WIRELESS INTERNET HANDBOOK

Technologies, Standards, and Applications

INTERNET and COMMUNICATIONS

This new book series presents the latest research and technological developments in the field of internet and multimedia systems and applications. We remain committed to publishing high-quality reference and technical books written by experts in the field.

If you are interested in writing, editing, or contributing to a volume in this series, or if you have suggestions for needed books, please contact Dr. Borko Furht at the following address:

Borko Furht, Ph.D., Director
Multimedia Laboratory
Department of Computer Science and Engineering
Florida Atlantic University
777 Glades Road
Boca Raton, FL 33431 U.S.A.

E-mail: borko@cse.fau.edu

WIRELESS INTERNET HANDBOOK

Technologies, Standards, and Applications

Edited by
Borko Furht, Ph.D.
Mohammad Ilyas, Ph.D.

CRC PRESS

Boca Raton   London   New York   Washington, D.C.

Library of Congress Cataloging-in-Publication Data

Wireless internet handbook : technologies, standards, and applications / editors, Borko Furht, Mohammad Ilyas.
    p. cm.
   Includes bibliographical references and index.
   ISBN 0-8493-1502-6 (alk. paper)
    1. Wireless Internet--Handbooks, manuals, etc. I. Furht, Borivoje. II. Ilyas, Mohammad, 1953-

   TK5103.4885 .W5714 2003
   004.67'8--dc21

                   2002038795

Visit the CRC Press Web site at www.crcpress.com

© 2003 by CRC Press LLC
Auerbach is an imprint of CRC Press LLC

No claim to original U.S. Government works
International Standard Book Number 0-8493-1502-6
Library of Congress Card Number 2002038795
Printed in the United States of America 1 2 3 4 5 6 7 8 9 0
Printed on acid-free paper

Preface

Just a few years ago, the only way to access the Internet and the Web was by using wireline desktop and laptop computers. Today, however, users are traveling between corporate offices and customer sites, and there is a great need to access the Internet through wireless devices. The wireless revolution started with wireless phones and continued with Web phones and wireless handheld devices that can access the Internet. Many nations and corporations are making enormous efforts to establish a wireless infrastructure, including declaring new wireless spectrum, building new towers, and inventing new handheld devices, high-speed chips, and protocols.

The purpose of the *Handbook of Wireless Internet* is to provide a comprehensive reference on advanced topics in this field. The *Handbook* is intended both for researchers and practitioners in the field, and for scientists and engineers involved in the design and development of the wireless Internet and its applications. The *Handbook* can also be used as the textbook for graduate courses in the area of the wireless Internet.

This *Handbook* is comprised of 24 chapters that cover various aspects of wireless technologies, networks, architectures, and applications. Part I, *Basic Concepts,* introduces fundamental wireless concepts and techniques, including various generations of wireless systems, security aspects of wireless Internet, and current industry trends.

Part II, *Technologies and Standards,* covers multimedia and video streaming over the wireless Internet, voice service over the wireless Internet, and wireless standards such as IEEE 802.11 (for wireless LANs) and Wireless Application Protocol.

Part III, *Networks and Architectures,* consists of chapters dealing with issues such as user mobility in IP networks, location-prediction techniques, wireless local access techniques, multiantenna technology, Bluetooth-based wireless systems, *ad hoc* networks, and others.

Part IV, *Applications,* includes chapters describing typical applications enabled by wireless Internet, including M-commerce, telemedicine, delivering music, and others.

We would like to thank the authors, who are experts in the field, for their contributions of individual chapters to the *Handbook.* Without their expertise and effort, this handbook would never have come to fruition. CRC Press editors and staff also deserve our sincere recognition for their support throughout the project.

Borko Furht and Mohammad Ilyas
Boca Raton, Florida

The Editors-in-Chief and Authors

 **Borko Furht** is a professor and chairman of the Department of Computer Science and Engineering at Florida Atlantic University (FAU) in Boca Raton, Florida. Before joining FAU, he was a vice president of research and a senior director of development at Modcomp, a computer company of Daimler Benz, Germany, and a professor at the University of Miami in Coral Gables, Florida. Professor Furht received Ph.D. degrees in electrical and computer engineering from the University of Belgrade. His current research is in multimedia systems, Internet computing, video coding and compression, video databases, and wireless multimedia. He is the author of numerous books and articles in the areas of multimedia, computer architecture, real-time computing, and operating systems. He is a founder and editor-in-chief of the *Journal of Multimedia Tools and Applications* (Kluwer). He has received several technical and publishing awards, and has consulted for many high-tech companies including IBM, Hewlett-Packard, Xerox, General Electric, JPL, NASA, Honeywell, and RCA. He has also served as a consultant to various colleges and universities. He has given many invited talks, keynote lectures, seminars, and tutorials.

 **Mohammad Ilyas** received his Ph.D. degree from Queens' University in Kingston, Ontario, Canada in 1983. His doctoral research was about switching and flow control techniques in computer communications networks. Since September 1983, he has been with the College of Engineering at Florida Atlantic University, Boca Raton, Florida, where he is currently Associate Dean for Graduate Studies and Research. From 1994 to 2000, he was chair of the department. During the 1993–1994 academic year, he was on sabbatical leave with the Department of Computer Engineering, King Saud University, Riyadh, Saudi Arabia. Dr. Ilyas has conducted successful research in various areas including traffic management and congestion control in broadband/high-speed communications networks, traffic characterization, wireless communications networks, performance modeling, and simulation. He has published 1 book and over 130 research articles. He has supervised 10 Ph.D. dissertations and 32 Master's theses to completion. He has been a consultant to several national and international organizations. Dr. Ilyas is an active participant in several IEEE technical committees and activities.

Contributors

Kalyan Basu
Center for Research in Wireless Mobility
and Networking
Department of Computer Science and
Engineering
The University of Texas at Arlington
Arlington, Texas

Nitish Barman
Department of Computer Science and
Engineering
Florida Atlantic University
Boca Raton, Florida

Amiya Bhattacharya
Center for Research in Wireless Mobility
and Networking
Department of Computer Science and
Engineering
The University of Texas at Arlington
Arlington, Texas

Jill Boyce
Corporate Research
Thomson Multimedia
Princeton, New Jersey

Stefano Cacciaguerra
Department of Computer Science
University of Bologna
Bologna, Italy

Jonathan Chan
CSIRO Centre for Networking
Technologies for the Information
Economy
Collingswood, Australia

Christine Cheng
Department of Electrical Engineering
and Computer Science
University of Wisconsin-Milwaukee
Milwaukee, Wisconsin

Igor D.D. Curcio
Nokia Corporation
Tampere, Finland

Sajal K. Das
Center for Research in Wireless Mobility
and Networking
Department of Computer Science and
Engineering
The University of Texas at Arlington
Arlington, Texas

Ahmed K. Elhakeem
Concordia University
Montreal, Quebec, Canada

Stefano Ferretti
Department of Computer Science
University of Bologna
Bologna, Italy

Mark L. Ferrey
Minnesota Pollution Control Agency
Site Remediation Section
St. Paul, Minnesota

David Furuno
Advanced Wireless Group
General Atomics, Photonics
Division
San Diego, California

Borko Furht
Florida Atlantic University
Department of Computer Science and
  Engineering
Boca Raton, Florida

José Antonio Garcia-Macias
CICESE Research Center
Esenada, Mexico

Vittorio Ghini
Department of Computer Science
University of Bologna
Bologna, Italy

David Goodman
Department of Electrical and Computer
  Engineering
Polytechnic University
Brooklyn, New York

Kevin Hung
Joint Research Center for Biomedical
  Engineering
Department of Electronic Engineering
The Chinese University of Hong Kong
Shatin, Hong Kong

Mohammad Ilyas
Florida Atlantic University
Department of Computer Science and
  Engineering
Boca Raton, Florida

Ravi Jain
DoCoMo USA Labs
San Jose, California

Sanjay Jha
School of Computer Science and
  Engineering
University New South Wales
Sydney, Australia

Björn Landfeldt
School of Information Technologies and
  School of Electrical and Information
  Engineering
The University of Sydney
Sydney, Australia

Dennis Seymour Lee
Forest Hills, New York

Andres Llana, Jr.
Vermont Studies Group, Inc.
King of Prussia, Pennsylvania

Angel Lozano
Wireless Communication Research
  Department
Bell Laboratories (Lucent Technologies)
Holmdel, New Jersey

Oge Marques
Department of Computer Science and
  Engineering
Florida Atlantic University
Boca Raton, Florida

Archan Misra
IBM T.J. Watson Research Center
Hawthorne, New York

Amitava Mukherjee
IBM Global Service
Calcutta, India

Gopal Racherla
Advanced Wireless Group
General Atomics, Photonics Division
San Diego, California

Sridhar Radhakrishnan
School of Computer Science
University of Oklahoma
Norman, Oklahoma

G. Radhamani
Faculty of Information Technology
Multimedia University
Cyberjaya Campus
Selangor D.E., Malaysia

Mahesh S. Raisinghani
Center for Applied Information
  Technology
Graduate School of Management
University of Dallas
Dallas, Texas

Marco Roccetti
Department of Computer Science
University of Bologna
Bologna, Italy

Valerie A. Rosenblatt
W Style
Burlingame, California

Abhishek Roy
Center for Research in Wireless Mobility
  and Networking
Department of Computer Science and
  Engineering
The University of Texas at Arlington
Arlington, Texas

Debashis Saha
Indian Institute of Management
Calcutta, India

Paola Salomoni
Department of Computer Science
University of Bologna
Bologna, Italy

Aruna Seneviratne
School of Electrical Engineering and
  Telecommunications
The University of New South Wales
Kensington, Australia

Mohammad Umar Siddiqi
Faculty of Engineering
Multimedia University
Cyberjaya Campus
Selangor D.E., Malaysia

Sirin Tekinay
Department of Electrical and Computer
  Engineering
New Jersey Institute of Technology
Newark, New Jersey

Binh Thai
School of Electrical Engineering and
  Telecommunications
The University of New South Wales
Kensington, Australia

Leyla Toumi
LSR-IMAG
CNSR/INPG
Grenoble, France

Eric van den Berg
Applied Research
Telcordia Technologies
Morristown, New Jersey

Yuan-Ting Zhang
Joint Research Center for Biomedical
  Engineering
Department of Electrical Engineering
The Chinese University of Hong Kong
Shatin, Hong Kong

Haitao Zheng
Bell Laboratories
Lucent Technologies
Holmdel, New Jersey

Table of Contents

Part I

Basic Concepts

1   The Fundamentals of the Wireless Internet

Borko Furht and Mohammad Ilyas

CONTENTS

ABSTRACT

This chapter presents a comprehensive introduction to the field of wireless systems and their applications. We begin with the fundamental principles of wireless communications, including modulation techniques, wireless system topologies, and performance elements. Next, we present three generations of wireless systems based on access techniques, and we introduce the basic principles of frequency division multiple access, time division multiple access, and code division multiple access techniques. We discuss various wireless Internet networks and architectures, including wireless personal area networks, local area networks, and wide area networks. We present common wireless devices and their features, as well as wireless standards such as Wireless Application Protocol. A survey of present and future wireless applications is given, from messaging applications to M-commerce, entertainment, and mobile Web services. We discuss briefly the future trends in wireless technologies and systems.

1.1 INTRODUCTION

The wireless Internet is coming of age! Millions of people worldwide already are using Web phones and wireless handheld devices to access the Internet. Nations and corporations are making enormous efforts to establish a wireless infrastructure, including declaring new wireless spectrum, building new towers, inventing new handset devices and high-speed chips, and developing protocols.

The adoption of the wireless Internet strictly depends on the mobile bandwidth, the bandwidth of access technologies. The current 2G wireless access technologies transmit at 9.6 to 19.2 kbps. These speeds are much slower than the dial-up rates of desktop PCs connecting to the Internet. However, 2.5G wireless technologies already in use provide speeds of 100 kbps, and 3G technologies with speeds of 2 to 4 Mbps will allow wireless connections to run much faster than today's wired cable and DSL services. Figure 1.1 illustrates the transmission speeds of wired networks and their applications. This figure includes wireless access networks, showing that 2G networks are basically used for voice and text messaging, but 2.5G networks and particularly 3G networks will open doors for many new wireless applications that use streaming video and multimedia.

Today, the number of subscribers with fixed Internet access is much higher than those with mobile Internet access. However, according to a forecast from Ericsson, in several years the number of mobile subscribers to the Internet will reach 1 billion and will be higher than those having fixed access (see Figure 1.2).

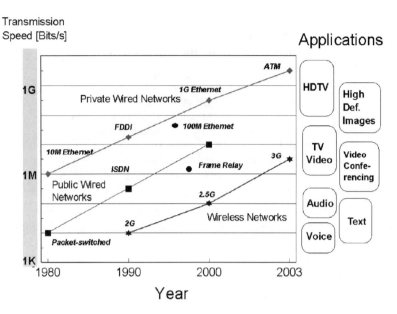

FIGURE 1.1 Wired and wireless networks and their applications.

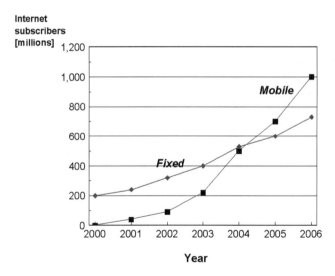

FIGURE 1.2 Mobile Internet access. (Source: Ericsson, Basic concepts of WCDMA radio access network, White Paper, www.ericsson.com, 2002.)

In this chapter, we introduce the fundamental concepts of wireless Internet. In Section 1.2, we describe the basic principles of wireless communications, including wireless network technologies. Section 1.3 presents the modulation techniques and basic access technologies. Wireless Internet networks are described in Section 1.4,

TABLE 1.1
Radio Spectrum and Applications

Applications	Frequency Spectrum
AM	535 to 1700 kHz
FM	88 to 108 MHz
TV	54 to 88, 174 to 220 MHz
GPS	1200 to 1600 MHz
Cell phones	800 to 1000 MHz
	1800 to 2000 MHz

while wireless devices and their functionality are presented in Section 1.5. Section 1.6 gives an overview of current and potential wireless Internet applications, while some future trends in wireless technologies are discussed in Section 1.7. Concluding remarks are given in Section 1.8.

1.2 PRINCIPLES OF WIRELESS COMMUNICATIONS

In this section, we describe fundamental principles of wireless communications and related wireless technologies, including wireless radio and satellite communications. We introduce basic modulation techniques used in radio communications and two fundamental wireless system topologies: point-to-point and networked topologies. We discuss performance elements of wireless communications.

1.2.1 WIRELESS TECHNOLOGIES

Today, there are many wireless technologies that are used for a variety of applications. Wireless radio communications are based on transmission of radio waves through the air. Radio waves between 30 MHz and 20 GHz are used for data communications. The range lower than 30 MHz could support data communication; however, it is typically used for FM and AM radio broadcasting, because these waves reflect on the Earth's ionosphere to extend the communication. Radio waves over 20 GHz may be absorbed by water vapor, and therefore, they are not suitable for long distance communication. Table 1.1 shows radio frequencies used for wireless radio applications in AM and FM radio, TV, GPS, and cell phones.[1]

Microwave transmission is based on the same principles as radio transmission. The microwave networks require a direct transmission path, high transmission towers, and antennas. Microwave equipment in the United States operates at 18 to 23 GHz. There are 23,000 microwave networks in the United States alone.

Satellite communications are used for a variety of broadcasting applications. The two most-popular frequency bands for satellite communications are C-band (frequency range 5.9 to 6.4 GHz for uplink and 3.7 to 4.2 GHz for downlink) and Ku-band (frequency range 14 to 14.5 GHz for uplink and 11.7 to 12.2 GHz for downlink). Recently, the Ku-band spectrum has been opened up to U.S. satellite communication, which receives at 30 GHz and sends at 20 GHz.

The radio transmission system consists of a transmitter and a receiver. The main components of a radio transmitter are a transducer, an oscillator, a modulator, and an antenna. A transducer, e.g., a microphone or a camera, converts the information to be transmitted to an electrical signal. An oscillator generates a reliable frequency that is used to carry the signal. A modulator embeds the voice or data signal into the carrier frequency. An antenna is used to radiate an electrical signal into space in the form of electromagnetic waves.

A radio receiver consists of an antenna, an oscillator, a demodulator, and an amplifier. An antenna captures radio waves and converts them into electrical signals. An oscillator generates electrical waves at the carrier frequency that is used as a reference wave to extract the signal. A demodulator detects and restores modulated signals. An amplifier amplifies the received signal that is typically very weak.

1.3 MODULATION TECHNIQUES

Modulation techniques embed a signal into the carrier frequency. They can be classified into analog and digital modulations. Traditional analog modulations include amplitude modulation (AM) and frequency modulation (FM). In digital modulations, binary 1s and 0s are embedded in the carrier frequency by changing its amplitude, frequency, or phase. Subsequently, digital modulations, called keying techniques, can be amplitude shift keying (ASK), frequency shift keying (FSK), and phase shift keying (PSK).

Some new popular keying techniques include Gaussian minimum shift keying (GMSK) and differential quadrature phase shift keying (DQPSK). GMSK is a type of FSK modulation that uses continuous phase modulation, so it can avoid abrupt changes. It is used in GSM (*Groupe Speciale Mobile*) systems, and DECT (digital enhanced cordless telecommunications). DPSK is a type of phase modulation, which defines four rather than two phases. It is used in TDMA (time division multiple access) systems in the United States.

A significant drawback of traditional radio frequency (RF) systems is that they are quite vulnerable to sources of interference. Spread spectrum modulation techniques resolve the problem by spreading the information over a broad frequency range. These techniques are very resistant to interference. Spread spectrum techniques are used in code division multiple access (CDMA) systems, and are described in more detail in Section 1.4.

1.3.1 WIRELESS SYSTEM TOPOLOGIES

Two basic wireless system topologies are point-to-point (or *ad hoc*) and networked topology. In the point-to-point topology, two or more mobile devices are connected using the same air interface protocol. Figure 1.3a illustrates the full mesh point-to-point configuration, where all devices are interconnected. Limitations of this topology are that the wireless devices cannot access the Web, send e-mail, or run remote applications.

In the networked topology, there is a link between wireless devices connected in the wireless network and the fixed public or private network. A typical configuration, shown in Figure 1.3b, includes wireless devices (or terminals), at least one

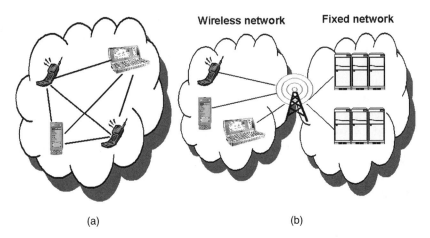

Wireless network   **Fixed network**

(a)   (b)

FIGURE 1.3 Wireless topologies: (a) point-to-point topology and (b) networked topology.

bridge between the wireless and the physical networks, and the numbers of servers hosting applications used by wireless devices. The bridge between the wireless and the physical networks is called the base station or access point.

1.3.2 PERFORMANCE ELEMENTS OF WIRELESS COMMUNICATIONS

Wireless communication is characterized by several critical performance elements:

- Range
- Power used to generate the signal
- Mobility
- Bandwidth
- Actual data rate

The range is a critical factor that refers to the coverage area between the wireless transmitter and the receiver. The range is strongly correlated with the power of the signal. A simplified approximation is that for 1 milliwatt of power, the range is one meter in radius. For example, 1 watt of power will allow the range of 1 kilometer in radius. As the distance from the base station increases, the signal will degrade, and data may incur a high error rate. Using part of the spectrum for error correction can extend the range; also, the use of multiple base stations can extend the range.

Mobility of the user depends on the size of the wireless device. Miniaturization of the wireless device provides better mobility. This can be achieved by reducing the battery size and consequently by minimizing power consumption; however, this will cause the generated signal to weaken, giving reduced range. In summary, there should be a trade-off between the range and the mobility: the extended range will reduce the mobility, and better mobility will reduce the range of wireless devices.

Bandwidth refers to the amount of frequency spectrum available per user. Using wider channels gives more bandwidth. Transmission errors could reduce the available bandwidth, because part of the spectrum will be used for error correction.

Actual data rate mostly depends on the bandwidth available to the user; however, there are some other factors that influence it, such as the movement of the transceiver, position of the cell, and density of users. The actual data rate is typically higher for stationary users than for users who are walking. Users traveling at high speed (such as in cars or trains) have the lowest actual data rate. The reason for this is that part of the available bandwidth must be used for error correction due to greater interference that traveling users may experience.

Similarly, interference depends on the position of the cell; with higher interference, the actual data rate will be reduced. Optimal location is where there is direct line-of-sight between the user and the base station and the user is not far from the base station. In that case, there is no interference and the transmission requires minimum bandwidth for error correction.

Finally, if the density of users is high, there will be more users transmitting within a given cell, and consequently there will be less aggregate bandwidth per user. This reduces the actual data rate.

1.3.3 GENERATIONS OF WIRELESS SYSTEMS BASED ON WIRELESS ACCESS TECHNOLOGIES

From the late 1970s until today, there were three generations of wireless systems based on different access technologies:

1. 1G wireless systems, based on FDMA (frequency division multiple access)
2. 2G wireless systems, based on TDMA and CDMA
3. 3G wireless systems, mostly based on W-CDMA (wideband code division multiple access)

In Section 1.7, we introduce the future efforts in building the 4G wireless systems.

1.3.3.1 The 1G Wireless Systems

The first generation of wireless systems was introduced in the late 1970s and early 1980s and was built for voice transmission only. It was an analog, circuit-switched network that was based on FDMA air interface technology. In FDMA, each caller has a dedicated frequency channel and related circuits. For example, three callers use three frequency channels (see Figure 1.4a). An example of a wireless system that employs FDMA is AMPS (Advanced Mobile Phone Service).

1.3.3.2 The 2G Wireless Systems

The second generation of wireless systems was introduced in the late 1980s and early 1990s with the objective to improve transmission quality, system capacity, and range. Major multiple-access technologies used in 2G systems are TDMA and CDMA. These systems are digital, and they use circuit-switched networks.

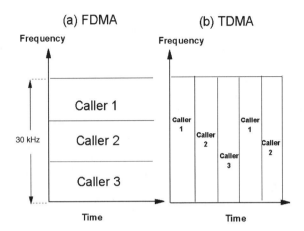

FIGURE 1.4 FDMA versus TDMA: (a) In FDMA, a 30-kHz channel is dedicated to each caller. (b) In TDMA, a 30-kHz channel is timeshared by three callers.

1.3.3.2.1 TDMA Technology

In TDMA systems, several callers timeshare a frequency channel. A call is sliced into a series of time slots, and each caller gets one time slot at regular intervals. Typically, a 39-kHz channel is divided into three time slots, which allows three callers to use the same channel. In this case, nine callers use three channels. Figure 1.4 illustrates the operation of FDMA and TDMA access technologies.

The main advantage of the TDMA system is increased efficiency of transmission; however, there are some additional benefits compared to the CDMA-based systems. First, TDMA systems can be used for transmission of both voice and data. They offer data rates from 64 kbps to 120 Mbps, which enables operators to offer personal communication services such as fax, voice-band data, and Short Message Services (SMS). TDMA technology separates users in time, thus ensuring that they will not have interference from other simultaneous transmissions. TDMA provides extended battery life, because transmission occurs only part of the time. One of the disadvantages of TDMA is caused by the fact that each caller has a predefined time slot. The result is that when callers are roaming from one cell to another, all time slots in the next cell are already occupied, and the call might be disconnected.

1.3.2.2 GSM

GSM (*Groupe Special Mobile* or Global System for Mobile Communications) is the best-known European implementation of services that uses TDMA air interface technology. It operates at 900 and 1800 MHz in Europe, and 1900 MHz in the United States. European GSM has been exported also to the rest of the world.

GSM has applied the frequency hopping technique, which involves switching the call frequency many times per second for security.

The other systems that deploy TDMA are DECT (digital enhanced cordless telephony), IS-136 standard, and iDEN (integrated Digital Enhanced Network).

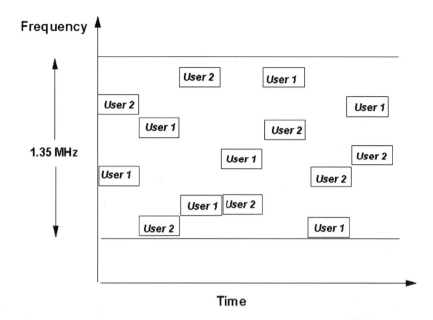

FIGURE 1.5 Frequency hopped spread spectrum applied in CDMA air interface.

1.3.2.3 CDMA Access Technology

CDMA is a radically different air interface technology that uses the frequency hopping (FH) spread spectrum technique. The signal is randomly spread across the entire allocated 1.35-MHz bandwidth, as illustrated in Figure 1.5. The randomly spread sequences are transmitted all at once, which gives higher data rate and improved capacity of the channels compared to TDMA and FDMA. It gives eight to ten times more callers per channel than FDMA/TDMA air interface. CDMA provides better signal quality and secure communications. The transmitted signal is dynamic bursty, ideal for data communication. Many mobile phone standards currently being developed are based on CDMA.

1.3.3 THE 3G WIRELESS SYSTEMS

The 3G wireless systems are digital systems based on packet-switched network technology intended for wireless transmission of voice, data, images, audio, and video. These systems typically employ W-CDMA and CDMA 2000 air interface technologies.

1.3.3.1 Packet Switching versus Circuit Switching

In circuit-switching networks, resources needed along a path for providing communication between the end systems are reserved for the entire duration of the session. These resources are typically buffers and bandwidth. In packet-switching networks,

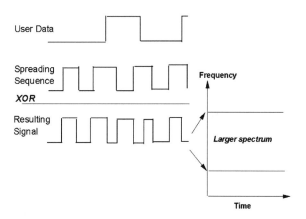

FIGURE 1.6 Direct sequence spread spectrum applied in W-CDMA air interface.

several users share these resources, and various messages use the resources on demand. Therefore, packet switching offers better sharing of bandwidth; it is simpler, more efficient, and less costly to implement. On the other hand, packet switching is not suitable for real-time services, because of its variable and unpredictable delays.

1.3.3.2 W-CDMA Access Technology

W-CDMA uses a direct sequence (DS) spread spectrum technique. DS spread spectrum uses a binary sequence to spread the original data over a larger frequency range, as illustrated in Figure 1.6. The original data is multiplied by a second signal, called spreading sequence or spreading code, which is a pseudorandom code (PRC) of much wider frequency. The resulting signal is as wide as the spreading sequence, but carries the data of the original signal.

1.3.4 2.5G WIRELESS SYSTEMS

An intermediate step in employing full packet-switching 3G systems is the 2.5 wireless systems. They use separate air interfaces – circuit switching for voice and packet switching for data – designed to operate in 2G network spectrum. The 2.5G provides an increased bandwidth to about 100 kbps, much larger than 2G systems, but much lower than the expected bandwidth of 3G systems. General Packet Radio Service (GPRS) is the 2.5G implementation of Internet protocol packet switching on European GSM networks.[2] It is an upgrade for the IS-136 TDMA standard, used in North America and South America. GPRS combines neighboring 19.2-kbps time slots, typically one uplink and two or more downlink slots per GPRS tower. The rate can potentially reach 115 kbps.

Enhanced Data for Global Enhancement (EDGE) is another packet-switched technology that is a GPRS upgrade based on TDMA. The theoretical pick of this technology is 384 kbps, but the tests show that the practical rates are in the range of 64 to 100 kbps. EDGE is a standard of AT&T Wireless in the United States.

TABLE 1.2
Basic Characteristics of Generations of Wireless Systems

Features	1G	2G	2.5G	3G
Air interfaces	FDMA	TDMA CDMA	TDMA	W-CDMA TD-CDMA CDMA 200
Bandwidth		~10 kbps	~100 kbps	~2 to 4 Mbps
Data traffic	No data	Circuit switched	Packet switched	Packet switched
Examples of services	AMPS	GSM IS-136 PDC IS-95	GPRS EDGE	UMTS cdma 2000
Modulation	Analog	Digital	Digital	Digital
Voice traffic	Circuit switched	Circuit switched	Circuit switched	Packet switched (VoIP)

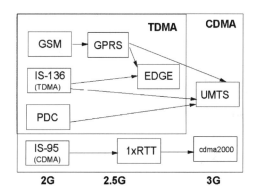

FIGURE 1.7 Migration path from 2G to 3G wireless systems.

1.3.5 UMTS

Universal Mobile Telecommunications System (UMTS) is a 3G wireless standard that supports two different air interfaces: wideband CDMA (W-CDMA) and time division CDMA (TD-CDMA). W-CDMA will be used for the cellular wide area coverage and high mobility service, while TD-CDMA will be used for low mobility, local in-building services, asymmetrical data transmission, and typical office applications. GSM, IS-136, and PDC (Personal Digital Cellular) operators have all adopted the UMTS standard, but Qualcomm has developed a similar standard, CDMA 2000, which could attract existing IS-95 carriers. Basic concepts of W-CDMA radio access network are described in the Ericsson white paper.[3]

Table 1.2 presents the characteristics of three generations of wireless systems, and Figure 1.7 shows the most-possible migration path from 2G to 3G wireless systems.

In summary, the target features of 3G wireless systems include:

- High data rates, which are expected to be 2 to 4 Mbps for indoor use, 384 kbps for pedestrians, and 144 kbps for vehicles
- Packet-switched networks, which provide that the users will be always connected
- Voice and data network will be dynamically allocated
- The system will offer enhanced roaming
- The system will include common billing and will have user profiles
- The system will be able to determine the geographic position of the users via mobile terminals and networks
- The system will be well suited for transmission of multimedia and will offer various services such as bandwidth on demand, variable data rates, quality sound, etc.

1.4 WIRELESS INTERNET ARCHITECTURES

The general wireless system architecture, which includes connections to the Internet, is shown in Figure 1.8.[4] A wireless device is connected to a base station through one of the wireless Internet networks (see Section 1.4.1); the base station is wired to a telecommunications switch. In 2.5G systems, the telecommunication switch is used to send voice calls through the circuit-switched telephone network, and data through the packet-switched Internet. However, 3G systems use the packet-switched Internet for both voice and data.

1.4.1 WIRELESS INTERNET NETWORKS

The wireless part of the Internet architecture, shown in Figure 1.8, is referred to as wireless Internet network. Wireless Internet networks can be classified as:

- Wireless personal area networks (PANs)
- Wireless local area networks (LANs)
- Wireless wide area networks (WANs)

The main difference between these networks is in the range they cover. Wireless PANs and LANs operate on unlicensed spectrum; wireless WANs are licensed, well-regulated public networks. They can all be used as access networks to the Internet, as discussed in Section 1.4.2.

1.4.1.1 Wireless PANs

Wireless PANs have a very short range of up to 10 meters. They are used to connect mobile devices to send voice and data in order to perform transactions, data transfer, or voice relay functions. They are used in personal computers to replace keyboard and printer cables and connectors. Two popular technologies for wireless PANs are infrared (IR) and Bluetooth technologies. Infrared devices use IRDA standard and

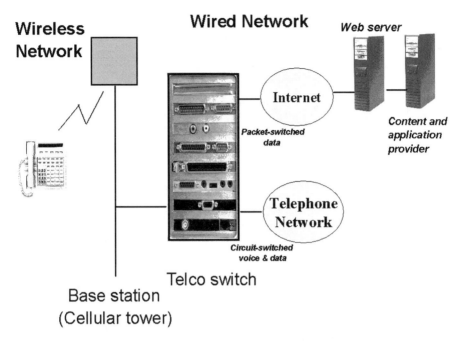

FIGURE 1.8 Wireless system architecture. (Adapted from Beaulieu, M., *Wireless Internet Applications and Architecture,* Addison-Wesley, Reading, MA, 2002.)

are used to transmit data among a variety of devices, including cell phones, notebooks, personal digital assistants, digital cameras, and others.

The Bluetooth network, called a *piconet,* is used to connect up to eight devices. It uses frequency hopping spread spectrum technique implemented with Gaussian frequency shift keying (GFSK). The Bluetooth network is intended for wireless connection between mobile devices, fixed computers, and cellular phones.

1.4.1.2 Wireless LANs

Wireless LANs are used to substitute fixed LANs in the range of about 100 meters. They are used in office buildings and homes to connect devices using a wireless LAN protocol. Typically, wireless LANs have a fixed transceiver, which is a base station that connects the wireless LAN to a fixed network. Popular wireless LANs include DECT, home RF, and 802.11 networks.

DECT is a standard for cordless phones that operate in the frequency range from 1880 to 1900 MHz in a range of 50 meters. It is based on TDMA technology. Home RF network is used to connect home appliances. It uses SWAP (Shared Wireless Access Protocol), which is similar to DECT, but carries both data and voice. It supports up to 127 devices in the range of about 40 meters. 802.11 is a standard developed for wireless LANs that cover an office building or a group of adjacent buildings. Standard 802.11b (a revision of an original 802.11 standard) subdivides its frequency band of 2.4 to 2.483 GHz into several channels. Its specification supports direct sequence spread spectrum technique.

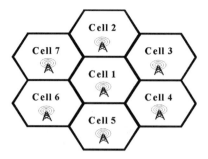

FIGURE 1.9 Cellular network is a wireless LAN that has multiple base stations positioned in a hexagon.

TABLE 1.3
Wireless Internet Networks

Wireless Networks	Range	Frequency Spectrum	Examples of Networks
PAN	~10 m	Unlicensed	IRDA
			Bluetooth
LAN	~100 m	Unlicensed	DECT
			HomeRF
			802.11b
WAN	~2500 m: One transceiver	Licensed	Cellular networks
	Unlimited: Multiple transceivers		GSM
			IS-95
			IS-136
			PDC

Adapted from Rhoton, J., *The Wireless Internet Explained,* Digital Press, 2002.

1.4.1.3 Wireless WANs

Wireless WANs are licensed public wireless networks that are used by Web cell phones and digital modems in handheld devices. With a single transceiver (also called base station or cellular tower), the range is about 2500 meters; however, wireless LANs usually have multiple receivers that make their range practically unlimited. The most popular wireless WANs are cellular networks that consist of multiple base stations positioned in a hexagon (see Figure 1.9). Cellular networks can be classified as mobile phone networks that primarily carry voice, and they typically use circuit switching technology, and packet data networks that primarily carry data and use packet-switching technology.

Table 1.3 summarizes basic features of three wireless networks.

1.4.2 WIRELESS INTERNET TOPOLOGIES

A typical wireless device that has one radio and one antenna can either connect to a public, cellular phone network (WAN), to a private wireless LAN, or to a PAN.

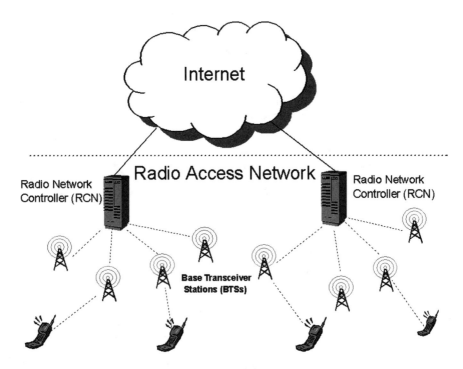

FIGURE 1.10 Wireless Internet architecture using star topology.

However, all these devices can connect to the wireless Internet. One of the recent trends is that some wireless devices have multiple antennas, and thus, multiple air interfaces. This approach allows the devices to connect to various wireless networks in order to optimize coverage.

Figure 1.8, presented earlier in this section, is a typical wireless Internet topology that consists of a wireless and fixed network. This architecture can be further expended into a star topology, shown in Figure 1.10.[5] In this topology, a centralized radio network controller (RNC) is connected by point-to-point links with the base stations that handle connectivity for a particular geographic area or cell. RNCs are interconnected to allow mobile users to roam between geographical areas controlled by different RNCs. RNCs are further connected to a circuit-switching network for voice calls (in 2G and 2.5G systems), and to a packet-switching network for data and access to the Internet. One of the drawbacks of this architecture is that the RNC presents a single-point-of-failure; therefore, if an RNC fails, the entire geographical region will lose service. This problem is addressed in Kempf and Yegani,[5] and some new architectures for future 4G generation of wireless systems are proposed.

Figure 1.11 illustrates a network topology that includes a combination of wireless PANs, LANs, and WANs, all connected to the Internet through base stations and fixed networks. Some devices, such as one denoted in Figure 1.11 as the MAI (multiple air interfaces) device, can be connected to several wireless networks, including a satellite network. Multiple air interfaces in this case can complement each other to provide optimized coverage of a particular area.

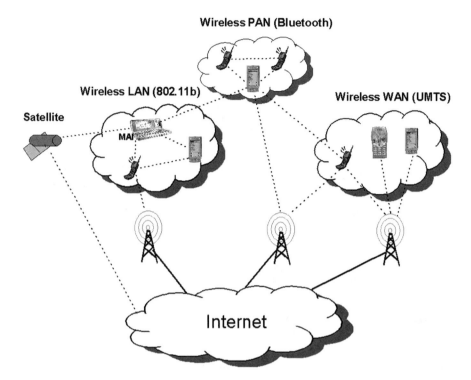

FIGURE 1.11 A network topology with various wireless networks connected to the Internet. A wireless device with multiple air interfaces (MAI) can be connected to the Internet through W-LAN, W-PAN, or through a satellite network.

1.5 WIRELESS DEVICES AND STANDARDS

In this section, we introduce the most-common wireless devices and their applications. We discuss the Wireless Application Protocol, which is a common standard for presenting and delivering services on wireless devices. We describe Java-enabled wireless devices, which use Java technology to run applications on wireless devices.

1.5.1 WIRELESS DEVICES

Wireless (or mobile) devices can be classified into six groups:[4]

1. Web phones. A Web phone is most commonly a cellular phone device with an Internet connection. The three major Web phones are the HDML&WAP phone in the United States, the WAP phone in Europe, and the i-mode phone in Japan. Web phones can exchange short messages, access Web sites with a minibrowser, and run personal service applications such as locating nearby places of interest. Web phones operate only when they have a network connection; however, advanced Web phones can run their own applications.

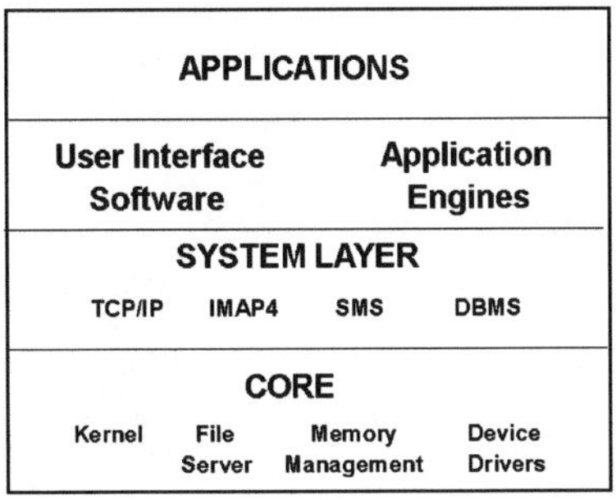

```
┌─────────────────────────────────────────────┐
│                 APPLICATIONS                 │
├─────────────────────────────────────────────┤
│   User Interface          Application        │
│     Software                Engines          │
├─────────────────────────────────────────────┤
│               SYSTEM LAYER                   │
│   TCP/IP    IMAP4      SMS        DBMS        │
├─────────────────────────────────────────────┤
│                    CORE                      │
│   Kernel    File      Memory       Device    │
│            Server   Management     Drivers    │
└─────────────────────────────────────────────┘
```

FIGURE 1.12   The architecture of the Symbian OS.

2. Wireless handheld devices. A wireless handheld device (such as Palm) is another common device that can exchange messages and use a mini-browser to access the Internet. Industrial handheld devices, such as Symbol and Psion, can perform complex operations such as completing orders.

3. Two-way pagers. A two-way pager allows users to send and receive messages and provides the use of a minibrowser. They are typically used in business applications.

4. Voice portals. Voice portals allow users to have a conversation with an information service using a kind of telephone or mobile phone.

5. Communication appliances. Communication appliances are electronic devices that use wireless technology to access the Internet. Examples include wireless cameras, watches, radios, pens, and others.

6. Web PCs. Web PCs are standard PCs connected to the Internet that can access mobile services wirelessly.

Wireless devices typically use an embedded real-time operating system. The most-common operating systems for wireless devices include Palm OS® (used in Palm handheld devices), Windows® CE and Windows NT Embedded by Microsoft (used in a variety of devices such as handheld PCs, pocket PCs, WebTV, Smart Phone, etc.), and Symbian OS.

We present a brief description of the Symbian OS (renamed from Epoc OS) that was used for many years in Psion handheld devices. It is currently used in many wireless devices, including the Nokia 9200 Communicator Series. The architecture of the Symbian OS, shown in Figure 1.12, consists of four layers. The Symbian core is common for all devices and consists of a kernel, a file server, memory management, and device drivers. The system layer consists of data service enablers that provide communications and computing services, such as TCP/IP, IMP4, SMS, and database management. User interface software is made and licensed by manufacturers, e.g., for the Nokia 9200 platform. Application engines enable software developers to create user interfaces. Various applications are at the last layer.

Figure 1.13 shows several representative wireless devices: the Palm VII, the Sony-Ericsson R520, and Nokia's 9210 and 9290.

Web phone - Palm VII   Nokia 9210 for GSM

SonyEricsson   Nokia 9290 Communicator

FIGURE 1.13 Contemporary wireless phones and handheld devices.

The Nokia 9290 Communicator is a wireless device that combines wireless phone and handheld device. The user can send and receive e-mail messages with attachments, and can an access the Internet. It has many applications built-in, such as MS Word, PowerPoint, and Excel. An interesting feature is that the user can take notes using the keyboard while conference calling on a built-in, hands-free speak-erphone.

1.5.2 WAP

Wireless Application Protocol (WAP) is a *de facto* standard for presenting and delivering wireless services on mobile devices. It is developed by mobile and wire-less communication companies (Nokia, Motorola, Ericsson, and Unwired Planet) and includes a minibrowser, scripting language, access function, and layered com-munication specification. Most wireless device manufacturers as well as service and infrastructure providers have adopted the WAP standard.

There are three main reasons why wireless Internet needs a different protocol:

1. Transfer rates
2. Size and readability
3. Navigation

The 2G wireless systems have data transfer rates of 14.4 kbps or less, which is much less than 56 kbps modems, DSL connections, or cable modems. Therefore, loading existing Web pages at these speeds will take a very long time.

Another challenge is the small size of the screens of wireless phones or handheld devices. Web pages are designed for desktops and laptops that have a resolution of 640×480 pixels. Wireless devices may have a resolution of 150×150 pixels, and the page cannot fit on the display.

Navigation is quite different on wireless devices. On desktops and laptops, navigation is performed using point-and-click action of a mouse, while typical wireless devices (specifically phones) use the scroll keys.

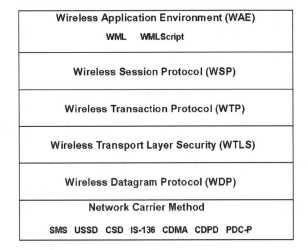

FIGURE 1.14 WAP stack consisting of six layers.

Therefore, WAP is created to provide Web pages to typical wireless devices, having in mind these limitations. Instead of using HTML, WAP uses Wireless Markup Language (WML), which is a small subset of XML (Extensible Markup Language). WML is used to create and deliver content that can be deployed on small wireless devices. It is scalable and extensible, because, like XML, it allows users to add new markup tags.

1.5.2.1 WAP Stack

The WAP stack consists of six layers, as illustrated in Figure 1.14.

1. The Wireless Application Environment (WAE) consists of the tools for wireless Internet developers. These tools include WML and WMLScript, a scripting language (similar to JavaScript or VBScript) that provides interactivity of Web pages presented to the user.
2. The Wireless Session Protocol (WSP) specifies a type of session between the wireless device and the network, which can be either connection-oriented or connectionless. Typically, a connection-oriented session is used in two-way communications between the device and the network. A connectionless session is commonly used for broadcasting or streaming data to the device.
3. The Wireless Transaction Protocol (WTP) is used to provide data flow through the network. WTP determines each transaction request as reliable two ways, reliable one way, or unreliable one way.
4. The Wireless Transport Layer Security (WTLS) provides some security features, similar to the Transport Layer Security (TLS) in TCP/IP. It checks data integrity, provides data encryption, and performs client and server authentication.

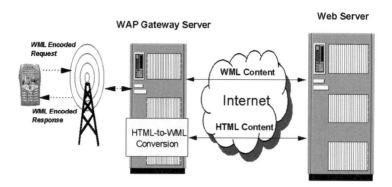

FIGURE 1.15 The WAP topology.

5. The Wireless Datagram Protocol (WDP) works in conjunction with the network carrier layer and provides WAP to adapt to a variety of bearers.
6. The Network Carrier Method. Network carriers or bearers depend on current technologies used by the wireless providers.

1.5.2.2 WAP Topology

Figure 1.15 shows a typical WAP topology. The wireless device, which is a WAP client, sends a radio signal searching for service through its minibrowser. A connection is established with the service provider, and the user selects a Web site to be viewed. The URL request from the WAP client is sent to the WAP gateway server, which is located between the carrier's network and the Internet. The WAP gateway server retrieves the information from the Web server. It consists of the WAP encoder, script compiler, and protocol adapters to convert the HTML data into WML. The WAP gateway server operates under two possible scenarios:

1. If the Web server provides content in WML, the WAP gateway server transmits this data directly to the WAP client.
2. If the Web server delivers content in HTML, the WAP gateway server first encodes the HTTP data into WML and then transmits to the client device.

In both cases, the WAP gateway server encodes the data from the Web server into a compact binary form for transmission over low-bandwidth wireless channels.

With the development of 3G wireless systems, there is a question whether WAP will be still needed. WAP was primarily developed for 2G systems that provide limited data rates of 9.6 to 14.4 kbps. The UMTS network, a 3G wireless system with expected data rates of 2 to 4 Mbps, will resolve the problem of limited bandwidth.

On the other hand, the WAP Forum argues that, even in 3G systems, bandwidth will play a crucial role and that WAP will be beneficial for the UMTS network as well. The WAP features that could be useful for the UMTS network include screen size, low power consumption, carrier independence, multidevice support, and intermittent coverage. Another argument is that new applications will require higher bandwidth and data rates, so WAP will still play a crucial role.

1.5.3 Java-Enabled Wireless Devices

New wireless devices, referred to as Java-enabled wireless devices, have recently emerged. While WAP wireless devices run new applications remotely using WAP, Java-enabled wireless devices allow users to download applications directly from the Internet. In addition, these devices allow users to download Java applets that can customize their devices. Another benefit of Java-enabled wireless devices is that they run applications and services from different platforms.

Java-enabled wireless devices use J2ME (Java 2 Platform Mobile Edition) that allows Java to work on small devices. J2ME includes some core Java instructions and APIs (application programming interfaces); however, its graphics and database access are less sophisticated than in J2SE and J2EE.

Java technology can be implemented either in software or in hardware. In a software implementation, the CPU of the wireless device runs the Java code, while hardware implementation is based on either a specialized Java acceleration chip or a core within the main processor. The hardware approach typically increases the performance of Java applications by running more efficiently and thus reduces power demands. Several companies are currently developing hardware chips that run Java or can be used as Java coprocessors, including ARC Cores, ARM Ltd., Aurora VLSI, and Zucotto Wireless.

Korea's LG Telecom developed the first Java-enabled phone in 2000. Java phones are presently produced by Nextel in the United States, NTT BoCoMo in Japan, and British Telecom. Nokia planned to ship 50 million Java phones in 2002 and 100 million in 2003.[6]

1.6 WIRELESS INTERNET APPLICATIONS

The wireless Internet will keep a large number of people in motion. Four wireless applications drive the wireless Internet: messaging, browsing, interacting, and conversing.[4] In messaging applications, a wireless device is used to send and receive messages. The device uses Short Message Service (SMS) and other e-mail protocols. In browsing applications, a wireless device uses a minibrowser to access various Web sites and receives Web services. In interacting applications, the applications run on wireless devices and include business and personal applications, and stand-alone games. In conversing applications, a wireless device calls voice portals (such as Wildfire®) to get voice information from Web services.

However, there are still a number of challenges in the development of wireless applications. The desktop computer will continue to be a dominant platform for generating content; however, professionals and consumers will increasingly use wireless devices to access and manage information. The great challenge for developers is to tailor content to the unique characteristics of wireless devices. The main objective is to provide quick and easy access to the required information rather than

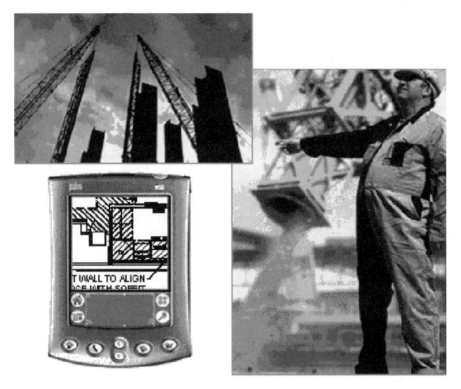

FIGURE 1.16 Firepad software comprises a high-speed vector rendering engine that can be used in CAD drawings.

to provide a complex directory tree where the user will easily get lost. Another challenge for developers is the design of user interfaces, which should be simple because of the limited size of the wireless devices.

The 2.5G and 3G wireless systems will allow new applications to include rich graphical content. Software vendors have been developing authoring tools for creating WAP-compatible WAP sites that include rich graphical content and animations. Examples include Macromedia and Adobe that are offering WAP and i-mode versions of their products. Macromedia Spectra, a product for creating dynamic, interactive, and content-rich Web sites, has been extended so a developer can easily add wireless Internet by creating WML code rather than HTML.

Firepad developed a vector-based graphics application for mobile devices. This application uses a high-speed vector rendering engine for complex applications such as geographic information systems and CAD drawings, as illustrated in Figure 1.16.

In the next section, we present several wireless applications that, in our opinion, are a major force in further driving the development of wireless Internet.

1.6.1 Messaging Applications

Messaging in mobile networks today mainly involves short text using the SMS protocol. The GSM has estimated that 24 billion SMS messages are sent each month.[7]

However, it is expected that soon wireless devices will support pictures, audio, and video messages. At the same time, the popular messaging services on the Internet, such as e-mail, chat, and instant messaging, are extending to wireless environments.

1.6.2 MOBILE COMMERCE

M-commerce applications refer to conducting business and services using wireless devices. These applications can be grouped into (1) transaction management applications, (2) digital content delivery, and (3) telemetry services.

Transaction management applications include online shopping tailored to wireless devices with online catalogs, shopping carts, and back-office functions. Other transaction applications include micro transactions and low cost purchases for subway or road tolls, parking tickets, digital cash, and others.

Digital content delivery includes a variety of applications:

- Information browsing for weather, travel, schedules, sport scores, stock prices, etc.
- Downloading educational and entertainment products
- Transferring software, images, and video
- Innovative multimedia applications

According to the recent study by HPI Research Group,[7] the following are the top ten mobile entertainment features:

1. Sending SMS messages
2. Checking local traffic and weather information
3. Using a still camera
4. Getting latest news headlines
5. Sending photos to a friend
6. Using a video camera
7. Booking and buying movie tickets
8. Getting information on movies
9. Listening to radio
10. Requesting specific songs

Entertainment on mobile devices is attractive because it is almost always with the user, whether commuting, traveling, or waiting.

Telemetry services include a wide range of new applications:

- Transmission of status, sensing, and measurement information
- Communications with various devices from homes, offices, or in the field
- Activation of remote recording devices or service systems

1.6.3 CORPORATE APPLICATIONS

Banks and transport companies were among the first businesses to deploy wireless applications based on WAP for their customers and employees. In banks, the goal was to reduce consumer banking transaction costs, while transport companies wanted to track transportation and delivery status online.

Gartner Research Group expects most corporations to implement wireless applications in four overlapping phases:[7]

1. The first group of applications is readily justifiable and includes high-value, vertical niche solutions, such as field force automation.
2. The second phase includes horizontal applications such as e-mail and personal information management applications.
3. The third wave of applications consists of vertical applications, such as mobile extensions to CRM (Customer Relationship Management), sales force automation, and enterprise resource planning systems.

In the long term, Gartner expects that 40 to 60 percent of all corporate systems will involve mobile elements.

1.6.4 WIRELESS APPLICATION SERVICE PROVIDERS

WASPs allow wireless access to various software products and services. Business WASP applications are targeted to mobile business people, field personnel, and sales staff. Other WASP applications include:[8]

- Mobile entertainment services
- Wireless gaming
- Wireless stock trading
- In-vehicle services, such as traffic control, car management, etc.

1.6.5 MOBILE WEB SERVICES

Web services include well-defined protocol interfaces through which businesses can provide services to customers and business partners over the Internet. Web services specify a common and interoperable way for defining, publishing, invoking, and using application services over networks. They are built on emerging technologies such as XML, SOAP (Simple Access Object Protocol), WSDL (Web Service Description Language), UDDI (Universal Description, Discovery, and Integration), and HTTP.

Mobile Web Services provide content delivery, location discovery, user authentication, presence awareness, user profile management, data synchronization, terminal profile management, and event notification services. Initially, wireless terminals are likely to access Mobile Web Services indirectly, through application servers. The application server will manage the interactions with the required Web services.

1.6.6 WIRELESS TEACHING AND LEARNING

Web-based distance learning could be extended to wireless systems. For example, the project Numina at the University of North Carolina–Wilmington is intended to

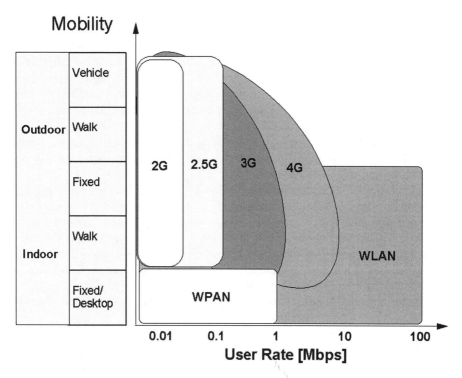

FIGURE 1.17 User mobility and data rates for wireless PANs and wireless LANs. (Adapted from Pahlavan, K. and Krishnamurthy, P., *Principles of Wireless Networks,* Prentice Hall, Englewood Cliffs, NJ, 2002.)

explore how wireless technology can be used to facilitate learning of abstract scientific and mathematical concepts.[9] Students use handheld computers (with appropriate software) which are connected to the wireless Internet. The system provides interactive exercises, and integrates various media and hypertext material.

1.7   FUTURE OF WIRELESS TECHNOLOGY

The major trend that is already emerging is the migration of mobile networks to fully IP-based networks. The next generation of wireless systems, 4G systems, will use new spectrum and emerging wireless air interfaces that will provide a very high bandwidth of 10+ Mbps. It will be entirely IP-based and use packet-switching technology. It is expected that 4G systems will increase usage of wireless spectrum. According to Cooper's law, on average, the number of channels has doubled every 30 months since 1985.

Figure 1.17 shows the user mobility and data rates for different generations of wireless systems, and for wireless PANs and LANs. The 3G and later 4G systems will provide multimedia services to users everywhere, while WLANs provide broadband services in hot spots, and WPANs connect personal devices together at very short distances.

Spread spectrum technology is presently used in 3G systems; however, there are already research experiments with Multicarrier Modulation (MCM), which is a step further from spread spectrum. MCM transmits simultaneously at many frequencies.

New types of smart antennas are currently under development. Most current antennas are omnidirectional, which means that they transmit in all directions with similar intensity. New directional antennas transmit primarily in one direction, while adaptive antennas vary direction in order to maximize performance.

New generations of software radios will dynamically adapt to wireless technology. They apply digital signal processors, so they can update the software with new versions of transmission techniques.

The transition from circuit-switched to packet-switched networks provides increased efficiency of the network and higher overall throughput. However, packet-switched networks operate on a best-effort basis, and therefore, cannot guarantee the service (specifically when the load is high). This will require the development of new QoS (quality of service) approaches to handle various network scenarios.

New wireless multimedia applications will require new solutions related to error resilience, network access, adaptive decoding, and negotiable QoS.

Error resilience solutions should enable delivery of rich digital media over wireless networks that have high error rates and low and varying transmission speeds. Network access techniques should provide the delivery of rich media without adversely affecting the delivery of voice and data services. Innovative adaptive decoding techniques should optimize rich media for wireless devices with limited processing power, limited battery life, and varying display sizes. New negotiable QoS algorithms should be developed for IP multimedia sessions, as well as for individual media components.

1.8 CONCLUSIONS

In this chapter, we presented fundamental concepts and technologies for wireless communications, and introduced various architectures and three generations of wireless systems. We are currently at the transition between 2G and 3G systems (2.5G systems). The 3G systems will soon offer higher data rates suitable for a variety of applications dealing with multimedia. Services and applications are driving 3G systems. With 3G systems, users will be able to send graphics, play games, locate a restaurant, book a ticket, read news updates, check a bank statement, watch their favorite soap operas, and many other exciting applications.

In July 2002, Ericsson delivered 15 real-life 3G applications, including real-time sport applications, face-to-face video calling, and exciting team games, to 40 operators so they can demonstrate to their customers what the wireless Internet is all about.

In the meantime, researchers are already working on 4G systems that will provide even higher data rates, will be entirely IP-based, and will include many other new features.

References

1. Rhoton, J., *The Wireless Internet Explained,* Digital Press, 2002.
2. Park, J.-H., Wireless Internet access for mobile subscribers based on GPRS network, *IEEE Communication Magazine,* 40 (4), 38–49, 2002.
3. Ericsson, Basic concepts of WCDMA radio access network, White Paper, www.ericsson.com, 2002.
4. Beaulieu, M., *Wireless Internet Applications and Architecture,* Addison-Wesley, Reading, MA, 2002.
5. Kempf, J. and Yegani, P., OpenRAN: a new architecture for mobile wireless Internet radio access network, *IEEE Communication Magazine,* May 2002, 118–123.
6. Lawton, G., Moving Java into mobile phones, *IEEE Comput.,* June 2002, 17–20.
7. Nokia, Mobile terminal software — markets and technologies for the future, White Paper, www.nokia.com, 2002.
8. Steemers, P., Critical success factors for wireless application service providers, White Paper, Cap Gemini Ernst & Young, 2002.
9. Shotsberger, P.G. and Vetter, R., Teaching and learning in the wireless classroom, *IEEE Comput.,* 110–111, 2001.
10. Buracchini, E., The software radio concept, *IEEE Communications Magazine,* September 2000, 138–143.
11. Hanzo, L., Cherriman, P.J., and Streit, J., *Wireless Video Communications,* IEEE Press, New York, 2001.
12. Krikke, J., Graphics applications over the wireless Web: Japan sets the pace, *IEEE Comput. Graphics Appl.,* May/June 2001, pp. 9–15.
13. Pahlavan, K. and Krishnamurthy, P., *Principles of Wireless Networks,* Prentice Hall, Englewood Cliffs, NJ, 2002.

2   Wireless Internet > Wireless + Internet

Sirin Tekinay and David Goodman

CONTENTS

ABSTRACT

The technical and business communities view a "wireless Internet" as an inevitable sequel to the spectacular growth of cellular communications and the World Wide Web in the 1990s. The prevailing wisdom is that without the nuisance of wired connections to consumer equipment, Internet access will be more convenient and enjoyable. While this is true, it is only part of the picture because it fails to acknowledge the fact that information services shaped by the needs and characteristics of people on the move and the nature of the information they send and receive will be qualitatively different from services delivered to people in fixed locations. In the long run, a wireless Internet will offer far more than the negative benefit of an Internet with some of its wires removed. However, to realize the full potential of a wireless Internet, it will be necessary to transcend the technical assumptions that nurtured cellular communications and the Web.

This chapter examines current industry trends in uniting wireless communications and the Internet. It describes the advances these trends will produce and the bottlenecks they do not address. It then surveys current research initiatives that go beyond the centralized topology of wireless systems and the client/server model of Internet information delivery.

2.1 INTRODUCTION

Since the late 1990s, the cellular industry and the business press have promoted wireless Internet as "the next big thing" in information technology. The idea was compelling in view of the huge public appetite for cellular telephones and the Web in the 1990s. The enthusiastic predictions of the growth of the wireless Internet were linked to two emerging technologies:

1. Third-generation (3G) cellular systems that would overcome the bit rate bottleneck of existing technology.
2. Internet-enabled cell phones that within a few years would be more numerous than personal computers.

As we write this chapter three years later, we are drawn to the adage "the future isn't what it used to be." Instead of cellular modems, the preferred mode of wireless access to the Internet in 2002 is a WLAN (wireless local area network) plug-in card or a WLAN modem built into a notebook computer. In limited coverage areas, WLANs give stationary (or slowly moving) users of notebook computers access to the two "killer apps" (mass-market applications) of the Internet: the Web and e-mail. In wide coverage areas (metropolitan and national), specialized wireless data networks and cellular networks transfer e-mail to and from PDAs (personal digital assistants) and specialized e-mail terminals. The most popular PDAs use the Palm operating system. Blackberry is a popular specialized e-mail terminal.

Simultaneous with the rapid growth of WLAN usage, the cellular industry is cautiously inaugurating 3G networks in Europe and Asia, and upgrading second-generation systems with "2.5G" (enhanced digital cellular) technology in many countries. In parallel with the cellular and WLAN radio developments, there is a high volume of activity in the Internet community focused on extending existing Internet protocols and introducing new ones with the aim of accommodating mobility and the characteristics of radio communications.

This chapter focuses on trends in wireless communications aimed at promoting a wireless Internet. The following section introduces a framework for comparing different wireless Internet radio technologies and describes the evolution of cellular systems and WLANs. The emerging technologies will overcome some of the deficiencies of mobile wireless communications relative to the transmission technologies

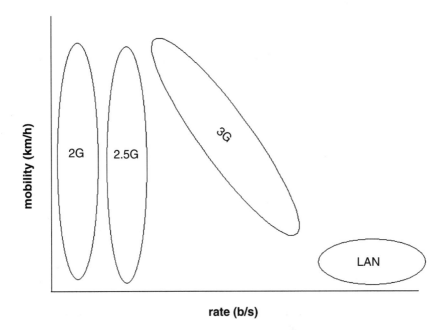

FIGURE 2.1 Bit rate and mobility in WLAN and cellular systems.

of the wired Internet. However, even after these evolutionary measures mature, many gaps will remain between expectations for a wireless Internet and what can be achieved in practice. The remainder of the chapter describes work in progress to overcome these gaps and create a wireless Internet that is more than the present Internet with some of the wires eliminated. Section 2.3 suggests an approach to advancing beyond the technologies emerging in 2002, and Section 2.4 describes research in progress that follows this approach.

2.2 WLANS AND CELLULAR NETWORKS: COMPARISON AND CONTRAST

Any practical communications system represents a compromise between a variety of technology and cost criteria. Some of the principal figures of merit for wireless communications systems are bit rate, mobility of terminals, signal quality, coverage area, service price, and demands on the power supplies of portable terminals.

Goals for third-generation wireless communication, enunciated in the early 1990s by the International Telecommunications Union Task Group IMT-2000, focused on the first two criteria, bit rate and mobility. Third-generation systems should deliver 2 Mbps to stationary or slowly moving terminals, and at least 144 kbps to terminals moving at vehicular speeds. Meanwhile, WLAN development has confined itself to communications with low-mobility (stationary or slowly moving) terminals, and focused on high-speed data transmission. The relationship of bit rate to mobility in cellular and WLAN systems has been commonly represented in two dimensions by diagrams resembling Figure 2.1. The principal goal of succeeding

TABLE 2.1
Figures of Merit for Wireless Internet Access Technologies

	Cellular	WLAN
Strong	Ubiquitous coverage	High bit rate
	High mobility	Low power
	Controlled signal quality	Low cost
	Infrastructure	
Weak	Low bit rate	Isolated coverage
	High power	Low mobility
	High cost	Vulnerable to interference

generations of cellular technology has been to move to the right in the bit rate/mobility plane. Coverage, the geographical area that a signal can reach, is a third figure of merit. The relationship between bit rate and coverage is similar to the relationship between bit rate and mobility. Cellular systems provide wide area ubiquitous coverage, while WLANs, as the name implies, cover only local areas, with large gaps between coverage areas. With respect to signal quality, a fourth figure of merit, cellular networks employ elaborate radio resources management technology to maintain high signal quality for the highest possible user population. Moreover, cellular network operators own expensive licenses, granting them exclusive use of radio spectrum in their service areas. By contrast, WLANs, operating in unlicensed spectrum bands, are vulnerable to interference from various sources, including other WLANs, cordless telephones, microwave ovens, and Bluetooth personal area networks.

In addition to the criteria of mobility, bit rate, coverage, and signal quality, Table 2.1 indicates that cellular terminals make greater demands on their batteries than WLAN modems. The radiated power in a cell phone can be as high as several hundred milliwatts, while WLANs transmit at a maximum of 100 milliwatts. In addition, cellular networks are far more expensive to establish and maintain than WLAN access points. As a consequence, WLAN service prices are considerably lower (in many situations, they are free) than cellular prices. Consider the fact that in a cellular network, a 1-MB file transfer uses comparable transmission resources to 500 seconds of a phone conversation (at 16 kbps speech transmission) and a few thousand short messages (at 200 characters per message). Consumers are accustomed to paying much more per bit for phone calls and short messages than for Internet access. It is a challenge to the cellular industry to establish prices for broadband services at a level high enough to compensate them for the radio resources consumed and simultaneously low enough to attract a large number of customers. The other cellular advantage in Table 2.1 is the network infrastructure of base stations, switches, routers, and databases that regulate access to a network and facilitate mobility.

All in all, we observe in Table 2.1 that, with respect to the figures of merit for wireless communications, cellular systems and WLANs are complementary; each one is strong where the other is weak. This suggests that both technologies will play

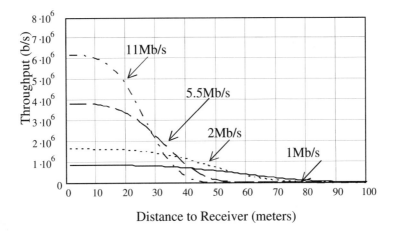

Distance to Receiver (meters)

FIGURE 2.2 Relationship of throughput to distance between transmitter and receiver in a WLAN.

important roles in a wireless Internet. As discussed in Section 2.2.3, coordinating WLAN and cellular access to a wireless Internet is a major task for industry and the research community. Meanwhile, the WLAN and cellular industries are moving ahead with technology advances in their own domains, as described in the following sections.

2.2.1 WLAN TRENDS

Although WLANs have been available commercially for more than a decade, their popularity as business and consumer devices dates from around 1999, when manufacturers converged on a technology referred to as 802.11b, published by the Institute of Electrical and Electronic Engineers (IEEE). The industry committed itself to interoperability, setting up the Wireless Ethernet Compatibility Alliance (http://www. wirelessethernet.org) to ensure that equipment produced by one company will communicate with equipment produced by other companies. An 802.11b WLAN operates in the 2.4-GHz unlicensed frequency band. The signaling rate is 11 Mbps, and terminals employ CSMA/CA (Carrier Sense Multiple Access with Collision Avoidance) to share the available radio spectrum.

At any particular time, the WLAN communicates at one of four possible bit rates (1 Mbps, 2 Mbps, 5.5 Mbps, and 11 Mbps), depending on the type of information it carries and the current channel conditions. The appropriate bit rate depends on channel quality, which can be measured as carrier-to-noise ratio (CNR). In a WLAN operating environment, the distance between transmitter and receiver has the greatest influence on CNR, which decreases with increasing distance. Accordingly, in order to maximize throughput, terminals transmit information at lower bit rates when they are far from the receiver and at higher bit rates when they are near the receiver. Figure 2.2, the result of a theoretical study,[1] predicts the relationship between user throughput and distance for the four transmission rates in 802.11b. The figure indicates that a terminal can achieve maximum throughput transmitting

TABLE 2.2
WLAN Bit Rates and Carrier Frequencies

	Spectrum Band (GHz)	Maximum Bit Rate (Mbps)
802.11a	5	54
802.11b	2.4	11
802.11g	2.4	20
HiperLAN1	5	20
HiperLAN2	5	54

at 11 Mbps when it is within 28 meters of the receiver. Between 28 and 38 meters, the throughput is highest at 5.5 Mbps, while 2 Mbps is preferred between 38 and 55 meters. At distances greater than 55 meters, transmission at 1 Mbps maximizes throughput.

Most WLANs transmit signals between an access point connected to an Ethernet and a laptop computer with a built-in WLAN modem or a modem contained in a plug-in card. WLANs also are capable of direct (peer-to-peer) communication between two terminals.

While the overwhelming majority of WLANs in operation conform to 802.11b, more-advanced technologies were on the drawing boards in 2002, and to a small extent marketed commercially. Two organizations guide the standardization of new WLAN technology, the IEEE (http://ieee802.org/11) and ETSI (European Telecommunications Standards Institute) (http://portal.etsi.org/bran/kta/Hiperlan/ hiperlan2.asp). IEEE efforts take place within the 802.11 Working Group, which consists of a number of task groups, each labeled with a lower case letter. ETSI activity, referred to as HiperLAN2 (high performance LAN), focuses on WLAN technology operating in the 5-GHz band at a bit rate of 54 Mbps. Table 2.2, a summary of bit rates and spectrum bands for existing and emerging WLANs, shows that 802.11a technology operates at the same bit rate and in the same part of the electromagnetic spectrum as HiperLAN2. This congruence has been the stimulus for discussions on harmonizing the two technologies.[2]

2.2.2 CELLULAR TRENDS

Progress in cellular communications technology has been measured by "generations." The principal characteristics of first-generation systems, introduced in the early 1980s, were analog speech transmission over radio channels and limited built-in roaming capability. Second-generation systems, transmitting digital speech signals, were introduced in the early 1990s and today account for the overwhelming majority of cellular telephone communications. Starting with a wide array of incompatible first-generation radio transmission technologies deployed throughout the world, the number converged to four in the second generation. GSM (Global System for Mobile Telecommunications), standardized by ETSI, has by far the largest

TABLE 2.3
Advanced Second-Generation (2.5G) and Third-Generation Cellular Systems

	Generation	Channel BW (Hz)	Channel Rate (bps)	Principal Information Format
GSM				
GSM	2	200 k	271 k	Voice and circuit data
EDGE	2.5	200 k	813 k	Voice and circuit data
GPRS	2.5	200 k	271 k	Packet data
E-GPRS	2.5	200 k	813 k	Packet data
W-CDMA/FDD	3	5 M	3.84 M	Multimedia
W-CDMA/TDD	3	5 M	3.84 M	Multimedia
CDMA				
CDMA1	2	1.25 M	1.2288 M	Voice and circuit data
1XRTT	2.5	1.25 M	1.2288 M	Voice and circuit data
HDR	2.5	1.25 M	Uplink 2.4 M Downlink 153 k	Packet data
CDMA2000	3	3.75 M	3.6864 M	Multimedia

subscriber base and the most-widespread adoption geographically. The most-salient characteristic of GSM radio transmission is its TDMA (time division multiple access) technique. A GSM signal occupies a bandwidth of 200 kHz. The transmission bit rate is 270 kbps with eight digital signals sharing the same carrier. The CDMA (code division multiple access) system, conforming to Interim Standard 95 published by the TIA (Telecommunications Industry Association), has the second-largest subscriber base. It is deployed throughout North America and in several Asian countries. CDMA signals occupy a bandwidth of 1.25 MHz with a binary signaling rate of 1,228,800 chips per second. The two other digital systems are similar to one another. NA-TDMA, the North American time division multiple access system conforming to TIA Interim Standard 136, operates with a bandwidth of 30 kHz per channel and a signaling rate of 48,600 bits per second. It is deployed throughout North America and in a few countries in Latin America. PDC (Personal Digital Cellular), with a signal bandwidth of 25 kHz, is a Japanese standard similar to NA-TDMA.

In 2002, the introduction of new radio technology proceeds in two streams, one based on GSM and the other on CDMA. Both streams contain 2.5G (advanced second generation) systems, with signals confined to the existing 2G bands (200 kHz for GSM and 1.25 MHz for CDMA) and 3G systems, with signals occupying 4 or 5 MHz bandwidth. Table 2.3 is a catalog of the systems in the GSM and CDMA streams. In North America, NA-TDMA operating companies have announced technology migration paths to the GSM stream. In Japan, PDC operating companies have introduced 3G systems based on W-CDMA (wideband code division multiple access).

The original second generation systems were designed to carry voice conversations, for the most part. They also carry circuit-switched data. Their enhancements

(2.5G) are segregated in two categories: EDGE and 1XRTT carry voice and circuit-switched data at higher bit rates than GSM and CDMA1, respectively. On the other hand, GPRS (General Packet Radio Service), E-GPRS, and HDR are designed for packet-switched data. A principal characteristic of the 3G systems is their ability to carry a variety of traffic types. While 2G and 2.5G systems classify information as either circuit or packet oriented, 3G systems' planners classify information according to latency requirements within four categories: background, interactive, streaming, and conversational.

Although the channel signaling rates are fixed for each system, only 2G systems specify constant user throughput. All of the other systems contain "rate adaptation" technology that matches the transmission rate available at each terminal to the current channel quality, as determined by network congestion and location-specific radio propagation conditions. For example, EDGE defines 12 "modulation and coding schemes," with user bit rates ranging from 8.8 to 88.8 kbps per time slot.[3] An application can use from one to eight time slots to exchange information.

2.2.3 UNITING WLANs AND CELLULAR

The complementary strengths and weaknesses of WLANs and cellular systems make it certain that a wireless Internet will contain both technologies. Recognizing this prospect, the technical community has turned its attention to coordination of cellular systems and WLANs. Short-term approaches to this coordination use existing network infrastructure, while more futuristic work anticipates new network architecture based on Internet protocols that inherently accommodate both types of radio access. One example based on existing infrastructure is an OWLAN (operator WLAN)[4] combining GSM subscriber management and billing mechanisms (authorization, authentication, and accounting) with WLAN radio access. A key aspect of the OWLAN is incorporation of a GSM SIM (subscriber identity module) in the subscriber equipment containing a WLAN modem. Another example uses a cellular data modem as a bridge linking the Internet with a cluster of laptop computers, all communicating with a WLAN access point.[5] The cellular modem relays data between the access point and the cellular network infrastructure operating a suite of Internet protocols.

In contrast to cellular–WLAN coordination using existing infrastructure, there is intense industry effort devoted to specification of a core network based on Internet protocols. Such a core network would serve terminals that communicate by means of WLAN, cellular, and a variety of other wired and wireless access technologies. Section 2.4.2 describes examples of work in progress on network architectures that address a broad range of technical challenges including roaming, handoff, security, and quality of service (QoS).

2.2.4 PERSONAL AREA NETWORKS

Although cellular telephones and WLANs have attracted the greatest consumer acceptance to date, other wireless networks have a role to play in a wireless Internet. Among them personal area networks (PANs) using Bluetooth technology are the most prominent.[6] The original aim of Bluetooth was to provide low-cost, low-power connections between a variety of consumer products. One example is a Bluetooth link between a laptop computer and a 3G cell phone enabling the computer to gain access to the Internet by means of the 3G packet data infrastructure. Another example is a cordless headset linked to a cell phone or a personal stereo device. In the context of these applications, Bluetooth appears as a low-cost alternative to WLAN modems. In addition, Bluetooth also contains sophisticated *ad hoc* networking capabilities. These capabilities are contained in technologies built into the Bluetooth standard for creating piconets and scatternets that use Bluetooth modems to create networks linking a large number of wireless devices.

2.2.5 TECHNOLOGY GAPS

Each of the emerging advances in the cellular, WLAN, and PAN domains works within a region of the six-dimensional figure of merit volume (mobility, bit rate, coverage, signal quality, power, and price) described at the beginning of Section 2.2. All of them address the "last mile" or "last five meters" problem of linking devices to the Internet. An examination of the details of each of these technologies reveals that in sum they will remain inferior to wired connections consisting of Ethernets, digital subscriber lines, or cable modems connected to a 10 Gbps Internet backbone. The result will be

wireless Internet = Internet − some of the wires < Internet with wires

To get beyond these limitations, it will be necessary to create new communications paradigms that are matched directly to the requirements and constraints of the users, the information, and the operating environment of a wireless Internet. The next section adopts the theme of "geography" to formulate a framework for technology creation, and Section 2.4 describes current research within this framework.

2.3 FRAMEWORK FOR TECHNOLOGY CREATION

The incremental evolution of a wireless Internet, described in Section 2.2, takes a bottom-up approach of augmenting existing wireless technology and Internet protocols in order to provide a smoother interface between the two marriage partners. On the wireless side, the principal aim is to increase channel bit rate. On the Internet side, extended protocols aim to accommodate mobility, the variable quality of wireless signals, and vulnerability of wireless systems to eavesdropping and unauthorized access. In this section, we introduce a top-down approach that aims for a wireless Internet that is more than the sum of the two existing communications systems. This approach begins with a three-dimensional analysis, including the characteristics of (1) the endpoints of communication, (2) the information transferred, and (3) the physical nature of wireless signals. In this analysis, we find significant differences from the wired Internet on all three dimensions. The common aspect of all three dimensions can be summed up in the word "geography." Cellular systems aim to be the same "anytime, anywhere," and the name "World Wide Web"

carries a similar suggestion. By contrast, our analysis of wireless Internet requirements in the following paragraphs reveals a fundamental dependence on location,
including (1) locations of information terminals (geography of users), (2) the location-dependent relevance of information (geography of information), and (3) location-dependent quality of signals (geography of signal transmission). Section 2.4
refers to examples of research in progress that aligns technology with the geography
of users, the geography of information, and the geography of signal transmission.

2.3.1 THE GEOGRAPHY OF WIRELESS INTERNET USERS

The Internet was originally designed to move data packets carrying many types of
information between host computers in stationary, known locations. By contrast,
cellular networks were originally designed to carry telephone calls and short messages in systems that are matched to the geographical distribution of subscribers,
their mobility patterns, and the temporal distribution their service needs. Technology
creation and deployment are considerably more complicated in a wireless Internet
because mobile terminals with different capabilities will transmit and receive multimedia information in a variety of formats, with widely different quality-of-service
requirements that place varying demands on network resources.

Wireless Internet technology needs to be sensitive to the characteristics of the
sources and destinations of information, which will often be groups that share
information. Groups form and dissolve as clusters in time and space. The formation
and disintegration of such groups may or may not be initiated by the users involved.
Key characteristics of the geography of users are location, mobility state (speed and
direction), timing of information needs, and demographics of individuals and user
groups. An example of a group formed spontaneously is the population of mobile
callers in an unexpected traffic jam. In this case, the defining characteristics of the
group are the locations and mobility states of the group members.

The endpoints of a wireless Internet will include familiar information devices
carried by people (telephones, PDAs, laptop computers). There will be an increasing
number of autonomous devices such as wireless sensors with specialized tasks of
acquiring, transmitting, and receiving diverse types of data. A few examples are
geolocation information, biomedical measurements, and surveillance pictures. Pervasive computing anticipates a proliferation of cooperating autonomous wireless
terminals.

Multicasting, an increasingly popular mode of Internet information transfer, is
likely to be even more attractive in a wireless Internet. In a multicast, the "end user"
is a group comprising a variable population of members defined on a per-session
basis. In a wireless Internet, multicasting is likely to be just as popular but, owing
to the mobility of terminals and variability of transmission conditions, it will present
challenges that do not arise in the wired Internet.[7]

Geocasting is a form of multicast that that can add to the value of a wireless
Internet.[8] Geocasting defines a multicast group with reference to a target area. The
members of the group are terminals with geographical coordinates within the target
area. In addition to location, mobility states (velocity and direction) and demograph
ics can be major factors in the definition of geocast groups. The geocast membership

can be specified by the sender of information, the recipient, or by a service provider. A geocast session may consist of one or more messages that are sent to the geocast group. A message can originate with a group member or outside the group. For example, in the action "send a reminder to all students and faculty within 3 km of the campus that a seminar will begin in 30 minutes," the originator of a message defines a geocast group by location and demographic category. In the action "get information about all shoe stores that I can reach in 30 minutes," the information recipient defines a group by location and mobility. Finally, in the action "notify everyone within a radius of 10 km of a traffic jam," the service provider uses an arbitrary criterion defined by location to specify a group.

Geolocation (discussed in further detail in Section 2.4.2.1), the process of determining the geographical coordinates of an information device, is a technology that supports geocasting. The construction and maintenance of the geocast group are nontrivial tasks for mobile networks. Most studies assume that geolocation information is continuously available to mobile nodes via the Global Positioning System (GPS). While this is generally a viable assumption, the manner in which the geolocation information is acquired and disseminated has significant impact on network capacity and performance.

2.3.2 THE GEOGRAPHY OF INFORMATION[9]

For twenty years, the expression "anytime, anywhere" has been a cellular technology mantra. At first only a lofty goal, the combination of satellite telephones and terrestrial cellular systems have made "anytime, anywhere" a reality for telephone calls and short text messages; it is also a good description of the World Wide Web paradigm in which content seems pervasive, contained in Web pages that can be summoned to any computer in the world at the click of a mouse. Although this paradigm is appealing, the geography of signal transmission, described in Section 2.3.3, makes it difficult and expensive to achieve with wireless technology, even for the simple task of delivering telephone calls and short messages. With the added complexity of multimedia wireless Internet information and the diversity of user characteristics, "anytime, anywhere" becomes prohibitively demanding. Thus, we would do well to examine the nature of the information conveyed in a wireless Internet to determine the conditions in which ubiquitous, instantaneous coverage is essential, not merely a convenience to be weighed against its costs. Rather than impose the burden of "anytime, anywhere" on all communications in a wireless Internet, we examine the temporal, spatial, and demographic coordinates of information. Matching communication technology to information geography promises gains in efficiency and quality of a wireless Internet.

In examining the geography of information, we classify services according to where, when, and to whom the information is relevant. We represent the classification in each of these dimensions — space, time, and personal — in a range from specific to general. At one extreme we have information that is useful to only one person, at a particular time, when the person is in a particular place. For example, a message generated while you are on your way to the airport that "you are urgently requested to deal with an emergency in your home" is localized in all three dimensions. Unless

TABLE 2.4
Localized and General Information

Location	Time	Personal	Information Example
Specific	Specific	Specific	Emergency dispatch message
Specific	Specific	General	Traffic conditions
Specific	General	Specific	Alert us when a friend is nearby
Specific	General	General	Local maps, directories
General	Specific	Specific	Horoscope
General	Specific	General	Stock market prices
General	General	Specific	Message containing family news
General	General	General	Music recording

you receive the message very soon and you are near home, the information is not very useful to you. Information at the other extreme is a popular music recording. It has no time localization and it is of interest to a large population of people throughout the world. Table 2.4 gives examples of information in the corners of the three-dimensional cube in which spatial relevance, temporal relevance, and personal relevance range from specific to general.

The first example, at the top of the table, requires "anytime, anywhere" message delivery through a network with ubiquitous coverage, while by contrast the music recording at the bottom of the table can be downloaded at a time and place that are convenient, economical, and conducive to reliable information transfer. If the recording is very popular, multicasting would make sense. Local maps and directories in the middle of the table lend themselves to geocasting by wireless information kiosks.

2.3.3 THE GEOGRAPHY OF SIGNAL TRANSMISSION

It is well known that the signals transmitted by wireless modems are subject to a variety of transmission impairments, the most prominent of which are:

- Attenuation that depends on the distance between transmitter and receiver
- Fading that depends on the physical characteristics of the transmission environment and the motion of wireless terminals
- Additive noise in modem receivers
- Interference due to transmissions by other modems

Attenuation and fading effects are highly dependent on the locations of transmitters and receivers and interference varies with both time and the locations of the interfering transmitters and the location of the signal receiver.

Engineers have devised a vast array of modulation, reception, coding, signal processing, and network control techniques to mitigate the effects of these impairments. To use them effectively, network managers devote high levels of effort and expense to address the geography of signal transmission and the geography of users in determining the locations of base stations and access points and precisely orienting their antennas. They aim for highly reliable signal reception in the greatest possible coverage area at all times.

In spite of the effort and expense devoted to erasing the inherent time and location dependence of signal quality, the goal of "anytime, anywhere" communications remains elusive in all WLANs and new (2.5G and 3G) cellular systems. All of these technologies prescribe radio modems that can operate with a collection of modulation and coding schemes, each with its own transmission rate and immunity to impairments. They employ rate adaptation to find at any time and place the best compromise between signal quality and transmission rate. As in the example of Figure 2.2, this trade-off depends on the locations of transmitters and receivers and on network activity.

The nature of the compromise resembles that of a telephone modem built into a personal computer in that the modem operates at a bit rate matched to the characteristics of each dial-up connection. However, the effects on applications are quite different. A dial-up modem operates at one rate for the duration of a connection. By contrast, the mobility of wireless terminals and the time-varying nature of the interference will cause wireless modems to change their rates far more frequently. Managing quality of service of applications in the presence of location-dependent and time-dependent transmission rates and signal quality levels is a major challenge that remains to be addressed.

2.4 RESEARCH INITIATIVES

The industry trends described in Section 2.2 follow evolutionary paths within the framework of established cellular and local area networks. In doing so, they fail to address directly the essential characteristics of a wireless Internet as represented by the geography of users, the geography of information, and the geography of signal transmission described in Section 2.3.3. To fill the gap, the research community is exploring novel network architectures supported by an IP-based core network. The network architectures include proximity-based communications, *ad hoc* networking, and hybrids incorporating these with cellular networks, WLANs, and other approaches. The IP core network will require a new service support sublayer between the transport layer and the applications layer. It will make use of geolocation information to facilitate network control and optimize the use of network resources. This section describes examples of work in progress on adaptive network architecture and on the core network.

2.4.1 ADAPTIVE NETWORK ARCHITECTURES

A wireless Internet presents new possibilities of adaptive networking solutions. Instead of simply cutting the wires in the last mile and viewing air as the hostile medium of transmission whose shortcomings need to be combated, the absence of wires can provide novel means of disseminating control and user information by taking advantage of mobility. The client/server model that governs the cellular approach, where all radio resource allocation is determined centrally, attempts to erase the effects of mobility rather than take advantage of it. As mentioned in Section 2.3.1, the diversity of users accessing a wireless Internet will proliferate with the future deployment of autonomous devices that collect, measure, process, query, and relay information. The growth in pervasive computing devices will make it impractical for fixed access points to provide centralized mobility management and ensure bandwidth efficiency. Therefore, in the confines of the client/server model of the cellular architecture, mobile users would experience limited quality of service and limited data access.

The radio links of a wireless Internet are at the perimeter of a complex information network. The interface between the core Internet and the radio links of a wireless Internet comprise a *radio access network,* which needs to respond to the changing wireless landscape. Radio access networks for cellular and WLAN radio links differ significantly. Cellular access networks consist of a sophisticated infrastructure linking base stations, routers, servers, and databases. WLAN access networks come in many varieties. They can have a substantial infrastructure or none at all, relying on *ad hoc* connections between WLAN modems for network control. The networks can have a hierarchical or peer-to-peer topology. However, no radio access network has yet emerged as the clear winner. The next-generation wireless network will need to be a network of networks where the boundaries between different modes of radio access are transparent to the user. The quest for this capability is reflected in the research-and-development efforts toward WLAN/cellular coordination discussed in Section 2.2.3. More radical examples of seamless transition between modes of access in real-time are emerging in the research community.

Future radio access networks will need to promote efficient use of electromagnetic resources by all transceivers, mobile and fixed. This mindset immediately points to the flexibility offered by proximity-based and peer-to-peer communications augmenting the conventional infrastructure networking. Such flexibility would present adaptation capability to overall spatial and temporal variation of traffic, as well as the mobility states of user groups and individual users. Further, mobility of nodes can now be viewed as an advantage in information dissemination, routing, and cooperation for improved quality of service. In an adaptive network of mobile nodes, each mobile device enriches the web of communication by contributing to the network density. Data can move from car to car, among people passing each other in the streets, in the hallways of an office building, in a park, or in an airport. Military communications development efforts have for some time taken these concepts into account. Proximity-based communications will promote the efficiency and resilience of a wireless Internet.

Solutions that take advantage of proximity and peer-to-peer cooperative communications include Infostations, multihop systems that extend the range of fixed wireless systems, *ad hoc* networks of various hierarchy levels, and hybrid systems. The common thread in this seemingly diverse set of architectures is the motivation to adapt to the geography of users, information, and signal transmission in a locally optimal manner. This section is a survey of exemplary new architectures.

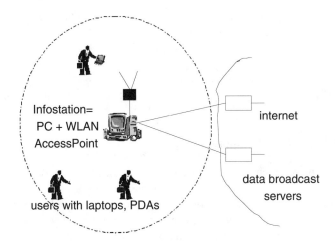

FIGURE 2.3 Infostation system elements.

2.4.1.1 Proximity-Based Systems

An Infostation is an example of a proximity-based system that takes advantage of user mobility;[10] it provides wireless information services to users located in or traversing a limited coverage area. As people with notebook computers or PDAs pass by an Infostation, they receive useful information, with little or no human interaction. This could be information that is most relevant near the Infostation, such as local maps, restaurant listings, or information about courses at a university; or it could be information of general interest, such as news articles or music.

As shown in Figure 2.3, an Infostation consists of a radio transceiver (such as a WLAN access point) that provides high bit rate, low-cost, low-power network connections to portable terminals in a restricted coverage area, along with computer hardware and software that caches relevant data and schedules transmissions. Because a subscriber to an Infostation service may spend a short time in the service area of each Infostation, the information transfer should be organized in advance and should take place at the speed of electronic processes rather than the speed of human–computer interactions. A network of Infostations consists of several isolated coverage areas separated by large gaps. The cellular network serving the region containing all the Infostations can enhance the operation of the Infostation network by observing the changing locations of users. The Infostation system can use this location knowledge to move information needed by a user to an Infostation before the user arrives.[11] The information can be quickly downloaded to an information terminal in the short time that the user is in the Infostation coverage area.

The Infostation paradigm is motivated by our earlier observation that no single "one size fits all" technology is suitable for all wireless information services, and WLANs and cellular networks both will be prominent in a wireless Internet. The best way to deliver information depends on various facets of the geography of the information, including spatial and temporal aspects, as well as characteristics of the users. For many types of information, "many-time, many-where" coverage offered by Infostations is sufficient to serve a mobile population. For services such as e-mail, voice mail, maps, restaurant locations, and many others, there is no penalty incurred by waiting until a terminal arrives in range of an Infostation, provided the user is sufficiently mobile. If information delivery becomes urgent, the cellular network is available to deliver it, albeit at lower bit rates and higher cost in fees and power dissipation.

2.4.1.2 Cooperative Communications

A large body of research in progress anticipates that devices will cooperate with one another to deliver information. The cooperation can occur at different protocol layers. At the physical layer, one device can provide diversity transmission and reception for another one. At the application layer, a user can receive a Web page from the cache of a nearby user and avoid the need to communicate with the Web server where the page originated. Multihopping is an example of cooperation at the network layer.

Cooperation naturally relies on proximity of network devices that can assist one another. Mobility enhances cooperation by increasing the probability that a device will be able to receive assistance from other devices.[12] The following paragraphs refer to work in progress on cooperative systems.

The large body of current research on *ad hoc* networks anticipates cooperative communications in many forms. Demon Networks (http://www.winlab.rut-gers.edu/~crose) envision an *ad hoc* local area network that takes advantage of mobility in order to route information. The network is not necessarily fully connected at any given time. Therefore, changes in network topology are essential for packet delivery rather than a complication to be overcome. Mobile stations can keep packets they receive and each packet to be delivered will almost certainly have many copies in the system at a given instant. The dissemination of information in this case resembles an epidemic, in which useful information is the contagious disease. The destination can be a single node or a multicast group. Each node is responsible for managing its memory allocation by making timely decisions regarding the deletion of packets it carries and disseminates.

The Terminode project applies this idea in a metropolitan area.[13] It uses mobility of users to disseminate information throughout a city. Each Terminode contains a map of the city and uses the map to make routing decisions.

Other forms of cooperation in *ad hoc* networks have been formulated as power combining and cooperative coding where diversity is exploited for purposes of maximizing bandwidth efficiency, extending coverage, network lifetime, and battery life.[14,15]

2.4.1.3 Hybrid Architectures

In addition to the limitations of cellular infrastructure in terms of quality of service and data rates, there are situations (for example shadowing and equipment failure) in which no communication infrastructure is available to terminals. The 7DS system

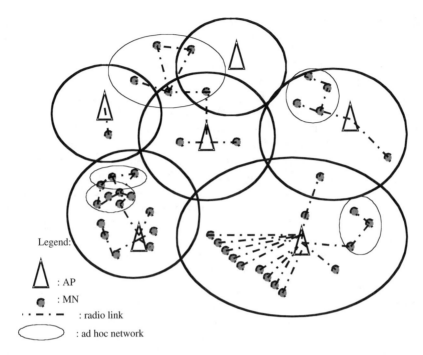

FIGURE 2.4 An exemplary snapshot of a CAHAN.

(seven degrees of separation) is motivated by the limitations of dependence on a network infrastructure.[16] In 7DS, mobile and stationary terminals cooperate to share information, help maintain connectivity to the network, and relay messages for one another. They can serve as *ad hoc* gateways into the Internet.

The 7DS architecture allows peer nodes to communicate via a WLAN, forming a flat *ad hoc* network, where some nodes have connectivity to the Internet. The connection to the infrastructure can be achieved by any access mode, such as Infostations, cellular base stations, or WLAN access points. For purposes of information sharing, peers query, discover and disseminate information. When the network connection sharing is enabled, the system allows a host to act as an application-based gateway and share its connection to the Internet. For message relaying, hosts that do have access to the Internet forward messages on behalf of other hosts. The system is an example of adaptive networking that adjusts its routing, cooperation, and power control based on the availability of energy and bandwidth. Furthermore, 7DS inherently exploits host mobility. Currently, the peer-to-peer portion of the network is implemented in a WLAN environment.

The hybrid Cellular Ad Hoc Augmented Network (CAHAN) has been developed with the above influences.[17] The goal of CAHAN is to make the best use of the cellular infrastructure where the centralized control and the fixed reference points provided by base stations are advantageous, and to incorporate peer-to-peer communications to optimize radio resource allocation, resilience, and power consumption. Figure 2.4 depicts an exemplary snapshot of CAHAN.

2.4.2 THE IP-BASED CORE NETWORK

It is widely accepted that future networks will converge to an IP-based core at the transport layer. In this event, an additional mobility layer between the transport and application layers is needed to ensure locally optimal wireless access to Internet services and applications. This layer will provide the intelligence for location and context awareness, media conversion, scaling, and seamless transition between modes of access in a manner that is transparent to the application or service. This layer introduces the true spirit of pervasive networking, where distributed computing and wireless access combine to make the network virtually disappear in the eyes of the user. In Yumiba et al.,[18] the intelligent mobility support layer is referred to as service support middleware consisting of several functions grouped in two sublayers. The *service support sublayer* performs location management, media conversion, and user profile management. The *network management* sublayer performs billing, security, and QoS provisioning.

The interaction between the two sublayers and the individual function blocks in these sublayers are under investigation by several working groups. The implementation of mobility management functions in this new context is the subject of intensive, ongoing research all over the world. The evolution of GSM systems into the IP-based future cellular network is discussed in Park.[19] Li et al.[20] present an architecture that supports delivery of advanced services through WLAN access points.

The service support sublayer needs input from the radio access network on the geolocation and mobility state of the user, in addition to the mode of access. Combined with prior user information maintained in user profiles, the coordinates, speed, and direction of a user as reported by the radio access network will be translated into immediate requirements for serving a user, as well as predictions of future needs. The flow of information between the radio access network and the service support sublayer will ensure appropriate service delivery on demand or in a proactive fashion. The enablers of such pervasive networking are new technologies in geolocation, prefetching, caching, and radio resource allocation.

2.4.2.1 Geolocation

Geolocation is the term coined for determining the geographic coordinates of a mobile node. Many methods of using the radio access network with or without the aid of specialized mobile terminals have been proposed in recent years primarily with the objective of locating emergency callers.[21] Geolocation also tracks terminals by measuring speed and direction. All geolocation mechanisms consist of acquisition, computation, and storage of measurements or computed coordinates for averaging or tracking. The distribution of these functions is a design decision reflecting the distribution of functions between mobile terminals and fixed network elements.

Most studies of location-based services assume that geolocation information is continuously available to mobile nodes via the Global Positioning System (GPS);[22] while this is generally a viable assumption, the manner in which the geolocation information is utilized has significant impact on network capacity and performance. We characterize the impact of the geolocation method by its error region size. The error region is the area around the actual position of the mobile that the computed geolocation will lie in with a given high probability. The optimization of the geolocation and update intervals will mostly depend on the mobility patterns of the mobile nodes, as well as the target area size.

As an enabler of wireless Internet, a geolocation method of choice should be dictated by the requirements of the location-aware application or service. Furthermore, the signaling flow for geolocation will be different in different network architectures. Measurement, computation, and storage functions can be performed by different network components, in a hierarchy or in a cooperative, peer-to-peer fashion. The geolocation method, along with the mode of communication, should be optimized locally, so that it will work in harmony with the adaptive network architectures. The role of geolocation with respect to the geographical framework in Section 2.3 is summarized as follows: Geolocation enables the adaptation to the geography of information and users by facilitating location-aware services and applications. It helps the radio access network cater to the geography of users and the geography of signal transmission.

Facilitation of location-aware services and applications prompts signaling between the radio access network and the service support sublayer. Media conversion, content, and location management components need to interpret the position, speed, and direction of the mobile obtained from the radio access network by cross-referencing these with target area and subscriber profile information, which is maintained in the service support sublayer. The amount of signaling between the service support sublayer and the radio access network needs to be optimized. Mobility modeling and trajectory prediction methods are found to be helpful in assigning the maximum geolocation update interval subject to the quality of service requirements of the particular application or service.[23]

Along with mobility modeling and trajectory prediction, geolocation and tracking mechanisms make it possible for the network to prefetch and cache information proactively. An example of file prefetching in a drive-through Infostation system is given in Iacono and Rose.[24] Other studies consider the file prefetching in base stations of the cellular infrastructure.[25]

2.4.2.2 Resource Management

The *ad hoc* networks investigated in Section 2.4.1 raise new issues related to management of radio resources and management of battery energy in terminals. In these networks, terminals sometimes function as endpoints of communication links and other times as relays, receiving and forwarding packets moving to and from other terminals. Each terminal therefore will use some of its energy for sending and receiving its own data and another portion of its energy assisting other terminals. Routing algorithms have a strong effect on overall energy consumption in a network and in individual terminals. They influence the proportion of energy each terminal expends for itself relative to energy used to assist other terminals. Research on energy-efficient routing in *ad hoc* networks considers total energy consumption in transmitting a message as well as average energy consumed by the terminals that participate in the transmission. Most of this work examines a stationary network with terminals in random positions. In this situation, there is considerable variation from terminal to terminal in the proportion of energy used for the tasks of relaying and communicating. By contrast, a recent study shows that when terminals are mobile, the variation is considerably diminished.[14] However, truly adaptive techniques that will optimize routing decisions based on the instantaneous remaining battery power of each node need to be devised.

Game theory appears to be a promising approach to the management of battery energy in the terminals of a network of cooperating nodes. A game theory formulation defines a utility function for each terminal and an overall ("social") utility function for the network. The utility function relates to the amount of data sent and received by the terminal and the energy consumed. Each terminal adopts a strategy for maximizing its utility. Because many terminals in a network share radio resources, the strategy adopted by one terminal affects the utility obtained by the others. This situation is similar to the one addressed in research on power control in cellular systems,[26] in which game theory strategies led to the design of efficient algorithms for power control for cellular data. In applying game theory to resource management in networks of cooperating nodes, a major issue is the nature of the cooperation that will promote effective distribution of radio resources and fair expenditure of battery energy across the terminals in a network. It is clear that a completely noncooperative game produces suboptimum results. In the studies of cellular data systems, the base station can coordinate the cooperation among terminals.[27] On the other hand, the terminals in the *ad hoc* networks under investigation are not in communication with a single coordinating device. Therefore, the cooperation must be distributed among the terminals in the network.

2.5 CONCLUSIONS

In this chapter, we have presented state-of-the-art wireless Internet technologies and their shortcomings. Prioritizing formulation of problems before solutions, we redefined wireless Internet as a range of opportunities fueled by the lack of wires and mobility rather than an array of obstacles in competing with wired Internet. Viewing wireless Internet as such, we introduced emerging network architectures and enabling technologies that pave the way toward this vision. Wireless Internet, if fully realized, will be the pervasive network that will disappear from the point of view of the end user. The network will sense the present state of its users, as well as predict their future needs and adapt, react, and plan proactively. In order to have these abilities, wireless Internet will need to be a network of networks, whose boundaries will be as transparent as possible to the user.

References

1. Fainberg, M. and Goodman, D., Maximizing performance of a wireless LAN in the presence of Bluetooth, Proc. 3rd IEEE Workshop on Wireless LANs, 2001.
2. Grass, E. et al., On the single-chip implementation of a Hiperlan/2 and IEEE802.11a capable modem, *IEEE Pers. Commun.,* 8 (6), 48–57, 2001.

3. Eriksson, M. et al., System performance with higher level modulation in the GSM/EDGE radio access network, *IEEE Globecom*, 5, 3065–3069, 2001.

4. Ala-Laurila, J., Mikkonen, J., and Rinnemaa, J., Wireless LAN access network architecture for mobile operators, *IEEE Communications Magazine*, 39(11), 82–89, 2001.

5. Noerenberg, J., Bridging wireless protocols, *IEEE Communications Magazine*, 39(11), 90–97, 2001.

6. Haartsen, J.C., The Bluetooth radio system, *IEEE Pers. Commun.*, 7 (1), 28–36, 2000.

7. Gossain, H., de Morais Cordeiro, C., and Agrawal, D.P., Multicast: wired to wireless, *IEEE Communications Magazine*, 40(6)116–123, June 2002.

8. Navas, J.C. and Imielinski, T., Geographic addressing and routing, Proc. Mobicom '97, Budapest, Hungary, September 1997.

9. Goodman, D.J., The wireless Internet: promises and challenges, *Computer*, 33 (7), 36–41, 2000.

10. Goodman, D.J. et al., Infostations: a new system model for data and messaging services, Proc. IEEE Vehicular Technology Conference, 969–973, 1997.

11. Iacono, A.L. and Rose, C., Bounds on file delivery delay in an Infostation system, Proc. IEEE Vehicular Technology Conference, 2295–2299, 2000.

12. Grossglauser, M. and Tse, D., Mobility increases the capacity of *ad hoc* wireless networks, *IEEE/ACM Transactions on Networking*, 10(4), 477–486, 2002.

13. Blazevic, L., Giordano, S., and Le Boudec, J.-Y., Self organized routing in wide area mobile *ad hoc* networks, *Proc. IEEE Globecom*, 5, 2814–2818, 2001.

14. Catovic, A. and Tekinay, S., A New Approach to Minimum Energy Routing for Next Generation Multihop Wireless Networks, *J. Communications and Networks*, 4(4), 351–362, 2002.

15. http://eeweb.poly.edu/~elza/Publications.htm.

16. http://www.cs.columbia.edu/~maria/project/.

17. Tekinay, S., Adaptive networks for next generation wireless communications: the growing role of peer-to-peer communications, in *Wireless Communications and Networking*, Sunay, O., Ed., Kluwer, Dordrecht, Netherlands, in press.

18. Yumiba, H., Imai, K., and Yabusaki, M., IP-based IMT network platform, *IEEE Pers. Commun.*, 8 (6), 18, 2001.

19. Park, J.-H., Wireless Internet access for mobile subscribers based on the GPRS/UMTS network, *IEEE Communications Magazine*, 40(4), 38–49, 2002.

20. Li, J. et al., Public access mobility LAN: extending the wireless Internet into the LAN environment, *IEEE Wireless Commun.*, 9(3), 22–30, 2002.

21. S. Tekinay, Guest Ed., Wireless geolocation systems and services, *IEEE Communications Magazine*, 36, 36(4), 28, 1998.

22. Sarikaya, B., Ed., *Geographic Location in the Internet*, Kluwer, Dordrecht, 2002.

23. Choi, W.-J. and Tekinay, S., Mobility modeling and management for next generation wireless networks, *Proc. Symp. on Wireless Personal Multimedia Communications* 2001, Aalborg, Denmark, 2001.

24. Iacono, A.L. and Rose, C., Infostations: a new perspective on wireless data networks, in *Next Generation Wireless Networks*, Tekinay, S., Ed., Kluwer, Dordrecht, Netherlands, 2001.

25. Kobayashi, H., Yu, S.-Z., and Mark, B.L., An integrated mobility and traffic model for resource allocation in wireless networks, *Proc. Workshop on Wireless Mobile Multimedia*, Boston, 3–63, 2000.

26. Saraydar, C.U., Mandayam, N.B., and Goodman, D.J., Efficient power control via pricing in wireless data networks, *IEEE Trans. Commun.*, 50(2), 291–303, 2002.

27. Goodman, D.J. and Mandayam, N.B., Network Assisted Power Control for Wireless Data, *Mobile Networks and Applications*, 6(5), 409–418, 2001.

3 Wireless Internet Security

Dennis Seymour Lee

CONTENTS

3.1 INTRODUCTION

Recalling the early days of the Internet, one can recount several reasons why the Internet came about:

- A vast communications medium to share electronic information
- A multiple-path network that could survive localized outages
- A means for computers from different manufacturers and different networks to talk to one another

0-8493-1502-6/03/$0.00+$1.50
© 2003 by CRC Press LLC

Commerce and security at that time were not high on the agenda (with the exception of preserving network availability). The thought of commercializing the Internet in the early days was almost unheard of. In fact, it was considered improper etiquette to use the Internet to sell products and services. Commercial activity and its security needs are more-recent developments on the Internet, having come about strongly in the past few years.

Today, in contrast, the wireless Internet is being designed from the very beginning with commerce as its main driving force. Nations and organizations around the globe are spending millions, even billions of dollars to buy infrastructure, transmission frequencies, technology, and applications in the hopes of drawing business. In some ways, this has become the "land rush" of the new millennium. It stands to reason, then, that security must play a critical role early on as well: where money changes hands, security will need to accompany this activity.

Although the wireless industry is still in its infancy, the devices, infrastructure, and applications development for the wireless Internet are rapidly growing on a worldwide scale. Those with foresight will know that security must fit into these designs early. The aim of this chapter is to highlight some of the significant security issues in this emerging industry that need addressing. These are concerns that any business wishing to deploy a wireless Internet service or application will need to consider to protect itself and its customers, and to safeguard investments in this new frontier.

Incidentally, the focus of this chapter is not about accessing the Internet using laptops and wireless modems; that technology, which has been around for many years, in many cases is an extension of traditional wired Internet access. Neither will this chapter focus on wireless LANs and Bluetooth, which are not necessarily Internet based, but deserve chapters on their own. Rather, the concentration is on portable Internet devices, such as cell phones and PDAs (personal digital assistants), which inherently have far less computing resources than regular PCs. Therefore, these devices require different programming languages, protocols, encryption methods, and security perspectives to cope with the technology. It is important to note, however, that despite their smaller sizes and limitations, these devices have a significant impact on information security, mainly because of the electronic commerce and intranet-related applications that are being designed for them.

3.2 WHO IS USING THE WIRELESS INTERNET?

Many studies and estimates are available today that suggest the number of wireless Internet users will soon surpass the millions of wired Internet users. The assumption is based on the many more millions of worldwide cell phone users who are already out there, a population that grows by the thousands every day. If every one of these mobile users chooses to access the Internet through a cell phone, indeed that population could easily exceed the number of wired Internet users by several times. It is this very enormous potential that has many businesses devoting substantial resources and investments in the hopes of capitalizing on this growing industry.

The wireless Internet is still very young. Many mobile phone users do not yet have access to the Internet through their cell phones. Many are taking a "wait-and-see" attitude toward available services. Most who do have wireless Internet access are early adopters experimenting with the potential of what this service could provide. Because of the severe limitations in the wireless devices — the tiny screens, the extremely limited bandwidth, as well as other issues — most users who have both wired and wireless Internet access will admit that, for today, the wireless devices will not replace their desktop computers and notebooks anytime soon as their primary means of accessing the Internet. Many admit that "surfing the Net" using a wireless device today could become a disappointing exercise. Most of these wireless Internet users have expressed the following frustrations:

- It is too slow to connect to the Internet.
- Mobile users can be disconnected in the middle of a session when they are on the move.
- It is cumbersome to type out sentences using a numeric keypad.
- It is expensive to use the wireless Internet, especially when billed on a per-minute basis.
- There is very little or no graphics display capabilities on wireless devices.
- The screens are too small and users have to scroll constantly to read a long message.
- There are frequent errors when surfing Web sites (mainly because most Web sites today are not yet wireless Internet compatible).

At the time of this writing, the one notable exception to these disappointments is found in Japan. The telecommunications provider NTT DoCoMo has experienced phenomenal growth in the number of wireless Internet subscribers, using a wireless application environment called i-mode (as opposed to the wireless application protocol, or WAP). For many in Japan, connection using a wireless phone is their only means of accessing the Internet. In many cases, wireless access to the Internet is far cheaper than wired access, especially in areas where the wired infrastructure is expensive to set up. i-mode users have the benefit of "always online" wireless connections to the Internet, color displays on their cell phones, and even graphics, musical tones, and animation. Perhaps Japan's success with the wireless Internet will offer an example of what can be achieved in the wireless arena, given the right elements.

3.3 WHAT TYPES OF APPLICATIONS ARE AVAILABLE?

Recognizing the frustrations and limitations of today's wireless technology, many businesses are designing their wireless devices and services not necessarily as replacements for wired Internet access, but as specialized services that extend what the wired Internet could offer. Most of these services highlight the attractive convenience of portable informational access, "anytime, anywhere," without having to sit in front of a computer; essentially, Internet services one can carry in one's pocket. Clearly, the information would have to be concise, portable, useful, and easy to access. Examples of mobile services available or being designed today include:

- Shopping online using a mobile phone; comparing online prices with store prices while inside an actual store
- Getting current stock prices, trading price alerts, trade confirmations, and portfolio information anywhere
- Performing bank transactions and obtaining account information
- Obtaining travel schedules and booking reservations
- Obtaining personalized news stories and weather forecasts
- Receiving the latest lottery numbers
- Obtaining the current delivery status for express packages
- Reading and writing e-mail "on the go"
- Accessing internal corporate databases such as inventory, client lists, etc.
- Getting maps and driving directions
- Finding the nearest ATM machines, restaurants, theaters, and stores, based on the user's present location
- Dialing 911 and having emergency services quickly triangulate the caller's location
- Browsing a Web site and speaking real-time with the site's representative, all within the same session

Newer and more-innovative services are in the works. As any new and emerging technology, wireless services and applications are often surrounded by much hope and hype, as well as some healthy skepticism. But as the technology and services mature over time, yesterday's experiments can become tomorrow's standards. The Internet is a grand example of this evolving progress. Development of the wireless Internet will go through the same evolutionary cycle, although probably at an even faster pace.

Like any new technology, however, security and safety issues can damage its reputation and benefits if they are not included intelligently into the design from the very beginning. It is with this in mind that this chapter is written.

Because the wireless Internet covers much territory, the same goes for its security as well. This chapter discusses security issues as they relate to the wireless Internet in a few select categories, starting with transmission methods to the wireless devices and ending with some of the infrastructure components themselves.

3.4 HOW SECURE ARE THE TRANSMISSION METHODS?

For many years, it has been public knowledge that analog cell phone transmissions are fairly easy to intercept. It has been a known problem for as long as analog cell phones have been available. They are easily intercepted using special radio-scanning equipment. For this reason, as well as many others, many cell phone service providers have been promoting digital services to their subscribers and reducing analog to a legacy service.

Digital cell phone transmissions, on the other hand, are typically more difficult to intercept. It is on these very same digital transmissions that most of the new wireless Internet services are based.

However, there is no single method for digital cellular transmission. In fact, there are several different methods for wireless transmission available today. For example, in the United States, providers such as Verizon and Sprint primarily use CDMA (Code Division Multiple Access), whereas AT&T primarily uses TDMA (Time Division Multiple Access) and Voicestream uses GSM (Global Systems for Mobile Communications). Other providers, such as Cingular, offer more than one method (TDMA and GSM), depending on the geographic location. All these methods differ in the way they use the radio frequencies and the way they allocate users on those frequencies. This chapter discusses each of these in more detail.

Cell phone users who want wireless Internet access are generally not concerned with choosing a particular transmission method, nor do they really care to. Instead, most users select their favorite wireless service provider when they sign up for service. It is generally transparent to the user which transmission method their provider has implemented. It is an entirely different matter for the service provider, however. Whichever method they implement has significant bearing on its infrastructure. For example, the type of radio equipment they use, the location and number of transmission towers to deploy, the amount of traffic they can handle, and the type of cell phones to sell to their subscribers are all directly related to the digital transmission method chosen.

3.4.1 FREQUENCY DIVISION MULTIPLE ACCESS TECHNOLOGY

All cellular communications, analog or digital, are transmitted using radio frequencies that are purchased by or allocated to the wireless service provider. Each service provider typically purchases licenses from the appropriate authority to operate a spectrum of radio frequencies.

Analog cellular communications typically operate on what is called Frequency Division Multiple Access (FDMA) technology. With FDMA, each service provider divides its spectrum of radio frequencies into individual frequency channels. Each channel has a width of 10 to 30 kilohertz (kHz) and is a specific frequency that supports a one-way communication session. For a regular two-way phone conversation, every cell phone caller is assigned two frequency channels: one to send and one to receive.

Because each phone conversation occupies two channels (two frequencies), it is not too difficult for specialized radio scanning equipment to tap into a live analog phone conversation once the equipment has tuned into the right frequency channel. There is very little privacy protection in analog cellular communications if no encryption is added.

3.4.2 TIME DIVISION MULTIPLE ACCESS TECHNOLOGY

Digital cellular signals, on the other hand, can operate on a variety of encoding techniques, most of which are resistant to analog radio frequency scanning. (Note: the term *encoding* in wireless communications does not mean encryption and is here used to refer to converting a signal from one format to another e.g., from a wired signal to a wireless signal.)

One such technique is called time division multiple access, or TDMA. Similar to FDMA, TDMA typically divides the radio spectrum into multiple 30-kHz frequency channels (sometimes called frequency carriers). Every two-way communication requires two of these frequency channels: one to send and one to receive. But in addition, TDMA further subdivides each frequency channel into three to six time slots called voice/data channels, so that now up to six digital voice or data sessions can take place using the same frequency. With TDMA, a service provider can handle more calls at the same time, compared to FDMA. This is accomplished by assigning each of the six sessions a specific time slot within the same frequency. Each time slot (or voice/data channel) is approximately seven milliseconds in duration. The time slots are arranged and transmitted over and over again in rapid rotation. Voice or data for each caller is placed into the time slot assigned to that caller and then transmitted. Information from the corresponding time slot is quickly extracted and reassembled at the receiving cellular base station to piece together the conversation or session. Once that time slot (or voice/data channel) is assigned to a caller, it is dedicated to that caller for the duration of the session, until it terminates. In TDMA, a user is not assigned an entire frequency, but shares the frequency with other users, each with an assigned time slot.

As of the writing of this chapter, there have not been many publicized cases of eavesdropping of TDMA phone conversations and data streams as they travel across the wireless space. Access to special types of equipment or test equipment would probably be required to perform such a feat. It is possible that an illegally modified TDMA cell phone also could do the job.

However, this does not mean that eavesdropping is unfeasible. With regard to a wireless Internet session, consider the full path that such a session takes. For a mobile user to communicate with an Internet Web site, a wireless data signal from the cell phone will eventually be converted into a wired signal before traversing the Internet itself. As a wired signal, the information can travel across the Internet in clear text until it reaches the Web site. Although the wireless signal itself may be difficult to intercept, once it becomes a wired signal, it is subject to the same interception vulnerabilities as all unencrypted communications traversing the Internet. As a precaution, if there is confidential information being transmitted over the Internet, regardless of the method, it is always necessary to encrypt that session from end-to-end. Encryption is discussed in a later section.

3.4.3 GLOBAL SYSTEMS FOR MOBILE COMMUNICATIONS

Another method of digital transmission is Global Systems for Mobile Communications (GSM). GSM is actually a term that covers more than just the transmission method alone. It covers the entire cellular system, from the assortment of GSM services to the actual GSM devices themselves. GSM is primarily used in European nations.

As a digital transmission method, GSM uses a variation of TDMA. Similar to FDMA and TDMA, the GSM service provider divides the allotted radio frequency spectrum into multiple frequency channels. This time, each frequency channel has a much larger width of 200 kHz. Again, similar to FDMA and TDMA, each GSM cellular phone uses two frequency channels: one to send and one to receive.

Like TDMA, GSM further subdivides each frequency channel into time slots called voice/data channels. However, with GSM, there are eight time slots, so that now up to eight digital voice or data sessions can take place using the same frequency. As for TDMA, when that time slot (or voice/data channel) is assigned to a caller, it is dedicated to that caller for the duration of the session until it terminates.

GSM has additional features that enhance security. Each GSM phone uses a subscriber identity module (SIM). A SIM can look like a credit-card sized smart card or a postage-stamp sized chip. This removable SIM is inserted into the GSM phone during usage. The smart card or chip contains information pertaining to the subscriber, such as the cell phone number belonging to the subscriber, authentication information, encryption keys, directory of phone numbers, and short saved messages belonging to that subscriber. Because the SIM is removable, the subscriber can take this SIM out of one phone and insert it into another GSM phone. The new phone with the SIM will then take on the identity of the subscriber. The user's identity is not tied to a particular phone, but to the removable SIM itself. This makes it possible for a subscriber to use or upgrade to different GSM phones without changing phone numbers. It is possible also to rent a GSM phone in another country, even if that country uses phones that transmit on different GSM frequencies. This arrangement works, of course, only if the GSM service providers from the different countries have compatible arrangements with each other.

The SIM functions as an authentication tool because the GSM phones are useless without it. When the SIM is inserted into a phone, users are prompted to put in their personal identification numbers (PINs) associated with that SIM (if the SIM is PIN-enabled). Without the correct PIN number, the phone will not work.

In addition to authenticating the user to the phone, the SIM also is used to authenticate the phone to the phone network itself during connection. Using the authentication (or Ki) key in the SIM, the phone authenticates to the service provider's Authentication Center during each call. The process employs a challenge-response technique, similar in some respects to using a token card to remotely log a PC onto a network.

The keys in the SIM have another purpose in addition to authentication. The encryption (or Kc) key generated by the SIM can be used to encrypt communications between the mobile phone and the service provider's transmission equipment for confidentiality. This encryption prevents eavesdropping, at least between these two points.

GSM transmissions, similar to TDMA, are difficult but not impossible to intercept using radio frequency scanning equipment. A frequency can have up to eight users on it, making the digital signals difficult to extract. By adding encryption using the SIM card, GSM can add yet another layer of security against interception.

However, when it comes to wireless Internet sessions, this form of encryption does not provide end-to-end protection; only part of the path is actually protected. This is similar to the problem mentioned previously with TDMA Internet sessions.

A typical wireless Internet session takes both a wireless and a wired path. GSM encryption protects only the path between the cell phone and the service provider's transmission site — the wireless portion. The remainder of the session through the wired Internet — from the service provider's site to the Internet Web site — can still travel in the clear. One would need to add end-to-end encryption if there is a need to keep the entire Internet session confidential.

3.4.4 CODE DIVISION MULTIPLE ACCESS TECHNOLOGY

Another digital transmission method is called code division multiple access (CDMA). CDMA is based on spread spectrum, a transmission technology that has been used by the U.S. military for many years to make radio communications more difficult to intercept and jam. Qualcomm is one of the main pioneers incorporating CDMA spread spectrum technology into the area of cellular phones.

Instead of dividing a spectrum of radio frequencies into narrow frequency bands or time slots, CDMA uses a very large portion of that radio spectrum, also called a frequency channel. The frequency channel has a wide width of 1.25 megahertz (MHz). For duplex communications, each cell phone uses two of these wide CDMA frequency channels: one to send and one to receive.

During communication, each voice or data session is first converted into a series of data signals. Next, the signals are marked with a unique code to indicate that they belong to a particular caller. This code is called a pseudorandom noise (PN) code. Each mobile phone is assigned a new PN code by the base station at the beginning of each session. These coded signals are then transmitted by spreading them out across a very wide radio frequency spectrum. Because the channel width is very large, it has the capacity to handle many other user sessions at the same time, each session again tagged by unique PN codes to associate them to the appropriate caller.

A CDMA phone receives transmissions using the appropriate PN code to pick out the data signals that are destined for it and ignores all other encoded signals.

With CDMA, cell phones communicating with the base stations all share the same wide frequency channels. What distinguishes each caller is not the frequency used (as in FDMA), nor the time slot within a particular frequency (as in TDMA or GSM), but the PN noise code assigned to that caller. With CDMA, a voice/data channel is a data signal marked with a unique PN code.

Intercepting a single CDMA conversation would be difficult because its digital signals are spread out across a very large spectrum of radio frequencies. The conversation does not reside on just one frequency alone, making it difficult to scan. Also, without knowledge of the PN noise code, an eavesdropper would not be able to extract the relevant session from the many frequencies used. To further complicate interception, the entire channel width is populated by many other callers at the same time, creating a vast amount of noise for anyone trying to intercept the call.

However, as seen earlier with the other digital transmission methods, Internet sessions using CDMA cell phones are not impossible to intercept. As before, although the CDMA digital signals themselves can be difficult to intercept, once these wireless signals are converted into wired signals, the latter signals can be intercepted as they travel across the Internet. Without using end-to-end encryption, wireless Internet sessions are as vulnerable as other unencrypted communications traveling over the Internet.

3.4.5 OTHER METHODS

There are additional digital transmission methods, many of which are derivatives of the types already discussed, and some of which are still under development. Some of these that are under development are called third-generation or 3G transmission methods. Second-generation (2G) technologies, such as TDMA, GSM, and CDMA, offer transmission speeds of 9.6 to 14.4 kbps, which is slower than today's typical modem speeds. 3G technologies, on the other hand, are designed to transmit much faster and carry larger amounts of data. Some will be capable of providing high-speed Internet access, as well as video transmission. Below is a partial listing of other digital transmission methods, including those in the 3G category.

- iDEN (integrated Digital Enhanced Network) is a 2G transmission method based on TDMA. In addition to sending voice and data, it can be used also for two-way radio communications between two iDEN phones, much like walkie-talkies.
- PDC (Personal Digital Communications) is based on TDMA and is a 2G transmission method widely used in Japan.
- GPRS (General Packet Radio Service) is a 2.5G (not quite 3G) technology based on GSM. It is a packet-switched data technology that provides "always online" connections, which means that the subscriber can stay logged on to the phone network all day but uses it only if there is actual data to send or receive. Maximum data rates are estimated to be 115 kbps.
- EDGE (Enhanced Data rates for Global Evolution) is a 3G technology based on TDMA and GSM. Like GPRS, it features "always online" connections using packet-switched data technologies. Maximum data rates are estimated to be 384 kbps.
- UMTS (Universal Mobile Telecommunications System) is a 3G technology based on GSM. Maximum data rates are estimated at 2 Mbps.
- CDMA2000 and W-CDMA (wideband CDMA) are two 3G technologies based on CDMA. CDMA2000 is more of a North American design, whereas W-CDMA is more European and Japanese oriented. Both provide maximum data rates estimated at 384 kbps for slow-moving mobile units and at 2 Mbps for stationary units.

Regardless of the methods or the speeds, the need for end-to-end encryption will still be a requirement if confidentiality is needed between the mobile device and the Internet or intranet site. Because wireless Internet communications encompass both wireless and wired-based transmissions, encryption features covering just the wireless portion of the communication is clearly not enough. For end-to-end privacy protection, the applications and the protocols have a role to play, as discussed later in this chapter.

3.5 HOW SECURE ARE WIRELESS DEVICES?

Internet security, as many have seen it applied to corporate networks today, can be difficult to implement on wireless phones and PDAs for a variety of reasons. Most of these devices have limited CPUs, memory, bandwidth, and storage abilities. As a result, many have disappointingly slow and limited computing power. Robust security features that can take less than a second to process on a typical workstation can take potentially many minutes on a wireless device, making them impractical or inconvenient for the mobile user. Because many of these devices have merely a fraction of the hardware capabilities found on typical workstations, the security features on portable devices are often lightweight or even nonexistent from an Internet security perspective. However, these same devices are now being used to log onto sensitive corporate intranets, or to conduct mobile commerce and banking. Although these wireless devices are smaller in every way, their security needs are just as significant as before. It would be a mistake for corporate IT and information security departments to ignore these devices as they start to populate the corporate network. After all, these devices do not discriminate; they can be designed to tap into the same corporate assets as any other node on a network. Some of the security aspects as they relate to these devices are examined here.

3.5.1 AUTHENTICATION

The process of authenticating wireless phone users has gone through many years of implementation and evolution. It is probably one of the most reliable security features digital cell phones have today, given the many years of experience service providers have had in trying to reduce the theft of wireless services. Because the service providers have a vested interest in knowing who to charge for the use of their services, authenticating the mobile user is of utmost importance.

As previously mentioned, GSM phones use SIM cards or chips that contain authentication information about the user. SIMs typically carry authentication and encryption keys, authentication algorithms, identification information, phone numbers belonging to the subscriber, etc. They allow users to authenticate to their own phones and to the phone network to which they are subscribed.

In North America, TDMA and CDMA phones use a similarly complex method of authentication as in GSM. Like GSM, the process incorporates keys, Authentication Centers, and challenge-response techniques. However, because TDMA and CDMA phones do not generally use removable SIM cards or chips, these phones rely instead on the authentication information embedded into the handset. The user's identity is therefore tied to the single mobile phone itself.

The obvious drawback is that for authentication purposes, TDMA and CDMA phones offer less flexibility when compared to GSM phones. To deploy a new authentication feature with a GSM phone, in many cases, all that is needed is to update the SIM card or chip. On the other hand, with TDMA and CDMA, deploying new authentication features would probably require users to buy new cell phones — a more expensive way to go. Because it is easier to update a removable chip than an entire cell phone, it is likely that one will find more security features and innovations being offered for GSM.

It is important to note, however, that this form of authentication does not necessarily apply to Internet-related transactions. It merely authenticates the mobile user to the service provider's phone network, which is only one part of the transmission if one is talking about Internet transactions. For securing end-to-end Internet transactions, mobile users still need to authenticate the Internet Web servers they are connecting to, to verify that indeed the servers are legitimate. Likewise, the Internet Web servers need to authenticate the mobile users that are connecting to it, to verify that they are legitimate users and not impostors. The wireless service providers, however, are seldom involved in providing full end-to-end authentication service, from mobile phone to Internet Web site. That responsibility usually falls to the owners of the Internet Web servers and applications.

Several methods for providing end-to-end authentication are being tried today at the application level. Most secure mobile commerce applications are using IDs and passwords, an old standby, which of course has its limitations because it provides only single-factor authentication. Other organizations are experimenting with GSM SIMs by adding additional security ingredients such as public/private key pairs, digital certificates, and other public key infrastructure (PKI) components into the SIMs. However, because the use of digital certificates can be process intensive, cell phones and handheld devices typically use lightweight versions of these security components. To accommodate the smaller processors in wireless devices, the digital certificates and their associated public keys may be smaller or weaker than those typically deployed on desktop Web browsers, depending on the resources available on the wireless device.

Additionally, other organizations are experimenting with using elliptic-curve cryptography (ECC) for authentication, digital certificates, and public key encryption on the wireless devices. ECC is an ideal tool for mobile devices because it can offer strong encryption capabilities, but requires less computing resources than other popular forms of public key encryption. Certicom is one of the main pioneers incorporating ECC for use on wireless devices.

As more and more developments take place with wireless Internet authentication, it becomes clear that, in time, these Internet mobile devices will become full-fledged authentication devices, much like tokens, smart cards, and bank ATM cards. If users begin conducting Internet commerce using these enhanced mobile devices, securing those devices themselves from loss or theft now becomes a priority. With identity information embedded into the devices or the removable SIMs, losing these could mean that an impostor can now conduct electronic commerce transactions using that stolen identity. With a mobile device, the user, of course, plays the biggest role in maintaining its overall security. Losing a cell phone that has Internet access and an embedded public/private key pair can be potentially as disastrous as losing a bank ATM card with its associated PIN written on it, or worse. If a user loses such a device, contacting the service provider immediately about the loss and suspending its use is a must.

3.5.2 CONFIDENTIALITY

Preserving confidentiality on wireless devices poses several interesting challenges. Typically, when one accesses a Web site with a browser and enters a password to gain entry, the password one types is masked with asterisks or some other placeholder to prevent others from seeing the actual password on one's screen. With cell phones and handheld devices, masking the password could create problems during typing. With cell phones, letters are often entered using the numeric keypad, a method that is cumbersome and tedious for many users. For example, to type the letter "R," one must press the number 7 key three times to get to the right letter. If the result is masked, it is not clear to the user what letter was actually submitted. Because of this inconvenience, some mobile Internet applications do away with masking so that the entire password is displayed on the screen in the original letters. Other applications initially display each letter of the password for a few seconds as they are being entered, before masking each with a placeholder afterward. This gives the user some positive indication that the correct letters were indeed entered, while still preserving the need to mask the password on the device's screen for privacy. The latter approach is probably the more sensible of the two, and should be the one that application designers adopt.

Another challenge to preserving confidentiality is making sure that confidential information such as passwords and credit card numbers are purged from the mobile device's memory after they are used. Many times, such sensitive information is stored as variables by the wireless Internet application and subsequently cached in the memory of the device. There have been documented cases in which credit card numbers left in the memory of cell phones were reusable by other people who borrowed the same phones to access the same sites. Once again, the application designers are the chief architects in preserving the confidentiality here. It is important that programmers design an application to clear the mobile device's memory of sensitive information when the user finishes using that application. Although leaving such information in the memory of the device may spare the user of having to reenter it the next time, it is as risky as writing the associated PIN or password on a bank ATM card itself.

Yet another challenge in preserving confidentiality is making sure that sensitive information is kept private as it travels from the wireless device to its destination on the Internet, and back. Traditionally, for the wired Internet, most Web sites use Secure Sockets Layer (SSL) or its successor, Transport Layer Security (TLS), to encrypt the entire path end-to-end, from the client to the Web server. However, many wireless devices, particularly cell phones, lack the computing power and bandwidth to run SSL efficiently. One of the main components of SSL is RSA public key encryption. Depending on the encryption strength applied at the Web site, this form of public key encryption can be processor and bandwidth intensive, and can tax the mobile device to the point where the communication session itself becomes too slow to be practical.

Instead, wireless Internet applications that are developed using the Wireless Application Protocol (WAP) use a combination of security protocols. Secure WAP applications use both SSL and WTLS (Wireless Transport Layer Security) to protect different segments of a secure transmission. Typically, SSL protects the wired portion of the connection and WTLS primarily protects the wireless portion. Both are needed to provide the equivalent of end-to-end encryption.

WTLS is similar to SSL in operation. However, although WTLS can support either RSA or ECC, ECC is probably preferred because it provides strong encryption capabilities but is more compact and faster than RSA.

WTLS has other differences from SSL as well. WTLS is built to provide encryption services for a slower and less resource-intensive environment, whereas SSL could tax such an environment. This is because SSL encryption requires a reliable transport protocol, particularly TCP (Transmission Control Protocol, a part of TCP/IP). TCP provides error detection, communication acknowledgments, and retransmission features to ensure reliable network connections back and forth. But because of these features, TCP requires more bandwidth and resources than what typical wireless connections and devices can provide. Most mobile connections today are low bandwidth and slow, and not designed to handle the constant, back-and-forth error-detection traffic that TCP creates.

Realizing these limitations, the WAP Forum, the group responsible for putting together the standards for WAP, designed a supplementary protocol stack that is more suitable for the wireless environment. Because this environment typically has low connection speeds, low reliability, and low bandwidth in order to compensate, the protocol stack uses compressed binary data sessions and is more tolerant of intermittent coverage. The WAP protocol stack resides in layers 4, 5, 6, and 7 of the OSI reference model. The WAP protocol stack works with UDP (User Datagram Protocol) for IP-based networks and WDP (Wireless Datagram Protocol) for non-IP networks. WTLS, which is the security protocol from the WAP protocol stack, can be used to protect UDP or WDP traffic in the wireless environment.

Because of the differences between WTLS and SSL, as well as the different underlying environments that they work within, an intermediary device such as a WAP gateway is needed to translate the traffic going from one environment into the next. The WAP gateway is discussed in more detail in the infrastructure section of this chapter.

3.5.3 Malicious Code and Viruses

The number of security attacks on wireless devices has been small compared to the many attacks against workstations and servers. This is due in part to the very simple fact that most mobile devices, particularly cell phones, lack sufficient processors, memory, or storage that malicious code and viruses can exploit. For example, a popular method for spreading viruses today is by hiding them in file attachments to e-mail. However, many mobile devices, particularly cell phones, lack the ability to store or open e-mail attachments. This makes mobile devices relatively unattractive as targets because the damage potential is relatively small.

However, mobile devices are still vulnerable to attack and will become increasingly more so as they evolve with greater computing, memory, and storage capabilities. With greater speeds, faster downloading abilities, and better processing, mobile devices can soon become the equivalent of today's workstations, with all their exploitable vulnerabilities. As of the writing of this chapter, cell phone manufacturers were already announcing that the next generation of mobile phones will support languages such as Java so that users can download organizers, calculators, and games to their Web-enabled phones. However, on the negative side, this also opens up more opportunities for users to unwittingly download malicious programs (or "malware"). The following adage applies to mobile devices: "The more brains they have, the more attractive they become as targets."

3.6   HOW SECURE ARE THE NETWORK
INFRASTRUCTURE COMPONENTS?

As many of us who have worked in the information security field know, security is usually assembled using many components, but its overall strength is only as good as its weakest link. Sometimes it does not matter if one is using the strongest encryption available over the network and the strongest authentication at the device. If there is a weak link anywhere along the chain, attackers will focus on this vulnerability and may eventually exploit it, choosing a path that requires the least effort and the least amount of resources.

Because the wireless Internet world is still relatively young and a work in progress, vulnerabilities abound, depending on the technology one has implemented. This chapter section focuses on some infrastructure vulnerabilities for those who are using WAP (Wireless Application Protocol).

3.6.1   THE "GAP IN WAP"

Encryption has been an invaluable tool in the world of E-commerce. Many online businesses use SSL (Secure Sockets Layer) or TLS (Transport Layer Security) to provide end-to-end encryption to protect Internet transactions between the client and the Web server.

When using WAP, however, if encryption is activated for the session, there are usually two zones of encryption applied, each protecting the two different halves of the transmission. SSL or TLS is generally used to protect the first path, between the Web server and an important network device called the WAP gateway that was mentioned previously. WTLS (Wireless Transport Layer Security) is used to protect the second path, between the WAP gateway and the wireless mobile device.

The WAP gateway is an infrastructure component needed to convert wired signals into a less-bandwidth-intensive and compressed binary format, compatible for wireless transmissions. If encryption such as SSL is used during a session, the WAP gateway will need to translate the SSL-protected transmission by decrypting this SSL traffic and reencrypting it with WTLS, and vice versa in the other direction. This translation can take just a few seconds; but during this brief period, the data sits in the memory of the WAP gateway decrypted and in the clear before it is reencrypted using the second protocol. This brief period in the WAP gateway — some have called it the "gap in WAP" — is an exploitable vulnerability. How vulnerable one is depends on where the WAP gateway is located, how well it is secured, and who is in charge of protecting it.

Clearly, the WAP gateway should be placed in a secure environment. Otherwise, an intruder attempting to access the gateway can steal sensitive data while it transitions in clear text. The intruder also can sabotage the encryption at the gateway, or even initiate a denial-of-service or other malicious attack on this critical network component. In addition to securing the WAP gateway from unauthorized access, proper operating procedures also should be applied to enhance its security. For example, it is wise not to save any of the clear-text data to disk storage during the decryption and reencryption process. Saving this data to log files, for example, could create an unnecessarily tempting target for intruders. In addition, the decryption and reencryption should operate in memory only and proceed as quickly as possible. Furthermore, to prevent accidental disclosure, the memory should be properly overwritten, thereby purging any sensitive data before that memory is reused.

3.6.2 WAP GATEWAY ARCHITECTURES

Depending on the sensitivity of the data and the liability for its unauthorized disclosure, businesses offering secure wireless applications (as well as their customers) may have concerns about where the WAP gateway is situated, how it is protected, and who is protecting it. Three possible architectures and their security implications are examined: (1) the WAP gateway at the service provider, (2) WAP gateway at the host, and (3) pass-through from service provider's WAP gateway to host's WAP proxy.

3.6.2.1 WAP Gateway at the Service Provider

In most cases, the WAP gateways are owned and operated by the wireless service providers. Many businesses that deploy secure wireless applications today rely on the service provider's WAP gateway to perform the SSL-to-WTLS encryption translation. This implies that the business owners of the sensitive wireless applications, as well as their users, are entrusting the wireless service providers to keep the WAP gateway and the sensitive data that passes through it safe and secure. Figure 3.1 provides an example of such a setup, where the WAP gateway resides within the service provider's secure environment. If encryption is applied in a session between the user's cell phone and the application server behind the business' firewall, the path between the cell phone and the service provider's WAP gateway is typically encrypted using WTLS. The path between the WAP gateway and the business host's application server is encrypted using SSL or TLS.

A business deploying secure WAP applications using this setup should realize, however, that it cannot guarantee end-to-end security for the data because it is decrypted, exposed in clear text for a brief moment, and then reencrypted, all at an external gateway, away from the business' control. The WAP gateway is generally housed in the wireless service provider's data center and attended by those who are not directly accountable to the businesses. Of course, it is in the best interest of the service provider to maintain the WAP gateway in a secure manner and location.

Sometimes, to help reinforce that trust, businesses may wish to conduct periodic security audits on the service provider's operation of the WAP gateways to ensure

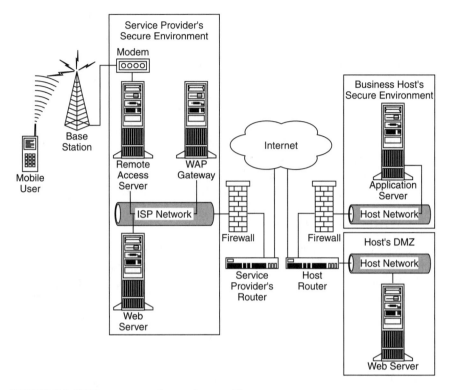

FIGURE 3.1 WAP gateway at the service provider.

that the risks are minimized. Bear in mind, however, that by choosing this path, the business may need to inspect many WAP gateways from many different service providers. A service provider sets up the WAP gateway primarily to provide Internet access to its own wireless phone subscribers. If users are dialing into a business' secure Web site, for example, from 20 different wireless service providers around the world, then the business may need to audit the WAP gateways belonging to these 20 providers. This, unfortunately, is a formidable task and an impractical method of ensuring security. Each service provider might apply a different method for protecting its own WAP gateway, if protected at all. Furthermore, in many cases the wireless service providers are accountable to their own cell phone subscribers, not necessarily to the countless businesses that are hosting secure Internet applications, unless there is a contractual arrangement to do so.

3.6.2.2 WAP Gateway at the Host

Some businesses and organizations, particularly in the financial, healthcare, and government sectors, may have legal requirements to keep their customers' sensitive data protected. Having such sensitive data exposed outside the organization's internal control may pose an unnecessary risk and liability. To some, the "gap in WAP" presents a broken pipeline, an obvious breach of confidentiality that is just waiting to be exploited. For those who find such a breach unacceptable, one possible solution

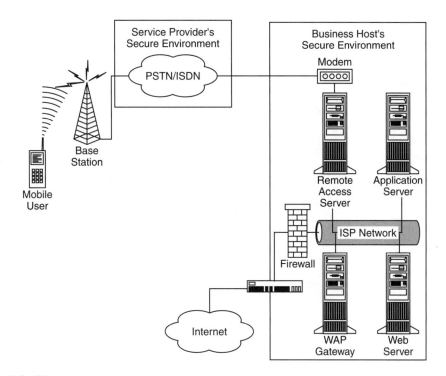

FIGURE 3.2 WAP gateway at the host.

is to place the WAP gateway at the business host's own protected network, bypassing the wireless service provider's WAP gateway entirely. Figure 3.2 provides an example of such a setup. Nokia, Ericsson, and Ariel Communications are just a few of the vendors offering such a solution.

This approach has the benefit of keeping the WAP gateway and its WTLS-SSL translation process in a trusted location, within the confines of the same organization that is providing the secure Web applications. Using this setup, users are typically dialing directly from their wireless devices, through their service provider's public switched telephone network (PSTN), and into the business' own remote access servers (RAS). Once they reach the RAS, the transmission continues onto the WAP gateway, and then onward to the application or Web server, all of these devices within the business host's own secure environment.

Although it provides better end-to-end security, the drawback to this approach is that the business host will need to set up banks of modems and RAS so users have enough access points to dial in. The business also will need to reconfigure the users' cell phones and PDAs to point directly to the business' own WAP gateway instead of (typically) to the service provider's. However, not all cell phones allow this reconfiguration by the user. Furthermore, some cell phones can point to only one WAP gateway, while others are fortunate enough to point to more than one. In either case, individually reconfiguring all those wireless devices to point to the business' own WAP gateway may take significant time and effort.

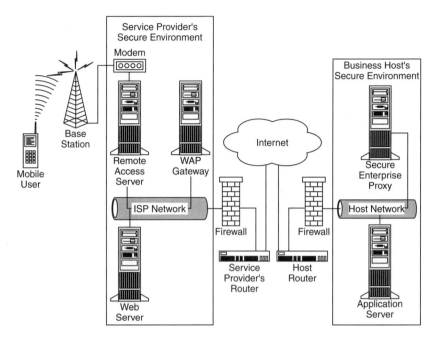

FIGURE 3.3 Pass-through from service provider's WAP gateway to host's WAP proxy.

For users whose cell phones can point to only a single WAP gateway, this reconfiguration introduces yet another issue. If these users now want to access other WAP sites across the Internet, they still must go through the business host's WAP gateway first. If the host allows outgoing traffic to the Internet, the host then becomes an Internet service provider (ISP) to these users who are newly configured to point to the host's own WAP gateway. Acting as a makeshift ISP, the host will inevitably need to attend to service- and user-related issues, which too many businesses can be an unwanted burden because of the significant resources required.

3.6.2.3 Pass-Through from Service Provider's WAP Gateway to Host's WAP Proxy

For businesses that want to provide secure end-to-end encrypted transactions and to avoid the administrative headaches of setting up their own WAP gateways, there are other approaches. One such approach, as shown in Figure 3.3, is to keep the WTLS-encrypted data unchanged as it goes from the user's mobile device and through the service provider's WAP gateway. The WTLS-SSL encryption translation will not occur until the encrypted data reaches a second WAP gateway-like device residing within the business host's own secure network. One vendor developing such a solution is Openwave Systems (a combination of Phone.com and Software.com). Openwave calls this second WAP gateway-like device the Secure Enterprise Proxy. During an encrypted session, the service provider's WAP gateway and the business' Secure Enterprise Proxy negotiate with each other, so that the service provider essentially passes the encrypted data unchanged to the business that is using this proxy. This solution utilizes the service provider's WAP gateway because it is still needed to provide proper Internet access for the mobile users, but it does not perform the WTLS-SSL encryption translation there and thus is not exposing confidential data. The decryption is passed on and occurs instead within the confines of the business' own secure network, either at the Secure Enterprise Proxy or at the application server.

One drawback to this approach, however, is its proprietary nature. At the time of this writing, to make the Openwave solution work, three parties would need to implement components exclusively from Openwave. The wireless service providers would need to use Openwave's latest WAP gateway. Likewise, the business hosting the secure applications would need to use Openwave's Secure Enterprise Proxy to negotiate the encryption pass-through with that gateway. In addition, the mobile devices themselves would need to use Openwave's latest Web browser, at least Micro Browser version 5. Although approximately 70 percent of WAP-enabled phones throughout the world are using some version of Openwave Micro Browser, most of these phones are using either version 3 or 4. Unfortunately, most of these existing browsers are not upgradable by the user, so most users may need to buy new cell phones to incorporate this solution. It may take some time before this solution comes to fruition and becomes popular.

These are not the only solutions for providing end-to-end encryption for wireless Internet devices. Other methods in the works include applying encryption at the applications level, adding encryption keys and algorithms to cell phone SIM cards, and adding stronger encryption techniques to the next revisions of the WAP speci-fications, perhaps eliminating the "gap in WAP" entirely.

3.7 CONCLUSION

Two sound recommendations for the many practitioners in the information security profession are:

1. Stay abreast of wireless security issues and solutions.
2. Do not ignore wireless devices.

Many in the IT and information security professions regard the new wireless Internet devices diminutively as personal gadgets or executive toys. Many are so busy grappling with the issues of protecting their corporate PCs, servers, and net-works that they cannot imagine worrying about yet another class of devices. Many corporate security policies make no mention of securing mobile handheld devices and cell phones, although some of these same corporations are already using these devices to access their own internal e-mail. The common fallacy is that these they can cause no harm.

Security departments have had to wrestle with the migration of information assets from the mainframe world to distributed PC computing. Many corporate attitudes have had to change during that evolution regarding where to apply security. With no exaggeration, corporate computing is undergoing yet another significant phase of migration. It is not so much that corporate information assets can be accessed through wireless means, because wireless notebook computers have been doing that for years; rather, the means of access will become ever cheaper and, hence, greater in volume. Instead of using a $3000 notebook computer, users (or intruders) can now tap into a sensitive corporate network from anywhere, using just a $40 Internet-enabled cell phone. Over time, these mobile devices will have increasing processing power, memory, bandwidth, storage, ease of use, and popularity. It is this last item that will inevitably draw upon corporate resources.

Small as these devices may be, once they access the sensitive assets of an organization, they can do as much good or harm as any other computer. Ignoring or disallowing these devices from an information security perspective has two probable consequences:

1. The business units or executives within the organization will push, often successfully, to deploy wireless devices and services anyway, shutting out any involvement or guidance from the information security department. Inevitably, information security will be involved at a much later date, but reactively and often too late to have a significant impact on proper design and planning.
2. By ignoring wireless devices and their capabilities, the information security department will give attackers just what they need: a neglected and unprotected window into an otherwise fortified environment. Such an organization will be caught unprepared when an attack using wireless devices surfaces.

Wireless devices should not be treated as mere gadgets or annoyances. Once they tap into the valued assets of an organization, they are indiscriminate and equal to any other node on the network. To stay truly informed and prepared, information security practitioners should stay abreast of the new developments and security issues regarding wireless technology. In addition, they need to work with the application designers as an alliance to ensure that applications designed for wireless take into consideration the many points discussed in this chapter. And finally, organizations need to expand the categories of devices protected under their information security policies to include wireless devices, because they are in effect yet another infrastructure component of the organization.

Bibliography

Appleby, T.P., WAP — the wireless application protocol, White paper, Global Integrity.
Blake, R., *Wireless Communication Technology,* Delmar Thomson Learning, 2001.
Certicom, Complete WAP Security from Certicom, http://www.certicom.com.
CMP Media, Wireless devices present new security challenges — growth in wireless Internet access means handhelds will be targets of more attacks, October 21, 2000.
DeJesus, E.X., Wireless devices are flooding the airwaves with millions of bits of information. Securing those transmissions is the next challenge facing E-commerce, White paper, http://www.infosecuritymag.com, October 2000.
Harte, L. et al., *Cellular and PCS: The Big Picture,* McGraw-Hill, New York, 1997.

Howell, R. et al., *Professional WAP,* Wrox Press Ltd., Birmingham, 2000.

Izarek, S., Next-gen cell phones could be targets for viruses, http://www.foxnews.com, June 1, 2000.

Muller, N.J., *Desktop Encyclopedia of Telecommunications,* 2nd ed., McGraw-Hill, New York, 2000.

Nobel, C., Phone.com plugs WAP security hole, *eWEEK,* September 25, 2000.

Nokia, Secure Corporate WAP services: Nokia Activ Server, White paper, http://www.nokia.com.

Radding, A., Crossing the wireless security gap, http://www.computerworld.com, January 1, 2001.

Saarinen, M.-J., Attacks against the WAP WTLS Protocol, University of Jvyskyl, Finland.

Saita, A., Case study: securing thin air, academia seeks better security solutions for handheld wireless devices, http://www.infosecuritymag.com, April 2001.

Schwartz, E., Two-zone wireless security system creates a big hole in your communications, http://www.infoworld.com, November 6, 2000.

Tulloch, M., *Microsoft Encyclopedia of Networking,* Microsoft Press, Redmond, WA, 2000.

Unstrung.com, Does Java solve worldwide WAP wait?, http://www.unstrung.com, April 9, 2001.

Van der Heijden, M. and Taylor, M., *Understanding WAP: Wireless Applications, Devices, and Services,* Artech House Publishers, 2000.

Part II

Technologies and Standards

4 Multimedia Streaming over Mobile Networks: European Perspective

Igor D.D. Curcio*

CONTENTS

4.1 INTRODUCTION

Mobile communications, Internet connectivity, and multimedia technologies are progressively merging in a single paradigm of personal communications. Mobile communications originate from the increasing need of users to have information

---

* The opinions expressed in this chapter are those of the author and not necessarily those of his employer.

0-8493-1502-6/03/$0.00+$1.50
© 2003 by CRC Press LLC

available "anytime, anywhere." Internet connectivity puts a huge quantity of information resources at users' disposal, including services such as searching, browsing, e-mail, and E-commerce. Multimedia technologies are emerging as users want to have more information in audio/visual form rather than in textual form.

Mobile networks have been developed in the past two decades to allow users to make phone calls in total mobility. These systems have evolved from the first generation (1G) of analog networks (such as AMPS, TACS, NMT, and NTT) in the 1980s, to the second generation (2G) of digital networks (such as GSM, PDC, D-AMPS, IS-95) in the late 1980s. Digital networks offer higher spectrum efficiency, better data services, and more-advanced roaming capabilities than the 1G systems. Furthermore, GSM has evolved to offer more-advanced services such as higher bit rates for circuit- and packet-switched data transmission. Those networks are commonly referred to as the 2.5G networks. HSCSD (High-Speed Circuit-Switched Data), GPRS (General Packet Radio Service) and EDGE (Enhanced Data rates for GSM Evolution) are extensions of the current GSM network, and allow reaching bit rates up to 64, 171.2, and 473.6 kbps, respectively. The new 2.5G networks are able to carry low and medium bit rate multimedia traffic, allowing the feasibility of applications requiring real-time video and audio.

A strong effort has been made by standardization bodies toward third generation (3G) networks that offer even higher bit rates (up to 2 Mbps), more flexibility, multiple simultaneous services for one user, and different quality-of-service (QoS) classes. For example, a user could establish a video streaming session or browse the World Wide Web while retrieving a file from a corporate intranet server as a background process. In the International Telecommunications Union (ITU), the 3G networks are called IMT-2000 (International Mobile Telecommunications year 2000). IMT-2000 represents the joint effort of merging the European, Asian/Japanese, and North American standards into a unique common platform for mobile communications. IMT-2000 specifications are defined by the Third Generation Partnership Project (3GPP), which has written standards for UMTS (Universal Mobile Telecommunications System) 3G networks and services since December 1998. The specifications have evolved through different releases, from Release '99, to Releases 4, 5, and 6. The different releases have been planned to enable the transition to packet-switched (PS), all-IP (Internet Protocol) mobile networks.

Recent advances in video compression technology have made possible the transmission of real-time video over low-bit-rate links. H.263 and MPEG-4 are two examples of video compression algorithms. However, the deployment of mobile video is a challenging issue. First, video processing, including compression and decompression, is CPU intensive; this and the constraints of a mobile device mean that the digital signal processing (DSP) platform must be of limited size and weight, but still capable of processing a large quantity of data, possibly in real-time. Second, efficient error-resilience techniques must be developed in order to recover from bit errors and packet losses inherently present in the air interface during data transmission.

A typical video application is multimedia streaming, which has been widely deployed over the Internet for many years. Mobile multimedia streaming is enabled by the capacity of the current 2.5G or 3G networks, and by the multimedia capabilities of current and next-generation mobile phones.

This chapter is an introduction to mobile multimedia streaming. It is organized as follows: Section 4.2 describes the end-to-end architecture for mobile streaming systems. Section 4.3 includes a review of the current mobile networks that enable streaming applications, and Section 4.4 introduces the current standards for mobile streaming. Section 4.5 contains some performance and QoS considerations for streaming. Section 4.6 concludes this chapter.

4.2 END-TO-END SYSTEM ARCHITECTURE

A streaming system is a real-time system of the nonconversational type. It is real-time because the playback of continuous media, such as audio and video, must occur in an isochronous fashion. A streaming application is different from a conversational application because it has the following properties:

1. *One-way data distribution.* The media flow is always unidirectional, from the streaming server to the mobile client (in the downlink direction). Normally, the user has limited control over a streaming session, and there is not a high level of interactivity between mobile client and streaming server. Typical user control commands in the uplink direction include PLAY, PAUSE, STOP, FAST-FORWARD, and REWIND.
2. *Offline media encoding.* A streaming system is similar to a Video On Demand system, where the user can play only prestored content. This content is not encoded in real-time, but in an offline fashion using specific content creation tools.
3. *Not highly delay sensitive.* Because high interactivity is not a requirement of a streaming system, end-to-end delays can be relaxed. For example, the time required by the streaming client to execute a command issued by the user (such as PLAY) does not need to be on the order of milliseconds. Media can be streamed after an initial latency period. This allows the mobile client to smooth out eventual network jitter without compromising user QoS.

Figure 4.1 describes the high-level architecture of a typical mobile multimedia streaming system over an IP-based mobile network. We will follow an end-to-end approach, analyzing the system in its different parts. A mobile streaming system consists mainly of three components: (1) the streaming server, (2) the mobile network, and (3) the mobile streaming client.

The *streaming server* is connected to a fixed IP network and can reside either within the mobile operator's domain or outside it (e.g., the Internet). The location of the streaming server is important when considering the end-to-end quality of service of a streaming service. In fact, if the server is located in the public Internet, the QoS of the network trunk between the streaming server and the mobile network is not usually controlled by the mobile operator, and it can be of the best-effort type in the worst case. This may have impact on the perceived streaming service user quality.

The content created offline is loaded onto the streaming server before a user can actually request its playback. The content that is estimated to be highly requested

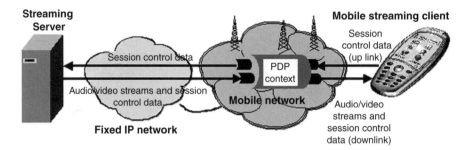

FIGURE 4.1 A typical mobile multimedia streaming system.

can be replicated or cached in proxy servers using appropriate techniques that make use of usage patterns.

The *mobile network* carries multimedia streaming traffic mainly between the streaming server and the mobile streaming client. A logical connection established between the network and the mobile client addresses is called PDP (Packet Data Protocol) context. This uses physical transport channels in the downlink and uplink directions to enable the data transfer in the two directions.

The *mobile streaming client* keeps a radio connection with the mobile network, utilizing the allocated PDP context for data transfer. The mobile client has the possibility to roam (i.e., upon mobility, change the network operator without affecting the received service), provided there is always radio coverage to guarantee the service. The data flows received by the mobile client in the downlink direction are, for example, audio and video plus additional information for session establishment, control, and media synchronization. The data flows sent by the client in the uplink direction are mainly session control data and QoS reports. The streaming server may react accordingly upon reception of the QoS reports, taking appropriate actions for guaranteeing the best-possible media quality at any instant.

4.3 THE CHALLENGES OF MOBILE NETWORKS

This section analyzes the differences between fixed IP networks and mobile networks, from the streaming service perspective. Deploying a streaming service over mobile networks is a challenging task, because all the constraints and properties of a mobile network must be taken into account when developing a streaming application.

In a fixed IP environment such as the Internet, the main obstacles for achieving a good QoS are packet losses and delays. Losses are mainly caused by congestion in the routers along the end-to-end path between the streaming server and the streaming client. If a router is congested, it starts to drop packets. These packets are normally not retransmitted by the network protocols, unless *ad hoc* retransmission techniques are used at the application layer or reliable transport protocols as employed. Delays may depend on congestion issues, out-of-sequence packet reordering and on the physical capacity of the network trunks between streaming server and client. A variable delay over time is called jitter. Whenever the inter-arrival time of the media packets at the streaming client is variable, delay jitter occurs. Normally, a good buffer management at the streaming client would help in de-jittering the incoming data flow.

In a mobile network environment some new factors must be considered:

- *Radio link quality.* In mobile networks the air interface is inherently affected by bit errors that can be up to 10^{-3} after channel coding. High bit error rates (BERs) can be caused, for example, by a weak radio signal in a determined area (such as under bridges, behind buildings or hills) or because of handover due to movement of the user. This factor may cause packet corruption or packet losses that can produce noticeable impairment of the streamed media.
- *Mobility.* As users become truly mobile, handover is an important issue. Handover (i.e., switching the mobile client from a cell to another cell of the same or another operator's network) is an operation that may cause service interruption for a certain amount of time, and it might cause delay and packet losses at the streaming client. When moving to a new cell, the capacity that was available in the old cell might no longer be available to the streaming user. This factor means that the bandwidth may be subject to change all the time with user mobility. The management of network bandwidth variation is one of the key points for a successful mobile streaming service.

4.3.1 MOBILE NETWORKS FOR STREAMING

In this section we will describe the suitable mobile channels for multimedia streaming. There are essentially two types of connections that allow streaming: circuit-switched and packet-switched connections.

4.3.1.1 Circuit-Switched Mobile Channels

4.3.1.1.1 High Speed Circuit-Switched Data (HSCSD)

HSCSD is a technology derived by the GSM (Global System for Mobile Communications) standard, and defined in the GSM Release '96 specifications. The limit of GSM networks in terms of capacity is 9.6 kbps. This speed is suitable for voice calls and nonreal-time data connections at very low bit rates, such as Web access. The minimum capacity requirement of a multimedia streaming application of acceptable quality is about 20 kbps (i.e., considering 5 kbps of audio and 15 kbps for video at four frames per second).

An HSCSD network is an enhancement of the GSM network. The basic idea is to allow one user to simultaneously allocate several TDMA (Time Division Multiple Access) time slots of a carrier. To achieve this, a new functionality is introduced in the network and mobile station (MS) for splitting and combining data into several data streams, which will then be transferred via n ($n = 1,2,\ldots,8$) channels over the radio interface. Once split, the data streams are carried by the n full-rate traffic channels as if they were independent of each other, up to the point in the network where they are combined (see Figure 4.2).[1]

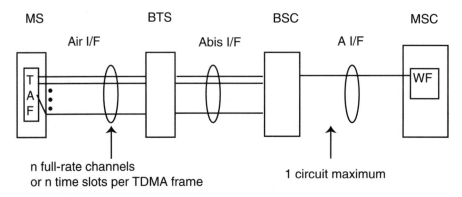

n full-rate channels
or n time slots per TDMA frame

1 circuit maximum

FIGURE 4.2 Network architecture for supporting HSCSD. (Source: ETSI, High speed circuit-switched data [HSCSD], Stage 2 [Release '96], GSM 03.34, v.5.2.0 [1999–05].)

TABLE 4.1
Bit Rates for HSCSD Networks

Number of Time Slots	Bit Rate per Time Slot (kbps)	Total User Bit Rate (kbps)
1	9.6	9.6
2	9.6	19.2
3	9.6	28.8
4	9.6	38.4
1	14.4	14.4
2	14.4	28.8
3	14.4	43.2
4	14.4	57.6

The data rate of a single time slot can be increased up to 14.4 kbps by puncturing (i.e., by deleting) certain error correction bits of the existing 9.6 kbps. In theory, up to 8 time slots can be allocated at the same time; therefore, the available user bit rate could be as high as 115.2 kbps (8 × 14.4 kbps). In practice, however, the maximum bit rate per user is limited to 64 kbps, because only one ISDN B-channel is reserved per user in the A interface of GSM network infrastructure.[1] Table 4.1 summarizes the possible bit rates achievable in uplink and downlink with HSCSD networks.

HSCSD has both transparent and nontransparent types of services. Transparent mode offers error protection at the channel coding level only. In this mode retransmission of packets hit by errors is not used. As a result, the bit rate and network delay are constant,[2] but the BER is variable (up to 10^{-3}), depending on the channel condition. Nontransparent mode offers retransmission of erroneous frames, using the GSM Radio Link Protocol (RLP) (see Figure 4.3),[1] in addition to error correction made by channel coding. Typical BER values in nontransparent mode are less than 10^{-6}. The available throughput and transmission delay vary with the channel quality

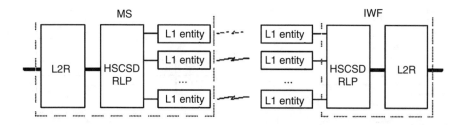

FIGURE 4.3 The HSCSD concept in nontransparent mode. (Source: ETSI, High speed circuit-switched data [HSCSD], Stage 2 [Release '96], GSM 03.34, v.5.2.0 [1999–05].)

TABLE 4.2
Bit Rates for ECSD Networks

Number of Time Slots	Bit Rate per Time Slot (kbps)	Total User Bit Rate (kbps)
1	28.8	28.8
2	28.8	57.6
2	32.0[a]	64.0
1	43.2	43.2

The 32 kbps configuration is available only for multiple slot transparent service.[56]

(the higher the BER, the lower the throughput and the higher the network delay). Typical delay values range from 400 milliseconds up to 1 second in case of mobile-to-mobile connections.[3]

HSCSD services can be further classified in symmetrical and asymmetrical services. Symmetrical service allows allocating equal bit rates to both the uplink and downlink connections. Asymmetrical service can provide different data rates in the uplink and downlink direction. Asymmetrical services are only applicable in nontransparent mode,[2] and are most suitable for multimedia streaming. In fact, in this case most of the data flow goes from the network to the MS; only a fraction of the traffic goes in the opposite direction. An example of a streaming connection over asymmetrical HSCSD is a 3+1 time slot service (43.2 kbps for downlink direction and 14.4 kbps for uplink direction).

4.3.1.1.2 Enhanced Circuit-Switched Data (ECSD)

ECSD is a network technology defined within EDGE in Release '99 specifications, and it has the same principle as HSCSD. The fundamental enhancement consists of a new modulation technique used in the air interface. This modulation is called 8-PSK (octagonal Phase Shift Keying) and it triples the data rate per time slot. However, the limitation of the A interface to 64 kbps is always in place. Table 4.2 summarizes the bit rates achievable with ECSD (in addition to those available with HSCSD).[4]

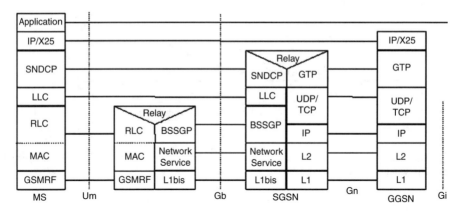

FIGURE 4.4 GPRS user plane protocol stack.

TABLE 4.3
Bit Rates for GPRS Networks (kbps)

	1 TS	2 TS	3 TS	4 TS	5 TS	6 TS	7 TS	8 TS
CS-1	9.05	18.1	27.15	36.2	45.25	54.3	63.35	72.4
CS-2	13.4	26.8	40.2	53.6	67.0	80.4	93.8	107.2
CS-3	15.6	31.2	46.8	62.4	78.0	93.6	109.2	124.8
CS-4	21.4	42.8	64.2	85.6	107.0	128.4	149.8	171.2

4.3.1.2   Packet-Switched Mobile Channels

4.3.1.2.1   General Packet Radio Service (GPRS)

GPRS networks introduce the concept of packet data in Release '97 specifications. Packet data is suitable for applications that exploit a bursty traffic, for which it is not needed to allocate a circuit-switched channel permanently, but resources are allocated from a common pool. The access time to GPRS networks is lower, and charging is done based on traffic volumes.

A great advantage of GPRS networks is that they are built to support packet-switched traffic based on IP and X.25 protocols. In this way, it is easy to connect GPRS networks to IP-based backbones, such as the public Internet. Figure 4.4 shows the GPRS protocol stack for the user plane.[5]

A GPRS MS can use up to eight channels or time slots (TS), that are dynamically allocated separately for downlink and uplink when there is traffic to be transferred. The allocation depends on the resource availability. In GPRS, different channel coding schemes are defined in the radio interface. They are named CS-1, CS-2, CS-3, and CS-4, and offer decreasing error protection. Depending on the number of time slots and the coding scheme used, the maximum bit rate achievable with GPRS networks can be as high as 171.2 kbps, as shown in Table 4.3.

Between the MS and the BSS (Base Station Subsystem) transmission can occur in Unacknowledged or Acknowledged mode at the Radio Link Control (RLC) layer.[6] Unacknowledged mode is a transparent transmission mode. In acknowledged mode, the RLC layer provides the ability to retransmit the erroneous frames that have been corrupted by errors in the air interface. Acknowledged mode is appropriate for mobile multimedia streaming.

The SNDCP (SubNetwork Dependent Convergence Protocol)[7] layer provides TCP/IP header compression[8] and V.42 bis[9] data compression to enhance the capacity of the network. SNDCP allows a reduction of the packet header size from 40 to 3 bytes.

GPRS introduces the concept of QoS profile for a PDP context. A QoS profile defines a set of attributes that characterize the expected quality of the connection. These attributes are described in Table 4.4.[5,10] For real-time traffic, such as multimedia streaming traffic, the QoS profile must be set with appropriate combination of values, in order to guarantee the best user QoS. It must be noted that the throughput values can be renegotiated by the network at any time.

4.3.1.2.2   Enhanced General Packet Radio Service (E-GPRS)

E-GPRS is defined in the EDGE framework of Release '99 specifications.[58] In E-GPRS, as in ECSD, the 8-PSK modulation is used to increase the network capacity. This modulation scheme is used in addition to the GMSK (Gaussian Minimum Shift Keying) already employed in GPRS. The major impacts of E-GPRS, compared to GPRS, are in layers 1 and 2 of the protocol stack. In layer 1 a new set of modulation and coding schemes (MCS) are defined. The GPRS GMSK coding schemes (CS-1 through CS-4) are replaced with four new GMSK modulation and coding schemes (MCS-1 through MCS-4) offering decreasing error protection. In addition, five 8-PSK coding schemes (MCS-5 through MCS-9) are defined, which yield decreasing error protection as well.

The E-GPRS coding schemes support incremental redundancy (IR), which is a physical layer performance enhancement for the RLC acknowledged mode of layer 2. Whenever a request for a retransmission is triggered by the RLC protocol, the IR mechanism dynamically adjusts the code rate to the actual channel conditions by incrementally redundant information, until the reception of the lost RLC block is successful. This effectively increases the probability of data reception at the RLC peer entity.[11] This feature is a great benefit in nonconversational real-time applications, such as mobile multimedia streaming.

In E-GPRS, bit rates can be increased up to 473.6 kbps, as illustrated in Table 4.8, which shows the bit rates for different combinations of time slots and coding schemes.

Another improvement offered by E-GPRS is the TCP and UDP (over IPv4 and IPv6) header compression capability in the SNDCP layer. This allows the packet headers to reduce from a maximum size of 60 to 4 bytes.[12]

The QoS profile for E-GPRS is essentially similar to that for UTRAN. Please refer to the information provided in Tables 4.9, 4.10, and in the next section.

4.3.1.2.3   UMTS Terrestrial Radio Access Network (UTRAN)

The IMT-2000 specifications for 3G networks are written by 3GPP, which has defined standards for UMTS networks. The air interface technology for UMTS is

TABLE 4.4
QoS Profile for GPRS Release '97 Networks

QoS Profile Attribute	Description
Precedence class	The precedence class indicates a priority in case of abnormal network behavior. For example, in case of congestion, the precedence class determines which packet to discard first. Values: [1…3] in decreasing order of precedence
Delay class	The delay class defines the maximum and 95-percentile of mean transfer delay within a GPRS network end-to-end (it does not include transfer delays in external networks). Examples for packet sizes of 128 and 1024 bytes are shown in Table 4.5. Values: [1…4]. A GPRS network must support at least the Class 4 (best effort)
Reliability class	Data reliability is defined in terms of residual probabilities of data loss, out-of-sequence delivery, duplicate data delivery and data corruption. These probabilities are defined for three classes in Table 4.6. The reliability class specifies the requirements of the various network protocol layers. The combinations of the GTP (GPRS Tunneling Protocol), LLC (Logical Link Control), and RLC transmission modes support the reliability class performance requirements. The combinations are shown in Table 4.7. Values: [1…5]. A GPRS network may support only a subset of the defined reliability classes
Mean throughput class	It specifies the average rate at which data is expected to be transferred across the GPRS network during the remaining lifetime of an activated PDP context. The rate is measured in bytes per hour. Values: [1…18, 31], where the value 31 means best effort, and the values from 1 to 18 define discrete rates in the range $[100, 50 \times 10^6]$ bytes per hour, i.e., in the range $[0.22, 111 \times 10^3]$ bits per second
Peak throughput class	It specifies the maximum rate at which data is expected to be transferred across the GPRS network for an activated PDP context. There is no guarantee that this peak rate can be achieved or sustained for any time period, and this depends on the MS capability and the available radio resources. The rate is measured in bytes per second. Values: [1…9], which define discrete rates in the range $[1 \times 10^3, 256 \times 10^3]$ bytes per second, i.e., in the range [8, 2048] kbps

W-CDMA (Wideband Code Division Multiple Access). The main objectives and features for UMTS networks can be summarized as:

- Full area coverage and mobility for 144 kbps (at vehicular speed), preferably 384 kbps (at pedestrian speed). Limited area coverage and mobility for 2 Mbps.
- Multiplexing of services with different QoS requirements on a single bearer (e.g., a speech call, a multimedia streaming session, and a Web session). This is one of the key features of W-CDMA. Power is the common shared resource for users. As the bit rate changes, the power

TABLE 4.5
Delay Classes for GPRS Release '97 Networks

	Delay (Maximum Values)			
	Packet Size: 128 Bytes		Packet Size: 1024 Bytes	
Delay Class	Mean Transfer Delay (sec)	95th Percentile Transfer Delay (sec)	Mean Transfer Delay (sec)	95th Percentile Transfer Delay (sec)
1. (Predictive)	< 0.5	< 1.5	< 2	< 7
2. (Predictive)	< 5	< 25	< 15	< 75
3. (Predictive)	< 50	< 250	< 75	< 375
4. (Best Effort)		Unspecified		

TABLE 4.6
Residual Error Probabilities for Reliability Classes in GPRS Release '97 Networks

	Probability				
Reliability Class	Packet Loss	Duplicate Packet	Out of Sequence Packet	Packet Corruption	Example of Application Characteristics
1	10^{-9}	10^{-9}	10^{-9}	10^{-9}	Error sensitive, no error-correction capability, limited error-tolerance capability
2	10^{-4}	10^{-5}	10^{-5}	10^{-6}	Error sensitive, limited error-correction capability, good error-tolerance capability
3	10^{-2}	10^{-5}	10^{-5}	10^{-2}	Not error sensitive, error-correction capability, very good error-tolerance capability

allocated to the channel is adjusted so that the continuity of service is guaranteed at any instant of the connection. The relative transmitted power during a 10-millisecond radio frame is a function of the bit rate: the higher the bit rate, the higher the transmitted power. In this way there is no waste of resources. For example, when a short voice segment has low information content, it can be encoded with few bits that will be transmitted using a relatively small amount of power, thus minimizing interference with other users.

- Delay requirements that range from the most stringent values for real-time traffic to more relaxed ones for best-effort traffic.
- Coexistence of 2G, 2.5G, and 3G networks through intersystem handover capability.

TABLE 4.7
Reliability Classes in GPRS Release '97 Networks

Reliability Class	GTP Mode	LLC Frame Mode	LLC Data Protection	RLC Block Mode	Traffic Type
1	Acknowledged	Acknowledged	Protected	Acknowledged	Nonreal-time traffic, error-sensitive application that cannot cope with data loss.
2	Unacknowledged	Acknowledged	Protected	Acknowledged	Nonreal-time traffic, error-sensitive application that can cope with infrequent data loss.
3	Unacknowledged	Unacknowledged	Protected	Acknowledged	Nonreal-time traffic, error-sensitive application that can cope with data loss, GMM/SM, and SMS.
4	Unacknowledged	Unacknowledged	Protected	Unacknowledged	Real-time traffic, error-sensitive application that can cope with data loss.
5	Unacknowledged	Unacknowledged	Unprotected	Unacknowledged	Real-time traffic, error nonsensitive application that can cope with data loss.

TABLE 4.8
Bit Rates For E-GPRS Networks (kbps)

	1 TS	2 TS	3 TS	4 TS	5 TS	6 TS	7 TS	8 TS
MCS-1	8.8	17.6	26.4	35.2	44.0	52.8	61.6	70.4
MCS-2	11.2	22.4	33.6	44.8	56.0	67.2	78.4	89.6
MCS-3	14.8	29.6	44.4	59.2	74.0	88.8	103.6	118.4
MCS-4	17.6	35.2	52.8	70.4	88.0	105.6	123.2	140.8
MCS-5	22.4	44.8	67.2	89.6	112.0	134.4	156.8	179.2
MCS-6	29.6	59.2	88.8	118.4	148.0	177.6	207.2	236.8
MCS-7	44.8	89.6	134.4	179.2	224.0	268.8	313.6	358.4
MCS-8	54.4	108.8	163.2	217.6	272.0	326.4	380.8	435.2
MCS-9	59.2	118.4	177.6	236.8	296.0	355.2	414.4	473.6

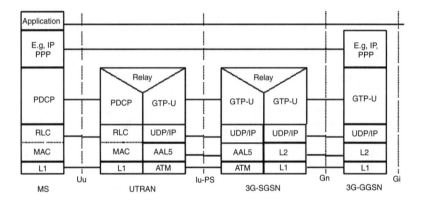

FIGURE 4.5 User plane protocol stack for UTRAN networks (Iu mode)

- Fast transmit power control (TPC). Because W-CDMA networks are interference limited, fast TPC based on the measurement of signal-to-interfer-
  ence ratio (SIR) can always minimize the transmitted power according to
  the traffic load, and thus interference to other users can be reduced.

3GPP has gone through different releases of the UTRAN specifications, namely
Release '99, Release 4, and Release 5. At the time of writing this chapter, the Release
6 specifications were in the process of being defined. UTRAN networks offer both
circuit-switched and packet-switched services. Release 5 specifications define the
IP Multimedia Subsystem (IMS)[13] in the Core Network (CN) that makes all-IP
networks a reality. IMS is based on the Session Initiation Protocol (SIP) defined in
the IETF (Internet Engineering Task Force).[14–16] IMS in the CN enables more
Internet-based multimedia services that are not available in the Release 4 CN.

Figure 4.5 shows the user plane protocol stack for UTRAN networks.[17] Between
the MS and the UTRAN, the RLC Protocol can operate in transparent, unacknowledged,

and acknowledged modes. In transparent mode no protocol overhead is added to the higher layer data, while in unacknowledged and acknowledged modes a certain RLC layer overhead is added (sequence numbers, length indication, and other information). Acknowledged mode is suitable for a mobile multimedia streaming application, because it provides retransmissions of the lost blocks. Retransmission can be configured in the RLC protocol in different ways:[18]

- *Retransmissions for* n *times.* The RLC layer tries to retransmit the lost block up to *n* times. If it does not reach the RLC peer entity within *n* retransmission attempts, the block is considered lost. This option tries to achieve a constant BLER (block error rate) at the cost of a variable delay at the RLC peer entity.
- *Retransmission with a timer.* The RLC tries to retransmit the lost block an undefined number of times until a timer fires. Afterwards, the block is considered lost. This option defines implicitly an upper bound on the delay at the RLC peer entity; however, the BLER is variable.
- *Fully persistent retransmission.* The RLC layer retransmits the lost block an undefined number of times until the block is received by the RLC peer entity. The RLC block is discarded only when the RLC layer buffer is full. This option defines an upper bound on the BLER, but it may produce the highest variations of delay at the receiver (jitter).

Other functions of the RLC layer include segmentation/reassembly, concatenation, padding, error correction, in-sequence delivery, duplicate detection, flow control, sequence number check, and ciphering.

The PDCP (Packet Data Convergence Protocol) layer[57] is located immediately below the IP layer, and it exists only for services from the packet-switched domain. Its main functionality is that of compressing higher-layer protocol headers for the purpose of reducing the bit rate toward the radio interface. For Release '99 networks the compression algorithm is the same as the one included in the E-GPRS specifications,[12] while from Release 4 onward the ROHC (Robust Header Compression) algorithm[19] is supported also to compress RTP/UDP/IP or UDP/IP (under IPv4 or IPv6 environment) headers from a maximum of 60 to 3 bytes. In the PDCP, differently than the SNDCP layer in GPRS, no data compression is supported. The reason for this choice is to achieve a higher protocol speed, and also because many types of data encapsulated using the Real-Time Transport Protocol (RTP) are already compressed (speech, audio, video, images), making a second compression step unnecessary.

To guarantee end-to-end quality of service, the UMTS specifications define a new, important parameter: the traffic class. This is considered as a fundamental way to distinguish services of different types and their respective quality. Table 4.9 summarizes the four traffic classes defined for UMTS networks.[20] The practical differences between the four classes is in terms of delay and error rates. While conversational and streaming classes guarantee low delays at the cost of higher error rates, interactive and background traffic classes guarantee lower error rates at the cost of higher delays.

TABLE 4.9
UMTS Traffic Classes

Traffic Class	Conversational Class	Streaming Class	Interactive Class	Background
Fundamental characteristics	• Preserve time relation (variation) between information entities of the stream • Conversational real-time pattern (very delay sensitive)	• Preserve time relation (variation) between information entities of the stream • Nonconversational real-time (not highly delay sensitive)	• Request response pattern (best effort) • Preserve payload content	• Destination is not expecting the data within a certain time (best effort) • Preserve payload content
Example application	• Voice over IP, video telephony	• Video streaming	• Web browsing	• Background download of e-mails

The QoS profile for an UMTS PDP context is defined in a slightly different way compared to the GPRS QoS profile. Table 4.10 contains the QoS profile attributes and values for the streaming traffic class.[20]

The rules to map UMTS and GPRS Release '97 QoS profile attributes (and vice versa) are not described here, but they are available (see reference 20).

4.3.1.2.4   GSM/EDGE Radio Access Network (GERAN)

GERAN networks in Release 5 specifications originate from the possibility of integrating UMTS and GSM/EDGE network technologies to provide more benefits to the end users. The two technologies have many things in common; for example, UMTS has adopted most of the functionalities of the GSM/EDGE networks. On the other hand, GERAN has adopted the Iu interface, which is the same interface between UTRAN and CN. The Iu interface enables the interfacing to UTRAN networks, allowing also the provisioning of the same IMS services as UTRAN. GERAN also makes use of the Gb and A interfaces to communicate with GSM/EDGE networks at the maximum speed of 473.6 kbps. GERAN implements a separation of radio-related and nonradio-related functionalities. For example, one operator could run a CN and deploy the same services using two different radio technologies seamlessly (e.g., W-CDMA, GSM/EDGE, WLAN).

One of the peculiarities of UTRAN and GERAN is the fact that their protocol stacks are aligned. Figure 4.6 shows the GERAN user plane protocol stack.[21]

In this architecture, the SNDCP and LLC protocols of E-GPRS are replaced by the PDCP layer, when communicating through the Iu interface. One of the features standardized in GERAN is the capability to efficiently handle RTP/UDP/IP traffic by using header compression[19] or header removal in the PDCP layer.

TABLE 4.10
QoS Profile for UMTS Networks

QoS Profile Attribute	Description
Traffic class	Type of application for which the UMTS bearer service is optimized. UMTS can make assumptions about the traffic source and optimize the transport for that traffic type. Values: [Conversational, Streaming, Interactive, Background]
Maximum bit rate	Upper limit a user or application can accept or provide. All UMTS bearer service attributes may be fulfilled for traffic up to the maximum bit rate depending on the network conditions. Its purpose is (1) to limit the delivered bit rate to applications or external networks with such limitations, and (2) to allow a maximum user bit rate to be defined for applications that are able to operate with different rates (e.g., applications with adapting codecs). Values: up to 2048 kbps
Guaranteed bit rate	Describes the bit rate the UMTS bearer service shall guarantee to the user or application. Guaranteed bit rate may be used to facilitate admission control based on available resources, and for resource allocation within UMTS. Values: up to 2048 kbps
Delivery order	Indicates whether the UMTS bearer shall provide in-sequence packet delivery. Values: [Yes, No]
Maximum SDU size	The maximum allowed SDU (packet) size. It is used for admission control and policing. Values: up to 1502 bytes
SDU format information	List of possible exact sizes of SDUs (packets). UTRAN needs packet size information to operate in transparent RLC mode. Thus, if the application can specify packet sizes, the bearer is less expensive. Values: specific values in bits
SDU error ratio	Indicates the fraction of packets lost or detected as erroneous. Values: $[10^{-1}, 10^{-2}, 7*10^{-3}, 10^{-3}, 10^{-4}, 10^{-5}]$
Residual bit error ratio	Indicates the undetected bit error ratio in the delivered packets. If no error detection is requested, residual bit error ratio indicates the bit error ratio in the delivered packets. Values: $[5*10^{-2}, 10^{-2}, 5*10^{-3}, 10^{-3}, 10^{-4}, 10^{-5}, 10^{-6}]$
Delivery of erroneous SDUs	Used to decide whether error detection is needed and whether frames with detected errors shall be forwarded to the upper layers. Values [Yes, No, —], where "Yes" means that error detection is employed and erroneous packets are delivered together with an error indication; "No" means that error detection is employed and that erroneous packets are discarded; "—" means that packets are delivered without considering error detection.
Transfer delay	Used to specify the delay tolerated by the application; in other words, it is the maximum delay for the 95th-percentile of the distribution of delay for all delivered packets during the lifetime of a bearer service, where delay for a packet is defined as the time from a request to transfer a packet at one SAP (service access point) to its delivery at the other SAP. A good maximum delay value would take into account delays produced by the RLC layer when the acknowledged mode is used. Values: [288, maximum value] milliseconds; the maximum value can be defined at bearer setup

TABLE 4.10 (continued)
QoS Profile for UMTS Networks

Traffic handling priority	Specifies the relative importance for handling of all the packets belonging to the bearer compared to the packets of other bearers. This parameter is not used for the streaming traffic class. Values: [1, 2, 3]
Allocation/ retention priority	Used for differentiating between bearers when performing allocation and retention of a bearer. In situations where resources are scarce, the network can use this attribute to prioritize bearers with a high priority over bearers with a low priority when performing admission control. This is a subscription attribute, which is not negotiated from the mobile terminal. Values: [1, 2, 3]
Source statistics descriptor	Specifies characteristics of the source traffic. Conversational speech has a well-known statistical behavior. By being informed that the packets are generated by a speech source, the network and the mobile station may, based on experience, calculate a statistical multiplex gain for use in admission control on the relevant interfaces. Values: [Speech, unknown].

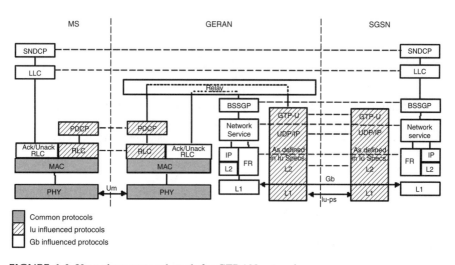

FIGURE 4.6 User plane protocol stack for GERAN networks.

The RLC layer is similar to the one defined for UTRAN, but also it can benefit from incremental redundancy of E-GPRS. All the other protocol details are similar to those for UTRAN. The same is valid for the QoS profile parameters.

In this section we have surveyed the standards for 2.5G and 3G mobile networks. These networks allow the deployment of mobile multimedia streaming. The next section is about standardized protocols, codecs, and issues related to multimedia streaming applications.

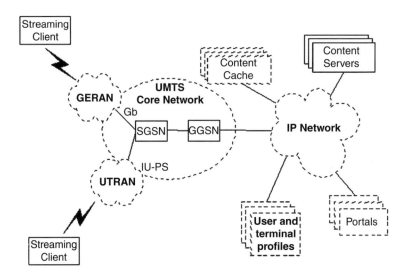

FIGURE 4.7 Network elements involved in a 3G packet-switched streaming service.

4.4 STANDARDS FOR MOBILE STREAMING

The need to have an optimized end-to-end multimedia streaming service over mobile networks has pushed the 3GPP organization to standardize such service within the Service Aspects Codec Working Group. The first standardized service is in Release 4 specifications of transparent end-to-end packet-switched streaming service (PSS). Release 5 specifications define additional capabilities to the streaming service. At the time of writing this chapter, Release 6 specifications were being written by 3GPP.

4.4.1 RELEASE 4 PSS

The specifications for PSS in Release 4 are basically defined in two documents.[22,23] Here we describe the main features and architecture of a mobile streaming service that is implemented according to Release 4 specifications. No details on the file formats for PSS are included in this chapter.

Figure 4.7 shows the end-to-end architecture for PSS service.[22] In this architecture only content servers and streaming clients are required. The dashed elements may not be present. The content server resides behind the Gi interface, possibly within the operator's network. However, streaming servers also may reside on the public Internet. Content cache servers can be used for service optimization, and may keep the replicated copies of the most-popular items streamed by the users. Portals servers are engines that allow an easier access facility to the end user, often through the use of searching and browsing capabilities. User and terminal profile servers store user preferences and terminal capabilities data that can be used to control the streaming session to the end user.

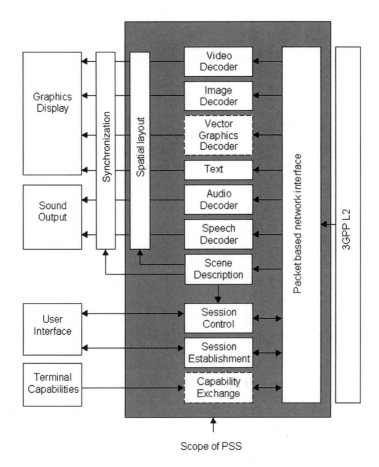

FIGURE 4.8 System architecture of a Release 4 PSS client.

While a streaming server can be designed primarily for use over the Internet and also serve mobile streaming users, eventually in an nonoptimized fashion, a streaming client must be designed merely for mobile use. A system view of a Release 4 PSS client is shown in Figure 4.8,[23] while the protocol stack for layer 3 and above is shown in Figure 4.9.[23] A PSS client can be either extended or simple, depending on whether or not it implements the functional components in the dashed blocks (vector graphics and capability exchange). The features for extended PSS clients have been defined in Release 5 specifications.

4.4.1.1 Control and Scene Description Elements

Three control elements are included in PSS (see Figure 4.8):

1. Session control (and setup) of streaming sessions between a streaming client and one or more streaming servers, with the possibility of VCR-like operations such as PAUSE, PLAY, STOP, FAST-FORWARD, and REWIND. Session control and setup is made with two protocols: HTTP

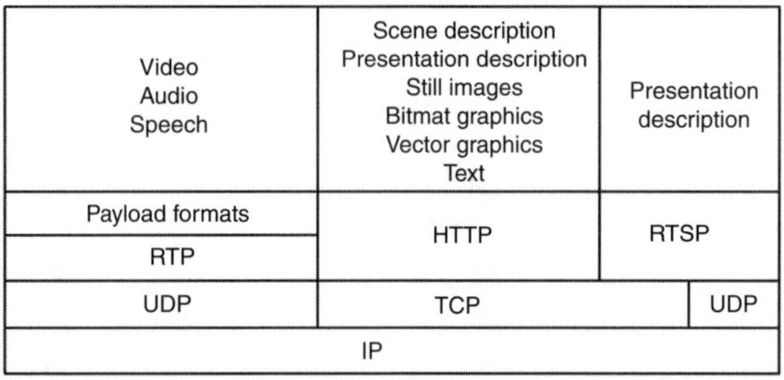

Video Audio Speech	Scene description Presentation description Still images Bitmat graphics Vector graphics Text	Presentation description	
Payload formats	HTTP	RTSP	
RTP			
UDP	TCP		UDP
IP			

FIGURE 4.9 Protocol stack for Release 4 PSS.

(Hypertext Transfer Protocol)[24] for reliable (i.e., over TCP/IP) transport
of discrete media (still images, bitmap graphics, text, scene or presentation
description) or RTSP (Real-Time Streaming Protocol)[25] for reliable or
unreliable transport of session setup (i.e., over TCP/IP or UDP/IP, respectively) and control of continuous media (speech, audio, or video).
2. Session establishment refers to the methods to invoke a PSS session from
a browser or directly by entering the URL[26] in the terminal's user interface;
in other words, the ways to obtain the initial session description (e.g., an
SDP[27] presentation description, a SMIL[28] scene description or directly the
RTSP URL of the content).
3. Capability exchange enables the adaptation of the streamed content to the
user's device, depending on its characteristics and capabilities. No explicit
protocol is defined for the simple PSS. This makes the assumption that
the user is aware of the requirements of the content to be streamed (e.g.,
the screen size). Protocols for capability exchange can be specified for
extended PSS.

The scene description component consists of spatial layout of the different media
and the description of the temporal relationship (i.e., synchronization) of the media
that is included in a media presentation.

4.4.1.2 Media Elements

3GPP PSS can support a rich set of media, either continuous or discrete. Continuous
media are media flows that must be displayed/played preserving their temporal
relationship. They are transported via RTP/UDP/IP packet encapsulation, using
appropriate payload formats defined for each codec. Discrete media have no time
element among their properties. They are transported using HTTP/TCP/IP packet
encapsulation.

Continuous media in Release 4 PSS are speech, audio, and video. The following
decoders can be supported in a streaming client:

- AMR (Adaptive MultiRate) narrowband is the mandatory speech decoder for PSS.[29] Speech is encoded at 8 kHz sampling frequency at 8 different bit rates ranging from 4.75 to 12.20 kbps. An AMR speech flow is packetized by a streaming server using the RTP payload format described in reference 30.
- AMR wideband decoder is mandatory for a PSS client when speech encoded at 16 kHz sampling frequency is supported.[31] It supports nine different bit rates ranging from 6.60 to 23.85 kbps. An AMR wideband flow is packetized using the RTP payload format described in reference 30.
- MPEG-4 AAC. If audio is supported, AAC Low Complexity[32] with a maximum sampling frequency of 48 kHz in mono (1/0) and stereo (2/0) is the decoder supported. However, the AAC Long-Term Prediction decoder can also be supported. An AAC flow is packetized using the payload format as described in reference 33.
- H.263 Video. Profile 0 Level 10 is the mandatory decoder for video streams.[34] It supports video at a maximum bit rate of 64 kbps at QCIF picture size (176 × 144 pixels). Optionally, Profile 3 Level 10 can also be supported[35] to provide better error resilience. An H.263 video stream is packetized using the payload format defined in reference 36.
- MPEG-4 Visual is an optional decoder that can be supported at Simple Profile Level 0.[37,38] An MPEG-4 video stream is packetized using the payload format described in reference 33.

Discrete media in Release 4 PSS are still images, bitmap graphics, vector graphics, and text. The following discrete media decoders can be supported in a streaming client:

- JPEG is the mandatory format for still images.[39] It is supported in the baseline and progressive DCT mode (nondifferential Huffman encoding).
- GIF is the format supported for bitmap graphics. Two formats can be decoded by a PSS client: GIF87a and GIF89a.[40,41]
- Vector graphics. No decoders are specified for a simple PSS client. However, decoders can be specified for extended PSS clients.
- The text decoder is supposed to be used in a SMIL presentation with the text formatted following XHTML Mobile Profile.[42] The supported character coding formats are UTF-8[44] and UCS-2.[43]

4.4.2   RELEASE 5 PSS

The specifications for Release 5 PSS are defined in three documents.[45-47] Release 5 enhances PSS with new features. The major features are described here (file format issues will not be covered here).

4.4.2.1   Control Elements

The most important enhancement of the application control plane is a new mechanism for capability exchange, which means the functionality of PSS servers to provide content for a wide set of mobile devices. The device capabilities and preferences are described in a device profile using attributes into an RDF[48] document that follows the structure of the CC/PP[49] framework and the CC/PP application, UAProf.[50] Following are some of the possible attributes that can characterize a mobile device:

- Audio channels
- Max polyphony capabilities when supporting scalable polyphony MIDI sounds
- List of MIME types that the PSS device accepts
- Screen size
- SMIL modules supported
- Decoding video byte rate
- Size of predecoder buffer
- Size of postdecoder buffer
- Number of bits per pixel
- Color capability
- Pixel aspect ratio
- List of supported character sets

Whenever this mechanism for capability negotiation is used, device capability profiles are stored on a device profile server that is inquired by a PSS server (via HTTP or RTSP) before delivering multimedia content to the mobile device.

4.4.2.2 Media Elements

The new media formats that are introduced in Release 5 of PSS specifications are all of the discrete type (i.e., transport over HTTP/TCP/IP):

- Scalable polyphony MIDI (SP-MIDI) is the format to support synthetic audio.[51]
- PNG (Portable Network Graphics) is yet another format to support bitmap graphics.[52]
- SVG (Scalable Vector Graphics) is the mandatory format used to support vector graphics. Optionally, the Basic profile also can be supported.[53,54]
- Timed text is supported as downloadable text (not streamed via RTP/UDP/IP). Timed text can define color, font, writing direction (i.e., left to right, or other directions), Karaoke highlighting, and other attributes.[47]

This section discussed the main features of PSS in Release 4 and Release 5 specifications. The next section covers some QoS and performance issues of mobile multimedia streaming.

4.5 PERFORMANCE ISSUES OF MOBILE STREAMING

When implementing a PSS services, the underlying mobile network brings requirements and constraints that must be taken into consideration. To understand how a

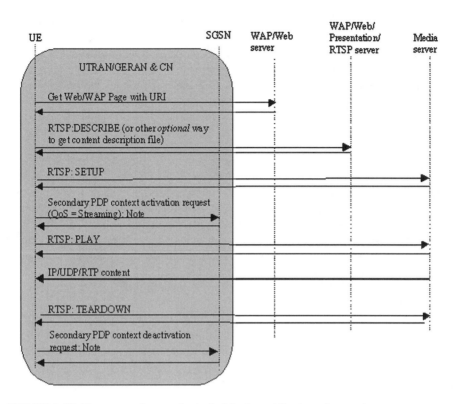

FIGURE 4.10 Message exchange of a typical basic mobile streaming session.

multimedia session is started and handled end-to-end, Figure 4.10 shows a possible scenario for an example session (here and in the following, we consider only the usage of packet switched bearers, rather than circuit switched bearers).[45]

The session is started by a mobile user getting a URI to the specific content that suits the user's terminal. The URI comes from a WAP (Wireless Application Protocol)/Web browser, or it is typed in. The URI specifies the streaming server and the address of the content on that server. A PSS application that establishes the multimedia session must understand an SDP file. The SDP file may be obtained in a number of ways. In the example described here, it is obtained through RTSP signaling via the DESCRIBE message. The SDP file contains the description of the session (session name, author, and other parameters), the type of media to be presented, and the bit rate.

The session establishment is the process in which the mobile user invokes a streaming client to set up the session with the server. The mobile device is expected to activate a PDP context that enables IP packet transmission, with an appropriate QoS profile for streaming media.

The setup of the streaming service is done by sending an RTSP SETUP message for each media stream chosen by the client. This returns, among other information, the UDP and TCP port to be used for the respective media stream. The client sends an RTSP PLAY message to the server that starts to send one or more streams over the IP network.

At the end of the session a TEARDOWN message is sent by the client and the PDP context can be deallocated.

The following sections introduce some QoS issues in mobile multimedia streaming.

4.5.1 BEARER CONSIDERATIONS

The PDP context activation described in Figure 4.10 must be activated with an adequate set of parameters. SDU error rates and transfer delay are key parameters and must be tuned according to the RLC mode chosen. If acknowledged mode is selected, the streaming client can rely on lower error rates because the lost blocks are retransmitted. As a downside, the delay jitter perceived at the PSS client may be rather high, especially if fully persistent acknowledged mode transmission is chosen. On the other hand, when unacknowledged mode is selected, the delay jitter is supposedly limited, but no retransmissions are occurring at the RLC layer. Consequently, the PSS client cannot benefit from a much higher level of data delivery reliability.

Particular attention must be paid to RTP packet sizes. When using RLC acknowledged mode, using bigger packets produces higher delay jitters at the PSS client, because there is a higher probability that more RLC blocks have to be retransmitted in order to deliver a single RTP packet. On the other hand, when using RLC unacknowledged mode, larger packets are more susceptible to losses than smaller packets, because the loss of a single RLC block produces the loss of the entire RTP packet. In this case using small packets is recommended. Too-small packets would cause too much RTP/UDP/IP header overhead. It is clear that the best settings are found when considering all the mentioned issues at the same time.

4.5.2 RTCP

The RTP (Real-Time Transport Protocol) specifies its control protocol RTCP[55] that allows monitoring of the data delivery or, in other words, the quality of service. The main function of RTCP is to convey feedback information about the participants in an ongoing session. The information provided includes packet losses and interarrival jitter information. In a point-to-point connection, such as a multimedia streaming session, RTCP can offer a valid means for adjusting the error resilience properties of the media streams carried over mobile networks. In such networks the quality of the connection may vary all the time, and prompt action is crucial for guaranteeing the best possible quality of service at any instant. An example of action to be taken by a streaming server is the dynamic change of the packetization parameters to provide an increased level of error resilience. However, the utility function for repairing media is decreasing over time, because an action taken too late can be useless or it may even worsen the quality of service. Normally, RTCP feedback is sent at an interval of at least 5 seconds. The recommended fraction bandwidth reserved for RTCP is 5 percent of the RTP session bandwidth. However, at low bit rates (up to 64 kbps), if the minimum RTCP transmission interval is 5 seconds, it is impossible to reach the full 5 percent bandwidth.

The PSS specifications of Release 5[47] define two working modes for PSS clients that intend to send more-frequent feedback to the PSS server:

1. Mode 1 (normal feedback) uses the rule defined in reference 55, where feedback is sent at intervals of at least 5 seconds.
2. Mode 2 (more frequent feedback) fills the 5 percent bandwidth reserved for RTCP to send feedback information.

Mode 2 allows feedback to be sent at a higher rate (the RTCP transmission interval is much smaller than 5 seconds), and it allows QoS-enabled PSS servers to take appropriate actions to guarantee good QoS at the PSS client. If 5 percent bandwidth is costly in terms of bearer allocation, Mode 1 can be used. It consumes less network resources, but the feedback capability is limited (on average) to one RTCP feedback report every 5 seconds.

4.5.3 RTSP Signaling Issues

Session setup time is one of the most important factors to determine the efficiency of a PSS service. RTSP is the protocol that handles session setup, and it supports both TCP and UDP transport. TCP is the mandatory transport mechanism for RTSP.

For TCP, two types of connections are possible:

1. Persistent, where a connection is used for several RTSP request/response pairs
2. Nonpersistent, where a connection is used for a single RTSP request/response pair

Every nonpersistent connection starts with a three-way handshake (SYN, ACK, SYN) before any RTSP message can be sent. This increases signaling time considerably. For this reason, the use of persistent TCP connections is recommended in order to keep the signaling time as low as possible.[47]

4.5.4 Link Aliveness

In mobile networks, connection can be lost because of low network coverage, fading, shadowing, loss of battery power, or turning off the PSS client even if the streaming session is active. In order for the streaming server to understand the client aliveness status, the PSS client should send periodic wellness information to the PSS server. The default period to send this information is 1 minute, as described in reference 25. For this purpose RTCP or RTSP can be used to signal link aliveness to the PSS server.

4.6 CONCLUSIONS

Current mobile networks, such as GPRS, E-GPRS, GERAN, and UTRAN, are designed and optimized to transport IP-based traffic. Consequently, mobile multimedia streaming can become one of the key applications for mobile users.

Packet-switched streaming will offer end users the ability to receive content such as audio and video directly to their mobile device. This service not only enables new content providers to distribute media content to mobile users, but also enables connection to the huge amount of content available on the Internet, making the transition toward the Wireless Internet feasible.

References

1. ETSI, High speed circuit-switched data (HSCSD). Stage 2 (Release '96), GSM 03.34, v.5.2.0 (1999–05).
2. ETSI, High speed circuit-switched data (HSCSD). Stage 1 (Release '98), GSM 02.34, v.7.0.0 (1999–08).
3. Nieweglowski, J. and Leskinen, T., Video in mobile networks, *European Conference on Multimedia Applications, Services and Techniques (ECMAST '96)*, 28–30 May 1996, Louvain-la-Neuve, Belgium, pp. 120–133.
4. 3GPP TSGS-SA, High speed circuit-switched data (HSCSD). Stage 1 (Release 4), TS 22.034, v.4.1.0 (2001–03).
5. 3GPP TSGS-SA, General packet radio service (GPRS), service description. Stage 2 (Release '97), TS 03.60, v.6.11.0 (2002–06).
6. 3GPP TSG-GERAN, General packet radio service (GPRS), mobile station (MS)–base station system (BSS) interface, radio link protocol/medium access control (RLC/MAC) protocol (Release '97), TS 04.60, v.6.14.0 (2001–07).
7. 3GPP TSG-CN, General packet radio service (GPRS), mobile station — serving GPRS support node (MS-SGSN), subnetwork dependent convergence protocol (SNDCP) (Release '97), TS 04.65, v.6.7.0 (2000–03).
8. Jacobson, V., Compressing TCP/IP headers for low-speed serial links, IETF (Internet Engineering Task Force) RFC 1144, February 1990.
9. ITU-T, Data compression procedures for data circuit-terminating equipment (DCE) using error correcting procedures, Recommendation V.42 bis, January 1990.
10. ETSI, General packet radio service (GPRS), service description (Stage 1) (Release '98), GSM 02.60, v.7.5.0 (2000–07).
11. Halonen, T., Romero, J., and Melero, J., Eds., *GSM, GPRS and EDGE Performance*, John Wiley & Sons, New York, 2002.
12. Degermark, M., Nordgren, B., and Pink, S., IP header compression, IETF (Internet Engineering Task Force) RFC 2507, February 1999.
13. 3GPP TSG SSA, IP Multimedia subsystem (IMS). Stage 2 (Release 5), TS 23.228, v.5.7.0 (2002–12).
14. 3GPP TSG-CN, Signaling flows for the IP multimedia call control based on SIP and SDP. Stage 3 (Release 5), TS 24.228, v.5.3.0 (2002–12).
15. 3GPP TSG-CN, IP multimedia call control protocol based on SIP and SDP. Stage 3 (Release 5), TS 24.229, v.5.3.0 (2002–12).

16. Rosenberg, J. et al., SIP: Session Initiation Protocol, IETF (Internet Engineering Task Force) RFC 3261, March 2002.

17. 3GPP TSGS-SA, General packet radio service (GPRS), service description. Stage 2 (Release 5), TS 23.060, v.5.4.0 (2002–12).

18. 3GPP TSG-RAN, Radio link control (RLC) protocol specification (Release 5), TS 25.322, v.5.3.0 (2002–12).

19. Bormann, C., Ed., Robust header compression (ROHC): framework and four profiles: RTP, UDP, ESP and uncompressed, IETF (Internet Engineering Task Force) RFC 3095, July 2001.

20. 3GPP TSGS-SA, QoS concept and architecture (Release 5), TS 23.107, v.5.7.0 (2002–11).

21. 3GPP TSG-GERAN, Overall description. Stage 2 (Release 5), TS 43.051, v.5.8.0 (2002–11).

22. 3GPP TSGS-SA, Transparent end-to-end packet-switched streaming service (PSS). General description (Release 4), TS 26.233, v.4.2.0 (2002–03).

23. 3GPP TSGS-SA, Transparent end-to-end packet-switched streaming service (PSS). Protocols and codecs (Release 4), TS 26.234, v.4.5.0 (2002–12).

24. Fielding, R. et al., Hypertext transfer protocol – HTTP/1.1, IETF (Internet Engineering Task Force) RFC 2616, June 1999.

25. Schulzrinne, H., Rao, A., and Lanphier, R., Real-time streaming protocol (RTSP), IETF (Internet Engineering Task Force) RFC 2326, April 1998.

26. Berners-Lee, T., Masinter, L., and McCahill, M., Uniform resource locators (URL), IETF (Internet Engineering Task Force) RFC 1738, December 1994.

27. Handley, M. and Jacobson, V., SDP: Session description protocol, IETF (Internet Engineering Task Force) RFC 2327, April 1998.

28. W3C, Synchronized multimedia integration language (SMIL 2.0), Recommendation, August 2001.

29. 3GPP TSGS-SA, Mandatory speech codec speech processing functions. AMR speech codec. General description (Release 5), TS 26.071, v.5.0.0 (2002–06).

30. Sjoberg, J. et al., RTP payload format and file storage format for the Adaptive Multi-Rate (AMR) and Adaptive Multi-Rate Wideband (AMR-WB) audio codecs, IETF (Internet Engineering Task Force) RFC 3267, March 2002.

31. ITU-T, Wideband coding of speech at around 16 kbits/s using Adaptive Multi-Rate Wideband (AMR-WB), Recommendation G.722.2, January 2002.

32. ISO/IEC, Information technology – Coding of audio-visual objects – Part 3: Audio, 14496–3, 2001.

33. Kikuchi, Y. et al., RTP payload format for MPEG-4 audio/visual streams, IETF (Internet Engineering Task Force) RFC 3016, November 2000.

34. ITU-T, Video coding for low bit rate communication, Recommendation H.263, February 1998.

35. ITU-T, Video coding for low bit rate communication. Profiles and levels definition, Recommendation H.263 Annex X, April 2001.

36. Bormann, C. et al., RTP payload format for the 1998 version of ITU-T Rec. H.263 (H.263+), IETF (Internet Engineering Task Force) RFC 2429, October 1998.

37. ISO/IEC, Information technology – Coding of audio-visual objects – Part 2: Visual, 14496–2, 2001.

38. ISO/IEC, Streaming video profile, 14496–2, 2001/Amd 2, 2002.

39. ISO/IEC, Information technology – Digital compression and coding of continuous-tone still images. Requirements and guidelines, 10918–1, 1992.

40. Compuserve Incorporated, GIF graphics interchange format: a standard defining a mechanism for the storage and transmission of raster-based graphics information, Columbus, OH, 1987.

41. Compuserve Incorporated, Graphics interchange format: version 89a, Columbus, OH, 1990.

42. WAP Forum Specification, XHTML Mobile Profile, October 2001.

43. ISO/IEC, Information technology – Universal multiple-octet coded character set (UCS) – Part 1: architecture and basic multilingual plane, 10646–1, 2000.

44. The Unicode Consortium, The Unicode standard, Version 3.0, Addison-Wesley, Reading, MA, 2000.

45. 3GPP TSGS-SA, Transparent end-to-end packet-switched streaming service (PSS). General description (Release 5), TS 26.233, v.5.0.0 (2002–03).

46. 3GPP TSGS-SA, Transparent end-to-end packet-switched streaming service. Service aspects. Stage 1 (Release 5), TS 22.233, v.5.0.0 (2002–03).

47. 3GPP TSGS-SA, Transparent end-to-end packet-switched streaming service (PSS). Protocols and codecs (Release 5), TS 26.234, v.5.3.0 (2002–12).

48. W3C, Resource description framework (RDF) schema specification 1.0, Candidate Recommendation, March 2000.

49. W3C, CC/PP structure and vocabularies, Working Draft Recommendation, June 2001.

50. WAP Forum, WAP UAProf specification, Specification, October 2001.

51. MIDI Manufacturers Association, Scalable polyphony MIDI specification version 1.0, RP-34, Los Angeles, February 2002.

52. Boutell, T., et al., PNG (Portable Network Graphics) specification version 1.0, IETF (Internet Engineering Task Force) RFC 2083, March 1997.

53. W3C, Scalable vector graphics (SVG) 1.1 specification, Working Draft Recommendation, February 2002.

54. W3C, SVG mobile specification, Working Draft Recommendation, February 2002.

55. Schulzrinne, H. et al., RTP: a transport protocol for real-time applications, IETF (Internet Engineering Task Force) RFC 1889, January 1996.

56. 3GPP TSG-CN, High Speed Circuit Switched Data (HSCSD), Stage 2 (Release '99), TS 23.034, v.3.3.0 (2000-12).

57. 3GPP TSG-RAN, Packet Data Convergence Protocol (PDCP) Specification (Release 4), TS 25.323, v.4.6.0 (2002-09).

58. 3GPP TSGS-SA, General Packet Radio Service (GPRS), Service Description, Stage 2 (Release '99), TS 23.060, v.3.14.0 (2002-12).

5   Streaming Video over Wireless Networks

Haitao Zheng and Jill Boyce

CONTENTS

5.1   INTRODUCTION

Video streaming is becoming ubiquitous on the wired Internet, as broadband Internet access is more commonly available. With the advent of higher-bandwidth wireless Internet access enabled by 3G wireless networks, video streaming over wireless networks also is likely to become common and enable new services and applications.

In this chapter, we provide an introduction to intelligent video streaming over wireless networks. We begin by providing a brief background on digital video compression standards that are frequently used for video streaming. Because many of the same problems exist and protocols are used for both wired and wireless packet networks, we first describe the protocols used for streaming video over IP networks, and the problems and solutions associated with video streaming over lossy packet networks. Then we describe the characteristics of wireless networks and the particular challenges associated with video streaming over wireless networks. We proceed to describe a cross layer design framework that enables adaptation to continuously changing wireless environments. Finally, we analyze proposed solutions that improve the quality of video streamed over wireless networks.

5.2 VIDEO COMPRESSION STANDARDS

Streaming of video over today's wired and wireless networks depends heavily on international video compression standards. There are numerous video compression systems that do not use open standards, such as Real Network's RealVideo and Microsoft's Windows Media Player, but they are not discussed in this chapter, as details of their inner workings are not publicly available. Standardization in the video compression space has been done primarily by two different standards bodies, the International Organization for Standardization (ISO) and the International Telecommunications Union (ITU), previously CCITT. The video communications standards of the highest past, current, and future interest are H.261, H.263, MPEG-1, MPEG-2, MPEG-4, and JVT. These video compression standards aree described briefly here, with their particular features relevant to wireless streaming highlighted. More details on the video compression standards themselves can be found in Puri and Chen[1] and Rao and Hwang.[2]

5.2.1 H.261

ITU-T H.261, "Video codec for audiovisual services at $p \times 64$ kbps," is the ancestor of all of the popular video compression standards in use today. H.261 was designed for video telephony and video conferencing, for use over one or more dedicated ISDN lines. The standardization effort for H.261 began with the establishment in December 1984 of CCITT Study Group XV, Specialist Group on Coding for Visual Telephony. In March 1989, the $p \times 64$ kbps specification was frozen. Final standardization was established in December 1990.

Like the other video compression standards that follow it, H.261 uses block-based motion estimation and compensation and block-based transform and quantization. Intracoded frames and intercoded frames are allowed in H.261. Intercoded frames are encoded with respect to a prediction formed from a previously coded frame. A Discrete Cosine Transform is applied to 8×8 pixel blocks, and the resulting transform coefficients are quantized and entropy coded using variable length coding (VLC) techniques. Macroblocks are arranged into Group of Blocks (GOBs). Pictures and GOBs contain unique start codes, which can be used as resynchronization points when transmission errors occur.

5.2.2 MPEG-1

Work on MPEG began in 1988. ISO IEC/JTC1 SC29 IS 11172, "Coded representation of picture, audio, and multimedia/hypermedia information," became an international standard in November 1992. MPEG was originally designed for digital storage applications with a target bit rate of about 1.5 Mbps, but has been applied to a wide spectrum of application, including video streaming over the Internet.

Like H.261, MPEG allows intracoded frames ("I" frames) and intercoded frames ("P" frames); also, MPEG introduced bidirectionally coded frames ("B" frames). B frames are predicted using a frame before and after the coded frame, and can be coded using relatively fewer bits. In MPEG, B frames are never used in coding other pictures. This disposable property of B frames can be important when MPEG is streamed over lossy networks. MPEG improved intracoding also by adding a quantization matrix, and improved intercoding by allowing motion estimation at half pel resolution. Any number of consecutive macroblocks, in scan order, can be grouped into a slice. Slices are begun with unique slice start codes, which can serve as resynchronization points.

In addition to providing a video compression standard, MPEG provided also an audio compression standard, and a systems standard. MPEG video can be carried either as a video elementary stream or as a program stream.

5.2.3 MPEG-2

Work on MPEG-2 began in 1990 and the video coding portion became an international standard in November 1994, entitled "Generic coding of moving pictures and associated audio," and standardized as ISO/IEC Committee Draft 13818 and ITU-T H.262. MPEG-2 was targeted at higher bit rate applications than MPEG-1, including standard definition television (SDTV) and high definition television (HDTV).

MPEG-2 builds on MPEG-1 coding techniques by adding tools for interlaced picture coding and methods of scalability. MPEG-2 was the first standard to introduce the concept of profiles and levels, to describe interoperability points. Each profile includes a group of tools that compliant decoders must support. Each level provides limitations of pixel dimensions and frame rates that a decoder must support. MPEG-2 defined seven profiles: Simple, Main, SNR, Spatial, High, 4:2:2, and Multi. MPEG-1 defined four levels: High, High1440, Main, and Low.

The methods of scalability that MPEG-2 provides are spatial scalability, SNR scalability, temporal scalability, and data partitioning. Scalable video encoding techniques can be of great use for video streaming when used in conjunction with Unequal Error Protection (UEP), as described in Section 5.4 of this chapter. The bit rates used in MPEG-2 video coding are generally higher than are used for Internet or wireless video streaming.

5.2.4 H.263

Design of ITU-T H.263, "Video coding for low bit rate communication," began in 1993, and the Version 1 standard was published in March 1996. H.263 was designed as an extension of H.261, and greatly increased compression efficiency over H.261.

H.261 added some of the tools from MPEG-1 and MPEG-2, as well as some original tools. The tools added to H.261 that improve coding efficiency include half pel motion compensation, median prediction of motion vectors, improved entropy coding, unrestricted motion vectors, and more efficient coding of Macroblock and block signaling overhead.

Version 2, also called H.263+, was standardized in September 1997. Version 3, or H263++, was standardized in January 1998. Version 2 added several features for error resilience, including a slice-structured mode, reference picture selection, and temporal, spatial, and SNR scalability. Version 3 added data partitioning and reversible variable length coding for additional error resilience.

H.263 is commonly used in videoconferencing over dedicated telecommunications lines, as well as over IP.

5.2.5 MPEG-4

Design of the MPEG-4 standard, "Coding of audio-visual objects," began in 1993. Its initial version, ISO/IEC 14496, was finalized in October 1998 and became an international standard in the first months of 1999. The fully backward compatible extensions under the title of MPEG-4 Version 2 were frozen at the end of 1999, and achieved formal international standard status in early 2000.

Relative to the preexisting video compression standards, MPEG-4 added object-based coding and improved video compression efficiency. According to Koenen,[3] MPEG-4 provides standardized ways to:

1. Represent units of aural, visual, or audiovisual content, called "media objects." These media objects can be of natural or synthetic origin, which means they could be recorded with a camera or microphone, or generated with a computer.
2. Describe the composition of these objects to create compound media objects that form audiovisual scenes.
3. Multiplex and synchronize the data associated with media objects, so that they can be transported over network channels providing a QoS appropriate for the nature of the specific media objects.
4. Interact with the audiovisual scene generated at the receiver's end.

MPEG-4 provides many profiles; for natural video alone, there are 11 profiles:

1. Simple Visual Profile
2. Simple Scalable Visual Profile
3. Core Visual Profile
4. Main Visual Profile N-Bit Visual Profile
5. Advanced Real-Time Simple Profile
6. Core Scalable Profile
7. Advanced Coding Efficiency
8. Advanced Simple Profile

9. Fine Granularity Scalability Profile
10. Simple Studio Profile
11. Core Studio Profile

Because of the large number of profiles for MPEG-4, interoperability has been difficult. The most commonly used profile is the Simple Profile.

MPEG-4 has several tools to improve error resilience, including reversible variable length coding and several methods of scalability. Fine Grain Scalability, in particular, is well suited for use with Unequal Error Protection for video streaming over lossy networks. Li[4] describes MPEG-4 Fine Grain Scalability in detail, and compares its use with SNR scalability and simulcast.

MPEG4IP[5] is an open source package designed to enable developers to create streaming servers and clients that are standards-based and free from proprietary technology. MPEG4IP uses the MPEG-4 Simple Profile.

5.2.6 JVT

In 2001, ISO and ITU-T joined forces to develop the JVT (Joint Video Team) standard. This effort was originally begun in the ITU-T as H.26L. Committee Draft status was reached in May 2002. JVT is scheduled to become an international standard in February 2003, and called H.264 by the ITU and MPEG-4 Part 10 by ISO.

JVT provides many of the tools found in H.263 and its extended versions H.263+ and H.263++, but at an improved coding efficiency. JVT is claimed to provide the same visual quality as MPEG-4 Advanced Simple Profile at half the bit rate.[6] JVT uses 4×4 block integer transform and motion blocks of a variety of sizes. JVT's May 2002 Committee Draft defines two profiles: Baseline and Main.

JVT's May 2002 Committee Draft does not include scalability, although it is intended for use in video streaming applications. Flexible Macroblock Ordering can improve performance over lossy networks, by allowing slices to be formed from nonneighboring macroblocks; in other words, to put neighboring macroblocks into different slices. Therefore, if one slice is unavailable at the decoder due to packet loss, neighboring macroblocks from other slices can be used to perform spatial concealment of the missing data. In JVT, pictures not used to predict other pictures are known as disposable pictures and are indicated in picture headers. In previous coding standards, B pictures were the only pictures to have this characteristic, while in JVT bipredictively coded pictures are not required to be disposable. Indication of the disposable nature of a picture in the picture header effectively allows temporal scalability, which can be used with Unequal Error Protection.

Table 5.1 provides a list of the video compression standards and the bit rate ranges that they were originally designed for. All of these video compression standards share the property that they use interframe prediction. A video frame is predicted from a previous frame, and only the differences are transmitted. This means that if transmission errors occur, the errors will persist for many frames. In general, macroblocks or entire frames are intracoded at regular intervals to limit the length of time an error can persist.

TABLE 5.1
Video Compression Standards

Standard	Bit Rate Range
H.261	64 to 384 kbps
H.263	64 kbps to 1 Mbps
MPEG-1	1 to 1.5 Mbps
MPEG-2	2 to 15 Mbps
MPEG-4	64 kbps to 2 Mbps
JVT	32 kbps to ?

5.3 PROTOCOLS

Transmission Control Protocol (TCP)[7,8] and User Datagram Protocol (UDP)[9] are two Internet Protocol transport protocols that can be used for transmitting compressed video over the Internet. TCP, a reliable protocol, guarantees delivery of all packets and in order, while UDP does not guarantee delivery of packets or the ordering of received packets. TCP uses retransmissions to guarantee that all packets arrive.

Most streaming video applications do not require guaranteed in-order arrival of all packets, and cannot tolerate the unbounded delay of using TCP to send compressed video data. So UDP is the transport protocol generally used for video streaming over IP networks.

The Real-Time Transport Protocol (RTP) is frequently used with UDP for streaming of video over IP networks. RTP provides functionality suited for carrying real-time content and for synchronizing different streams with timing properties. RTP specifies a header at the beginning of each packet that includes fields for payload type, time stamp, and sequence number. The RTP specification was published as RFC 1889[10] by the Audio/Video Transport Working Group of the Internet Engineering Task Force (IETF).

RFC 1889 defines also the Real-Time Transport Control Protocol, RTCP, which works in conjunction with RTP. RTCP defines a syntax for providing feedback of quality-of-service (QoS) parameters to the participants of an RTP session.

RTP can be used with many different audio or video compression standards. The Audio/Video Transport Working Group also has published several RFCs that specify carriage of specific video compression standards over RTP, in general by adding standard-specific RTP header extensions. For example, RFC 2032[11] "RTP Payload Format for H.261 Video Streams," describes a recommended syntax for an H.261 specific header to be included in an RTP packet, after the basic RTP header. In order to be error resilient, higher layer syntax elements from the H.261 bit stream are redundantly repeated in each packet header, in a fixed length format.

RFC 2038,[12] "RTP Payload Format for MPEG1/MPEG2 Video," similarly describes a recommended syntax for MPEG video data to be streamed using RTP. RFC 2038 applies only to MPEG elementary streams. RFC 2038 requires that coded pictures be fragmented into separate packets. New pictures must be at the start of a packet. Certain picture layer parameters are repeated in the MPEG specific RTP header extension.

RFC 2429,[13] "RTP Payload Format for the 1998 Version of ITU-T Rec. H.263 Video (H.263+)," describes a syntax for streaming H.263 over RTP. In addition to providing syntax for an H.263+ payload header, it provides an optional Video Redundancy Coding Header that works with H.263+'s reference picture selection to improve error resilience.

RFB 3016, "RTP Payload Format for MPEG-4 Audio/Visual Streams," does not provide an MPEG-4-specific RTP header extension. It does provide rules for fragmenting the MPEG-4 Visual Bitstream into RTP packets. An IETF Internet Draft, draft-ietf-avt-mpeg4-multisl-04.txt, "RTP Payload Format for MPEG-4 Streams," provides an MPEG-4-specific RTP header extension.

The Real-Time Streaming Protocol (RTSP) is a session control protocol for initiating and direction streaming of multimedia over IP. RTSP provides VCR-like control functions, such as PLAY, PAUSE, RESUME, FAST-FORWARD, and FAST-REWIND. RTSP is not used to deliver compressed video data itself, but is used in conjunction with other protocols such as RTP.

5.4 STREAMING VIDEO OVER THE INTERNET

Because the protocols typically used for streaming of compressed video over the Internet, UDP and RTP, do not guarantee end-to-end delivery of compressed video data, packet losses introduce errors into the decoded video, which reduces the perceived video quality by viewers. Because interframe coding is used in all of the common video compression standards, those errors propagate and hence can have a large impact on video quality.

Consider a typical application, with video encoded at 30 fps, and an intracoded frame occurring every 15 frames or every half a second. If packet loss occurs in the transmission of the intracoded (I) frame, a visible error can persist for half a second, until a new I frame is transmitted. An error persisting for half a second is quite noticeable to a viewer. As shown in Boyce,[14] packet loss rates as low as 3 percent can translate into frame error rates as high as 30 percent. Figure 5.1 shows frame error rates from sample traces of MPEG video data transmitted over the public Internet at 384 kbps, with I frames occurring every 15 frames. Frame error rate is defined by counting the percentage of decoded frames that are affected by a packet loss.

Error concealment techniques applied at the decoder can reduce the visual impact of packet losses. An overview of error concealment techniques for video compression was provided in Wang and Zhu.[15] These techniques generally copy information from spatial or temporal neighbors to reduce the visual effect of packet losses. Error concealment techniques are most effective at relatively low error rates. To protect video quality from higher loss rates, it is necessary to involve the transmitting as well as the receiving end. A good overview of error control techniques involving both the send and receiver ends was provided also in Wang and Zhu.[15] A summary of approaches to streaming video over the Internet can be found in Wu et al.[16]

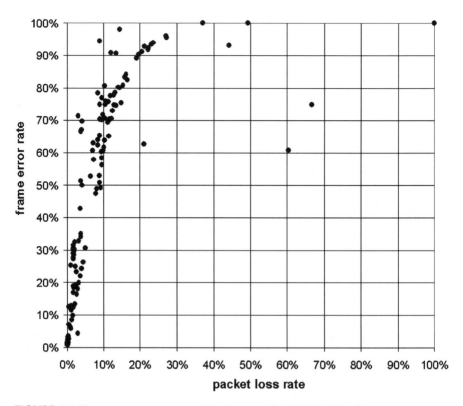

FIGURE 5.1 Frame error rate versus packet loss rate for MPEG video data.

Because the visual effects of packet losses persist until an intracoded Macroblock is received, an encoder can choose to perform intracoding more frequently to protect against packet loss. However, this comes with a visual-quality penalty, as intercoding is generally considerably more efficient than intercoding. More-sophisticated techniques can reduce the coding efficiency penalty by allowing the intra update rates for different image regions to vary according to various channel conditions and image characteristics.[17]

Alternatively, reference picture selection, such as that available in H.263+, can be used in networks with NAK feedback capability.[18] Instead of encoding a picture using intracoding after detecting a network transmission error, this approach eliminates the persistence of the error effects by intercoding the picture with respect to a previously coded picture, which has been decoded and stored at the decoder.

Scalable video coding can be used to improve the quality of video streamed over lossy networks. With scalable video coding, a base and one or more enhancement layers are encoded, and it is expected that the base layer alone should provide at least a minimally acceptable quality representation of the video. For networks that possess paths with different levels of QoS, the base layer is transmitted with a higher level of QoS than the enhancement layer. In Aravind and coworkers,[19] the performance of different types of MPEG-2 scalability over lossy networks was described. In Receiver Driven Layered Multicast,[20] scalable video coding is used with IP multicast, and each layer of video is transmitted in a separate multicast group. Clients can join as many multicast groups as may fit in their available bandwidth.

For streaming applications, where a small amount of additional delay can be tolerated, the use of Forward Error Correction (FEC) or Forward Erasure Correction (FXC) can protect against packet loss Using media-independent FEC, well-known information theory techniques can be applied to streaming video. In Rosenberg and Schulzrinne,[21] several variations of XOR operations are used to create parity packets from one or more data packets. More-complex techniques such as Reed Solomon (RS) coding also can be used. In RS coding, the original information bytes are transmitted, as well as additional parity bytes. When an RS(n,k) codeword is constructed from byte data, h parity bytes are created from k information bytes, and all $n = k + h$ bytes are transmitted. Such a Reed Solomon decoder can correct up to any $h/2$ byte errors, or any h byte erasures, where an erasure is defined as an error in a known position. Because in wired IP networks packets are generally lost completely rather than being transmitted with bit errors, when FEC is applied to video streaming over IP networks, the FEC is applied across packets. When RS coding is applied, k information packets of length l bytes are coded using l RS codewords. For each RS codeword, k information bytes are taken from k different packets (one from each packet), and the constructed parity bytes are placed into separate parity packets, and all $n = k + h$ packets are transmitted. Because RTP sequence numbers make it possible to determine if a given packet is lost, an RS(n,k) code can protect against up to any $h = n - k$ packet losses. Figure 5.2 shows an example of an RS(5,3) code applied to IP data. For this example, three information packets are RS encoded, yielding two parity packets and the $3 + 2 = 5$ packets are transmitted. The three original information packets can be recovered perfectly if no more than two of the five transmitted packets are lost.

Because RS coding is systematic, i.e., the original information bytes themselves are transmitted, if all k information bytes are received, no computations are needed at the receiver to reconstruct the original information bytes. A key advantage of RS coding over simple parity is its ability to protect against several consecutive errors, depending on the parameter choices.

Varying amounts of packet loss protection can be achieved by varying the RS(n,k) parameters. The trade-off between delay and error protection capability affects the choice of the n, k parameters. As n and k increase for protection against a burst of length h, the overhead rate h/k decreases, but the delay in the system increases. In Rizzo,[22] any code parameter values of n, k up to 255 can be generated using the same generator polynomial, such that as the value of n increases, the parity bytes generated for lower values of n are unchanged. For example, the first 9 bytes of a (10,5) code are the same as would be used in a (9,5) code. The type of FEC code with multicast was used in Rhee et al.[23] to achieve variable levels of error protection for different users. Several multicast groups transmit different numbers of parity packets, and individual receivers join as many of the multicast groups as needed to achieve the level of error protection appropriate for their network connection. FEC is well suited to multicast, because the same parity packets can be used to protect against different losses in the separate multicast transmission paths.

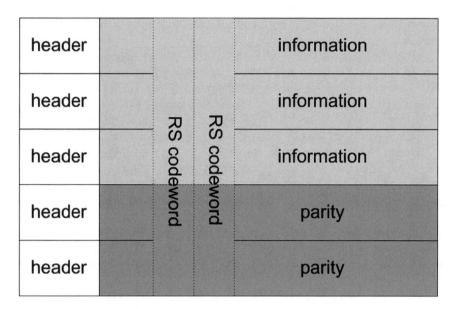

FIGURE 5.2 Reed Solomon (5,3) code applied to IP data.

FEC and scalability can be combined to achieve Unequal Error Protection (UEP). The overhead rates can be reduced by applying more error protection to the more-important layers of a scalable video stream than to the less-important layers, while maintaining the best possible received video quality in the presence of channel loss. In Priority Encoding Transmission (PET),[24] different layers of scalable video compressed data can be placed in the same packets and given different levels of protection.

In the High Priority Protection method (HiPP),[14] UEP is accomplished using an MPEG-2-like data partitioning to divide a compressed video stream into two partitions, a high-priority partition and a low-priority partition. Overhead parity data for the video stream is created by applying forward erasure correction coding to only the high-priority partition of the video stream. The high- and low-priority data and parity data are arranged into the same packets and are sent over a single channel. The packetization method used maximizes resistance to burst losses, while minimizing delay and overhead. The HiPP method is discussed in more detail in Section 5.6.

5.5 WIRELESS NETWORKS AND CHALLENGES

Before we study the application of streaming video to wireless networks, it is conducive to gain a historic perspective on the wireless industry. The cellular concept was first conceived and developed in the late 1970s. When the first wireless systems, the Advanced Mobile Phone System (AMPS) and its variations, were deployed in the early 1980s, they were built strictly for voice communications. Generally, these analog cellular networks were considered as the first-generation (1G) wireless technologies. The advent of voice coding and digital modulation technologies brought the evolution to the second generation or 2G wireless networks. The leading technologies included Global Systems for Mobile (GSM) in Europe, the IS-95 CDMA and IS-136 in the United States, and Pacific Digital Cellular (PDC) in Japan. Similar to the 1G wireless networks, the 2G networks are mostly used for low data rate, circuit-switched voice applications.

In the past few years, the explosion of Internet traffic has inevitably increased the need for packet-based wireless networks. As a result, the circuit-switched 1G and 2G wireless networks have gradually evolved into packet-switched 2.5G technologies such as GPRS, EDGE, and 1X-EVDV to provide packet data services and further improve voice capacity, which will eventually be phased out by the 3G wireless technologies. Employing increased spectrum, highly sophisticated air interfaces, and packet switching at the core, the 3G wireless networks further improve the capability to provide advanced data services. The high data rate (up to 2 Mbps) provided by 3G networks is much higher than that of today's wireline networks. In addition, 3G technologies provide seamless roaming across global networks. With these advantages, the 3G networks can support a wide variety of data services, including real-time, streaming multimedia and fast Internet access. In the end, the evolution of the 3G networks will bridge the gap between the wireline and the wireless worlds.

Given that the Internet traffic increases dramatically and users desire ubiquitous Internet access, the next generation of networking systems will be data-centric with the addressed mobility consideration. IP-based communications systems, which enable much-higher data rates and network flexibility, will gradually predominate over the traditional circuit-switched systems. In recent years, enormous effort has been made to support IP in wireless networks. Protocols and programming languages, including WAP, WML, and J2ME, have been developed to adapt Web content to the limitations of handheld devices by reducing the amount of transmitted data with minimum sacrifice of information. Mobile IP networks have been designed to maintain consistent transport-layer quality as the remote terminal is constantly in motion. However, in developing IP-based wireless data networks, significant difficulties remain to be addressed. They are summarized next.

5.5.1 DYNAMIC LINK CHARACTERISTICS

The process of a mobile device transmitting and receiving radio signals through the air makes wireless transmission vulnerable to noise and interference. The shadowing effect, multipath fading, and interference from the other devices make channel conditions vary unpredictably over time. Changing the transmission rate as the channel varies does improve efficiency but results in data rate oscillation. Furthermore, mobility introduces difficulty in channel estimation and prediction, thus raises error rate. Two approaches have commonly been used to address this problem. The first approach employs sophisticated channel coding and interleaving technologies. For example, turbo coding, despite its complexity, is now standard channel coding technique in 3G UMTS.[25] This approach, however, heavily relies on the quality of channel estimation. The second approach, the link layer ARQ mechanism, performs error control by retransmitting lost frames.[26] Although insensitive to the quality of channel estimation, this approach introduces latency and delay jitters to IP packet flow. The trade-off between latency and reliability depends on the ARQ persistence, which defines the willingness of the protocol to retransmit lost frames to ensure reliable transmission.[27] The persistence can be expressed in terms of time or the maximum number of retransmissions.

5.5.2 ASYMMETRIC DATA RATE

A mobile terminal has limited power so that the uplink data rate is usually less than the downlink data rate. This limitation is less stringent because most data applications are asymmetric.

5.5.3 RESOURCE CONTENTION

As in wireline networks, users share channel resources in wireless networks. When multiple users run a variety of applications, the most salient issue is the significant variability in terms of QoS requirements such as error rate, latency, and bandwidth. The resource contention problem is already quite challenging in wireline networks. As the result of mobility and unpredictable link variation, dynamic network topology makes wireless networks even harder to coordinate. The Medium Access Control (MAC) layer uses a scheduler to determine the next user to be served based on an individual user's channel condition and QoS requirement.[28–30] Currently, this scheduler is developed only for downlink transmission because only the base station gathers all the user information. The uplink transmission is typically made through contention, yielding high delay jitters.

Overall, high transmission errors and variable latency are the major causes of data loss in wireless networks. In the past, IP-based data applications have been designed mostly for wireline networks, where links and subnetworks normally have relatively stable transmission rates at low error rates. Data loss is primarily due to network congestion and buffer exhaustion. As described earlier in this chapter, many techniques have been developed to support efficient packet transmission over wireline networks. Unfortunately, they are not applicable to wireless networks. For example, in wireline networks, adding bandwidth can solve latency problems because bandwidth is not a paramount concern. However, in a wireless environment, this is quite difficult due to adverse channel condition and limited battery life of the mobile device.

5.6 ADAPTATION BY CROSS LAYER DESIGN

An important aspect of wireless networks is dynamic behavior. The conventional protocol structure is inflexible as various protocol layers can only communicate in a strict manner. In such a case, the layers are designed to operate under the worst conditions, rather than adapting to changing conditions. This leads to inefficient use of spectrum and energy.

Adaptation represents the ability of network protocols and applications to observe and respond to the channel variation. Central to adaptation is the concept of cross layer design.[31,32] Cross layer design for the three key layers in the overall protocol stack (i.e., application layer, transport layer, and network and link layer) are reviewed in this section. An example framework is illustrated in Figure 5.3 in terms of streaming video over wireline-to-wireless networks.

5.6.1 APPLICATION TRANSMISSION ADAPTATION

Application Transmission Adaptation refers to the application's capability to adjust its behavior to changing network and channel characteristics. Wireless networks often have to deal with adverse conditions where handoffs, deep fading, and bad carrier signals result in a high rate of packet losses. Only adaptive applications can cope with these challenging circumstances. For multimedia delivery, a media server can track packet losses and adjust media source rate accordingly.[33–37] To reduce information loss, the media server can employ packet FEC coding and UEP, as described in Section 5.4.

Whereas this level of adaptation is system independent and application specific, an application is able to reconfigure itself accurately only if it identifies the underlying network and channel variations.

5.6.2 TRANSPORT LAYER TRANSMISSION ADAPTATION

Instead of application layer adaptation, the adaptation can be shifted to the underlying transport layer, making it transparent to the application layer, so that applications originally developed for wireline networks remain intact. One drawback of this level of adaptation is that it is impossible to implement a complete adaptation if part of it is application specific.

The protocol should differentiate various packet loss patterns (i.e., packet losses due to network congestion or from channel errors),[38–40] and invoke congestion control and rate adaptation accordingly. Several cross layer approaches, such as EBSN,[41] snoop,[42] and freeze TCP,[43] have been proposed as TCP alternatives to distinguish congestion loss from noncongestion loss and invoke different flow control mechanisms. TCP and its variants provide reliable connections by retransmitting the lost packets. However, the resulting latency is in general too large for real-time and streaming media applications. For this reason, most streaming applications use UDP protocol with an unreliable packet delivery. However, by discarding corrupted packets, UDP does not distinguish between packet losses due to congestion and corruption. Alternatively, UDP-Lite applies partial checksum to some parts of a packet (i.e., packet payload) and reduces packet loss rate.[44] It is explicitly designed for certain applications, multimedia for example, which can detect and even recover from certain level of errors. CUDP conducts a precise error detection and recovery through error location information from link-layer.[45]

Note that the transport layer can only adapt effectively if it can observe the network layer and link layer conditions, propagate the information to the application layer, and in the meantime, identify and accommodate the application layer's need.

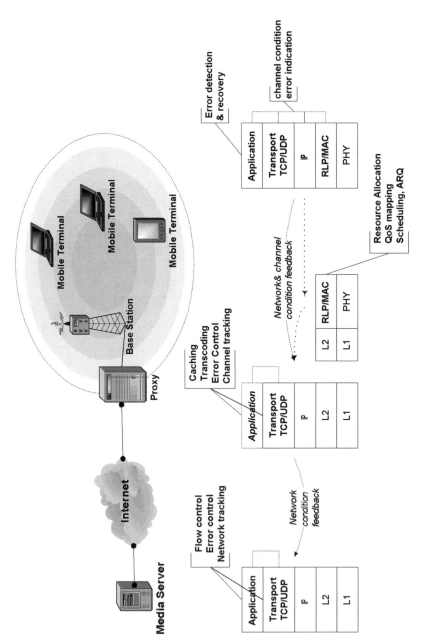

FIGURE 5.3 A video streaming architecture using cross layer design.

5.6.3 NETWORK LAYER AND LINK LAYER
TRANSMISSION ADAPTATION

The application characteristics, such as QoS requirement and packet priority, could be used in coordinating the network layer and link layer. In particular, the persistence level of the link layer ARQ mechanism should adapt to each application's latency and reliability requirements, while the link layer scheduler allocates radio resources to various packet flows based on their QoS priorities. The adaptation, however, requires the link layer and network layer to distinguish different packet flows, which in general can be achieved by an explicit indication of the QoS requirement associated with each packet flow.[27] Note that in some systems, the transport layer and link layer both conduct error recovery by using FEC coding and retransmissions. The balance between both schemes is important for the optimal usage of the overall communication resources.[46] Meanwhile, the network could operate efficiently by using the link layer and physical layer information, such as rate fluctuation and error condition, to distribute channel resources.

5.6.4 NETWORK AND CHANNEL CONDITION ESTIMATION
AND REPORT

The adaptation relies on each layer's ability to estimate current and even future network and channel conditions. The receiving entity evaluates current condition to invoke reception mechanism accordingly, while the sending entity uses current and future condition to adjust transmission flow. A condition report based on receiver feedback is normally more accurate than estimations at the transmitter.

Within a protocol stack, the link layer must detect its present status, including link availability, congestion, and error conditions, and signal it to upper layers for appropriate adaptation. In Zheng and Boyce,[45] the receiving link layer formats the location of channel errors in a meaningful manner, either implicitly or explicitly, so that the upper layers can identify and use it to detect and recover channel errors. Network layer and transport layer must propagate signals of the current conditions issued by lower layer(s) and themselves to upper layer(s). A proper form of the information exchange across multiple layers is crucial to the effectiveness of the adaptation.

5.6.5 PROXY SERVER

To allow efficient packet delivery through heterogeneous networks, a proxy server or gateway is placed between different networks. It provides seamless connection between the application server and the end users, regardless of their underlying network behavior. Using mobile streaming video as an example, a proxy server at the edge of the wireless network can virtually separate transmission path to server-to-proxy (e.g., wireline) and proxy-to-mobile-user (e.g., wireless). It can transcode media signal to a format suitable for low rate wireless transmission and limited mobile display,[47] add channel coding or perform retransmissions to maintain reliable transmissions,[48,49] and prefetch portions of media signals to allow continuous play-back during adverse channel conditions.[50] In addition, a proxy server can monitor network conditions in different paths and feedback them to the application server for appropriate adaptation.[51]

In general, in building an efficient wireless network, we strive to create a series of protocol layers that communicate, interact, and thus yield continuously improved applications and services. Next, we will highlight some of the innovative processes to improve the performance of streaming video over wireless network in terms of adaptation and cross layer design.

5.7 INTEGRATING THE ADAPTATION FOR STREAMING VIDEO OVER WIRELESS NETWORKS

UDP is generally used for video streaming; however, it is unable to distinguish between packet losses caused by network congestion and by channel errors. For this reason, it is more appropriate to use UDP over wireline networks than over wireless networks. UDP-Lite, on the other hand, ignores channel errors unless they corrupt a packet header. By doing so, it shifts the error-handling responsibility to the application. When packet-level FER coding is deployed to provide error control, CUDP is superior to UDP-Lite because CUDP utilizes error indication from the link layer. In general, the transmission unit at the link layer is smaller than that at the network and transport layers, so that link layer error indications provide a precise estimation of the error location. In general, a (n,k) FEC code can recover any $(n-k)/2$ errors or $(n-k)$ erasures per n data units. A packet FEC decoder can use the error locations to group erroneous data blocks to erasures and double the error recovery capability. Without packet FEC coding, the error indication is still beneficial because it can assist a video decoder to locate errors by formatting the corrupted link layer unit as all "1s."

The performance of UDP, UDP-Lite, and CUDP was compared in Zheng and Boyce[45] in terms of streaming MPEG video through a UMTS-similar system. The simulation flow chart is shown in Figure 5.4, where an MPEG video sequence of QSIF format (176 × 120 pixels) was coded at a bit rate of 288 kbps and a frame rate of 24 fps. The HiPP method[14] was used to provide packet FEC coding with an overhead of 25 percent, yielding a total source rate of 384 kbps. In particular, the MPEG video was split into two partitions with different priorities, HP and LP. The HP data contain more-important information. Video can be decoded with reduced quality, by using only the HP data. In regard to the trade-off between overhead and resilience, only the HP data was protected with an FEC coding by application server. Experimental IP packet loss traces and simulated wireless error traces[52] were applied to the packet flow. Transport packets were segmented to IP packets of 800 bytes and experimental IP packet loss traces were applied at the network layer. Link layer provides up to 384-kbps connections, and data units are 90 bytes each. The effect of link layer retransmissions and MAC layer scheduling is embedded in the error traces.

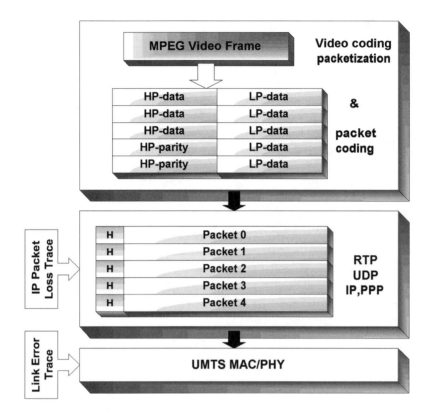

FIGURE 5.4 Simulation flow chart.

Figure 5.5 depicts video performance in terms of the averaged PSNR, where CUDP achieves 2 to 6 dB of PSNR improvement over UDP, and 5 to 10 dB over UDP-Lite. As congestion packet loss increases, the advantages diminish because network congestion becomes the dominant impairment.

Please note that the overall performance can be further enhanced by adjusting packet coding redundancy as well as source rate according to channel condition which has been extensively studied in Girod and coworkers,[33,34] Liu and Zarki,[35] Aramvith and coworkers,[36] Hsu and coworkers,[37] Chan et al.,[53] and Zhang and coworkers.[54]

5.8 CONCLUSIONS

This chapter provided an overview of intelligent video streaming over wireless networks. Background issues including video compression standards and protocols for streaming video were covered. Adaptation to continuously changing wireless environments was achieved through a cross layer design framework, which promotes communication and interaction across multiple protocol layers.

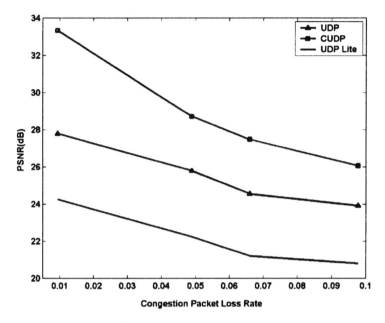

FIGURE 5.5 Video PSNR for UDP, UDP lite and CUDP.

References

1. Puri, A. and Chen, T., *Multimedia Systems, Standards, and Networks,* Marcel Dekker, New York, 2000.
2. Rao, K. and Hwang, J., *Techniques and Standards for Image, Video and Audio Coding,* Prentice Hall, New York, 1996.
3. Koenen, R., Overview of the MPEG-4 standard, ISO/IEC JTC1/SC29/WG11 N4668, March 2002.
4. Li, W., Overview of fine granularity scalability in MPEG-4 video standard, *IEEE Circuits Syst. Video Technol.,* 11 (3), 301–317, 2001.
5. MPEG4IP: open source, open standards, open streaming, http://www.mpeg4ip.net.
6. Wiegand, T., JVT coding, Workshop on multimedia convergence (IP Cable-com/MEDIACOM 2004/Interactivity in Multimedia), ITU Headquarters, Geneva, Switzerland, March 12–15, 2002, www.itu.int/itudoc/itu-t/workshop/converge/s6am-p3_pp4.ppt.
7. Postel, J., Transmission Control Protocol, RFC 793, 1981.
8. Allman, M., Paxson, V., and Stevens, W., TCP congestion control, RFC 2581, 1999.
9. Postel, J., User Datagram Protocol, Request for comments RFC 768, ISI, August 1980.
10. Schulzrinne, H. et al., RTP: A transport protocol for real-time applications, IETF RFC 1889, January 1996.
11. Turletti, T. and Huitema, C., RTP payload format for H.261 video streams, IETF RFC 2032, October 1996.
12. Hoffman, D. and Fernando, G., RTP payload format for MPEG1/MPEG2 video, IETF RFC 2038, October 1996.

13. Zhu, C., RTP payload format for H.263 video streams, IETF RFC 2190, September 1997.

14. Boyce, J., Packet loss resilient transmission of MPEG video over the Internet, Signal Processing: Image Communication, pp. 7–24, September 1999.

15. Wang, Y. and Zhu, Q.F., Error control and concealment for video communication: a review, *Proc. IEEE,* 86 (5), 974–997, 1998.

16. Wu, D. et al., Streaming video over the Internet: approaches and directions, *IEEE Trans. Circuits Syst. Video Technol.,* 11 (3), 282–300, 2001.

17. Liao J. and Villasenor, J., Adaptive intra block update for robust transmission of H.263, *IEEE Trans. Circuits Syst. Video Technol.,* 10 (1), 30, 2002.

18. Fukunaga, S., Nakai, T., and Inoue, H., Error resilient video coding by dynamic replacing of reference pictures, Proc. IEEE Global Telecommun. Config. (GLOBE-COM), Vol. 3, London, pp. 1503–1508.

19. Aravind, R., Civanlar, M., and Riebman, A., Packet loss resilience of MPEG-2 scalable video coding algorithms, *IEEE Trans. Circuits Syst. Video Technol.,* 6 (5), 426–435, 1996.

20. Jacobson, V., McCanne, S., and Vetterli, M., Receiver-driven layered multicast, Proc. ACM SIGCOMM '96, Stanford, CA, August 1996, pp. 117–130.

21. Rosenberg, J. and Schulzrinne, H., "An RTP payload format for generic forward error correction," RFC2733, http://www.faqs.org/rfcs/rfc2733.html.

22. Rizzo, L., Effective erasure codes for reliable computer communication protocols, *Comput. Commun. Rev.,* 27 (2), 24–36, 1997.

23. Rhee, I. et al., Layered multicast recovery, Technical report TR-99–09, NCSU, Computer Science Dept., February 1999.

24. Albanese, A. et al., Priority encoding transmission, Proc. 35th Ann. IEEE Symp. Foundations of Computer Science, November 1994, pp. 604–612.

25. 3rd Generation Partnership Project, Technical specification group radio access network, physical layer aspects of UTRA high speed downlink packet access (Release 2000), 3G Technical report (TR) 25.848.

26. Wang, Y. and Lin, S., A modified selective-repeat type-II hybrid ARQ system and its performance analysis, *IEEE Trans. Commun.,* COM31, 593–608, 1983.

27. Advice to link designers on link Automatic Repeat reQuest (ARQ), Internet Draft, March 2002, draft-ietf-pilc-link-arq-issues-04.txt.

28. Jalali, A., Padovani, R., and Pankaj, R., Data throughput of CDMA-HDR: a high efficiency high data rate personal communication wireless system, Proc. IEEE Vehicular Technology Conference, Tokyo, Japan, May 2000.

29. Andrews, M. et al., Providing quality of service over a shared wireless link, *IEEE Communications Magazine,* 39 (2), 150–154, 2001.

30. Tse, D., Forward link multiuser diversity through rate adaptation and scheduling, Bell Labs presentation, New Jersey, 1999.

31. A multilayered approach to mobile networking, Stanford University Project Report.

32. Tong, L., Zhao, Q., and Mergen, G., Multipacket reception in random access wireless networks: from signal processing to optimal medium access control, *IEEE Communications Magazine,* 39 (11), 108–112, 2001.

33. Girod, B. and Färber, N., Wireless Video, in *Compressed Video Over Network,* Reibman, A.and Sun, M.T., Eds., Marcel Dekker, New York, 2000.

34. Girod, B. et al., Advances in channel adaptive video streaming, Proc. IEEE International Conference on Image Processing (ICIP 2002), Rochester, September 2002.

35. Liu H. and Zarki, M.E., Adaptive source rate control for real-time wireless video transmission, *Mobile Networks Appl.,* 3, 49–60, 1998.

36. Aramvith, S., Pao, I.-M., and Sun, M.-T., A rate-control scheme for video transport over wireless channels, *IEEE Trans. Circuits Syst. Video Technol.,* 11, 569–580, 2001.

37. Hsu, C.Y., Ortega, A., and Khansari, M., Rate control for robust video transmission over burst error wireless channels, *IEEE J. Selected Areas Commun.,* 17 (5), 756–773, 1999.

38. Cen, S., Cosman, P.C., and Voelker, G.M., End-to-end differentiation of congestion and wireless losses, SPIE Multimedia Computing and Networking, (MMCN2002), San Jose, CA, January 18–25, 2002.

39. Samaraweera, N.K.G., Noncongestion packet loss detection for TCP error recovery using wireless links, *IEEE Proc. Commun.,* 146 (4), 1999.

40. Biaz, S. and Vaidya, N., Discriminating congestion losses from wireless losses using interarrival timers at the receiver, Technical report 98–014, Computer Science Department, Texas A&M University, June 1998.

41. Bikram, S. et al., Improving performance of TCP over wireless networks, Technical report 96–014, Texas A&M University, 1996.

42. Balakrishnan, H. et al., A comparison of mechanisms for improving TCP performance over wireless links, *IEEE/ACM Trans. Networking,* 5(6), 756–759, 1997.

43. Goff, T. et al., Freeze-TCP: A true end-to-end TCP enhancement mechanism for mobile environments, in Proc. of IEEE Infocom 2000.

44. Larzon, L., Degermark, M., and Pink, S., Efficient Use of Wireless Bandwidth for Multimedia Applications, MoMuc '99, San Diego, November 1999, pp. 187–193.

45. Zheng, H. and Boyce, J., An improved UDP protocol for video transmission over Internet-to-wireless networks, *IEEE Trans. Multimedia,* 3 (3), 356–364, 2001.

46. Chockalingam, A. and Bao, G., Performance of TCP/RLP protocol stack on correlated fading DS-CDMA wireless links, *IEEE Trans. Vehicular Technology,* 49, 28–33, 2000.

47. de los Reyes, G., Reibman, A.R., and Chang, S.F., Error resilient transcoding for video over wireless channels, *IEEE J. Selected Areas Commun.,* 18 (6), 1063–1074, 2000.

48. Vass, J. et al., Mobile video communications in wireless environments, IEEE International Workshop on Multimedia Signal Processing, Copenhagen, Denmark, Sept. 13–15, 1999.

49. Pei, Y. and Modestino, J.W., Robust packet video transmission over heterogenous wired-to-wireless IP networks using ALF together with edge proxies, Proc. European Wireless 2002, Feb. 25–28, Florence, Italy.

50. Fitzek, F.H.P, and Reisslein, M., A prefetching protocol for continuous media streaming in wireless environments, *IEEE J. Selected Areas Commun.,* 19 (10), 2015–2028, 2001.

51. Yu, F. et al., QoS adaptive proxy caching for multimedia streaming over the Internet, 1st IEEE Pacific Rim Conference on Multimedia (IEEE PCM 2000), December 2000, Australia.

52. Foschini, G.J., Layered space-time architecture for wireless communication in a fading environment when using multiple antennas, *Bell Labs Tech. J.,* 1 (2), 41–59, 1996.

53. Chan, C.W. et al., Eds., Special issue on error resilient image and video transmissions, *IEEE J. Selected Areas Commun.,* 18 (6), 2001.

54. Zhang, Q., Zhu, W., and Zhang, Y.Q., Network-adaptive rate control and unequal loss protection with TCP-friendly protocol for scalable video over Internet, *J. VLSI Signal Process.: Syst. Signal, Image Video Technol.,* 2001.

55. Sun, M. and Reibman, A. *Compressed Video over Networks,* Marcel Dekker, New York, September 2000.

56. Wiegand, T., ISO/IEC 14496–10: 2002, Joint Committee Draft, May 2002.
57. Larzon, L., Degermark, M., and Pink, S., UDP lite for real time multimedia applications, Proc. QoS Mini-Conference, IEEE International Conference of Communications (ICC '99), Vancouver, Canada, June 1999.
58. Budge, D. et al., Media-independent error correction using RTP, Internet Engineering Task Force, Internet Draft, May 1997.
59. Wenger S. and Côté, G., Using RFC2429 and H.263+ at low to medium bit-rates for low-latency applications, Packet Video '99, New York, April 1999.
60. Gallant, M. and Kossentini, F., Robust and efficient layered H.263 Internet video based on rate-distortion optimized joint source/channel coding, Packet Video '00, Italy, 2000.
61. Wenger, S. and Côté, G., Test model extension justification for Internet/H.323 video transmission, Document Q15-G-17, ITU Q15, Video Coding Experts Group, February 1999.

6   Clustering and Roaming Techniques for IEEE 802.11 Wireless LANs

Ahmed K. Elhakeem

CONTENTS

ABSTRACT

Clustering refers to the set of rules and algorithms that different nodes follow to group themselves into interconnected communications networks. Tactical, emergency, and rural communications have traditionally applied various clustering algorithms in fields where prior communications infrastructure does not exist. Recently, clustering gained wider attention due to the advent and wide deployment of IEEE 802.11 wireless LANs, Bluetooth, and other noncellular wireless platforms. This chapter surveys the various algorithms used in the IEEE 802.11 Standard and *ad hoc* wireless LANs for clustering, and describes also the close interaction between clustering and routing.

6.1 INTRODUCTION

Nodes should be grouped into clusters in a way that maintains maximum connectivity. Maximizing the stability and links connectivity as well as traffic intensities within each cluster leads to efficient clustering, routing, and overall communication efficiency. Minimizing the amount of clustering and routing information and numbers of intracluster links contributes also to overall communication efficiency.

Clustering helps also to manage the allocation of wireless channel resources, e.g., in the IEEE 802.11 Standard, the access point coordinates the transmission times of various nodes. Clustering leads also to formation of backbones and reduction of the state of the network. As an example, it is easier for cluster heads to exchange information in regard to 50 nodes in each cluster rather than allowing the 10,000 nodes of the whole wireless network to exchange information about the other 9,999 nodes. Clustering and subsequent formulation of backbone links of high quality and low delay may lead to overall end-to-end transmission delay in multihop networks.

Because of node movement, the establishment and maintaining of clusters in *ad hoc* wireless networks where there is no access point becomes a harder task compared to fixed nodes. More-frequent exchange of clustering information and routing tables may lead to less time available for information transmission and hence less communication efficiency. Added to this inefficiency is the nature of the wireless channel, which faces additive Gaussian noise, path loss due to shadowing and various obstacles, fading, sensitivity to distance, nodes transmission powers, etc.

In this chapter, some of the basic clustering techniques for wireless LANs and the various trade-offs involved are presented. While trying to explain the clustering/routing interrelationship, more details on the wireless LAN routing techniques may be found in Elhakeem[1] complementing the clustering techniques discussed herein. Other issues such as security and authentication are related but not investigated here.

6.2 WIRELESS LANS CLUSTERING

The IEEE 802.11 Standard[2] used in numerous wireless LAN products has few roaming and clustering facilities. This standard maintains two basic modes of operation: (1) access point (AP) and (2) *ad hoc*. In the AP mode, users (mobile or stationary) group around a typically stationary station, which then resembles the base station of the cellular radio system. The AP may be connected to other APs by radio or other ground-based networks, but 802.11 does not address such interconnection. The access point is assigned a basic service set (BSS) identification (BSSID, or network ID), which distinguishes between neighboring APs, as shown in Figure 6.1. The amalgamation of BSS will constitute the extended service set. A mobile station roaming from one BSS to another will have to know the IDs of the BSS it is passing through. Most APs will come with a default BSSID; however, an operator can easily change this, therefore no two APs will share the same BSSID.

When a mobile station is powered on, it tries first to see the availability of other APs to join. The AP is responsible for coordinating the sharing of the radio channel capacity among the nodes, authenticating the nodes, association with the nodes, and relaying various management and supervisory information to the nodes such as allowed transmission powers, etc. The AP determines when to switch the MAC access mode from contention mode (distribution function[3]) to the point coordination function where polling is used.

The mobile station goes through either active or passive scanning modes to check for such availability. In the passive scanning mode, it will listen to the pilot

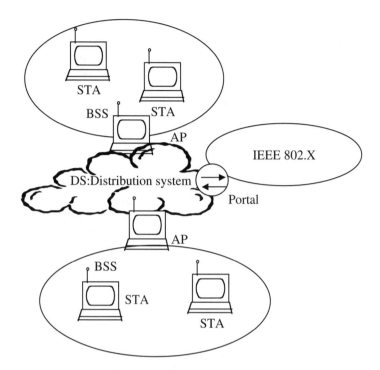

FIGURE 6.1 Infrastructure network.

signal (called beacon) of the expected APs (BSSIDs are preknown) in the area, one after another. These beacon signals, typically transmitted every 100 milliseconds, carry the identification of the BSSID or the extended service set ID and synchronization, among other information. In the active scanning mode, the mobile station sends its own probing signal carrying a specific BSSID it wishes to join, and waits for a response from the corresponding AP. If a response is received from the AP in case of active scan or if the AP beacon is heard in the passive scan, the mobile station will proceed to the authentication and association phases. Details of these phases can be found in Gier.[2] If no response or beacon is heard, the mobile station will proceed to claim itself as the AP, typically after 10 seconds of scanning and trying to find other APs.

The standard allows mobile stations to roam among APs. As the associated signal gets weaker, the mobile station will reassociate (register) with another AP from which it has received a stronger signal (while still being associated with the original AP).

Lucent's Wave Around is a protocol that facilitates roaming among similar vendor APs. A beacon signal which contains the domain ID, the BSSID, quality of communication, and cell search threshold values is transmitted at a certain repetition rate from each AP. The mobile station listens to beacons to find other APs. This is triggered by continuous measurements of the current AP beacon power and comparison with the search threshold values. If the beacon frequency is low, this is considered a relaxed condition where responsiveness of the mobile station to lower received signal power is slow. If the beacon frequency is normal, then it defines a normal response. If the beacon frequency is fast, responsiveness is higher, meaning faster reaction of the mobile station to declining received signal power.

The cell search thresholds contained in the AP beacons determine the times at which a mobile station will switch to another AP or remain in cell search mode, depending on the quality of service (QoS) of the received signal at the mobile station. When this QoS is less than or equal to the "regular cell search" threshold, the mobile station will start to look for another AP, but it will switch to the new AP if the QoS of that AP is higher than the "stop cell search" threshold. "Fast cell search" threshold corresponds to the level of QoS at which the mobile station should immediately switch to any AP that yields better quality of service.

"Stop cell search" corresponds to the acceptable QoS range, meaning the mobile station will stop looking for a new AP. However, in this case three sensitivity levels arise: (1) "low" defines the condition where the mobile station should stay associated with the AP as long as possible, e.g., due to unavailability of other APs; (2) "normal" means that the mobile station will stay an average amount of time; and (3) "high" means that the mobile station will try to switch to another AP as soon as possible.

6.2.1 IAPP

The previous mechanisms of the standard work well for overlapping or nonoverlapping BSS as long as all APs are made by the same manufacturer. However, Lucent and other companies came up with the IAPP (Interaccess Point Protocol) to facilitate roaming between different vendor APs. IAPP uses UDP or IP on top of the 802.11 Protocol, and consists of the "announce" and "hand-over" protocols. The "announce" protocol informs APs about new APs, and all APs about networkwide configuration. The "hand-over" protocol informs one AP that one of its mobile stations has moved to another AP. In this case, the bending files will be transmitted to the new AP from the old AP. Filter tables (bridging functionality) are provided also by the hand-over protocol to enable the extended LAN configuration (similar to terrestrial LAN bridging).

6.3 LOCATION-BASED CLUSTERING

The aforementioned clustering techniques apply to infrastructure networks where an AP is typically present. If all nodes (mobile stations) assume the same functionality and have similar processing capabilities, then the need arises for *ad hoc* clustering and routing techniques to enable roaming and forwarding of information and control packets to the appropriate destination. Although all nodes are similar in *ad hoc* networks, most routing and clustering protocols would assign certain functions to some nodes, i.e., cluster heads and gateways. Accurate position information could be available to the nodes as, for example, when GPS interfaces are available within the nodes, or when certain nodes transmit position aiding signals (pilots or beacons) so other nodes will know their location. Such position information could lead to facilitating the clustering and routing problems in *ad hoc* networks,[4] where a zone-based, two-level link state (called zone-based hierarchical LSR routing, ZHLS) is used for both clustering and routing.

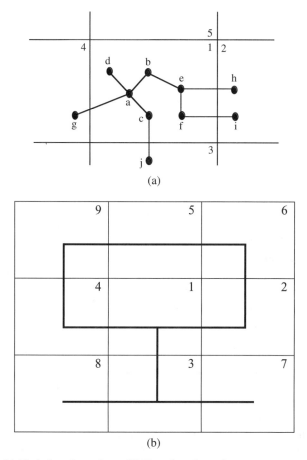

FIGURE 6.2 (a) Node-based topology. (b) Zone-based topology.

The network area is divided into nonoverlapping numbered zones, and a typical node address consists of node ID and zone ID (by reading from a memory table, the GPS coordinates of the node translate into its current zone ID). Although the emphasis here is on clustering and not routing,[1] the two functions are interleaved in most *ad hoc* networks.

The hierarchical address corresponds to the node-based and zone-based topologies, as shown in Figure 6.2. The zone size depends on the nodes' power, geography, application, etc.

The clustering and routing procedures consist of two phases: intrazone and interzone. In the first phase, neighboring nodes exchange their link state packets (LSP), each consisting of the node ID and zone ID. As shown in Figure 6.3, node a sends a link request, receives a link response from neighbors (at one-hop hearing distance), formulates its LSP (Table 6.1), and broadcasts it to all neighbors.

All nodes perform the same steps asynchronously. In Table 6.1, which shows all LSPs of zone 1, each table entry corresponds to all immediate neighbors of each node as well as the neighboring zone of that node.

TABLE 6.1
Node LSPs in Zone I

Source	Node LSP
a	b, c, d, 4
b	a, e
c	a, 3
d	a
e	b, f, m, 2
f	e, 2

As an example, node a has b, c, and d as immediate neighbors and can reach the neighboring zone 4 through node g, while node d has only a as a neighbor and cannot reach any neighboring zones. From Table 6.1, node a derives Table 6.2, which shows the intrazone (same zone) routing table of node a. As an example, for node a to reach node c, it will go directly to node c (same zone). For node a to contact any node in zone 2, it will go through node b [as in Figure 6.3d, the border node h in zone 2 hears node e, which in turn hears node b].

At the end of the first phase (intrazone) and following the exchange of all LSPs by all nodes of each zone, each node would then have the same table of zone LSPs.

In the second phase (interzone), the gateway nodes [those hearing messages from different zones such as nodes e, a, f, and c of Figure 6.3d] will propagate the zone LSPs through the network, as illustrated in Figure 6.4. This will enable each node to store a zone LSP similar to the one in Table 6.3. Taking the fourth entry of this table means the neighboring zones of 4 are 1, 9. Figure 6.4b shows the spanning tree or virtual links between the zones in this network.

Because each node receives all zone LSPs (Table 6.3), shortest-path algorithms are used at each node to find the best route to each zone, as shown in Table 6.4.

Routing of data messages is now proactive if the destination lies within the same zone as the source, according to Table 6.2, and reactive if the destination is not within the same zone. In the later case, a route search is conducted and the facilities of the zone LSPs of Table 6.3 and the various nodes interzone routing (Table 6.4) are used to find an end-to-end route. See Elhakeem[1] and Jao-Ng and Lu[4] for further routing details.

Needless to say, exchange of node LSPs and zone LSPs among nodes will cause some flooding, which can be minimized if repeated LSP messages are not transmitted again by any node. To have some idea about savings in the cost of control messages, one finds that in a flat topology, the number of control messages exchanged per clustering cycle is defined as:

$$C_f = U^2 \tag{6.1}$$

where U is the total number of nodes, because each node transmits one clustering message to every other node. A flat topology assumes that all nodes are considered to be in one large zone.

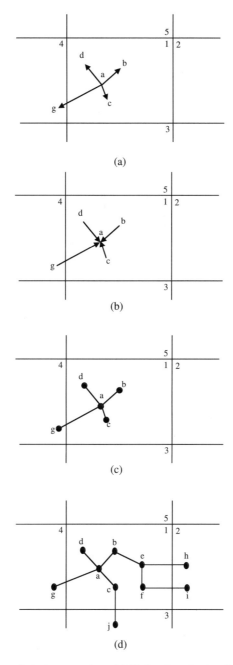

FIGURE 6.3 Intrazone clustering procedure. (a) Node a broadcasts a link request to its neighbors. (b) Node a receives link responses from its neighbors. (c) Node a generates its own node LSP and broadcasts it throughout the zone. (d) All nodes perform the previous steps asynchronously. (*Source:* Jao-Ng, M. and Lu, I.-T., A peer to peer zone-based two-level link state routing for ad hoc networks, *IEEE J. Selected Areas Commun.*, 17 (8), 1415–1425, 1999.)

TABLE 6.2
Intrazone Routing Table of Node a

Destination	Next Node
b	b
c	c
d	d
e	b
f	b
2	b
3	c
4	g

TABLE 6.3
Zone LSPs

Source	Node LSP
1	2, 3, 4
2	1, 6
3	1, 7, 8
4	1, 9
5	6, 9
6	2, 5
7	3
8	3
9	4, 5

TABLE 6.4
Interzone Routing Table of Node a

Destination Zone	Next Zone	Next Node
2	2	b
3	3	c
4	4	g
5	4	g
6	2	b
7	3	c
8	3	c
9	4	g

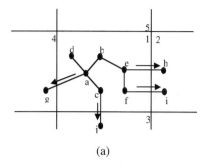 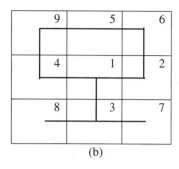

(a)   (b)

FIGURE 6.4 Interzone clustering procedure. (a) Gateway nodes broadcast zone LSPs throughout the network. (b) Virtual links between adjacent zones are established. (*Source:* Jao-Ng, M. and Lu, I.-T., A peer to peer zone-based two-level link state routing for ad hoc networks, *IEEE J. Selected Areas Commun.*, 17 (8), 1415–1425, 1999.)

In the location-based technique described previously, each node transmits one node LSP message to every other node in its zone, the average number of these nodes per zone is U/Z, where Z is the number of zones and the number of these messages become $Z(U/Z)^2 = U^2/Z$. This adds to the total number of zone LSP messages, UZ, to give the total number of control messages per clustering cycle in ZHLS, i.e.,

$$C_g = U^2/Z + UZ \qquad (6.2)$$

Clearly, Cg is less than Cf.

Although the selection of zone size is restricted by the radio powers of the nodes, channel conditions, and network deployment among other factors, it is possible to find the optimal number zones by differentiating C_g and equating to zero, thus obtaining

$$Z_{opt} = \sqrt{U} \qquad (6.3)$$

and optimal $C_g = 2N^{3/2}$ control messages. In the face of nodes' mobilities, and if the ratios of nodes generating node and zone LSPs due to mobility are p_a and p_b, respectively, then the control overhead per cycle of ZHLS due to mobility becomes

$$D_g = p_a U^2/Z + p_b UZ \qquad (6.4)$$

while for flat topology

$$D_f = p_a U^2 \qquad (6.5)$$

Because zone topology changes are less frequent than node-level changes, p_a is greater than p_b, and hence

$$D_g \leq p_a U^2/Z + p_a UZ \leq D_f \qquad (6.6)$$

Jao-Ng and Lu[4] present simulation results that display the superiority of zone clustering using GPS; also, Chen and coworkers[5] report a similar location-based technique, called geographical-based routing and clustering technique, which emphasizes radio range considerations in the process of zone formations. These clustering techniques are typically implemented within the application layers above the wireless IEEE 802.11 MAC layer. We notice that only gateways (not cluster heads or APs) are defined here, and that both gateways and ordinary nodes cooperate to route nodes packets from source to destination based on the routing tables mentioned previously.

6.4 GRAPH-BASED CLUSTERING

The coverage area of the network is to be divided into nonoverlapping clusters by means of the clustering algorithm. For clustering purposes, each node is assigned a unique ID number, and maintains a set of its single-hop neighbors. In one version of this clustering technique, called minimum ID, the node with the minimum ID among the neighbors is selected as the cluster head. The cluster head would have functionality similar to the AP mentioned previously.

A node turning on would listen to the beacons of the cluster head and configuration messages of the various nodes and join the cluster head with the minimum ID.

Each node transmits a clustering configuration message composed of the node ID and the thought-of cluster head ID to its neighbors. All neighboring nodes hear the message, adjust their conclusion in regard to cluster head election accordingly, but do not repeat this control message, but modify, and transmit a new configuration message. Each node in a cluster is at one hop away from the cluster head and, at most, two hops away from other nodes. A node, which is heard in two neighboring clusters, will belong to only one cluster.

A manager node also may exist in some systems,[5] where *ad hoc* network management is required, but this requirement is not a must for ID-based clustering, as shown in Figure 6.5.

It is possible also that one node (e.g., node 5 in Figure 6.5) can hear nodes from two different clusters, but will join only one. Clustering will take place according to cycles (timespan where topology is preserved, and through which a node keeps its thought-of cluster head ID). This cycle may vary from node to node. As an example, had node 1 been idle, node 2 would have been the cluster head of C1. But as soon as node 1 starts to transmit the clustering configuration messages, node 2 and other nodes in C1 will change their thought-of cluster head in the next clustering cycle.

If C1 and C2 clusters are formed in the same clustering cycle, then node 4 thinks of node 2 as the thought-of cluster head and thinks to join C1. However, after hearing

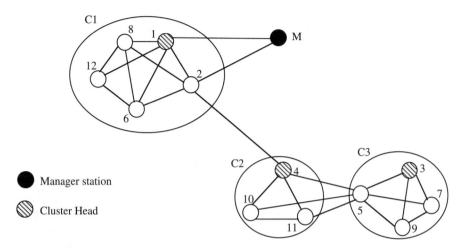

FIGURE 6.5 Clusters formed using graphical clustering.

from node 2 that node 1 is the thought-of cluster head, and because node 4 cannot hear node 1, node 4 will change its cluster configuration message so it becomes a cluster head of a new C2. Node 4 will not join C3 for a similar reason, i.e., it hears from node 5 that node 3 is the thought-of cluster head, but because it cannot hear node 3 directly, it will not accept node 3 as a cluster head and will start to form its own cluster.

Figure 6.5 assumes that clusters C2 and C3 were formed in the same clustering cycle; however, if C2 started to form before C3, then node 5 would have joined C2 instead (because nodes 3, 7, and 9 were idle, for example). Following the cluster formation and election of the cluster head, each node will keep the following database in regard to clustering:

- Cluster list (all nodes in the cluster), neighbor list (all nodes one hop away)
- A ping counter, which counts the time since the node last heard from the cluster head

As mentioned previously, when the cluster is formed each node is at one hop from the cluster head and two hops at most from other nodes in the cluster. As some of the nodes and cluster heads move in and out of the cluster, the cluster formation may be affected, as shown in Figure 6.6.

A good clustering algorithm should yield cluster node selection, cluster head selection, and possibly gateway nodes that do not change much in the face of nodes mobility. As shown in Figure 6.6a, in C1 nodes 2, 6, and 8 moved but remain in C1; cluster head node 1 moved, is now at two hops from nodes 6 and 12, but still remains in C1. In C2, cluster head node 4 moved, is now at two hops from nodes 5 and 10, but still remains in C2. In C3, nodes 3, 5, and 7 moved and the result is similar. The conclusion from Figure 6.6a is that there is no need for reconfiguration of the clusters. However, in Figure 6.6b, node 4 (the cluster head of C2) moved too far from C2, and accordingly joins C1 (through node 2). On the other hand, nodes 10

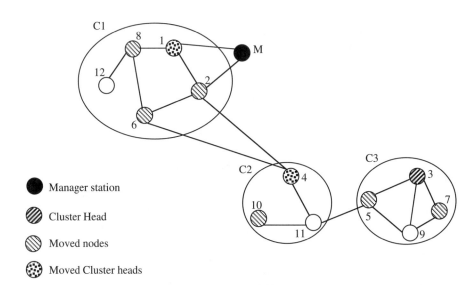

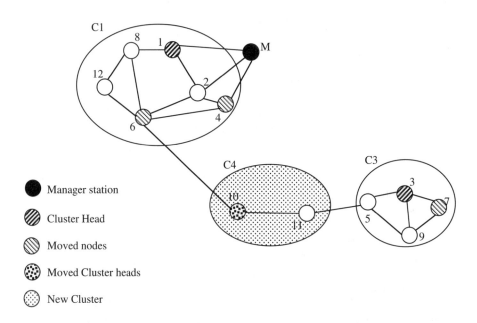

FIGURE 6.6 (a) Effect of node mobility on clusters. (*Source:* Chen, W., Jain, N., and Suresh, S., ANMP: Ad Hoc Network Management Protocol, *IEEE J. Selected Areas Commun.,* 17 (8), 1506–1531, 1999.) (b) Effect of node mobility on clusters. (*Source:* Chen, W., Jain, N., and Suresh, S., ANMP: Ad Hoc Network Management Protocol, *IEEE J. Selected Areas Commun.,* 17 (8), 1506–1531, 1999.)

and 11 find out that the cluster head has moved, and so form a new cluster C4, with node 10 becoming the cluster head. Because of node mobility as explained previously, some variations of the clustering rules[5] allow certain adaptability to mobility. One of those rules is the following:

> If a node detects that it has lost its links to the nodes in its cluster, it either joins another cluster or forms another cluster by itself.

In this case, the node may join a cluster head, which is two hops away. For further adaptability to mobility, at regular intervals of time each node detects and reports to the cluster head one or more of the following events:

- One neighbor in the same cluster moved out of the cluster.
- One node moved into the cluster, and wishes to join the cluster.
- One node became a neighbor and is a member of the cluster.
- The node was at one hop from the cluster head, but now is at 2 hops or more.
- The node was at two or more hops from the cluster head, but now is at one hop from the cluster head.

The cluster head will receive all those periodical reports from the nodes in its cluster, process them, and then broadcast a fresh list of cluster nodes to all nodes of the cluster.

The crucial function of the cluster head now becomes evident: if the cluster head moves, this leads to confusion among the member nodes. To protect against such an occurrence, a ping counter is incremented at each node at a regular interval of time. If the value of this counter exceeds a certain threshold, the cluster head broadcasts a ping message to all cluster nodes. On the other hand, if these nodes do not hear the ping message after their counters exceed the threshold, the nodes infer that the cluster head has been turned off or has moved out, in which case the voting process for establishing a new cluster head will commence again.

A number-crunching simulation involving 30 nodes[5] moving in a 1500×1500 unit area was conducted to test the resilience of the graph-based technique to mobility. The speed of each node varied in the range of 1 to 50 units per second, the transmission range was 450 units, and the nodes moved in random direction (uniformly distributed from 0 to 360 degrees). Figure 6.7 shows the number of messages exchanged to maintain the cluster in the face of mobility for different ping intervals, where no ping means infinite time steps, and 1 means one ping message every time interval. This number increases linearly with nodes' speeds.

Figure 6.8 shows the percentage of nodes unmanaged by cluster heads. Figure 6.8a outperforms those of Figure 6.8b due to the utilization of the available information from the 802.11 MAC layer.

A similar graph-based clustering appears in Lin and Gerla,[6] where it is proved that the ID-based algorithm guarantees that each node joins only one cluster. Figures 6.9 and 6.10 show some of the simulation results from Lin and Gerla.[6] Figure 6.9 shows the average node connectivity of the algorithm versus the transmission range for

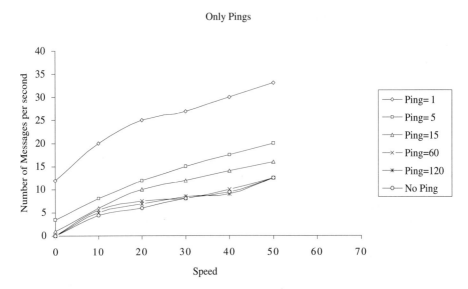

FIGURE 6.7 Control volume in graphical clustering.

different numbers of nodes N. Node connectivity refers to the number of nodes heard by a typical node. For a node to hear at least one neighbor, Figure 6.9 shows that the transmission range should be at most 40 for N = 20. Figure 6.10 shows that the average order of a typical repeater lies in the of range 2 to 3. This order is defined as the number of clusters this gateway node (repeater) can access.

Maximum connectivity clustering[7] assigns the cluster head rule to the node that hears the maximum number of its one-hop nodes. This clustering technique is formulated as follows:

1. A node becomes a cluster head if it has the maximum number of "uncovered" nodes within one hop (hearing distance). A tie is broken based on minimum ID.
2. A node is said to be "uncovered" if it has not yet elected a cluster head, otherwise it becomes a "covered" node.
3. A node that has elected another node as cluster head will give up the cluster head rule if elected by other nodes.

Figure 6.11 shows a typical clustering configuration based on maximum connectivity clustering, where nodes 5, 7, and 8 are elected as cluster heads, while nodes 2, 3, 9, and 10 are gateway nodes.

Figure 6.12 compares the two strategies, i.e., minimum ID clustering and maximum connectivity clustering and shows the average number of clustering changes per clustering unit time (time tick) as a function of the transmission range. The minimum ID clustering yields less cluster changes than the other policy and thus is more stable in the face of nodes mobility. The reason is simple: if a node with

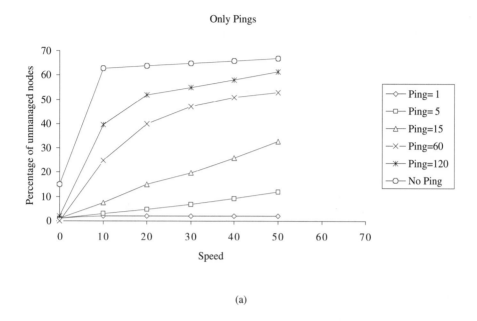

(a)

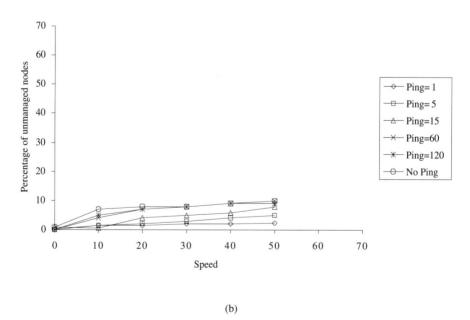

(b)

FIGURE 6.8 Percentage of nodes unmanaged by cluster heads. (*Source:* Chen, W., Jain, N., and Suresh, S., ANMP: Ad Hoc Network Management Protocol, *IEEE J. Selected Areas Commun.,* 17 (8), 1506–1531, 1999.)

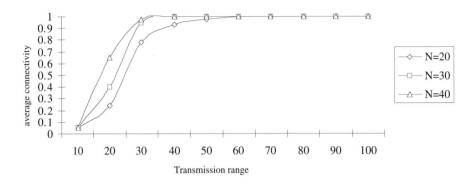

FIGURE 6.9 Connectivity property. (*Source:* Lin, C.R. and Gerla, M., Adaptive clustering for mobile wireless networks, *IEEE J. Selected Areas Commun.*, 15 (7), 1265–1275, 1997.)

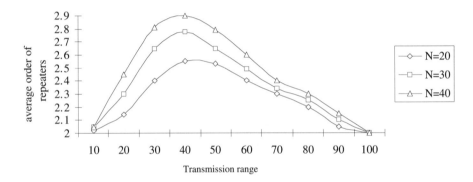

FIGURE 6.10 Average order of repeaters.

highest connectivity moves away, many links are broken; with minimum ID clustering fewer links are broken, and hence the cluster may remain intact.

6.5 QUASIHIERARCHICAL ROUTING

Quasihierarchical routing is mainly intended to provide routing efficiency in *ad hoc* and other networks. As shown in Figure 6.13, N nodes are arranged into m level hierarchy of clusters. Clusters at a certain level i ($0 \leq i < m$) are typically assumed to be disjointed. Each node is assigned an address of the form bcd, which is the succession of the ID of each cluster starting with b, i.e., the ID of (m − 1)th cluster, and ending with the zero-th level cluster ID, which is the node ID within its first cluster, and with the ID of the parent cluster.

A cluster head[8] generates and distributes routing information for the cluster. The cluster head summarizes the cluster condition and relays this to neighboring clusters. The mobility or downtimes of this cluster head may disrupt both routing and clustering functionality unless hot standbys are provided. In quasihierarchical routing,

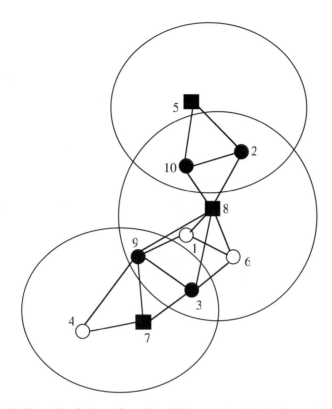

FIGURE 6.11 Example of cluster formation (highest connectivity). (*Source:* Gerla, M. and Tsai, J.T.C., Multiuser, mobile, multimedia radio network, *Wireless Networks J.*, 255–265, 1995.)

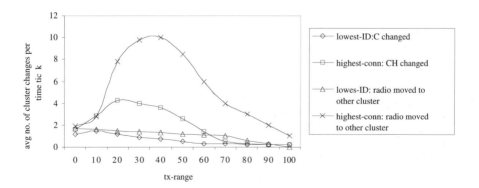

FIGURE 6.12 Comparisons of clustering (N = 30): random movements. (*Source:* Gerla, M. and Tsai, J.T.C., Multiuser, mobile, multimedia radio network, *Wireless Networks J.*, 255–265, 1995.)

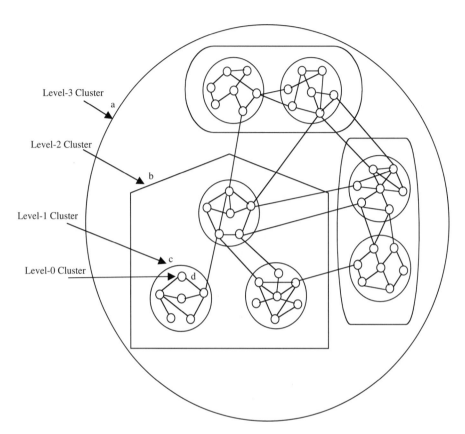

FIGURE 6.13 The nested cluster architecture.

a node seeks to minimize the number of hops by sending the message to the boundary of that highest level cluster that encloses the destination node.

Figure 6.14 shows an example with c_j the lowest level cluster having both source and destination nodes s_0, d_0, and $j = 3$. Source packets are sent directly from s_0 to d_2, which is the highest level cluster that includes the destination node, then to d_1, then d_0.

In order to build the forwarding tables based on which routing commences, the cluster head broadcasts to all neighbors its routing cost, which is the average it has to all nodes within its cluster.[9] This cost will be used distributively by all nodes to determine the least cost from a typical node to every level i cluster, $0 \leq i < m$, lying inside the node i + 1 cluster. This cost information is updated and propagated by nodes per the following steps, which pertain to two neighboring nodes, x and y, belonging to the same j + 1 level parent cluster but a different j-th cluster level. Node x receives from another neighbor, node z, a new cost in regard to cluster c, i.e., one of the i-th cluster levels, where $0 \leq i < m$. Node x has to determine if it needs to update its cost in regard to c:

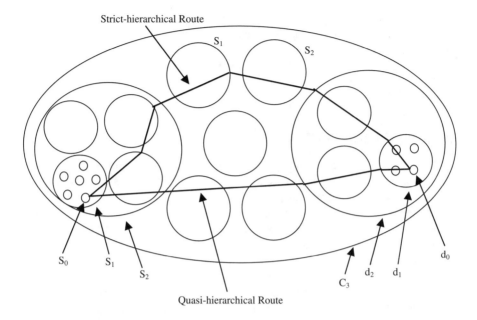

FIGURE 6.14 Quasihierarchical routing versus strict hierarchical routing. (*Source:* Perkins, C.E., *Ad Hoc Networks,* Addison-Wesley, Reading, MA, 2001.)

1. If c is not a parent cluster of x, then x will add the advertised cost received from node z to the cost of its link to z. If the sum is less than the cost that node x has stored for cluster c before, it will replace this cost with the new reduced cost, and update its forwarding tables accordingly. If not, the forwarding table entry in regard to cluster c remains the same and no change is broadcast from x to the neighbors.
2. If c is a parent cluster of x, x does not update its forwarding tables accordingly.
3. The new updated information at node x in Step 1 is broadcast to the neighbors of lowest level cluster of node x only if $i \geq j$, and sent to node y if the forwarding point from x to cluster c is not y, or looping may take place.

In a uniform hierarchical network of N nodes and m levels where every cluster has the same number of lower level clusters, the size of the forwarding tables at each node is of the order of $mN^{1/m}$, while for the quasihierarchical clustering this becomes of the order of mC_{max}, where C_{max} is the maximum number of inner clusters within the parent $(j + 1)$ level cluster.

Propagating the cost information according to the steps mentioned previously does not necessarily yield optimum shortest path to any node; in this regard, Kamoun and Kleinrock[10] and Lauer[11] try to treat this shortcoming of quasihierarchical clustering. In the literature, all clusters at a certain level were assumed to be nonoverlapping.

Overlapping clusters may provide better amenability to roaming because a node will not lose connection as it switches from cluster to cluster (if that node happened to belong to the overlapping clusters).

Allowing cluster overlap adds to the complexity of establishing clustering and routing tables for all nodes, not just nodes belonging to more than one cluster at the same level.[12]

6.6 STRICT HIERARCHICAL ROUTING

Similar to quasihierarchical routing, strict hierarchical routing helps routing objectives, and provides communications connectivity in mobile wireless networks at the expense of increased processing cost. The forwarding tables help to identify the clusters and boundaries enclosing both source and destination nodes. As illustrated in Figure 6.14, where the number of cluster levels is k = 3, the data packets are forwarded from s_0 to the boundary of s_1, then to level 1 clusters until they reach the boundary of s_2, then to the boundary of s_{k-1}. The packets hop then on k − 1 level clusters until they reach the boundary of d_{k-1}. The packets then hop to cluster levels k − 2, then k − 1, around the destination node until they reach the destination node d_0.

Building the clustering and routing tables at the cluster head of cluster c at level j ($0 \le j < m$) involves the following steps:

1. Calculation of the average cost of cluster c.
2. Determination if cluster c is at the boundary of a higher level cluster j + 1.
3. If 2 is true, determination of which clusters at level i, i = j + 1 are neighbors of c, these clusters at level j + 1 are directly linked to c and together with c lie within parent level j + 2 cluster. However, i can be larger than j + 1 if c happens to lie on the boundary of a higher level parent cluster.
4. Each cluster head within clustering level j relays the cost and neighbor information in 1 to 3 to cluster heads of neighboring clusters of level j, by which each cluster head would be able to compute its cluster minimum route cost to any level j cluster and the identity of the next cluster to take to reach another level j cluster (all within the next parent level j + 1).
5. In the sequel and once these pieces of information in 1 to 4 are exchanged, all cluster heads would know also the identity of level j clusters lying on the boundary of level j + 1, and the clusters these boundary clusters are linked to, as well as least cost routing information.

Strict clustering is not as accurate in calculating minimum routing cost as quasi clustering, because the strict clustering table contains mainly costs between clusters of various levels rather than between actual nodes. The details of the costs of various links of a certain cluster are averaged out.

This coarse granularity of strict clustering is a mixed blessing in the sense of providing less-frequent cost advertisement and hence savings in the precious radio channel capacity (cost is averaged over many links before being relayed to neighbors), while yielding higher routing cost by the same mechanism.

Clustering and routing tables in strict hierarchical routing may have the same number of entries as in quasihierarchical routing. However, the processing time to find the route to a certain node in strict clustering costs $2m - 1$[8] investigations of table entries, while in quasi clustering only one table entry may be consulted.

Routes computed based on strict routing are generally longer than their quasi routing counterparts, which are longer than routes based on nonhierarchical routing techniques.[11]

6.7 CONCLUSION

Clustering is still an ongoing research area, and the literature in this area is rich. For example, McDonald and Znati[13] introduce a clustering routing technique for *ad hoc* wireless LANs, where the network is partitioned into clusters of nodes mutually reachable with a certain specified probability for a certain time. Simulation results support the inherent adaptability and stability of the protocol. Hierarchical clustering and routing techniques have been and continue to be under investigation.[11,14] A strict hierarchical clustering and routing technique, which was designed for multimedia support in large mobile wireless networks, is presented in Ramanathan and Steenstrup.[15] Another hierarchical clustering and routing technique, NTDR,[16] was designed for the tactical environment, where a backbone network exists between cluster heads.

References

1. Elhakeem, A.K., Ad-hoc routing techniques for wireless LANs, in *The Communications Handbook, 2nd ed., Gibson, J.D., Ed.,* CRC Press, Boca Raton, FL, 2002, 92-1.
2. Gier, J., *Wireless LANs Implementing Interoperable Networks,* McMillan Technical Publishing, New York, 1999.
3. Crow, P.C. et al., IEEE 802.11 wireless local area networks, *IEEE Communication Magazine,* September 1997, p. 116–126.
4. Jao-Ng, M. and Lu, I.-T., A peer to peer zone-based two-level link state routing for ad hoc networks, *IEEE J. Selected Areas Commun.,* 17 (8), 1415–1425, 1999.
5. Chen, W., Jain, N., and Suresh, S., ANMP: Ad Hoc Network Management Protocol, *IEEE J. Selected Areas Commun.,* 17 (8), 1506–1531, 1999.
6. Lin, C.R. and Gerla, M., Adaptive clustering for mobile wireless networks, *IEEE J. Selected Areas Commun.,* 15 (7), 1265–1275, 1997.
7. Gerla, M., Baltzer, J.C., and Tsai, J.T.C., Multiuser, mobile, multimedia radio network, *Wireless Networks J.,* 255–265, 1995.
8. Perkins, C.E., *Ad Hoc Networks,* Addison-Wesley, Reading, MA, 2001.
9. Kleinrock, L. and Kamoun, F., Hierarchical routing for large networks: performance evaluation and optimization, *Comput. Networks,* 1 (1), 155–174, 1977.
10. Kamoun, F. and Kleinrock, L., Stochastic performance evaluation of hierarchical routing for large networks, *Comput. Networks,* 3 (5), 337–353, 1979.
11. Lauer, G.S., Hierarchical routing design for SURAN, IEEE International Communications Conference (ICC), June 1986, pp. 93–102.
12. Garcia-Luna-Aceves, J.J. and Shacham, N., Analysis of routing strategies for packet radio networks, IEEE Infocom Conference, March 1985, pp. 292–302.

13. McDonald, B. and Znati, T., A mobility based framework for adaptive clustering in wireless ad hoc networks, *IEEE J. Selected Areas Commun.,* 17 (8), 1–20, 1999.

14. Lee, W.C., Topology aggregation in hierarchical routing in ATM networks, ACM Sigcomm '95, *Corp. Commun. Rev.,* 25 (2), 82–92, 1995.

15. Ramanathan, R. and Steenstrup, M., Hierarchically-organized multihop mobile wireless networks for quality-of-service support, *ACM/Blatzer Mobile Networks Appl. J.,* 3 (1), 101–119, 1988.

16. Zavgren, J., NTDR mobility management protocols and procedures, Proc. IEEE Military Communications Conference (MILCOM) '97, November 1997.

7

VoIP Services in Wireless Networks

Sajal K. Das and Kalyan Basu

CONTENTS

7.1 INTRODUCTION

The success of wireless voice services has created opportunities for new information and entertainment services, leading to a ubiquitous information environment for the future. The potential for wireless data services has been demonstrated recently through the tremendous success of i-mode services in Japan. More and more people are getting accustomed to the concept of wireless appliances that can provide many attractive services by integrating content, voice, and text communications. Along

149

with this development, core global information networks are becoming a reality that converge multimedia (audio, video, and text) services on a single, seamless network infrastructure. Such convergence will not only reduce network complexity and infrastructure equipment, but also will be easy to maintain. The switching and routing technologies of core networks are evolving toward packet-based transmission such as IP (Internet Protocol), ATM (asynchronous transfer mode), frame relay, and optical routing. Thus, the migration of the core network toward packet-based transmission has created the opportunity for integrating voice and data traffic on the same wireless and IP network infrastructure.

Although the evolution of the core network to IP is enabling the migration of traditional circuit-switched voice- and call-signaling message traffic over the Internet using VoIP (Voice over IP) technology, there are many technical issues and challenges that need to be resolved for its successful commercial deployment. Before proceeding further, we analyze the benefits offered by such a unified end-to-end IP network:

- Cost reduction: The convergence of voice and data traffic can reduce operations costs and improve network efficiency.
- Simplification: An integrated infrastructure that supports all forms of communication allows more standardization and reduces network complexity.
- Consolidation: Because personnel are among the most-significant expense elements in a network, any opportunity to combine operations would eliminate points of failure and reduce expense.
- Advanced applications: Although telephony is the basic application for voice over all-IP networks, the long-term benefits are expected to be derived from multimedia and multiservice applications. For example, E-commerce solutions can combine World Wide Web access to information with a voice call button that allows immediate access to a call center agent from a personal computer.

Voice traffic transfer through the packet network involves the following important components:

- Coding of voice signals for packet mode transfer, and the resulting impact of the code on subjective voice quality
- Network impediments that are acceptable for voice packets and their allocation to different parts of the network
- Voice connection signaling mechanisms
- Core network quality of service (QoS) management for voice communication

There are three general areas of research that will influence the successful migration of wireless voice traffic to integrated voice, data, and text on the IP network:

1. Migration of traditional circuit-switched voice-call session signaling to the Internet-based signaling scheme that meets voice signaling performance requirements

2. Selection of appropriate voice coding and decoding methods, and the mechanism for assembling and dissembling wireless packet frames to transfer VoIP frames through the wireless link, to meet network performance requirements

3. Developing a successful micro-mobility management scheme to hand over VoIP frames from one base station to another without impacting voice quality

At present these issues are not fully resolved to provide a wireless VoIP service that can be compared with the voice quality offered on current wireless voice services. Many efforts are being made by the different standards bodies and research laboratories to address these challenging issues. In this chapter, we identify and explore some of these issues and describe the current state of the art in VoIP services in wireless networks. The focus of this discussion is mostly on the system issues of this problem, and hence all design- and architecture-related issues might not be covered.

This chapter presents a summary of our original research in VoIP services in wireless data networks. Section 7.2 deals with the basics of wireless networks, including GPRS (General Packet Radio Service) and the challenges in providing VoIP services. Section 7.3 briefly summarizes the principles of voice coding, while Section 7.4 analyzes the network quality requirements for VoIP implementation. Sections 7.5 and 7.6 give an overview of the H.323 Protocol and SIP (Session Initiation Protocol) for multimedia services, respectively, and Section 7.7 discusses the Radio Link Protocol (RLP) standard in wireless networks. Section 7.8 details the architecture implementation of H.323 using wireless links, presents the delay analysis for control messages and call set-up message with and without RLP, as well as the experimental verification using Microsoft® NetMeeting®. Media packets blocking and VoIP traffic blocking analysis in GPRS are discussed in Section 7.9. Section 7.10 concludes the chapter.

7.2 WIRELESS NETWORKS

Currently, voice and data traffic are treated separately in wireless networks such as GSM (Global Special Mobile)[1] and GPRS (General Packet Radio Service)[2] systems. The Release 1 definition of third generation (3G) wireless systems also kept this separation. Wireless voice traffic is routed through the circuit-switched infrastructure whereas data traffic is routed through the packet network. To explain this integration, we look to GPRS integration in the existing GSM network. As shown in Figure 7.1, two new GPRS dedicated nodes, SGSN (Serving GPRS Support Node) and GGSN (Gateway GPRS Support Node), are added in the existing GSM network. An SGSN is responsible for the delivery of data packets to and from the mobile station within its service area. Its tasks include packet routing and transfer, mobility management (attach/detach and location management), logical link management, and the authentication and charging functions. The location register of the SGSN stores the location information and user profiles of all GPRS users registered within the SGSN. GGSN acts as an interface between the GPRS backbone and the external packet data

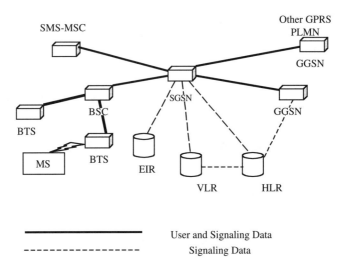

FIGURE 7.1 GPRS network. SMS-MSC: Short Messaging Services-Mobile Switch Center; PLMN: Public land mobile network; GGSN: Gateway GPRS Support Node; BSC: Base Station Controller; SGSN: Serving GPRS Support Node; MS: Mobile Station; BTS: Base Station; EIR: Equipment Identifying Register; VLR: Visiting Location Register; HLR: Home Location Register.

networks. It converts GPRS packets coming from the SGSN into appropriate PDP (Packet Data Protocol) format and sends them out on the corresponding packet data network. In the other direction, the PDP addresses of incoming data packets are converted to the GSM addresses of the destination. The addressed packets are sent to the responsible SGSN. For this purpose the SGSN stores the current SGSN address of the user and the user's profile in its location register. GGSN also performs the authentication and charging function.

Although wireless and Internet technologies perform quite well within their own domains, the integration of wireless links into the Internet exhibits considerable challenges. Unpredictability in the air-link conditions, such as rapid fading, shadowing, and intermittent disconnection, affect the radio link performance, and the frame error rate (FER) can be as high as 10^{-1}. This causes serious quality degradation to the data users. Packet traffic is very sensitive to the bit error rate (BER). To overcome the higher BER problem, the Wireless Air Interface Protocol has included the definition of a new protocol layer, RLP (Radio Link Protocol), on top of the MAC (medium access control) layer. The RLP layer brings higher reliability on the wireless link by using the ARQ (automatic repeat request) retransmission technique in conjunction with the cyclic redundancy check (CRC).

The signaling for VoIP should be able to gracefully migrate into the existing infrastructure to converge voice and data networks. Existing GSM voice networks and terminals use SS7 (Signaling System 7) and ISDN (Integrated Signaling Digital Network) protocols. The International Telecommunications Union (ITU) has defined H.323[3] as the key protocol that implements this migration to today's PSTN (public switched telephone network) signaling domain. New protocols such as SIP (Session

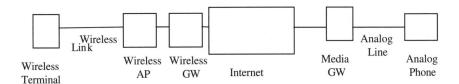

FIGURE 7.2 Schematic diagram for VoIP implementation. AP: Access points; GW: Gateway.

Initiation Protocol),[4] MGCP (Media Gateway Control Protocol),[5] and H.248[6] have been proposed recently to implement VoIP services to overcome the limitations of H.323 through gateway decomposition.

SIP, as defined by the IETF (Internet Engineering Task Force), is a signaling protocol for telephone calls over the Internet. Unlike H.323, SIP was specifically defined for the Internet and includes mobility management functions. It exploits the manageability of IP and makes telephony applications development relatively simple. It is basically used for setting up, controlling, and tearing down sessions on the Internet that include telephone calls and multimedia conferences. SIP supports various facets of telecommunications such as user location, user capabilities, user availability, call setup, and call handling.

In our view, multiple VoIP protocols will exist and interact among themselves for the foreseeable future. For our study in this chapter, we have considered H.323 and SIP as the protocols of choice for VoIP services implementation. The proposed analysis, although based on H.323 implementation, is general in nature and can be easily extended to SIP.

The introduction of VoIP service introduces new impediments because the IP network, as shown in Figure 7.2, replaces the network between the two ends. The previous assumption of negligible jitter in the synchronous digital network is no longer valid because of the packet-handling process in the best-effort IP network that will introduce variable amounts of delay between the subsequent packets. This variability must be eliminated at the destination so that the voice packets are offered to the decoder at a constant rate. To achieve this, the receiving end uses a dejitter buffer to delay packets so that all the packets are offered at a constant rate. To compensate for the high degree of network delay variability, the setting of the buffer delay can be high, thus introducing additional delay on the voice path. To reduce this delay jitter, the Internet is introducing new service types, such as Diff-Serv (differentiated service)[7] and MPLS (multiprotocol label switching),[8] to maintain the performance of VoIP traffic within only a small delay jitter tolerance.

Wireless VoIP introduces additional impediments within the network due to wireless data links in the access. In GSM and CDMA systems today, digitized voice packets are transmitted over wireless links without an ARQ-type protocol. Thus, if there is an error in the voice packets, the system accepts the packet or drops it. Voice quality is not significantly affected by this mechanism. Although the FER (frame error rate) in wireless access links can be as high as 10^{-1} to 10^{-3}, voice quality degradation is not significant. But for data communication, no channel error is acceptable unless it is protected by error correction code. To overcome channel error, the ARQ protocol is used to retransmit the erroneous packets. This retransmission currently happens at the TCP level because the original design of the IP protocol considered a very low error probability on the channel. Application of TCP retransmission for error correction in wireless networks only will significantly slow down the wireless link throughput and increase delay. To overcome this delay and throughput problem, MAC layer retransmission is considered for wireless data links. Wireless data access standards, such as GPRS and cdma2000, use MAC layer retransmission, called RLC/MAC, which causes additional delay jitter for wireless VoIP packets.

7.3 BASIS OF VOICE CODING

Voice signals in the telephone device are analog signals within the frequency range of 300 Hz to 3.4 kHz. The landline digital voice network converts these analog signals into digital signals with the help of the PCM (Pulse Code Modulation)[9] scheme. In this scheme, the voice band analog signals are sampled at 8 kHz speed to meet the Nyquist sampling rate, $f_s > 2 \times$ bandwidth. The samples are digitized by the different quantization techniques to 8 bits of data. Thus, the wireline PCM system is a 64-kbps stream where one voice sample of 8 bits is generated every 125 microseconds. The quantization of the samples creates error. Successive quantization errors of voice samples can be assumed uncorrelated random noise. Therefore, the quantization error is viewed as noise and expressed as the signal-to-quantization noise ratio (SQR). It is expressed as

$$SQR = \frac{E\{X^2(t)\}}{E\{[Y(t) - X(t)]^2\}}$$

where E[.] denotes the expectation value, X(t) is the analog input signal at time t, and Y(t) is the decoded output signal at time t. The error [Y(t) − X(t)] is limited in amplitude to q/2, where q is the height of the quantization interval. If all quantization intervals have equal lengths and the input analog signal is a sinusoidal, the SQR in dB is expressed as

$$SQR(dB) = 10.8 + 20\log 10[v/q]$$

where v is the RMS (root mean square) value of the amplitude of the input. The SQR values of the signals increase with the sample amplitude and penalize the small sample-size signals. A more-efficient coding can be achieved by not having uniform sample size, but rather by having sample size vary. The process of compounding[9] is used to achieve this nonuniform sampling. The compression algorithm used in North America and Japan for PCM is called μ-law, and a compounding formula recommended by the ITU for Europe and the rest of the world is called A-law. The 64-kbps voice-coding standard is issued from the ITU as Recommendation G.711.[10] In addition, the ITU developed Recommendations G.726 and G.727 on 40, 32, 24, and 16-kbps adaptive differential PCM (ADPCM) coding standards and G.722 on 56 to 64 kbps, 7-kHz wideband ADPCM standard.

The introduction of digital wireless communication in the early 1980s introduced a new challenge: G.711 and G.721 are not usable for wireless networks due to very limited bandwidth of wireless links. The early digital wireless access standard can support only 108 kbps over 200-kHz bandwidth in GSM and about 24 kbps over 30 kHz for North American IS-54 TDMA digital standards. It was felt that lower rate voice coders are needed for wireless voice communications. In addition, the voice coder should be able to provide robust communication under fading channel behavior. The initial answer came from many different solutions such as the code excited linear prediction (CELP) codec at 6.5 kbps, called the half-rate codec. For example, CDMA introduced the system with a full-rate 13-kbps CELP codec. The ETSI (European Telecommunications Standards Institution) introduced residual excited linear predictive speech coding (RPE-LTP) that is 13 kbps with a frame size of 20 milliseconds and no look-ahead delay. The ANSI (American National Standards Institution) proposed vector-sum excited linear predictive coding (VSELP) at 7.95 kbps for TDMA IS-54 system that has a frame size of 20 milliseconds and look-ahead delay of 5 milliseconds. This RPE-LTP codec of GSM allows supporting 8 voice calls simultaneously within the 200-kHz bandwidths and three voice calls within the 30-kHz bandwidth of the North American IS-54 TDMA system. New coders with improved performance were introduced in the subsequent GSM networks. They include full rate,[11] half rate, enhanced full rate,[12] adaptive multirate,[13] and RECOVC (recognition-compatible voice coding) speech transcoding. Other coders include G.728 16-kbps speech coding using low delay code excited linear prediction and G.729[14] 8-kbps speech coding using conjugate structure algebraic code excited linear prediction. The Qualcom code excited linear prediction (QCELP) is a variable bit rate codec of 8.5, 4.0, 2.0, and 0.8 kbps speeds, a frame size of 20 milliseconds, and a look-ahead delay of 5 milliseconds. Further to this bit rate reduction, a single call also can be provisioned for half-rate coding by taking advantage of the silence period during the conversation. At present a number of lower bit rate coding techniques are under consideration for wireless voice communications.

7.4   NETWORK QUALITY REQUIREMENTS

The measurement of voice quality is rather difficult. A subjective rating scale of 1 to 5, called mean opinion score (MOS),[15] is used to state voice quality. Wireline voice quality is normally within 4 to 4.5 MOS, the current wireless voice quality lies between 3.5 and 4 MOS. To improve the quality of voice signal for wireless networks, G.729 adopted 10-millisecond frame times. The computation time is 10 milliseconds and look-ahead delay is 5 milliseconds. This results in a total one-way codec delay of 25 milliseconds. In addition to the delay criteria, speech performance depends also on the bit error rate. The objective of performance under random bit error rate $< 10^{-3}$ is recommended not to be worse than that of G.726 under similar conditions.[16] To introduce VoIP, the appropriate selection of coding technology is necessary to meet the criteria of delay and bit error rate of the network.

In network applications of speech coding, coded voice signals are transmitted through multiple nodes and links, as shown in Figure 7.2. All these network links and nodes cause impediments to the coded voice signals. The ITU Recommendations G.113 and G.114[17] specify several system requirements, including:

- End-to-end noise accumulation is limited to 14 QDU (quantization distortion unit), where each QDU is equal to the noise of a single 64-kbps PCM device
- End-to-end transmission delay budget is 300 milliseconds
- G.114 limits the processing delay for codec at each end to 10 milliseconds

Among these network requirements, the most important one in the design of VoIP using wireless is the end-to-end delay budget of 300 milliseconds. In digital networks, because of the synchronous nature of transmission, this delay budget is mostly used for the switching and transmission delay. There is very little variation of this delay, called jitter, within the synchronous digital network for voice signals. In the current wireless voice links, the air interface introduces additional delay because of the air link multiple access standards. Most of the wireless link designs attempt to meet the delay requirements of 300 milliseconds for single-link voice calls in a national network. So single-link wireless calls without intermediate satellite links perform with a reasonable MOS rating today, however, if two ends of the connection are wireless links, the speech quality deteriorates below MOS 3.0.

There are three important network performance parameters for wireless VoIP service:

1. Performance to set up and tear down the call
2. Quality of voice payload packets during conversation
3. Performance of the voice session handoff

These parameters will depend on the selection of VoIP service protocols to set up the voice session, the voice signal coding and transporting scheme, and the micro/macro mobility protocols used for VoIP services.

The call setup delay depends on the successful transfer of the current Q.931 and ISDN-type messages and the additional message sets for capability check of the terminals and media packet synchronization in the IP network. Both H.323 and SIP signaling protocol implement this function. The average call setup delay of the present wireless voice network is about 3 seconds, H.323 and SIP implementations will have to meet this delay requirement.

Network voice quality will depend on the contribution of the different components of a hypothetical connection of VoIP in wireless network. A hypothetical connection, as shown in Figure 7.2, includes network components such as a wireless terminal, a wireless link, an access point, a wireless gateway, the Internet, a media gateway, and a landline terminal. The landline phone is assumed connected through an analog line. The wireless terminal includes the codec and the function to map the coded frames into the wireless data channel for communication with the wireless access point. The common wireless channel is shared between multiple users, so to get access to the capacity of the channel is a delay process. The fading and propagation loss of the wireless channel causes the frame error that introduces additional impediments to the transfer of data. Once the packet is received at the wireless access point, it is transferred to the terminating media gateway through the Internet. Internet performance depends on the delay at the different routers and the propagation delay through the transport links. Due to the use of fiber transmission, the transmission related bit error rate or packet loss is almost nonexistent in the links and propagation delay is very small. The delay in the router depends on the long-range dependency of the traffic and link congestion. Using proper engineering techniques, this Internet delay can be maintained within strict limits. The DiffServ and MPLS protocols will be able to support the core Internet with minimum delay and jitter. At the media gateway, the coded voice packets are reconverted to analog voice signals and transferred to the terminating analog voice terminals using copper wire connection. The end-to-end delay of the voice packet for this hypothetical connection can be represented by the following equation:

$$D_{end-to-end} = d_{jitter} + d_{wt} + d_{wc} + d_{w1} + d_{wap} + d_{internet} + d_{mgw}$$

where

d_{jitter} = Delay introduced by the jitter buffer. To compensate for the fluctuating network conditions, it is necessary to implement a jitter buffer in voice gateways or terminals. This is a packet buffer that holds incoming packets for a specified amount of time before forwarding them to decoding. This has the effect of smoothing the packet flow, increasing the resiliency of the codec to packet loss, delayed packets, and other transmission effects. The downside of the jitter buffer is that it can add significant additional delay in the path. It is not uncommon to see jitter buffer settings approaching 80 milliseconds for each direction.

d_{wt} = Delay at the wireless terminal for coding and decoding voice packets and creating voice frames conforming to the Internet frame packet format (TCP or UDP). Each coding algorithm has certain built-in delay. For example, G.723 adds fixed 30-millisecond delay. To reduce the IP overhead, multiple voice packets may be mapped to one Internet frame and thus introduce bundling delay.

d_{wc} = Delay at the wireless terminal to get a wireless data channel and map the Internet voice packet to it. This includes delay for buffer allocation such as GSM TBF (temporary buffer flow)[2] allocation. In uplink transmission: Before sending the data to the base station, the mobile station must access the common channel in the uplink direction to send the request. Getting permission to send data in the uplink direction takes time, which increases the end-to-end delay.

d_{w1} = Delay to transfer the voice packets to the wireless access point, including the retransmission delay to protect the frame error during propagation.

d_{wap} = Delay at the wireless access point to assemble and reassemble the voice frame from wireless frame formats to the Internet format.

$d_{internet}$ = Delay to transfer the packet through the Internet to the media gateway. This is the delay incurred in traversing the VoIP backbone. In general, reducing the number of router hops minimizes this delay. Alternatively, it is possible to negotiate a higher priority to voice traffic than for delay-insensitive data.

d_{mgw} = Delay at the media gateway to convert Internet voice packets to analog voice signals and transfer to analog voice lines.

The TCP retransmission delay impacts the delay parameters of d_{wc} and d_{wap}. The end-to-end delay, $D_{end-to-end}$, of voice packets for conversation should be less than 300 milliseconds, as specified by ITU G.114; one-way delay greater than 300 milliseconds has an adverse impact on conversation, and the conversation seems like half duplex or push-to-talk.

The wireless VoIP session handoff between the two wireless access points is determined by the handoff mechanism supported by the wireless mobility protocol. There are two mobility functions for the wireless IP network: (1) micro–mobility and (2) macro–mobility. The consensus between the different standards bodies is that current Mobile IP may be suitable for macro–mobility, but a new technique is necessary for micro–mobility. The potential micro-mobility protocols are GPRS GMM, Cellular-IP, Hawaii, TIMIP,[18] IDMP,[19,20] etc. The challenge of these mobility protocols is to ensure that the voice packets can be routed to the new access point without any packet loss or significant additional delay on the voice path. Most of the protocols in their current form have difficulty meeting the micro-mobility requirements of the voice packets. The current soft handoff of CDMA, the make-before-break mechanism on the GSM, and hard handoff in TDMA maintain the continuity of voice packet flow during handoff. The method of GPRS and 3G packet handoff do not allow similar mechanisms at this time.

7.5 OVERVIEW OF THE H.323 PROTOCOL

This section presents a brief introduction to the H.323 and RTP/RTCP[21] protocols. The H.323 standard provides a foundation for audio, video, and data communications across IP-based networks for multimedia communications over LAN with no QoS guarantee. The standard is broad in scope and includes both stand-alone devices, embedded personal computer technology, and point-to-point and point-to-multipoint conferences. The standard specifies the interfaces between LANs and other networks, and addresses call control, multimedia management, and bandwidth management methods. It uses the concept of channels to structure information exchange between the communications entities. A channel is a transport layer connection (unidirectional or bidirectional).

The Real-Time Transport Protocol (RTP) is used in conjunction with H.323 to provide end-to-end data delivery services with real-time characteristics. RTP can handle interactive audio and video services over a connectionless network. At present, RTP/RTCP (Real-Time Control Protocol) along with UDP provides the bare-bones real-time services capability to IP networks with minimum reliability. A

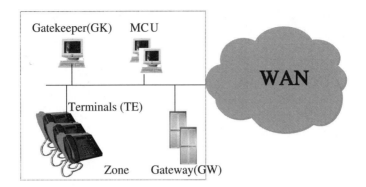

FIGURE 7.3 H.323 system components.

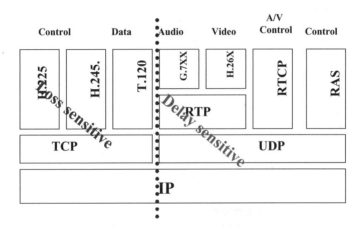

FIGURE 7.4 H.323 protocol relationships.

typical H.323 network, shown in Figure 7.3, consists of a number of zones inter-connected by a wide area network (WAN). The four major components of a zone are terminals, gateways (GW), gatekeepers (GK) and multipoint control units (MCU). An H.323 terminal is the client and endpoint for real-time, two-way com-munications with other terminals, GWs, or MCUs. The (optional) GK provides address translation-and-controls access and bandwidth-management functions to the H.323 network. The GW is an endpoint that interconnects the VoIP terminal to the PSTN network. The MCU is the network endpoint for multipoint conferencing. An H.323 communication includes controls, indications, and media packets of audio, moving color video pictures, and data. Thus, H.323 is a protocol suite, as shown in Figure 7.4, which includes separate protocol stacks for control and media packet transport. All H.323 terminals must support the H.245 protocol, which is used to negotiate channel usage and capabilities. Three additional components are required in the architecture:

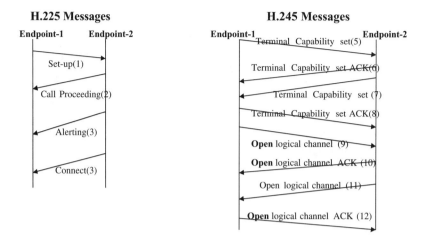

FIGURE 7.5 H.323 messages.

1. H.225 for call signaling and call setup (a variation of Q.931)
2. Registration/admission/status (RAS) protocol for communicating with a gatekeeper
3. RTP/RTCP for sequencing audio and video packets

A typical H.323 call setup is a three-step process that involves call signaling, establishing a communication channel for signaling, and establishing media channels. In the first phase of call signaling, the H.323 client requests permission from the (optional) gatekeeper to communicate with the network. Once the call is admitted, the rest of the call signaling will proceed according to one of several call models. Figure 7.5 describes the message flows in H.225 and H.245, where Endpoint-1 is the calling endpoint and Endpoint-2 is the called endpoint.

In the direct routing call model, the two endpoints communicate directly instead of registering with a gatekeeper. As shown in Figure 7.5, Endpoint-1 (the calling endpoint) sends the H.225 setup (1) message to the well-known call signaling channel transport identifier of Endpoint-2 (the called endpoint), which responds with the H.225 connect (3) message. The connect message contains an H.245 control channel transport address for use in H.245 signaling. The H.225 call proceeding (2) message is optional. Once the H.245 control channel (unidirectional) is established, the procedures for capability exchange and opening media channels are used, as shown in Figure 7.5. The first H.245 message to be sent in either direction is terminal capability set (5 and 7), which is acknowledged by the terminal capability set ACK (6 and 8) message. There can be an optional master–slave determination procedure invoked at this stage to resolve conflicts between the two endpoints trying to open a bidirectional channel. The procedures of H.245 are used to open logical channels for the various information streams (9 and 11). The open logical channel ACK (10 and 12) message returns the transport address that the receiving end has assigned to the logical channel. Both the H.225 and H.245 messages are transmitted over a reliable transport layer.

After the media channel has been set up and RTCP CNAMEs (canonical name) are exchanged, two TCP sessions are established, one for H.225 and the other for H.245 procedures, so that multiple media streams (e.g., audio and video) to the same user can be synchronized.[21] The setup delay is very large in H.323 for a regular call, which involves multiple messages from its underlying protocols such as H.225 and H.245. This delay is rather prominent on a low-bandwidth, high-loss environment. It is further aggravated by the higher delay margin on wireless access links. The fast call setup method is an option specified in H.323 that reduces the delay involved in establishing a call and initial media streams.

7.6 OVERVIEW OF SIP

SIP (Session Initiation Protocol) is an application layer protocol used for setting up and tearing down VoIP sessions. The major difference between SIP and H.323 is SIP is fully based on Internet context and thus does not support the Q.931 or ISUP messages that are currently used for telephony networks. But SIP extends the functionality of telephony signaling and supports mobility, and it is part of the overall IETF multimedia architecture framework that includes protocols such as RTP, RTSP (Real-Time Streaming Protocol), SDP (Session Description Protocol), SAP (Session Announcement Protocol), and others. SIP uses a text message format with an encoding scheme very similar to HTTP. Currently, SIP uses SDP[22] to establish the media session and the terminal capabilities, as H.245 in H.323. SDP messages are carried as the message body of a SIP message. A complete VoIP session includes a number of SDP messages for resource reservation, connection, and ringing in addition to SIP INVITE and BYE messages, as shown in Figure 7.6. All SIP messages are transported at the RTP layer, whereas H.323 control messages use TCP. SIP is based on client/server architecture with a SIP user agent and a SIP proxy server. The SIP user agent has two important functions: (1) it listens to the incoming SIP messages and (2) it sends SIP messages on receipt of an incoming SIP message or on user actions. The SIP proxy server relays SIP messages, so that it is possible to use a domain name to find a user. This simplifies the user location determination and allows scalability. The SIP server can be used as a redirect server, in which case it will provide the host location information without relaying the SIP messages, and the SIP user client will set up the session directly with the user.

The SIP mobility architecture components for VoIP are shown in Figure 7.7, where we assumed the mobile host and foreign network use DHCP (Dynamic Host Configuration Protocol)[23] or one of its variations for subnetwork configuration. An SIP-capable mobile host uses DHCP to register in the network. A mobile host broadcasts a DHCP_DISCOVER message to register to a network. Multiple DHCP servers will respond to this request with the IP address of the server and default gateway in the DHCP_OFFER message. The mobile host selects the DHCP server and sends the DHCP_REQUEST message to register. The registration is confirmed by DHCP_ACK at the DHCP server. As previously mentioned, SIP includes the mobility function, and the mobile host then uses its temporary IP address to register to the visiting register of the foreign network. The registration in a foreign network includes the authentication function, which uses AAA (authentication, accounting,

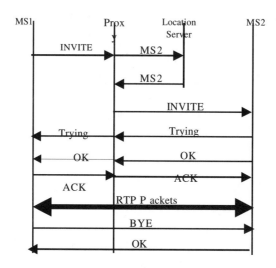

FIGURE 7.6 SIP signaling for VoIP.

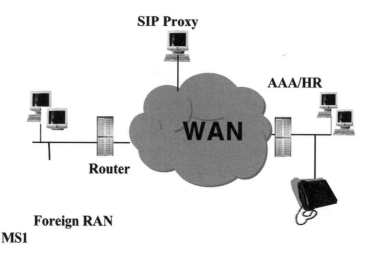

FIGURE 7.7 SIP mobility architecture components.

and administration)[21] servers. The foreign AAA server communicates with the home AAA server to get the confirmation from the home register about customer authenticity. Ultimately, the visiting register receives the authentication response message and, if it is accepted, it sends the 200 OK messages to the mobile host. If authentication fails, it sends 401 messages indicating unauthorized request for registration. After this registration, the mobile host initiates SIP registration for session start by sending an INVITE message to the SIP proxy server. In the case of micro–mobility, authentication with the AAA server is not necessary. The visiting register can authenticate the mobile host (expedited registration). The complete SIP registration sequence is required for macro–mobility. To reduce the macro-mobility registration delay of SIP, a quasiregistration concept is proposed in Schulzrinne.[23] In quasiregistration, whenever the mobile host hands off from an old visiting register to a new visiting register, it informs its home register of its location by sending a REGISTER message. When the home register replies its OK message, it will include the old visiting register's IP address along with the response. Thus, visiting registers will know the adjacent visiting register's address to use for fast registration.

7.7 RLP

ETSI and T-1 standards bodies defined the GPRS and 3G standards to carry packet data traffic in addition to voice in wireless access links. A new protocol layer, Radio Link Protocol (RLP), is defined on top of the MAC layer to improve bit error rate performance by using an ARQ retransmission technique. In addition to retransmission, GPRS RLC (Radio Link Control) performs block segmentation, reassembly, and buffering. As GPRS provides limited data capability, 3G standards are defined that extend the data capability of radio frequency. The main operation principles of GPRS and 3G RLP standards are very similar, although 3G provides some additional QoS management capability in RLP management at the MAC layer. In our analysis, we will use GPRS and 3G RLP capabilities interchangeably without significant impact on the results. cdma2000[24] is one of several 3G air-interface standards under consideration by standards bodies such as the 3GPP/3GPP2 (3G Partnership Project). The MAC layer of cdma2000 provides two important functions:

1. Best-effort delivery using the retransmission mechanism of RLP that provides reliability
2. Multiplexing and QoS management by mediating conflicting service requests

In addition, voice packets are directly given to the multiplex sublayer that bypasses the RLP function. Many transport channels have been defined for cdma2000 to provide services from physical to higher layers. These channels are unidirectional and either common (shared between multiple users) or fully dedicated to a user for the duration of the service. cdma2000 has defined many channels for its operation, the following transport channels are of interest:

1. Forward common control channel (F-CCCH): Communication from base station to mobile station for layer 3 and MAC messages
2. Forward supplemental channel (F-SCH): Operated in two modes, blind mode for data rate not exceeding 14.4 kbps, and explicit mode, where data rate information is explicitly provided, individual F-SCH target frame error rates can be different than other F-SCHs
3. Forward fundamental channel (F-FCH): Transmits at variable data rates, as specified in TIA/EIA-95-B

4. Reverse access channel (R-ACH) and reverse common control channel (R-CCCH): Used by a mobile station to communicate layer 3 and MAC messages
5. R-CCCH supports the low latency access procedure required for efficient operation of packet data suspend state
6. Reverse supplemental channel (R-SCH): Operates on two modes, blind and explicit
7. Reverse fundamental channel (F-SCH) supports 5- and 20-millisecond frames, the 20-millisecond frame structures provide rates derived from the TIA/EIA-95-B Rate Set-1 or Rate Set-2

The RLC and MAC layers are responsible for efficient data transfer of both real-time and nonreal-time services. The transfer of nonreal-time data includes the ARQ for low-level data to provide reliable transfers at higher levels. The network layer data PDUs (N-PDUs) are first segmented into smaller packets and transformed into link access control PDUs. The link access control overhead includes a service access point identifier, a sequence number for higher-level ARQ, and other data fields. The link access control PDUs are then transferred to SRBP (Signaling Radio Burst Protocol), a connectionless protocol for signaling messages. The data PDUs are segmented into smaller packet RLC PDUs corresponding to the physical layer transport blocks. Each RLC PDU contains a sequence number for lower-level ARQ and CRC fields for error detection. CRC is calculated and appended by the physical layer. When RLP at the receiving end finds a frame in error or missing, it sends back a NAK (negative acknowledgment) request for retransmission of this frame and starts a retransmission timer. When the timer expires for the first attempt, the RLP resets the timer and sends back a NAK request. This NAK triggers a retransmission of the requested frame from the sender. In this way, the number of attempts per retransmission increases with every retransmission trial. As noted in Bao,[25] the number of trials is usually less than four.

The GPRS structures of different radio channels and MAC/RLC are very similar to cdma2000. GPRS uses the same TDMA/FDMA structure as GSM to form the physical channels. For the uplink and downlink direction, many frequency channels with a bandwidth of 200 kHz are defined. These channels are further subdivided into the length of 4.615 milliseconds. Each TDMA frame is further split into eight time slots of equal size. As an extension to GSM, GPRS uses the same frequency bands as GSM and both share the same physical channels. Each time slot can be assigned to either GSM or GPRS. Time slots used by the GPRS are known as packet data channels (PDCH). The basic transmission unit of a PDCH is called a radio block. To transmit a radio block, four consecutive TDMA frames are utilized. Depending on the message type transmitted in one radio block, a sequence of radio blocks forms a logical channel.

- PRACH (packet random access channel, uplink): This common channel is used by the mobile stations to initiate the transfer in the uplink direction.
- PPCH (packet paging channel, uplink): The base station controllers (BSC) uses this channel to page the mobile station prior to downlink data transmission.

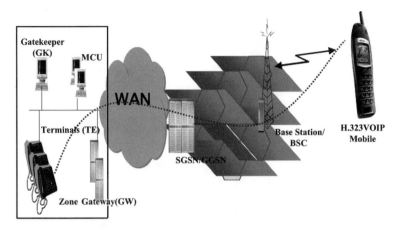

FIGURE 7.8 High-level view of VoIP implementation.

- PAGCH (packet access grant channel, downlink): Resource assignment on the uplink and downlink channel is sent on this channel.
- PBCCH (packet broadcast control channel): GPRS specific information is broadcast on this channel.
- PDTCH (packet data transfer channel) : Data packets are sent on this channel. A mobile station can use one or several PDTCHs at the same time.
- PACCH (packet associated control channel): This channel conveys signaling information related to a given mobile station and the corresponding PDTCHs.

7.8 H.323 IMPLEMENTATION ARCHITECTURE

Implementation of H.323 using a wireless link is shown in Figure 7.8. The wireless VoIP terminal is H.323 capable and uses a wireless air interface protocol such as GPRS or cdma2000 to connect to the base station. The other VoIP terminal connected to the Internet also is H.323 capable. The medium is voice packets. The terminals have two major functional planes:

1. The signaling plane that receives all messages coming to H.323 protocol supporting RAS (registration, admission, and status), H.225, and H.245.
2. The media plane that receives messages of H.323 belonging to RTP/RTCP. Recall that the control messages of H.225 and H.245 are carried on TCP, whereas RTP/RTCP messages are carried on UDP.

If the H.225 and H.245 messages are carried on UDP, there should not be significant differences in the analysis of our architecture. Note that TCP has three-way handshake and retransmission, while UDP has none. The characteristic difference between UDP and TCP over RLP (air link) is the three-way handshake, i.e.,

TABLE 7.1
Message Sizes Associated with the Regular Call Setup Procedure

Messages	Payload Size	Number of Frames (9.6 kbps)	Number of Frames (19.2 kbps)
Setup: H.225	254 octet	14	7
Alerting: H.225	97 octet	7	4
Connect: H.225	165 octet	10	5
TC Set: H.225	587 octet (×2)	29 (×2)	15 (×2)
TC Set ACK: H.245 (×2)	71 octet (×2)	6 (×2)	3 (×2)
OC: H.245 (×2)	115 octet (×4)	8 (×4)	4 (×4)
OC ACK: H.245 (×4)	64 octet (×4)	5 (×4)	3 (×4)
RTCP packet	120 octet	8	4

the beginning of the TCP session. The purpose of RLP is to provide reliable transmission, hence, TCP retransmission has no, if not zero, significant impact. The advantage of using UDP over TCP is (1) the UDP header overhead is lower (20 bytes as opposed to about 40 bytes for TCP), and (2) TCP has three-way handshake before each session. For an H.323 connection, typically, one handshake over air link is used. But there may be more than one if a gateway is being used, in that case, it is over land link communication only. If the air-link bandwidth is higher, so that 20 bytes vs. 40 bytes is not an issue, then the initial delay caused by the TCP three-way handshake is the only difference. The TCP retransmission could come in handy over a complex link (air and land) situation.

The performance of VoIP call setup using wireless links is influenced by three factors:

1. The number of messages exchanged to set up the call
2. The size of the messages
3. The number of TCP sessions that are set up

Reduction of any of these factors results in shorter call setup delay. Table 7.1 shows the size of the signaling and control messages for call setup using regular H.323 procedures. The frames indicate the number of air links. The wireless link FER (probability) can be as high as 0.1. With this high FER, the regular H.323 procedure call setup delay contribution can be as high as 40 seconds for a 9.6-kbps link and 30 seconds for a 19.2-kbps link. This high delay is unacceptable, compared to the current call setup delay (less than 3 seconds) of the PSTN system. To improve performance, we consider direct routing call model for this implementation. Table 7.2 shows the fastConnect messages and their sizes, which are significantly reduced.

The calling endpoint initiates the fastConnect procedure by sending an H.225 setup message containing the fastStart element to the called endpoint. When the called endpoint accepts the fastConnect procedure, it sends an H.225 message (call proceeding, alerting, progress or connect) containing a fastStart element selected

TABLE 7.2
Message Sizes of the fastConnect Procedure

Messages	Payload Size	Number of Frames (9.6 kbps)	Number of Frames (19.2 kbps)
Setup + fastStart	599 octet	30	15
Alerting	97 octet	7	4
Connect + fastStart	280 octet	15	8
RTCP packet	120 octet	8	4

from among the open logical channel proposals offered by the calling point. The channel accepted is considered opened once the H.245 open logical channel and open logical channel ACK procedures have been completed. Then the called endpoint may begin transmitting media (according to the channel opened) immediately after sending an H.225 message containing the fastStart element. The calling endpoint must therefore be prepared to receive media on any of the receive channels it proposed in the H.225 set message, because it is possible for the media to be received prior to the H.225 message indicating precisely which channels to use. Once an H.225 message containing the fastStart element is received by the calling endpoint, it may discontinue attempting to receive media on the channels for which proposals were not accepted by the called endpoint. The calling endpoint may begin transmitting media according to the channels opened immediately upon receiving an H.225 message containing the fastStart element. Therefore, the called endpoint must be prepared to receive media on the channels it accepted in the H.225 message containing the fastStart element. After establishing a call using the fastConnect procedure, either endpoint may initiate the H.245 procedures at any time during the call, using tunneling or a separate H.245 connection. If a call using the fastConnect procedure continues to completion without initiating the H.245 procedure, it may be terminated by either endpoint sending an H.225 Release Complete message

The challenge for wireless VoIP is the higher FER of the wireless links. The control messages require a very high level of integrity and a very low BER quality although it is tolerant to delay. On the other hand, the RTP voice messages are very sensitive to delay, but can tolerate higher FER. It was observed in the current wireless network that the FER of 10^{-1} to 10^{-3} is suitable for reasonably good voice quality. In the wireless air interface, the RLP layer brings the attributes that are needed for the control messages, whereas for voice packets, RLP will degrade voice quality due to its delay variations and subsequent jitter. Therefore, we propose separating these two packet streams and handling them through two different mechanisms of the air interface. The control messages will be routed through the MAC/RLP layer of the air interface. But RTP voice packets will bypass the RLP stage and will be transmitted without any ARQ protection. In addition, we propose taking advantage of the tunneling concept of H.323 to encapsulate H.245 messages within H.225 messages. This capability is supported in H.323 recommendations, thus, N-PDUs created by H.323 processing will be classified into two classes by using a classifier.

H.323 media streams for interactive traffic (particularly, VoIP) are treated differently than the H.323 control packet streams (e.g., H.225, H.245, RTCP messages). Because the VoIP packets can sustain higher FER than the control traffic, it is recommended that transparent RLP (i.e., no RLP retransmission) be used for VoIP packets, while nontransparent (regular) RLP be used for H.323 control packets. Due to the interactive nature of voice, VoIP packets should get higher priority over any in-band H.323 control packets within the session. Two types of subflows requiring differentiation have been identified for VoIP service using H.323:

1. H.323 control packets including H.225 and RTCP: These packets need higher reliability for better performance. They have less-stringent delay requirements compared to media packets. The QoS to be satisfied for these packets is
   • Call setup delay is the time required to set up the call after completion of sending the information for the call, i.e., after pressing the send button of the wireless terminal. The H.225 signaling will control this delay.
   • Connection delay is the time required to make the media connection after receipt of the answer signal. RTCP packet synchronization will control this delay.
2. H.323 media packets carried by RTP: These packets have lower reliability requirements than the control packets, but have stringent delay requirements. Handle these packets at highest priority within a delay restriction of 250 to 300 milliseconds. The two components of this QoS are the end-to-end delay of the voice packets and the blocking probability of the voice packets when multiple sources are sharing the same wireless resources.

7.8.1 Delay Analysis of H.323 Control Signaling over Wireless

We will assume the simple call setup message flows of H.323, as depicted in Figure 7.4. The total call setup delay will be the cumulative delay due to:

1. Setup time for two TCP sessions that include exchange of SYN, SYN-ACK, and ACK messages
2. Successfully transmitting all H.225 and H.245 messages
3. Successfully receiving an RTCP CNAME message

First, we present the analysis of RTCP packets based on mathematical models as suggested in Bao[25] and Sen et al.[26] for CDMA networks. Next, we analyze H.323 call setup flows and their interactions with TCP. Some information in an RTCP packet is more important than other information, e.g., CNAME and BYE. They are critical and need to be reliably transmitted to guarantee user-perceived QoS and network performance. We assume a maximum of three trials in our analysis for RLP, and restrict the maximum retransmission time for a packet to be much less than the

RTCP transmission interval T. (*Note:* We present here the results of the paper [see Das et al.[27,28]] without discussing the details of the analysis. Interested readers can refer to the original papers for the details.)

7.8.2 ANALYSIS OF RTCP:CNAME PACKET DELAY

Let
p = The probability of a frame being in error in the air link
k = The number of RLP frames in an RTCP packet transmitted over air and T $\geq$ 5 seconds is the RTCP transmission interval
D = The end-to-end frame propagation delay over the radio channel, with typical values on the order of 100 milliseconds
τ = The interframe time of RLP with typical values on the order of 20 milliseconds

Assume a user just joined the RTP session. The average delay of receiving the CNAME packet indicates the waiting period for the user to play the associated RTP streams properly. Thus

$$\text{The packet loss rate without RLP: } q = 1 - (1 - p)^k$$

$$\text{The packet loss rate with RLP: } q = 1 - (1 - p(p(2 - p))^6)^k$$

Thus, the average delay (T_1) associated with receiving the first RTCP CNAME packet after joining the session without RLP is given by

$$T_1 = [T\frac{1+q}{2(1-q)}] + D + (k-1)\tau = T[\frac{2-(1-p)^k}{2(1-p)^k}] + D + (k-1)\tau$$

If RLP is used, the average delay T_2 for a newly joined member to receive the first RTCP packet containing the CNAME item after joining the session with RLP can be approximated by

$$T_2 = T[\frac{1+q}{2(1-q)}] + D'' = T[\frac{2-P_f^k}{2P_f^k}] + D''$$

where the delay D is changed to D'' for RLP retransmission and is given by

$$D'' = D + (k-1)\tau + \frac{k[P_f - (1-p)]}{P_f^2} \times \left[\sum_{j=1}^{n}\sum_{i=1}^{j} P(C_{ij})(2jD + (\frac{j(j+1)}{2} + i)\tau]\right]$$

where C_{ij} represents that the first frame received correctly at destination is the i-th retransmission frame at the j-th retransmission trial, n denotes the maximum number of RLP retransmissions, and P_f denotes the effective packet loss seen at the RTCP layer (q) and is given by

$$P_f = 1 - P(B_n) = 1 - p[p(2-p)]^{\frac{n(n+1)}{2}}$$

$$q = 1 - P_f^k$$

7.8.3 H.323 CALL SETUP MESSAGE DELAY ANALYSIS

Because H.225 and H.245 messages are carried over TCP, an analysis of TCP transport delay over wireless will lead to an insight to the H.323 call setup delay performance. We have assumed a radio channel bandwidth of 9.6 kbps. The two endpoints are assumed to be in close proximity, hence any wireline network delay is assumed to be negligible. The following assumptions are made about the end-to-end TCP sessions carrying the H.323 messages:

1. TCP operates in an interactive mode.
2. The delayed acknowledgment mode of TCP operation is turned off.
3. TCP always times out whenever a packet is lost (i.e., it never does fast retransmit).
4. Round-trip delay is 200 milliseconds, because the one-way delay for message (D) is assumed to be 100 milliseconds (approx.).
5. The initial TCP round-trip timer (RTO) value is exactly equal to the round-trip delay. Subsequent variation of RTO is as specified for TCP.[29]
6. If initial capability exchange or master–slave determination procedures fail, no retry should be issued, as opposed to the standard that suggests at least two additional retransmissions before the endpoint abandons the connection attempt.

Following Karn's algorithm for TCP timer backoff, the RTO is multiplied by a constant factor after each retransmission due to time out. Hence, $RTO_{i+1} = c \times RTO_i$, where RTO_i is the i-th TCP retransmission timer. This causes RTO to grow exponentially after each retransmission. We let $c = 2$, as it is most commonly implemented. Initially, RTO = 100 milliseconds, as assumed previously. Hence, $RTO_i = 2 \times 100$ milliseconds, ..., $RTO_i = 2^i \times 100$ milliseconds. Furthermore, TCP will not allow infinite number of retransmissions. Hence, if TCP retransmission succeeds, after N_m attempts (without loss of generality, we assume $N_m = 10$) for example, the average delay to transmit a TCP packet is

$$T' + RTO_1 + RTO_2 + ... + RTO_m = T' + (2^{N_m+1} - 2) \times 100$$

where T' is the end-to-end propagation delay of the TCP packet. We shall establish the average delay for successful transmission of a TCP packet both with and without a radio link reliable transmission scheme (RLP).

For the purpose of our analysis, we stipulate in the following that the packet loss rate is q < 0.5. Recall that D(100 milliseconds) and τ(20 milliseconds) are the end-to-end frame propagation delay over the radio channel and the interframe time, respectively.

7.8.4 Average TCP Packet Transmission Delay

7.8.4.1 Average TCP Packet Transmission Delay without RLP

The TCP packet loss rate is $q = 1 - (1 - p)^k$, where p is the probability of a frame being in error in the air link and k is the number of air-link frames contained in a TCP segment. The probability of successfully transmitting a TCP segment is

$$(1-q)+(1-q)q+...+(1-q)q^{N_m-1}=1-q^{N_m}$$

The average delay for successfully transmitting a TCP segment with no more than N_m retransmission trials is

$$TN = (k-1)\tau + \frac{D}{(1-q^{N_m})(1-2q)} + \frac{1-q}{1-q^{N_m}} D\left[\frac{q^{N_m}}{1-q} - \frac{2^{N_m+1}q^{N_m}}{1-2q}\right]$$

where N_m denotes the maximum number of TCP retransmissions.

The TCP packets are going to carry the H.323 control messages in the payload. Hence, the total call setup delay is the cumulative addition of the delays for transmitting all the H.323 call setup messages and the RTCP:CNAME packet.

The total delay without RLP is

$$T_{noRLP} = \sum_{i=1}^{N_m} TN_i + TC$$

where TN_i is the average delay given above for i-th TCP segment (carrying one of the H.323 control messages in the payload) and TC is the average delay to receive the first RTCP:CNAME packet after joining the session.

7.8.4.2 Average TCP Packet Transmission Delay with RLP

Thus, the average delay to transmit a TCP segment successfully is given by

$$TR = D' + \frac{2Dq(1-q)}{1-q^{N_m}}\left[1 + \frac{4q(1-(2q)^{N_m-2})}{1-2q} - \frac{q(1-q^{N_m-2})}{1-q}\right]$$

where D′ denotes the effective transport delay for TCP and is represented by

$$D' = \frac{1}{P_f}\left[D(1-p) + \sum_{j=1}^{n}\sum_{i=1}^{j} P(C_{ij})\left[(2j+1)D + \left(\frac{j(j+1)}{2} + i \right)\tau \right] \right]$$

When the air-link FER is very high, too many RLP retransmissions may cause enough delay that TCP's retransmission timer always times out, thus, the average call setup delay with RLP is

$$T_{RLP} = \sum_{i=1}^{N_m} T_i + TC$$

where $T_i = \min\{TR_i, TN_i\}$.

7.8.5 AVERAGE H.323 CALL SETUP DELAY

We now compute the average call setup delay for a regular H.323 procedure and a fastConnect procedure. The models presented in the previous section imply that an average regular H.323 call setup delay increases exponentially as FER increases. Three major factors contribute to the delay — the number of message exchanges, size of message exchanges, and the number of TCP sessions set up. Reduction of any of these factors results in shorter call setup delay for any H.323 call. H.323 provides ways for addressing this or similar issues, including encapsulation of H.245 messages within H.225 messages (tunneling), and the fastConnect procedure. In order to conserve resources, either or both these mechanisms synchronize call signaling and control, and reduce call setup time. These mechanisms can be invoked by the H.323 calling endpoint. In this discussion, we will investigate the call setup delay incurred by the fastConnect procedure only.

H.323 endpoints may establish media channels in a call using either the regular procedures defined in Recommendation H.245 or the so-called fastConnect procedure. The fastConnect procedure allows the endpoints to establish a basic point-to-point call with as few as one round-trip message exchange, enabling immediate media stream delivery upon call connection.

Figures 7.9 and 7.10 compare the average call setup delays associated with a regular connect procedure and a fastConnect procedure for a 9.6- and a 19.2-kbps channel, respectively. Each procedure is further evaluated with and without support from RLP over the error-prone wireless channel. It can be observed that fastConnect with RLP support provides the minimum call setup delay. Furthermore, the call setup delay for the fastConnect procedure is consistently below 5 seconds for 9.6 kbps and 4 seconds for 19.2 kbps channels, if FER is less than 9%. This is close to the PSTN call setup time of 3 seconds.

7.8.6 EXPERIMENTAL VERIFICATION

Experiments for call setup delay at various FER rates between 1 and 10 percent over the wireless link emulator (WLE) were conducted using Microsoft NetMeeting.

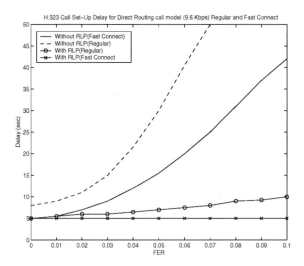

FIGURE 7.9 Call setup delay (9.6 kbps).

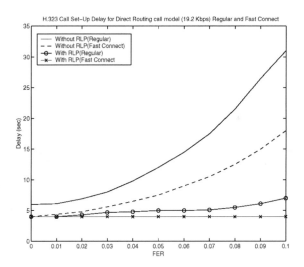

FIGURE 7.10 Call setup delay (19.2 kbps).

An end-to-end NetMeeting VoIP session from Endpoint A (Caller) to Endpoint B (Callee) over the WLE was created. NetMeeting 3.01 uses H.323v2 call signaling (Q.931/H.225/H.245 over TCP) to set up VoIP sessions and the default Microsoft codec G.723.1, 8-kHz mono, 6400 bps for audio compression. In these experiments, the call is considered successful only when the voice path is cut through (i.e., Endpoint B can hear the caller's voice). In addition, the maximum amount of time the called party waited for voice cut-through is 2 minutes, after which the call was marked as unsuccessful.

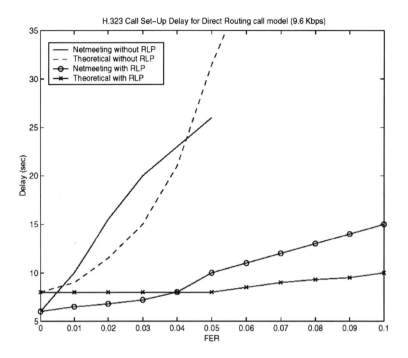

FIGURE 7.11 Comparison with NetMeeting results (9.6 kbps).

Two sets of experiments were conducted, one with the channel bandwidth fixed at 9.6 kbps and the others at 19.2 kbps. Thirty delay samples were collected at each FER with RLP and without RLP. Then, the sample mean was computed (see Figures 7.11 and 7.12). The success rate at each FER is shown in Table 7.3.

With RLP turned on, the result of the call setup success for both 9.6 and 19.2 kbps is perfect, at 100 percent success. With RLP turned off, at 1 percent air-link FER for 9.6 kbps and 1 to 2 percent air-link FER for 19.2 kbps, the call setup success rate is 100 percent. However, as the air-link FER rate increases, the average call setup delay increases (Figures 7.11 and 7.12) and the success rate declined (Table 7.3).

A fast call set-up time is considered a significant step toward providing QoS. Previous sections of this chapter illustrate that it is advantageous to transmit H.323 call setup messages and RTCP packets over air link with RLP, especially for VoIP clients that require exchange of several signaling and control messages before connection. For instance, the call setup procedure for H.323 involves exchange of H.225 and H.245 messages. Hence, the total delay to replay media is the sum of delays experienced by each of the messages. Therefore, the usage of RLP with either the regular or fastConnect call setup procedure enhances services without significantly sacrificing the limited bandwidth.

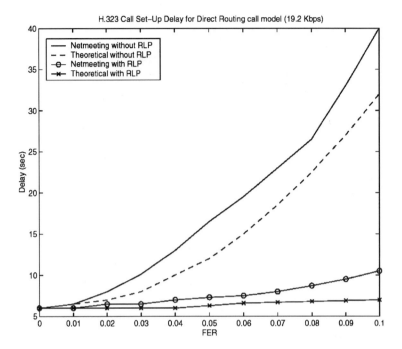

FIGURE 7.12 Comparison with NetMeeting results (19.2 kbps).

TABLE 7.3
Success Rate with NetMeeting

FER (percent)	Call Setup Success Rate (percent)			
	9.6 kbps (w/RLP)	19.2 kbps (w/RLP)	9.6 kbps (w/o RLP)	19.2 kbps (w/o RLP)
1	100	100	100	100
2	100	100	93	100
3	100	100	83	93
4	100	100	47	93
5	100	100	30	87
8	100	100	0	40
10	100	100	0	23

7.9 MEDIA PACKET-BLOCKING ANALYSIS IN GPRS

The media packet transport of VoIP consists of three important functions at the terminal:

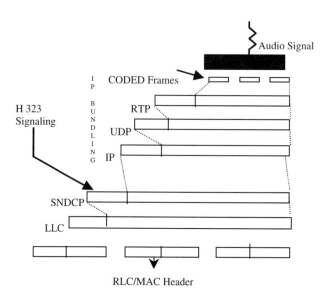

FIGURE 7.13 Voice payload design of GPRS VoIP.

1. The analog or PCM voice signals are decoded by the codec and codec frames are delivered to the bundling function at the codec rate. For example, AMR[13] codec outputs a coded frame every 20 milliseconds, whereas G.729[14] outputs a coded frame every 10 milliseconds. The bundling of the multiple voice frames of the coder is necessary, otherwise the overhead penalty becomes too significant.
2. The second is the selection of the Internet Protocol (IP). There are two options for IP at this stage: TCP (Transport Control Protocol) or UDP (Uniform Datagram Protocol) packet format. Most of the designs use the UDP packet protocol to transfer the media packets because the retransmission of the voice packets under BER is of no value to voice quality.
3. The next challenge is the mechanism of transporting the voice packets though the wireless access link to the Internet. The decoded UDP or TCP voice packet is to segment to the wireless link transport frame format, which depends on the wireless link standards such as GPRS, cdma2000, and IEEE 802.11 (wireless LAN), etc. The bundled coded frames are used as a payload for RTP and header for UDP and IP are added to construct a GPRS packet. The GPRS packet is then compressed by SNDCP and a LLC PDU is formed. This GPRS LLC PDU is broken down to GPRS MAC/RLC[30] layer frames that are transmitted every 20 milliseconds. The block diagram of these functions for GPRS is shown in Figure 7.13.

Thus, there are multiple design parameters in this voice payload assembly problem. We consider the case of a system using GPRS and UDP for transfer of voice packets. Let

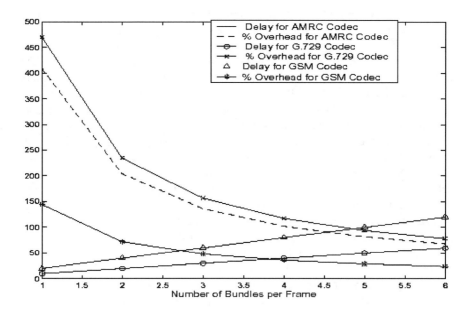

FIGURE 7.14 Overhead vs. delay due to packet bundling.

r   = Rate of voice coding in kilobits per second.
f   = Frame time of voice coding in milliseconds.
F   = Frame size of voice codec in bytes.
H_{ip} = IP header in bytes (normally 20 bytes).
H_{UDP} = UDP header in bytes (normally 8 bytes).
H_{CTP} = CTP header in bytes (normally 12 bytes, can be compressed to 2 to 4 bytes).
H_{GPRS} = GPRS header in bytes (normally 10 bytes).
β = Bundling factor. This bundling will result in additional delay on the voice
   path.

GPRS packet size:

$$P_{GPRS} = F.\beta + [H_{ip} + H_{UDP} + H_{CTP} + H_{GPRS}]$$

The overhead of this conversion of voice frames on UDP packets is $100[H_{ip} + H_{UDP} + H_{CTP} + H_{GPRS}]/P_{GPRS}$ (percent).

Figure 7.14 shows the overhead penalty of three different voice coders: AMR at 4.5 kbps, G.729 at 8 kbps, and GSM voice coder at 13.2 kbps. Depending on the different bundling parameter, additional delay will result in the system, although the overhead penalty will be reduced. The delay curve is linear to bundling factor, whereas the percent overhead reduction is high at bundling two or three codec frames. The VoIP designer will select the delay and the overhead penalty to get the appropriate value for the end to end delay of the service.

The codec voice packets after being bundled are segmented to fit into the GPRS frame size. The actual transmission of the packets through the GPRS channel in the upstream involves two functions. In the uplink direction, the GPRS traffic channel request is sent to the GPRS base station using uplink random access channel. The BSC transmits a notification to a mobile station terminal indicating temporary buffer flow (TBF) allocation. TBF allocation is basically a physical connection used to support the transfer of blocks. Each TBF connection is assigned one TFI (temporary flow identifier). TFI is included in each of the transmitted radio blocks so that multiplexing of different mobile stations can be done on the same packet data channel. As a matter of fact, GPRS uses dynamic bandwidth allocation. When a user needs more bandwidth, physical connection is established by the interaction between the mobile station and the base station to allow the mobile station to send the data in the uplink direction. This physical connection is considered TBF allocation. Every time the mobile station wants to send the data, it must establish this physical connection with the base station and tear down the connection at the end of the transmission. This whole process is time consuming, especially when the transmitting packet is too small, and the delay incurred by the TBF increases the mean delay. To support the packet-switched principle of GPRS, resources of the one packet data channel are assigned only temporarily to one mobile station.

In GPRS, uplink and downlink are carried out on different 200 kHz channels. The BSC handles the resource allocation in downlink as well as in uplink. All the packets are originated from the BSC (downlink), no concurrent access on one packet data channel can occur. While in the uplink, more than one mobile station can try to get hold of one packet data channel at the same time. To avoid the access conflict in the uplink direction, the BSC transmits the USF associated with each of the radio blocks in the downlink direction (USF identifies each of the mobile stations distinctly), indicating which mobile station has rights to send the data in the corresponding uplink block.

The PCU (process control unit) of the GPRS performs the allocation of capacity of the traffic channel for transfer of the voice packets to the base station. Because of the sharing of multiple GPRS terminals by the same GPRS traffic, the VoIP will encounter blocking when it requests the service from the GPRS system. This blocking depends on the number of active GPRS terminals in the cell and the intensity of the voice packets generated by the active voice terminals. The channel blocking probabilities are determined by the two-stage modeling technique. In the first stage, the probability of the number of active sources in the system is determined, and in the second step, out of these active terminals, probabilities of numbers of simultaneous traffic bursts generated by the terminals are determined. The probabilities of the simultaneous traffic bursts of the sources are modeled by considering the on–off burst source model of the individual source. Consider the case of an only-VoIP terminal in a GPRS system.

7.9.1 VoIP Traffic Blocking

The model of the GPRS VoIP user in a cell is considered as a finite-state Markov process with quasirandom arrival. The idle-state arrival rate decreases and departure rate increases, as more users are in the system. This is the traditional Engset Model[31] of telephony. This model will give the state probability of active users in the system. The actual design of the system will include a threshold of maximum number of users that the system will allow to enter (new users and handoff users). Let

n   = Total number of VoIP terminals in the cell coverage area.
α = Arrival rate of the free VoIP traffic source.
$1/\mu$ = Mean service time of one VoIP source during active voice session.
P_i = Probability of i-VoIP traffic source active in the system.

$$P_i = \frac{\binom{n}{i}\beta^i}{\sum_{k=0}^{n}\binom{n}{k}\beta^k}$$

where $\beta = \alpha/\mu$.

The accepted VoIP session will generate the voice traffic bursts depending on the coding speed and the bundling mechanism used in the system. The GPRS uses multiple coding schemes. Four GPRS coding schemes are proposed: CS-1 (9.05 kbps), CS-2, CS-3, and CS-4 (21.4 kbps). The CS-1 scheme has the highest error-correction capability, whereas CS-4 has no error-correction capability. In the typical GPRS system, most likely only CS-1 and CS-2 will be used. CS-2 has a user data rate of 13.4 kbps with some error-correcting capability. Any GPRS terminal can initiate a session and the talk bursts can be modeled as an on–off source. We can divide the entire holding time in the two regions: user is in on-state when it is active, and in off-state when there is no talk burst and a silent period during the conversation.

So from "i" users in the system, the probability that "k" users will be in on-state is given as

$$P_{ik} = \binom{i}{k}\rho^k(1-\rho)^{i-k}$$

where ρ =probability of being in on-state in one second (assuming the same for all the users), and $1 - \rho$ = probability of being in off-state.

The joint probability that the i users are active and k users are in on-state is given by $P_i \times P_{ik}$. Hence for a GPRS system with 8 slots in 200-kHz spectrum and assuming each active burst is using one slot at a time, the probabilities of burst level blocking of the system is given by

$$B = 1 - \sum_{i=0}^{n}\sum_{k=0}^{Min[i,8]} P_i \times P_{ik}$$

The PCU of the GPRS performs the allocation of the capacity of the traffic channel for transfer of the voice packets to the base station. The GPRS traffic channel

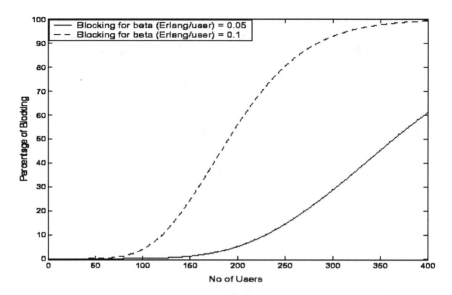

FIGURE 7.15 GPRS burst level blocking without silent detection.

is shared by the multiple voice sources within the cell-site coverage area. Because of this sharing of common GPRS resource by multiple GPRS terminals, the VoIP will encounter blocking. This blocking depends on the number of active GPRS terminals in the cell and intensity of the voice packets generated by the active voice terminals. The blocking performance with respect to number of users is shown in Figure 7.15. The system is using AMR coding with a bundling factor of 3 and a UDP transport mechanism with header compression and a GPRS CS-2 coding scheme. Here we have assumed that the VoIP terminal generates 50 milli-Erlang and 100 milli-Erlang of traffic and a burst occupancy of 0.5.

The burst level blocking performance can be improved by including the silent detection mechanism in the codec. In a two-way conversation, when the codec detects the silent period, the average number of talk bursts will reduce and thus bring additional capacity in the radio link. Normally, a user talks about 50 percent of the time and listens the remaining 50 percent of the time. If we use a 40-percent savings in capacity due to the unidirectional flow, more users can be supported in the system, as shown in Figure 7.16. Assuming a very low burst level blocking, the system can support more than 200 users for 50 milli-Erlang per user. The actual 200 kHz GSM system channel may not support more than 100 50-milli-Erlang users. There is an opportunity to gain voice traffic capacity in the GPRS VoIP.

7.10 CONCLUSION

In this chapter, we explained VoIP services in the wireless network. A two-imple-mentation architecture using H.323 and SIP is proposed, and architecture of H.323 was discussed in detail. We discussed the voice-quality issues in the implementation of the VoIP in the wireless network, and mentioned the areas where delay budgets

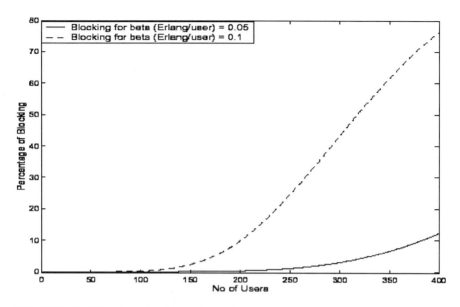

FIGURE 7.16 GPRS burst level blocking with silent detection.

can be improved. We proposed an implementation architecture of VoIP using the H.323 Protocol. The architecture recommends the use of direct connect with the fastConnect features of H.323 to keep the call setup delay low. We proposed also a concept of classifier to separate the control and media packets and exclude RLP function for media packet transmission. We have given a detailed analysis of the H.323 call-set model with variable frame error rate and RLP retransmission. An experimental NetMeeting setup was used to determine the call setup delay under various frame error rate conditions. The analytical results were compared with the experimental results. We proposed a model to determine the voice packet burst level blocking. This model is used to determine the capacity of VoIP using a GPRS system. The results of this model showed that using AMR coding with silent detection and CS2 coding and assuming a bundling factor of 3, considerable capacity gain is possible by using VoIP over GPRS, compared to the current GSM system. The end-to-end delay and the handoff performance of the VoIP in wireless networks are still open question. Until these two aspects are properly addressed, VoIP in wireless networks will not attain the quality comparable to current wireless voice services. But in the meantime, VoIP wireless services can be offered for business applications where the inferior voice-quality standard may be acceptable.

References

1. Mouly, M. and Putet, M.B., *GSM System for Mobile Communications,* Mouly and Putet, 1992.
2. GSM 03.60, Digital Cellular Telecommunications Systems (Phase 2+), General Packet Radio Service (GPRS) Service Description Stage-2, ETSI DTS/SMG-030360Q, May 1998.

3. International Telecommunications Union, Packet Based Multimedia Communications Systems, Recommendation H.323, Telecom Standardization Sector, Geneva, Switzerland, Feb. 1998.

4. Handley, M. et al., SIP: Session Initiation Protocol, Request for Comments (RFC) 2543, Internet Engineering Task Force, Mar. 1999.

5. Greene, N., Ramalho, M.A., and Rosen, B., Media Gateway Control Protocol Architecture and Requirements, RFC 2805, Apr. 2000.

6. International Telecommunications Union, Gateway Control Protocol, Recommendation H.248, Telecom Standardization Sector, Geneva, Switzerland, June 2000.

7. Blake, S. et al., An Architecture of Differentiated Services, RFC 2475, Dec. 1998.

8. Rosen, E., Viswanathan, A., and Callon, R., Multiprotocol Level Switching Architecture, RFC 3031, Jan. 2001.

9. Bellamy, J.C., *Digital Telephony,* Wiley Interscience, New York, 1991, pp. 98–142.

10. International Telecommunications Union, Pulse Code Modulation of Voice Frequencies, Recommendation G.711, Telecom Standardization Sector, Geneva, Switzerland, 1988.

11. GSM 06.10, Digital Cellular Telecommunications System (Phase 2+): Full Rate Speech Transcoding, Version 7.0.1, Release 1998.

12. GSM 06.60, Digital Cellular Telecommunications System (Phase 2+): Enhanced Full Rate (FER) Speech Transcoding, Version 7.0.1, Release 1998.

13. GSM 06.10, Digital Cellular Telecommunications System (Phase 2+): Adaptive Multi-Rate (AMR) Speech Transcoding, Version 7.1.0, Release 1998.

14. International Telecommunications Union, Coding of Speech at 8 kbits/sec Using Conjugate-Structure Algebraic Code-Excited Linear-Predictive (CS-ACLEP) Coding, Recommendation H.729, Telecom Standardization Sector, Geneva, Switzerland, Mar. 1996.

15. Lakaniemi, A. and Parantainen, J., On Voice Quality of IP Voice over GPTS, IEEE International Conference on Multimedia, *ICME,* 3, 751–754, 2000.

16. Cox, R.V., Three new speech coders from the ITU cover a range of applications, *IEEE Communications Magazine,* 39(9), 40–47, Sep. 1997.

17. International Telecommunications Union, G.114: Mean One-Way Propagation Time, Recommendation G.114, Telecom Standardization Sector, Geneva, Switzerland, Nov. 1988.

18. Grilo, A., Estrela, P., and Nunes, M., Terminal independent mobility for IP (TIMIP), *IEEE Communications Magazine,* 39(12), 34–46, Dec. 2001.

19. Das, S. et al., IDMP: An intra-domain mobility management protocol for next generation wireless networks, *IEEE Wireless Communications*, Special issue on Mobile and Wireless Internet: Architectures and Protocols, Agrawal, P., Omidyar, G., and Wolisz, A., Guest Eds., 9 (3), 38–45, 2002.

20. Misra, A. et al., IDMP-based fast handoffs and paging in IP-based 4G mobile networks, *IEEE Communications,* Special issue on 4G Mobile Technologies, Lu, W., Guest Ed., 40 (3), 138–145, 2002.

21. Schulzrinne, H. et al., RTP: A Transport Protocol for Real-Time Applications, RFC 1889, IETF, Jan. 1996.

22. Song, J. et al., MIPv6 User Authentication Support through AAA, Internet draft, draft-song-mobileip-mipv6-user-authentication-00.txt, Nov. 2001.

23. Schulzrinne, H., DHCP Option for SIP Servers, Internet draft, draft-ietf-sip-dhep-05.txt, Nov. 2001.

24. http://3GPP2.org.

25. Bao, G., Performance evaluation of TCP/RLP protocol stack over CDMA wireless link, *Wireless Networks,* 3 (2), 229–237, 1996.
26. Sen, S.K. et al., A call admission control scheme for TCP/IP based CDMA voice/data network, ACM/IEEE International Conference on Mobile Computing and Networking, Oct. 1998, Dallas, pp. 276–283.
27. Das, S.K. et al., Performance Optimization of VoIP Calls over Wireless Links Using H.323 Protocol, IEEE Infocom, June 2002.
28. Das, S.K. et al., Performance optimization of VoIP calls over wireless links using H.323 Protocol, *IEEE Trans. Computers,* Special issue on Wireless Internet, Lin, Y.B. and Tseng, Y.-C., Guest Eds., submitted, 2002.
29. Postel, J., Transmission Control Protocol, RFC 793, IETF, Sept. 1981.
30. GSM 04.60, Digital Cellular Telecommunications System (Phase 2+): General Packet Radio Service (GPRS): Radio Link Control/Media Access Control (RLC/MAC) Protocol, July 1998.
31. Kleinrock, L., *Queueing Systems,* Vol. 1: *Theory,* John Wiley and Sons, New York, 1975, pp. 99–110.
32. Kwon, T.T. et al., Mobility management for VoIP service: mobile IP vs. SIP, *IEEE Wireless Communications,* Special issue on IP Multimedia in Next-Generation Mobile Networks, Apostolis, S. and Merakos, L., Guest Eds., Oct. 2002.
33. Westberg, L. and Lindqvist, M., Realtime Traffic over Cellular Access Networks, IETF draft-westberg-realtime-cellular-01.txt, Oct. 1999.

8 Wireless Application Protocol (WAP) and Mobile Wireless Access

Andres Llana, Jr.

CONTENTS

8.1 INTRODUCTION

It is projected that in 2003 there will be over 400 million Internet subscribers and 600 million mobile phone users. As a result of this expansion, there will be a growing demand for wireless data services with a corresponding demand for quick access to information from any location; hence, the watchword of "anytime, anywhere." While WAP does provide access to the Internet, a "killer" application has not yet made an appearance.

FIGURE 8.1 Get Rid of the Phones

WAP-enabled pagers are fast becoming a way to send messages back and forth over the Internet. Today, there are many regional and national pager carriers that offer short messaging services using a Palm Pilot or one of the popular Motorola devices such as the Talkbout or Timeport. These must be a viable service because they have been very popular with teenagers since they found that they can send messages back and forth to their friends. Teachers have now banned them from classrooms because obviously they can be used to cheat on tests.

The costs vary from service carrier to service carrier. Generally, the devices range from $149 to $400, in addition to an access fee of about $18 per month for up to 25,000 characters; extra messages above this cost $0.01 per 100 characters. The subscriber is given an 800 number that can be accessed by other similar devices to send messages to the subscriber. For those who do not have such a device, an operator, who will manually key in any desired messages, can be accessed through an 800 number. The subscriber pays about $10 per month for this service for up to 30 such messages and $0.65 for additional messages above the 30-message rate.

Generally speaking, these services have become quite popular because the per-call costs are very low and the subscriber can be reached anywhere in the United States, and there are carriers overseas that can be used to extend services. The bottom line is that short messaging services are more cost effective than WAP-enabled telephones.

Behind this rapid growth in the Wireless Application Protocol (WAP) has been the WAP Forum. The WAP Forum began in December 1997 as an industry association to develop, support, and promote a world standard for wireless information and services accessed via a digital mobile telephone or similar wireless device. The WAP Forum was chartered to bring together service providers, handset manufacturers, Internet content providers, applications developers, and infrastructure manufacturers to ensure interoperability between devices and promote the growth of wireless Internet-based service (see Figure 8.1). The WAP Forum has over 300 regular members plus associate members, including handset manufacturers representing over 95 percent of the market, carriers with over 150 million customers, infrastructure providers, software developers, and other related industries.

The WAP Forum maintains liaison with other industry organizations to include the European Telecommunications Standards Institute, Cellular Telecommunications Industry Association, the Worldwide Web Consortium, and the Internet Engineering Task Force. All are actively working with the Forum to evolve the next-generation HTML (HTML-NG).

WAP also is being enhanced to address the 3G wireless networks that will support fully packetized information transmission.

8.2 WIRELESS APPLICATION PROTOCOL

Wireless Application Protocol (WAP) is the *de facto* standard for providing Internet communications and advanced telephony-based services over digital mobile telephones, pagers, personal digital assistants (PDAs), and other wireless terminals.

WAP is an open, global standard that empowers users of mobile telephones and wireless devices to securely access and instantly react with Internet information and services.

This single-industry, agreed-upon standard for wireless application interoperability uses an XML-compliant markup language called WML (Wireless Markup Language). The advantage of WML is that it provides a path for application developers as well as content providers to develop and deliver Web-based services. The WML user interface is a WAP microbrowser that maps into mobile phones and other wireless devices. Devices using WAP-based microbrowsers can access an array of innovative value-added services.

The basic concept of WAP is to specify the network server, the mobile telephone software, and the communications between them. Communication is established between the mobile handset (client) and a gateway that serves as the gateway to the Internet. The gateway supports protocol and format conversion between a network application server, enabling communication with a WAP-enabled handset or client.

WAP is designed to function over any wireless network, including CDPD, CDMA, GSM, PDC, Mobitex, and others. An application server on the Internet provides the information or data desired by the client while the network serves as the bearer for the data.

Microbrowser firmware is embedded into the mobile phone and the developer must be able to support that version of the microbrowser. However, because each manufacturer is different, there are subtleties that the developer must be able to support for each handset. Further, the programming language used to develop applications (WML or HTML) may have variants when deployed in the Asian market, where slightly different versions of these programming languages are used. This forces developers to be conversant in four versions of the markup language and to design their applications to interface with each of these language variants.

Design differences in the handset create additional problems for the developer because screen space varies, forcing the application designer to work toward the lowest common denominator. Because present-day WAP telephones are slow and do not always respond as expected, software applications may not perform as designed. This creates additional design problems affecting security and user privacy issues.

8.2.1 WAP SPECIFICATION

WAP specification is unique because it defines an open standard architecture and set of protocols intended to facilitate wireless Internet access. It provides solutions for problems not solved by other standards bodies (e.g., W3C, ETSI, TIS, IETF, etc.) and serves as a catalyst for wireless development and standardization. The specification's key elements include a definition of the WAP programming model, which is based on the existing World Wide Web programming model. This serves to benefit the developer community because it provides a familiar programming model, an established architecture, and the ability to leverage existing tools (i.e., Web servers, XML tools, etc.). A markup language adhering to XML standards is designed to enable powerful applications within the constraints and limitations of handheld devices. WML and WML Script do not assume the availability of a QWERTY keyboard or mouse for user input. Unlike the flat structure of HTML documents, WML documents are divided into a set of well-defined units of user interaction.

Another key element is the specification for a microbrowser in the wireless terminal that controls the user interface and is analogous to a standard Web browser. This specification defines how WML and WML Script should be interpreted in the handset and presented to the user.

In addition to the above, there is a lightweight protocol stack to minimize bandwidth requirements, guaranteeing that a variety of wireless networks can run WAP applications, a framework for Wireless Telephony Applications (WTA) that allows access to telephony functionality such as call control, phone book access, and messaging within WML Script applets. This allows the operator to develop secure telephony applications integrated into WML/WML Scripts.

8.3 WAP SOLUTION BENEFITS

8.3.1 BENEFITS TO THE SERVICE PROVIDER

It should not be a secret that service providers can add significant value to their service offerings by adding WAP-based services to their wireless networks. At present, many of the mobile handset and PDA manufacturers are starting to sell WAP-enabled devices. The only thing for the service providers to do is to package a product. By developing a WAP-based product line, the service providers will be able to market new services to the subscribers, which greatly increase network usage. By controlling the data connection through a WAP gateway, service providers can maintain strong relationships with their subscribers, forestalling customer churn. Many WAP developers are beginning to offer new content systems that provide the service provider with new easy-to-access subscriber services. This is much the same as with a new public page on the Internet that can be accessed globally by any Internet user. Because of the WAP open standard, many more options are available to the service providers for WAP gateways, WAP-enabled handsets, or Web-enabled content services. This flexibility of choice makes it possible for the service provider to choose from a wide array of vendor products, all at competitive price levels.

8.3.2 BENEFITS TO THE MANUFACTURER

Handset manufacturers are now beginning to see the advantage of integrating a microbrowser into their handsets that is low in cost and provides additional capability beyond voice access. Some vendors are including Bluetooth chips as well, which will enhance the value of their handsets over that of competitors. These microbrowser-enabled handsets will allow handsets to work on all WAP servers and all networks that offer WAP-based services. These new enhancements increase their value to the network service provider, who is now in a position to package these new handsets into a variety of new service offerings.

8.3.3 Developer Benefits

Applications developers are now in a position to reach a much-larger audience of end users who carry Web-enabled mobile handsets. Phone.com, a wireless Internet service provider, reported that its registered developers who create Web sites and applications grew from 62,000 to over 110,000. Another element that has encouraged developers is the fact that WML is based on XML and is an easy markup language for the developers to learn. Because WML has its basis in XML, it sets the stage for automatic content transformation. Information written in XML can be automatically translated into content for HTML or WML. As the technology for universal content continues to evolve, applications developers can feel secure in using present-day WML because there will always be a migration path upward from WML. WML serves as the common denominator for all developers, with no one having a unique advantage over competitors. WML provides a common thread among developers because any application written in WML will run on any network. WML allows developers to integrate applications with any device or telephony function.

8.4 SOME CONSTRAINTS OF A WAP-ENABLED WIRELESS NETWORK

8.4.1 Security Issues

Many of the applications destined for the Web require a secure connection between the client (mobile handset) and the application server. The WAP specification ensures that there is a secure protocol to support transactions between a wireless handset and the application server. This secure protocol is known as Wireless Transport Layer Security (WTLS) and is based on the industry-standard Transport Layer Security (TLS) Protocol, also known as Secure Sockets Layer (SSL). WTLS is designed to be used with WAP transport protocols and has been optimized for use over narrowband communications channels. WTLS is designed to ensure data integrity, privacy, authentication, and denial-of-service protection. Where Web applications employ standard Internet security techniques using TLS, the WAP gateway automatically and transparently manages wireless security.

In the WAP environment, the WAP gateway serves to translate WAP to Web protocols, thereby enabling WAP devices to access the Web. WTLS encrypts transmission from the mobile handset to the gateway. However, before the gateway can encrypt the transmission into TLS/SSL, it must first decrypt the WTLS packets. In this situation, all of the data is briefly in the clear before being encrypted for its journey to the application server. This results in a weak link in the WAP transmission process. To correct this problem, the WAP Forum is working on a fix that may well appear in WAP Version 1.2 or 1.x in the near future.

There have been some half solutions proposed to combat this situation, such as securing one's own gateway in a locked facility. There are a number of software vendors (e.g., Entrust Technologies) that offer software suites that will provide end-to-end security. Utilizing PKI (public key infrastructure) software modules, such systems can issue WAP server certificates as well as client certificates for complete user-to-server authentication.

Baltimore Telepathy offers a security gateway that supports end-to-end security from the mobile user to the WAP/Web server. This is a stand-alone solution for content service providers that requires digital signatures for authentication.

Hardware manufacturers are starting to announce secure WAP servers that can be placed online and provide immediate security. Hewlett Packard has recently announced its Praesidium Virtual Vault, which is aimed at the financial arena. This trusted WAP solution sits at the edge of the network between the outside world and the enterprise to connect mobile users to corporate applications and databases.

8.4.2 SECURE APPLICATIONS DEVELOPMENT

To date, there have been a number of products that support securing WAP-based operations. Many of these developments have been in the software arena.

Certicom and 724 Solutions have joined forces to develop a wireless PKI solution for the financial industry. This will be an open standards-based security solution that enables secure communications and digital signatures via a variety of Internet-enabled devices such as PDAs, mobile telephones, and pagers. This system will serve to support the new legislation that went into effect October 1, 2000, which allows businesses and consumers the ability to close contracts with digital signatures. The new wireless PKI solution will provide financial institutions with the ability to offer consumers the confidence and convenience of performing secure "anytime, anywhere" high-value transactions.

8.5 PREPARING FOR THE MOVE FORWARD

High-speed Internet access over circuit-switched wireless networks is not a very viable means for providing a base for data services that expect access to screens of information. Fortunately, circuit-switched wireless networks are undergoing change from their present form to one where all information will travel in the form of packets. At present, there are any number of GPRS tests underway in various GSM networks in Europe and Asia. Similarly in the United States, Sprint and Bell Atlantic (now Verizon) have moved to convert their wireless networks from circuit-switched to 2G IP packet-based networks. The impact of these migrations will serve to enhance WAP-based applications because the higher-speed IP packet networks will support greater throughput for all services.

Operators see the introduction of data as a way of addressing declining voice revenues. These operators will be only too glad to accommodate the WAP-enabled user who wishes access to the Internet for a variety of services. If nothing else, WAP-enabled handsets will more than ensure the operator of much-higher revenues over voice because data access will ensure long call-hold times and therefore much greater network occupancy.

Integrated access of both voice and enhanced WAP data services will ensure that the network operators will have more services to sell under a variety of pricing plans. Network operators will be more than a data delivery pipe; they will be purveyors of a range of upscale consumer services. These services will serve the needs of the subscriber in ways not thought of previously.

8.6 RECENT WAP DEVELOPMENTS AND APPLICATIONS

Numerous announcements were made during 2000, many coming during the PCIA 2000 conference. There are starting to emerge many content and value services that until now have been well developed for those who access the Web through a wired connection. In addition, another new application that will be coming along with WAP-based applications will be location services utilizing GPS. While GPS services have been around for some time, they will now be embedded in mobile phones. These new services will allow the end user to access a Web site that will provide directions to a specific location or service. This is very similar to the very same services that are available to a wired user accessing the Web with a browser. For example, GeePS announced that it has agreed with Advanced Internet, a creator of community-based Web sites, to provide a wireless version of its product. This new product will merge WAP and GPS technologies and will allow consumers to surf the Web to locate local merchants.

Visa and BT Cellnet, a U.K. service provider have announced a new WAP location service for Visa card holders who have WAP-enabled handsets. This new location service will allow card holders to use their WAP telephones to locate the nearest Visa ATM by entering the postal code for the area where they are located. BT Cellnet will extend this service to locate over 531,000 Visa ATMs located throughout the world. Future versions of this service will support mobile handsets equipped with GPS so that the service can locate the nearest ATM automatically without regard to a postal code.

8.6.1 INFORMATION SEARCH AND RETRIEVAL

There have been a number of WAP utility packages developed to search the Web for a specific information stream. MobileWAP.com is a good example of a search engine dedicated to finding WAP content on the Internet. A built-in electronic agent continuously searches the Web, seeking and indexing relevant Internet pages written for WAP-enabled devices using WML and adding these to its range of listings. MobileWAP.com can be accessed wirelessly with any WAP-enabled device or through the Internet using wired access.

8.6.2 E-MAIL AND MORE

Sheffield Dialogue Communications recently demonstrated in Europe a Windows e-mail attachment to a WAP-enabled device. Using the latest version of their Dialogue Expressway 2000 E-Mailconnector, users can read any document from the Microsoft Office Suite of software using their WAP-enabled telephone. In this system, documents using MS Word or Powerpoint are translated into simplified text that can be read on a mobile handset screen. This connector allows the user to read, reply, forward, or delete messages, as well as view attachments and have access to address books.

The Expressway 2000 acts as a broker for WAP-enabled devices providing a fully functional e-mail client on a phone. Access to personal or global address books is provided through LDAP support. Expressway 2000 employs advanced session handling and e-mail session spoofing to ensure that the user's e-mail remains intact even if the WAP device drops the connection to the network.

8.6.3 BANKING AND E-COMMERCE

There have been a number of initiatives undertaken in the banking industry. This effort has been reinforced with the introduction of PKI systems to ensure customer security. In Germany, Savings Bank Dortman has introduced WAP-based services using MATERNA Information & Communication's WAP-based software and their Anny Way WAP gateway. This system allows any user with a WAP-enabled mobile telephone to request account balances as well as view the financial status of all their accounts and deposits. End users can make transfers and payments through their WAP brokerage service. Customers can request stock exchange indices and stock values, as well as buy and sell securities. All of these services are available through any WAP-enabled telephone.

8.6.4 MANAGEMENT APPLICATIONS

Memorex Telex Ireland's field sales force is using WAP technology to update its customer management database using WAP-enabled handsets. The system uses the Esat Digifone network and software designed by eWARE Limited. This system allows the user to access content and then update relevant information to the Customer Relationship Management (CRM) system. The Memorex system is isolated from the rest of the Memorex network and users must sign on through a separate firewall. The eWARE has its own separate application-level security that serves to secure the entire application.

The Memorex sales staff are now able to dial up current customer histories, pricing, or any other information that they may have had access to back at the home office that is necessary to service their customers. This new WAP-based CRM system allows the Memorex salespeople to concentrate on selling without the burden of administrative details because all of the information needed is available to them via their WAP-enabled telephone.

Phone.com, a developer of WAP-based software, has announced a software package for service providers: Mobile Management Server (MMS) version 1.0. This WAP-based system will enable service operators to provision their WAP gateways, applications, and handsets "over the air." MMS uses WAP's WTLS secure protocol to communicate with a handset; also, it uses a trusted provisioning domain mechanism to authenticate MMS to a handset. This version of MMS allows the service operator to remotely alter specific software settings and configurations of handsets once they have been placed in service.

8.6.5 GPS POSITIONING-BASED LOCATION SERVICES

Landstar Systems (a transportation services company) and PhoneOnline.com (a wireless software development company) have launched a WAP-based vehicle location and intermodal transportation service for over 8000 independent trucking operators. Three applications were put online via a WAP-based solution using a WAP-enabled handset (Nokia 7190). The first application was the Balance Inquiry application. This application allows a driver to access his account to determine the balance in his debit account. The amount can be read on his handset screen. The next application is the Check Call application, in which a driver can call the Landstar system to update his arrival at a customer location. The driver can enter arrival information, tractor number, trailer number, freight bill, current date, time, and location using his Nokia 7190 WAP mobile telephone. The third application is the Available Load application. This allows the driver to access the Landstar system to identify available loads that the driver can elect to pick up for his return trip back to his point of origin. This is a very valuable service because it allows the driver to gain revenue from a return trip rather than driving back home "empty" or "dead heading" as it is known in the industry.

8.6.6 WAP MOBILE WIRELESS MOVES AHEAD

While some observers have felt that mobile handset manufacturers would continue to produce voice-only handsets, particularly for Third World users, this has not been the case. The PCIA 2000 show seemed to indicate that manufacturers are moving ahead with WAP-enabled phones, some equipped with Bluetooth chip sets. For some manufacturers, China has proven to be their best customer with major purchases of WAP-enabled phones. Perhaps the feeling among some developing nations is that while every village cannot be equipped with PCs, at least one mobile handset might be available for Internet access.

Some observers estimate that there are over four million WAP-enabled phones in the United States alone, and 12 million in Japan. In Japan, NTT DoCoMo's WAP-based i-mode service has proven to be very successful, due to the fact that i-mode uses a cut-down version of HTML (compact "cHTML") and employs an "always-on" link to the Internet.

To keep pace with this rapidly evolving future, mobile service providers are rapidly upgrading their networks to support future foolproof methods for delivering wireless data services while overcoming bandwidth and ergonomic obstacles associated with mobile communications.

8.7 SUMMARY

8.7.1 THE FUTURE EXPANSION OF TECHNOLOGY

While the present adoption of WAP technology is still evolving, future WAP telephone designs will provide the required WAP improvements. However, there is a lot of hype going full throttle that would have one believe that nearly everyone will be WAP-enabled next week. For example, the Strategis Group in Washington, D.C. predicts that the sales of handsets with microbrowsers will grow by more than 900 percent to 7.8 billion by the year 2005. They predict that by 2005, more than 9.6 million people will have subscribed to 3G networks or 2.5G mobile high-speed data services. While all of this is encouraging, we still must wait and see which WAP-based applications come to the top that will encourage the widespread use of the Internet for purely data-based applications.

There is no doubt that some form of short messaging services will continue to prevail. At present, many vendors have reported utter amazement at the sale of PDAs and other short messaging devices. For example, short messaging devices are being sold to a very large subscriber base in the teenage and young-adult market, both in the United States and in Europe. These devices have quickly replaced pagers because they now provide two-way capability.

In other information areas such as stock market quotes, weather, location services, etc., it remains to be seen how quickly these services will expand into convenience services such as banking, which is already taking hold as a Web-based service via wired access.

Another area that shows promise for the future is in vehicle systems such as wireless enhanced "smart vehicles." Such systems are already making their appearance in GPS/cellular location systems such as OnStar, which provides location and direction services, as well as basic vehicle security support.

There are a lot of novelty Web-based services that have made their appearance; however, as more WAP-based systems become available that empower the user to better navigate the business world, there will be an expanded use of WAP-based services. Further, as these services along with Bluetooth-based services become more affordable and prove to increase personal productivity, there will be an observable increase in wireless access services of all types.

Part III

Networks and Architectures

9 User Mobility in IP Networks: Current Issues and Recent Developments

Björn Landfeldt, Jonathan Chan, Binh Thai, and Aruna Seneviratne

CONTENTS

0-8493-1502-6/03/$0.00+$1.50
© 2003 by CRC Press LLC

9.1   INTRODUCTION

The introduction of wireless networks to the Internet infrastructure will bring many changes to the way we use and relate to computers. Wireless networks have many potential advantages such as lowering of installation and deployment costs, but the biggest impact will come from users becoming mobile.

The mobility factor has proven itself one of the most-successful features in modern telephony. The advent of cellular telephony clearly showed the demand from users, who have been willing to pay a considerably higher price for telephony services if only they could be mobile. There has been unprecedented growth in customer bases in most if not all rolled-out cellular networks, and regions such as Scandinavia have almost full geographic coverage, making telephony ubiquitous. A similar success story can be seen in Japan, where NTT Docomo's i-mode system has brought mobile data services to the Japanese public. With i-mode, similar to pure cellular telephony networks, the growth in customer base has been unprecedented for data services.

The deployment of GPRS and 3G networks will bring packet switching to cellular terminals. This will create an integration of mobility and data services, and lay the foundation of the mobile Internet. Cellular technologies, as is the case of 2G networks, provide a wide range of coverage from local to wide area. The data rates of these networks are modest at present, but are expected to increase considerably over the next few years. However, it is difficult for cellular technologies to compete with wireless LAN (WLAN) solutions in terms of providing high data rates. The costs involved with the two technologies also are very different. Cellular networks are inherently more complex than WLANs and use licensed spectrum. Therefore, it is more costly to run traffic through these networks.

This has lead to the emergence of a market for WLAN in so-called hot spots. This market segment consists of areas of predictably high mobile user densities such as hotels, airports, and conference centers. At these locations, wireless coverage through the IEEE 802.11 WLAN standard is being rolled out and offered to the general public. This standard is being widely deployed in enterprise networks, as well as gaining momentum in the home network market segment. Together, the cellular and WLAN technologies constitute the base needed for the emergence of the mobile Internet. Users will have ubiquitous access through a variety of access networks as they move around geographically.

There are many forms of mobility and all play a role in the mobile Internet. By *user mobility* we mean the ability of users to either move geographically or to change access points in the Internet by either updating the destination IP address or to change the routing of packets to the destination address. There are other forms of mobility as well. For example, teleporting[1] is an example of presentation mobility where the applications are executed on one host but the presentation (screen output) can be moved between computers as the user requires. Another example is applications that can maintain states while hosts are disconnected from the network. For example, many FTP clients can maintain states if the network connection disappears and resume the file transfer when another connection is available even if the terminal has a new address. In this chapter, we focus on user mobility because it is fundamental and critical for the success of the emerging mobile Internet.

9.2 A CONTEMPORARY VIEW OF USER MOBILITY

Current user mobility support mechanisms can be divided into two categories: personal mobility and terminal (device) mobility. Personal mobility refers to users' ability to access network services from any terminal at any location. Thus, personal mobility management schemes enable the network to identify end users, as their point and method of access change.

Terminal mobility refers to the networks' ability to provide support for handover between networks for mobile devices as they change point of access while still maintaining connectivity.

9.2.1 Terminal Mobility

Terminal mobility support can be handled at different layers in the DOD reference model, starting from the link layer and finishing at the application layer.

In the existing cellular networks and emerging 3G networks, mobility is handled at the link layer. In doing so, the mobility is handled entirely by the access network and transparent to the outside. However, managing mobility at this level can only be done as long as the terminal stays within the same access network technology. If it were to attach to a different type of access network, the mobility management would fail.

Generally speaking, mobility management can be better optimized the lower in the protocol stack. However, the lower the layer, the more specialization is required, with the ensuing increases in complexity and limitations in scope. Therefore, there is a trade-off between optimization, complexity, and functionality that has to be considered when deploying terminal mobility.

In future all-IP mobile networks, the trend is to use a combination of link layer and network layer mobility management to provide ubiquitous access and global roaming. The cellular industry is currently deploying link layer tunneling solutions for cellular networks and the IETF (Internet Engineering Task Force) currently supports a network layer solution, Mobile IP,[2] as the global mobility management scheme.

9.2.1.1 Network Layer Mobility

The current form of IP, IPv4, cannot support terminal mobility. The cause of the problem is the double meaning of IP addresses, which can be interpreted as both the endpoint identifier and the topological location of a terminal. Unfortunately, the initial IP design did not take mobility into account. The initial design was made with the idea that in connectionless IP networks, it is necessary to deploy a hierarchical addressing scheme[3] to make routing simpler and more manageable. This solution allows routers to only maintain routing information about the local topology and to use a fallback forwarding mechanism for all traffic destined outside this local topology. Although such an addressing scheme provides a scalable solution for routing data across large internetworks, there is a serious implication with locking the topological knowledge to only local routers. Should a host move from its local network to a foreign location, the foreign routers would not have any rules for forwarding traffic to the host. Therefore, it will use the fallback mechanism to forward the traffic toward the local network specified in the terminal's address, and even if the local routers know which foreign network a host is attached to, they cannot forward the traffic there because the foreign routers would only return the traffic back to them.

In mobile environments, users can freely move from one network to another and therefore this restriction on address usage is violated. One possible solution is to assign a new IP address to the mobile terminal when it arrives at a new subnet. This approach, however, can create problems in the transport and application layers, where an IP address serves as part of an endpoint identifier. For instance, a TCP connection is identified by the source/destination addresses and port numbers. If any one of these four components in the identifier changes, the ongoing TCP connection will be broken. For a UDP data session destined to a mobile host, it is possible to update its endpoint address whenever the mobile host moves. Despite its feasibility, such an arrangement implies that user applications need to be mobility aware. Unfortunately, it is unlikely that this global awareness of user mobility will become a reality in the near future.

Because IP addresses contain implicit location information, in order to support terminal mobility at the network layer, either the IP forwarding mechanism needs to be changed or the addressing scheme has to be modified. There are three primary alternatives to achieving this goal.

1. IP encapsulation (tunneling): This process involves the technique of encapsulating a packet, including the header, as data inside another packet. Because the header of the new packet, i.e., the tunnel header, has a topologically correct IP address, this packet can follow the standard IP routing mechanisms and reach the IP subnet where the mobile terminal is attached. This method has been widely studied and adopted as the data-forwarding mechanism in the IETF Mobile IP.[2]

2. Loose source routing: As an option in an IP packet header, loose source routing enables the sender to perform address translation operations. The source generates a list of addresses of intermediate routers it wants the packet to pass on its way to the destination, with the last entry being the current address of the mobile terminal. When building a normal IP packet, the destination field is filled with the address of the mobile terminal. When using loose source routing, this field is assigned instead the first entry in the address list. When the packet reaches this node, the destination address

is replaced with the next address from the list. This process is repeated until the packet reaches the mobile terminal. This function has been carefully integrated in IPv6 using a routing extension header, which avoids current problems in IPv4 with regard to security and performance.[4]

3. Dynamic per-host routing: In this routing scheme, the destination IP address is used as a terminal identifier only, removing its association with the terminal's current location. Packets are forwarded on a hop-by-hop basis from a gateway over special dynamically established paths to the terminal. The forwarding entries at each router along the path are refreshed periodically using update messages sent from the terminal. This category of packet forwarding has been proposed lately in some micro-mobility architectures, which we will discuss later in this chapter.

The main difference between these three routing schemes is the way location information is placed. In the case of tunneling, it is embedded within the packet payload; with loose source routing, it is provided in the packet header; and in dynamic host routing, it is maintained in the forwarding table of each intermediate router.

9.2.1.2 Mobile IP

The current supported IETF standard for terminal mobility in the Internet is called Mobile IP. In this approach, a fixed terminal (corresponding host, CH) that wants to communicate with another host (mobile host, MH) is unaware if the other host is in its home network or is away in a different (foreign) network. This transparency is provided by using two network agents, one located at the mobile host's home network (home agent, HA) and the other located on the visited network (foreign agent, FA). These mobility agents (i.e., HA and FA) and the MH cooperate with each other and perform mobility management without any other modifications to the network. The functionality of Mobile IP can be roughly divided into three components (Figure 9.1): (1) location registration, (2) packet forwarding, and (3) handover detection.

9.2.1.2.1 Location Registration

Mobile IP adopts a simplistic approach to location management. For example, in Mobile IP there is no terminal paging algorithm. Instead a location update is executed every time a mobile host arrives at a new subnet. In addition, Mobile IP does not use databases to store user location information, but performs location updates by creating or modifying a mobility binding at the HA. When a mobile host moves to a foreign network, it registers with the FA on that network and obtains a care-of address (COA). The mobile host then updates its current COA, possibly via the current FA, by sending a registration request message to its HA. After receiving this message, the HA associates the mobile host's home address with its current COA via a binding cache. This mobility binding is automatically deleted from the HA if the lifetime of the binding expires without receiving any new registration from the mobile host.

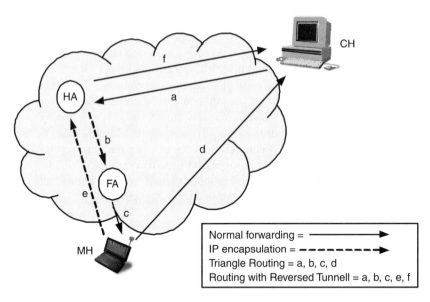

FIGURE 9.1 Mobile IP components.

9.2.1.2.2 Packet Forwarding

The packets destined to a mobile host are always forwarded to this mobile host's home network because the CH is not assumed to be mobility aware. In the case that the mobile host is currently away, the HA intercepts the packets, encapsulates them based on its binding cache and relays them to the FA currently serving the mobile host (the COA). When these packets arrive at the FA, it decapsulates them and forwards them to the mobile host via the local link layer technology.

It is worth noting that packets in the reverse direction can be delivered in two different ways, depending on the level of security the foreign network implements (see Figure 9.1). If routing is independent of source address within the foreign network, the mobile host can send packets directly to the CH. This routing asymmetry between the corresponding and mobile hosts is known as "triangular routing." On the other hand, if source-filtering routers are installed in the network (i.e., the routers check the source address of packets for correctness), they will drop all packets originating from the mobile host because its source address is topologically incorrect. One possible solution to this problem is to establish a reverse tunnel from the mobile host to its HA so that all tunneled packets bear a correct source address.[5] When these packets arrive at the HA, they will be decapsulated and forwarded to the CH.

9.2.1.2.3 Handover Detection

Using a tunneling mechanism, the HA can easily reroute mobile connections to the current location of a mobile host provided its binding cache is up-to-date. Mobile IP specifies some generic mechanisms for mobile hosts to discover FAs without assistance from the link layer. In essence, the FA advertises its availability through periodically transmitted router advertisements. The mobile host can detect that it has moved from one subnet to another in one of two ways. First, the mobile host can use the lifetime field of an FA advertisement to refresh its association with that FA. If the lifetime expires before receiving another advertisement, the mobile host will attempt to register with a new agent. Second, the mobile host can compare the subnet prefixes of agent advertisements. If the prefixes differ from the current care-of address, the mobile host will assume that it has moved.

It is worth noting that because agent advertisements are either broadcast or multicast in nature, they cannot be transmitted too often as they consume radio resources. In the older Mobile IP specification,[2] the maximum frequency of advertisement is only once per second, but in the revised version[6] this limit is lifted (i.e., unspecified), and FAs can be discovered by a link layer protocol.

9.2.2 PERSONAL MOBILITY

Presently, "all-in-one" mobile devices are being introduced in the marketplace. These devices integrate diverse network accesses and general purpose operating systems so they can be used both as cellular phones and as hyper-portable computers. Even if these devices are versatile and convenient, it is unlikely that they will satisfy a user's every need. Existing devices such as desktop PCs are difficult to completely replace with such a mobile device due to the difference in power constraints, screen size, and CPU capability, etc. Therefore, rather than relying on a single device for all activities, users will continue to use several devices for different purposes.

With a combination of fixed and mobile devices, terminal mobility support alone will not be sufficient to address user mobility as the user is moving and switching between different devices. Therefore, personal mobility will be of fundamental importance in future communications.

Personal mobility support has not been addressed as much as terminal mobility and, as a consequence, it is not as well defined as the role of terminal mobility. To date, two distinct roles have emerged together constituting a user's networked presence. The first area addresses user location and means of contact; the second area addresses the issues associated with personalizing a user's presence and describing the user's operating environment. Therefore, the area can be defined as personalization.

In traditional telecommunications systems, devices (or telephones) have been identified through a hierarchical numbering scheme. It has been possible to map the ID of the device to its current point of attachment, and therefore make the device reachable using this scheme, even if mobile. Similarly, personal mobility management schemes need a mechanism for allowing users to change their point of presence while still being reachable.

The user has to use a globally unique identifier for the scheme to work. The identifiers have to be resolvable to identify an anchor point for personal information retrieval. Possible candidates are, for example, telephone numbers, e-mail addresses or URLs. Consider an e-mail address such as user@domain.com. From this address it is possible to extract the user name (user) and the anchor point (domain.com).

With personalization, the presence information can be used for a number of purposes by the user, network services, and peers. For example, the presence information

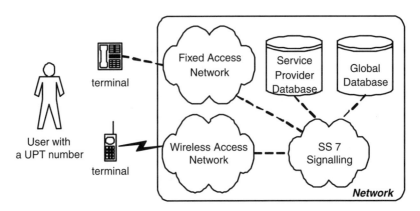

FIGURE 9.2 The architecture of UPT.

can be used to identify the device where the user currently can be reached. Different peers can obtain different results according to the user's preferences so that, for example, some peers can reach the user for personal communication whereas other users will obtain an alternative device such as an answering machine or e-mail inbox. Another use of presence information is to obtain the characteristics of the device the user is currently using to determine if some means of communication is possible. For example, capability information can be used to determine if the user device is capable of displaying video or whether only voice communication is possible.

Despite the fact that personal mobility and presence has not yet been thoroughly investigated, a couple of systems for supporting personal mobility have been standardized: Universal Personal Telecommunication (UPT)[7] by the ITU-T and Session Initiation Protocol (SIP)[8] by the IETF.

9.2.2.1 Universal Personal Telecommunication

UPT provides access to telecommunications services while allowing personal mobility. It enables each UPT user to initiate and receive calls on the basis of a personal, network-transparent UPT number across multiple networks, regardless of the type of terminal or geographical location.

In the architecture (Figure 9.2), the user has a personal UPT number subscribed to a UPT service provider. Each UPT service provider maintains a UPT service provider database that tracks the location of a user through registrations. A UPT user can register with the database at either a fixed or wireless terminal. To support service provider portability of UPT numbers, a UPT global database registers the UPT service provider for each assigned UPT number. When a UPT user changes UPT service provider, the new service provider can notify the UPT global database administrator to update the database.

A UPT user may register to separately receive incoming calls and originate outgoing calls at any specified terminal according to a service profile. The service profile includes identification and authorization information that can be used to allow or disallow making or receiving calls from a UPT terminal.

9.2.2.2 SIP

Session Initiation Protocol (SIP) was developed to assist in establishing, maintaining, and terminating advanced telephone services between two or more users across the Internet. It is part of the IETF standards process and is modeled on HTTP. The protocol supports personal mobility by providing the capability to each called party at a single, location-independent address.

SIP is based on client/server architecture. The main entities are user agent (UA), the SIP proxy server, the SIP redirect server and the registrar.

The user agents are the SIP endpoints. They operate as clients when initiating requests and as servers when responding to requests. A UA can communicate with another UA directly or via an intermediate server, and also store and manage the states of a call. SIP intermediate servers can act as proxy or redirect servers with the following functions: (1) proxy servers forward requests from the user agent to the next SIP server or user agent within the network; (2) proxy servers can maintain information for billing and accounting purposes; and (3) redirect servers respond to client requests and inform them of a requested server's address.

The final entity in the SIP architecture is the SIP registrar. The UA sends a registration message to the SIP registrar when the user location needs to be updated. The registrar stores the registration information in a location service via a non-SIP protocol and sends an appropriate response back to the user agent.

When a user (caller) wants to place a call to another user (callee), the process is initiated by the caller issuing an invite request. The request contains enough information for the callee to join the session. There are two possible sequences of events in issuing the invite request. If the caller knows the address of the callee, the invite request is sent directly the callee's UA; otherwise the invite request is sent to a SIP server (Figure 9.3).

The type of SIP server determines its response to the caller's invite request. If the server is a SIP proxy server, it will try to resolve the callee's location and forward the request to callee's UA; however, if it is a SIP redirect server it will return the location of the callee to the caller after the location resolution process, which enables the caller to send the invite request directly to the callee. When locating the callee, a SIP server can proxy or redirect the call to additional servers until the callee is located.

Once the request has arrived at the callee, several options are available. In the simplest case, the callee will be notified that a call has arrived. For example, the phone rings. If the callee answers the call, the callee's UA will respond to the invite and establish a connection; if the callee declines the call, the session can be redirected to other entities such as a voice mail server or another user.

SIP has two additional significant features. The first is a SIP proxy server's ability to split incoming calls so that several extensions can be run concurrently. This feature is useful if a callee may potentially be at two different locations. The second feature is the protocol's ability to return different media types in response to requests. For example, when a caller is contacting a collee, the SIP server that receives the connection request can return the caller an interactive Web page instead of a busy tone.

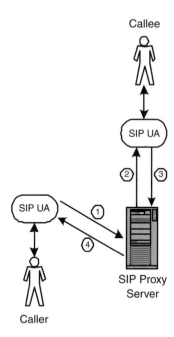

FIGURE 9.3 Call establishment with SIP servers.

In the following sections, we will give a brief overview of other proposals for supporting personal mobility.

9.2.2.3 Personal Mobility Systems that Support User Location

The following presence systems have been designed to support user location management, including UPT and SIP.

- The Mobile People Architecture (MPA)[9] attempts to enable users to be contacted from anywhere, using a variety of communications media, such as e-mail, telephones, and instant messages. This is facilitated by identifiers called personal online IDs (POID), together with a personal proxy located at the user's home network. The POID provides a way of uniquely identifying the user, and the personal proxy takes care of the user's movement, preferences, and any required protocol and content conversions. As all call signaling is required to go through the personal proxy, the location of the user and the device that she is currently using can be hidden from other users.
- ICEBERG[10] provides similar functionality to MPA. However, it was primarily designed to integrate different types of networks with the Internet. This enables, for example, the user with a cellular device to be contacted by others on the Internet and vice versa. Under such circumstances, content needs to be adapted to suit the characteristics of the user's network.

This is performed by ICEBERG Point of Presence (iPOP). Because it is based on the Ninja clustering platform,[11] it provides an execution environment, where adaptation libraries can be downloaded and configured on demand.

9.2.2.4   Personal Mobility Systems that Support Personalization

The following architectures were designed to support personalization:

- NetChaser[12] is a mobile-agent-based framework that is designed to support personal mobility in accessing three Internet services: HTTP, e-mail, and FTP. The mobile agents in NetChaser form a wrapper layer between the applications (Internet clients and servers) and the network. They assist the users by following them when they change working terminals. Interaction with the user is achieved by using a Web browser.
- Secure and Open Mobile Agent (SOMA) is an architecture where mobile agents are used as middleware to support mobile computing, which includes both terminal and personal mobility.[13] Personal mobility is supported in this architecture through the use of user virtual environment (UVE). The UVE service lets users connect to the Internet at different locations while maintaining the personal configurations indicated in their user profiles.
- In the Telecom Research Programme (TELEREP), a mobile-agent-based architecture to support mobility was introduced.[14] The architecture uses various mobile agents to perform different tasks. A user agent (UA) acts on the user's behalf when the user moves between different networks and controls the other agents. The application agent (AA) and the data agent (DA) maintain the user's applications and data by moving them close to the user. Finally, the profile agent (PA) stores the user's profiles. When a user moves from the home network, all the above-mentioned agents will migrate as well, carrying with them the applications, data, and profiles. This allows the user to access applications and data locally.
- Using the technique of application migration, Takashio and coworkers introduced a framework called follow-me Desktop, or f-Desktop.[15] In this framework, the context of so-called follow-me applications is mobile. The follow-me applications can be implemented using mobile agent technologies. When a user changes terminal, the user's applications are "frozen." The frozen applications are transmitted across the network to the user's new terminal and resume their execution there.
- None of the above-mentioned personal mobility architectures support both user location and personalization. The Integrated Personal Mobility Architecture (IPMOA)[16] addressed this issue by introducing an overlay network that caters for various personal mobility services such as interpersonal communications (location support), customized Internet services, remote application execution, and file synchronization (personalization support).

9.3 CHALLENGES AND RECENT DEVELOPMENTS OF TERMINAL MOBILITY

At present, we experience the early stages of the emerging mobile Internet and the road has not yet been mapped out. The work that has been proposed from different research groups has largely addressed the fundamental issues of user mobility and many areas still remain relatively untouched. As can be expected, the industry lags behind the research community and there is no universal consensus on standards for mobility management.

To date, the predominant picture is to use Mobile IP for global addressing and specific link layer technologies for certain access technologies, e.g., GPRS Tunneling Protocol (GTP) in UMTS networks. However, because the rollout of mobile data networks has just begun, there is little experience with the difficulties and properties of mobile computing. Therefore, over the next few years as the mobile experience grows, many new issues will be revealed and problems that only have been touched by researchers to date will be thoroughly investigated.

9.3.1 MOBILE IP ENHANCEMENTS

Although Mobile IP is a simple and scalable solution for IP mobility, it suffers from performance and security problems and has a number of drawbacks, especially when serving users with high mobility and quality of service (QoS) expectations. Currently, there are many enhancements being proposed to Mobile IP, and these proposals are summarized in the following sections.

9.3.1.1 Route Optimization

Because all packets destined to a mobile host have to be routed through its HA, the chosen path can be significantly longer than the direct route. In a QoS supported mobile network, a longer path and the ensuing delay will produce a higher probability of call dropping and service refusal. To rectify this problem, the extension of route optimization in Mobile IP[17] provides a means for corresponding hosts to cache the actual location of a mobile host so that their packets can be tunneled to the mobile host directly.

The IPv6 base specification incorporates this idea. In addition, it replaces the tunneling mechanism with loose source routing that incurs less delivery overheads.[18] Thus, Mobile IPv6 overcomes the previously mentioned shortcoming of Mobile IPv4.

Another approach to this problem is to introduce the concept of a location server. Through the location server, the mobile host can update its current location and the CH can then query the server for the current location of a mobile user before transmitting a packet. Because the CH knows the actual location of an MH, both triangle routing and tunneling can be avoided. The associated call setup and signaling protocols can be implemented by either changing the Mobile IP protocol (e.g., MIP-LR[19]), setting up user agents at the application layer (e.g., the SIP with mobility support[20] and session layer mobility management[21]), or modifying the upper layer protocol (e.g., the MSOCKS[22] and the end-to-end approach to host mobility[23]).

It is worth noting that in order to enable optimal routing, all the above proposals have to introduce mobility awareness in the corresponding hosts. In particular, it requires either modifications to the IP protocol stack (binding cache followed by tunneling or loose source routing), or the addition of a location server and associated signaling protocols. Unfortunately, a global deployment of these enhancements will not be available in the near future.

9.3.1.2 Frequent Handover and Fast Location Updates

Mobile IP does not mandate fast handover and location updates, and hence there are several problems when serving highly mobile users. Each time a mobile host moves from one subnet to another, it needs to register its new location with the HA. Thus, if the visited network is some distance away from the home network, the signaling delay for reregistrations can become large and, consequently, many packets could be misrouted.

In addition, Mobile IP does not require a mobile host to inform its previous FA when it moves. Therefore, the previous FA is not able to reroute packets to the current FA. The extension of routing optimization in Mobile IP[17] allows the misrouted packets to be tunneled from the old to the new foreign agent. However, a drawback is that the mobility agents need to deal with many associated security issues.

Another disadvantage is that when intersubnet handovers occur, the mobile host obtains a new COA and thus packets destined to the mobile host will be encapsulated with a different tunnel header. Even if we assume that network resources can be dynamically reallocated across a QoS-capable Internet backbone during handovers, such rerouting and resource allocation processes may take a long time when the HA and the FA are far apart (see Figure 9.4(a)).

Lately, in addition the work on Mobile IPv6, there have been attempts in the Internet community to resolve these problems by introducing the concept of micro mobility. This concept is a promising approach to efficiently manage high user mobility within a single administrative domain (see Figure 9.4(b)). In micro-mobility schemes, (1) location updates within a domain are handled locally, thus avoiding frequent reregistrations across the Internet to the HA; and (2) by using a specific packet delivery scheme within the region, it prevents the costly reestablishment of end-to-end routing between the HA and the FA. This regional framework is normally coordinated by a domain gateway, which serves as the interchange entity for mobility management within a domain (micro–mobility) and mobility management across different domains (macro–mobility).

9.3.1.2.1 Micro Mobility

So far, only Mobile IP has been considered as the solution for macro-mobility management. On the other hand, many types of regional network architectures have been proposed for micro-mobility management. The proposals can be divided into four categories:

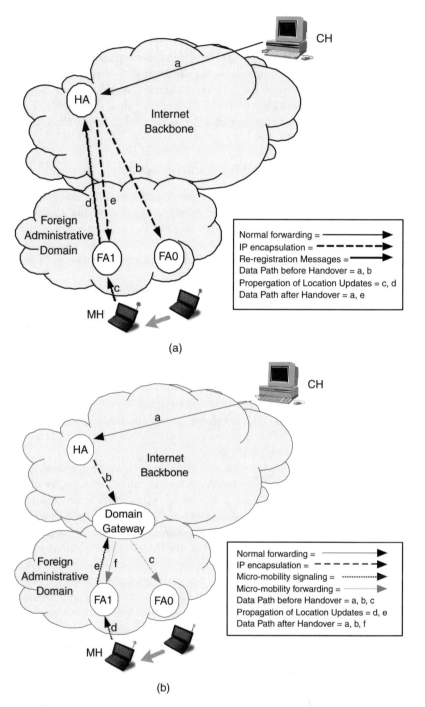

FIGURE 9.4 An illustration of the necessity of micro-mobility management in IP networking: (a) signaling and rerouting without the support of micro mobility; (b) signaling and rerouting with the support of micro mobility.

1. Cascade tunneling: This has been proposed as a solution to location registration latencies resulting from large distances between mobile hosts and servers. The solution is to perform registrations locally in the visited domain through a hierarchy of foreign agents with tunneling between them as described in Gustafsson and coworkers.[24] This approach reduces the number of signaling messages to the home network and shortens the signaling latency when moving from one foreign agent to another. However, the scheme requires the introduction of new registration messages for local mobility. Furthermore, a special gateway entity (gateway foreign agent) is required to handle and transform these regional registrations, and also to dynamically manage regional tunnels for the mobile host in both the forward and reverse directions. As a result, changes to both HA, FA, and MH are necessary. The changes ensure that the HA will always see the gateway foreign agent as the current location of a mobile host, regardless of which FA is serving the mobile host in the visited domain. The cascade tunneling approach also can be found in other proposals such as Regional Aware Foreign Agent[25] and Transparent Hierarchical Mobility Agents.[26] Instead of modifying those well-defined mobility agents of Mobile IP, these proposals introduce several new entities in their frameworks for handling the regional tunnel management.

2. Dynamic per-host routing: In a limited geographical area, i.e., between different subnets within the same management domain, the concept of dynamic per-host routing can be deployed as an alternative to regular IP routing. Problems associated with scalability and compatibility can be minimized because of the limited scope and single ownership of the management domain. With this scheme, IP addresses have no location significance inside the domain, and therefore neither tunneling nor address conversion is necessary during packet delivery. For example, HAWAII[27] uses path refresh messages to establish and update host-specific forwarding entries in routers between a gateway entity called the domain root router and the base station. In HAWAII, the role of a base station is twofold. First, it emulates an FA for replying to Mobile IP registration messages, thus making HAWAII entities transparent to mobile hosts that use Mobile IP. Second, it converts Mobile IP registration updates into HAWAII refresh messages, which in turn either revives or creates new forwarding entries in the routers, depending on whether a subnet handover has taken place. A similar approach is used in Cellular IP,[28] which also requires special routers that can set up, refresh, or modify host-specific forwarding entries using control packets. Besides control packets, routers utilize user data packets to refresh forwarding entries. To cater to large-scale deployment, special paging packets and paging caches are integrated in such a way that the gateway router can efficiently locate any idle mobile hosts.

3. Overlay routing: Regional overlay routing applies an overlay model where IP packets are either segmented or encapsulated into another packet format for local delivery. Because the data forwarding mechanism is no longer IP-based, address conversions need to be done at the gateway entity and

the base stations, and all mobility support issues have to be resolved by the overlay network. One example of this approach is IP over Mobile ATM,[29] where IP packets are segmented into ATM cells and delivered by virtual connections between the gateway and a base station across a mobility-enabled ATM network. When the MH roams from one base station to another, its ongoing virtual connections need to be rerouted to its latest location. Another example of overlay routing that can be considered is multiprotocol label switching (MPLS) for Mobile IP,[30] where IP packets are encapsulated with a label that directs the forwarding path to a base station. In fact, this is very similar to the cascade tunneling scheme mentioned earlier, except that the regional IP tunnel is now replaced by a label-switched path across the MPLS domain.

9.3.1.3 Tunneling across QoS Domains

Even though tunneling has been adopted as the standard mechanism for redirecting packets in Mobile IP, there are certain constraints if it is used in conjunction with the currently developed QoS frameworks. For example, when the protocol that has been adopted by the IETF Resource Reservation Protocol (RSVP) is applied to Mobile IP it is assumed that RESV messages follow the inverse path of PATH messages. However, this is not the case for the base specification of Mobile IP which results in triangle routing.

Another incompatibility comes from the IP-in-IP encapsulation. The insertion of a tunnel header offsets the packet payload, which prevents the fields in the transport and higher layers from being accessed normally. When RSVP signaling messages enter a tunnel, they are encapsulated with a tunnel header that carries an IP-in-IP encapsulation rather than a router-alert option. Consequently, RSVP-capable routers cannot recognize the packets, and resources are not reserved accordingly.

Moreover, even if the required resources could be reserved, the intermediate RSVP routers will not be able to access port numbers correctly in order to distinguish data packets belonging to different flows. Therefore, it will not be possible to honor the per-flow state resource reservations.

To resolve these problems, an RSVP tunneling algorithm has been proposed in Terzis et al.[31] This scheme passes end-to-end RSVP messages transparently (i.e., without reservations) between tunnel endpoints and instead the tunnel ingress and egress nodes are responsible for generating additional RSVP signaling to reserve resources between them. When data packets arrive at the tunnel ingress node, they are wrapped with extra IP and UDP headers such that intermediate routers can apply a standard RSVP filter specification to map a packet to the appropriate reservation. Because the source and destination IP address and the destination UDP port number (being assigned as a constant value of 363) are identical for every flow inside the tunnel, a unique source UDP port number is used to differentiate packets from various flows within the tunnel. However, this approach results in further complicating both the signaling and encapsulation at the tunnel endpoints. Moreover, it considerably increases the overhead of transferring small packet payloads such as voice data.

RSVP is proposed to be used with and tightly connected to the Integrated Services (IntServ) architecture, and it is considered as the signaling protocol to be used with Differentiated Services (DiffServ) as well.[32] The same shortcomings when combining QoS and tunneling will apply to both the IntServ and DiffServ environments. In the case of a DiffServ infrastructure without RSVP signaling, tunneling poses fewer problems because the DiffServ code point (DSCP) can be copied forwards and backwards between the tunnel header and the original IP header when encapsulation and decapsulation take place. However, in certain networking scenarios when path- or source-dependent services are desirable, multiple field (MF) classification has to be invoked at the ingress and egress DiffServ routers.[33] Similar to RSVP-compliant routers without modifications, these DiffServ edge routers will not be able to access the higher layer information in the packet payload due to the extra location offset created by the tunnel header, thus MF classification cannot be performed properly.

9.3.1.4 Link Layer Assisted Handover Detection

The base specification of Mobile IP was designed to be independent of the underlying link layer technology. However, because of its passive approach to handover detection, the registration process when moving from one FA to another can be long enough to cause problems, especially for real-time communications and reliable data transfers using TCP.[34,35]

Recently in the IETF, there have been at least two proposals that couple link layer functionality with Mobile IP in an attempt to minimize service disruption experienced by a mobile host when moving between foreign agents. In the Fast Handoffs draft,[36] the movement of an MH is anticipated from the link layer information, and simultaneous bindings are used to send multiple copies of packets to different potential foreign agents. In the FA Assisted Handoff draft,[37] the FA takes a proactive approach to manage handover events. When an FA is aware of a handover occurring at the link layer of its current cell, it sets up a mobile host's visitor entry and issues the handover messages on behalf of the mobile host to the next FA. As a result, packets can be forwarded from the current FA to the next FA prior to receiving a formal registration request at the network layer. Unfortunately, these handover proposals assume that the identity of the new FA is known to the MH. If there are multiple foreign agents appearing at the link layer, these proposals do not provide a solution to choose the one the mobile host will select.

9.3.1.5 Discussion of Mobile IP Enhancements

Future IP mobility frameworks need to consider the QoS constraints of active connections more closely when handling the usual requests of handover and rerouting. Several optimizations can be done to improve the overall mobility performance:

- The handover latency could be improved significantly by tightly coupling the IP layer with the link layer to give "hints" of potential handovers.
- It is beneficial to forward packets directly between the corresponding and mobile hosts, as it enables the resultant path to be optimized for quality

of service. However, it is unclear if and when all corresponding hosts would become mobility aware to provide this service.

- It would be advantageous to assign a gateway entity near the mobile host to handle micro–mobility. Techniques such as cascade tunneling, dynamic per-host routing, and overlay routing can be realized with such a gateway. However, it is still debatable as to which approach is the most appropriate. The most obvious consideration is that if tunneling is used to redirect packets in the future wireless Internet environment, its integration into the IntServ and DiffServ framework demands more attention and new solutions.

9.3.2 HIGHER-LAYER MOBILITY MANAGEMENT

One of the visions of future mobile communication is that of a multiaccess environment in which terminals will be attached to several different overlaid access networks simultaneously. In a multiaccess scenario, it is possible to select network interfaces for separate connections. It is likely that users will have access to several different networked devices. If this becomes a reality, it is likely that future services will emerge where it is possible to hand over sessions between different terminals.

These functions and others are, at the very best, cumbersome to solve in the network layer and therefore some proposals have been made that enable these functions by performing IP mobility management in layers above the network layer.

The MSOCKS proposal[22] places the mobility management function in the transport layer. This proposal is optimized for local mobility management within a single access network. The mobility management is carried out through splicing different socket endpoints together in the kernel of an intermediate node. The binding between different sockets is updated as clients move around and make address updates.

Another proposal, Session Layer Mobility Management (SLM)[21] places mobility management in a layer above the transport layer. This proposal lifts transport protocol dependency by maintaining states outside of sockets. This way, the data delivery can be totally separated from IP addresses and even socket endpoints. This in turn enables easy integration of intermediate nodes such as PEPs and handover between different devices. SLM uses a location server for both users and terminals; this way, it is possible to resolve user identifiers to terminals and to follow these terminals as they move and change addresses.

A final proposal[23] advocates changes to the TCP stack on end hosts to include mobility support. As is the case with MSOCKS, this proposal is transport-protocol-specific, but the mobility management is carried out end-to-end without requiring any intermediate node.

All these proposals require significant changes to end systems. However, it can be argued that software updates are necessary in any case, and that this is an ongoing process. If IPv6 is to be implemented or if RSIP is to be used, end hosts have to be updated. Similarly, if QoS support will be offered in access networks, end hosts will have to be upgraded with this support.

9.3.3 ENHANCEMENTS TO SUPPORT CONVERSATIONAL MULTIMEDIA

Traditional circuit switched telephony networks were designed and optimized to support voice services. Contrary to this, the packet-switched data network paradigm introduces a network that is able to support a variety of services and is optimized for maximum utilization. QoS management is an important building block to support classes of services with strict requirements in terms of parameters such as bandwidth, delay, and jitter from the network. In a fixed networking environment, resource reservations will suffice because the network behavior is predictable, but a wireless link will always suffer from varying conditions due to factors such as noise, handover latency, and overcrowded cells. However, interactive and especially conversational media require predictable network behavior even when the host is mobile.

The measures taken to alleviate these problems in mobile networks can be divided into two orthogonal classes, proactive and reactive. Proactive measures try to predict the needs of applications and make advance measures on their behalf. Network resource reservation is one of the measures in this category, but at handover QoS management has to be extended with additional functions to guarantee a smooth transition to the new cell.

9.3.3.1 Advance Resource Reservation

It is generally difficult to promise a specified level of QoS to mobile uses in mobile environments, because there may not be enough resources in the part of the network into which the mobile user is moving. Moreover, during handovers the ongoing traffic is likely to be disrupted, which can violate some of the previously agreed upon QoS parameters such as packet delay, jitter, and loss.

To date, there have essentially been two methods proposed to make QoS support more mobility aware. The first method focuses on the preallocation of network resources in locations where the mobile host is likely to visit. Preallocation can improve the continuity of a connection after a move, and reduce packet losses and latency of resource allocation. However, preallocation requests can fail under severe network congestion because there are simply no resources available for reservation. Moreover, even after preallocation, signal fluctuations of the wireless link can contribute to the failure of QoS guarantees. Under such conditions, it becomes necessary to take an alternative approach.

The second approach emphasizes the adaptivity of end systems, where their application, middleware or proxy, reacts to the changes of network resources caused by wireless channel fading and user mobility. This functionality can be achieved by means of session customization at various places. In the Internet environment, adaptive end systems have to rely on mechanisms that probe the network in order to avoid congestion. Unfortunately, this action is likely to cause traffic interruptions before congestion can be avoided on a long-term basis.

From this discussion, we believe that the first method (i.e., advance reservation) is a proactive approach to deliver some level of QoS guarantees (at least statistically) to the mobile users, whereas the second method (i.e., end system adaptivity) is a reactive approach to cope with changes of QoS and, as such, they are complementary.

9.3.3.1.1 Current Approaches of Advance Reservation

Advance reservation needs to deal with two nontrivial problems, namely, how to configure resources in advance, and where to preallocate resources for mobile devices.

How to Configure Resources in Advance when Mobile: Advance reservation was originally considered in Wireless ATM research.[38–41] Because of the recent interest in providing mobile QoS in an IP environment, the research community has addressed this issue using a combination of Mobile IP and IntServ models. Based on the topology used for reserved data paths, these proposals can be classified into three categories as shown in Figure 9.5.

1. Preconfigured anchor rerouting: The MRSVP protocol[42] is an example of preconfigured anchor routing. In MRSVP, an MH specifies and dynamically maintains a set of locations (known as the MSPEC), from which it wants to make advance reservations to its HA (i.e., the anchor point). Special routing entities, called RSVP proxy agents, are provided at the locations specified in the MSPEC to make reservations on behalf of the mobile host. To allow for better link utilization, reservations made by these proxy agents allow resources to be temporarily borrowed by lower priority services. The reservations are either classified as active or passive, depending on whether the reserved resources are strictly used for a data flow or if they can be temporarily borrowed by other services. Of all proxy agents associated with an MSPEC, only the one currently serving the mobile host is allowed to make active reservations. The others will remain capable of making passive reservations until the mobile host moves into their wireless region. Similar mechanisms also can be found in other proposals such as RSVP-A[43] and Mobile Extensions to RSVP.[44]

2. Preconfigured path extensions: An example of preconfigured path extensions is Advanced Reservation Signaling.[45] This scheme uses a concept of passive reservations similar to MRSVP. However, instead of making multiple reservation paths connecting the HA with other foreign agents, the advance reservation signaling simply extends the existing RSVP data path from the current position of a mobile host to all its adjacent base stations.

3. Preconfigured tunneling tree: The proposal in Terzis and coworkers[46] is an example of preconfigured tunneling tree schemes. It does not rely on the notion of passive reservations. Instead, it requires RSVP-capable tunnels[47] to be established between the HA and other foreign agents. Unlike ordinary IP encapsulations, these RSVP tunnels are preprovisioned with certain levels of resources while accommodating multiple end-to-end RSVP sessions. When the resources consumed by mobile hosts visiting an FA exceed the reserved amount, the FA can request an incremental block of resources to be added to the RSVP tunnel.

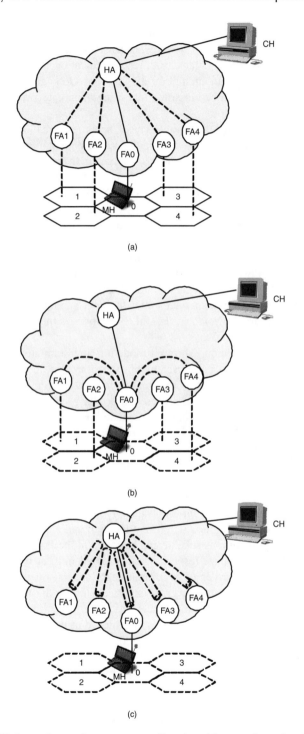

FIGURE 9.5 Various schemes for resource preallocation: (a) preconfigured anchor rerouting; (b) preconfigured path extensions; (c) preconfigured tunneling tree.

Where to Preallocate Network Resources for Mobile Devices: It is difficult to determine where to preallocate resources, because of the difficulties of predicting user movements. Despite this, many resource preallocation algorithms have been proposed in the literature as an attempt to safeguard the QoS agreements for mobile services. The algorithms can be classified into three main categories.

1. Neighborhood-based allocation: These schemes preallocate network resources between an anchor node and a set of base stations surrounding the MH. The number of base stations involved in the preallocation process depends on how far ahead in time the network is willing to support a mobile service. For example, Virtual Connection Tree[48] configures resources in advance between a root switch and each base station in the management domain upon the admission of a call. This implies that the network is willing to support this mobile host as long as it stays within the domain, but the network also may have low utilization of resources. In contrast, Advanced Reservation Signaling[45] reserves resources only between a, MH's current location and all adjacent cells. Thus the network guarantees continuity of services after the next handover, but its further commitments are subject to successful reservations at the new neighboring cells.

2. History-based allocation: Through modeling and simulation, many proposals have shown that the user mobility history can be helpful in predicting the future movements of a mobile host. Depending on the service commitment to mobile users, these proposals preallocate resources at various levels in advance along the predicted path. For instance, in order to obtain mobility-independent service guarantees, MRSVP and other similar protocols attempt to make resource reservations at each location a mobile host may visit during the lifetime of a session. The Shadow Cluster concept,[49] on the other hand, estimates an MH's future location in the short term rather than the long term. Based on the probabilities of previous visits and the current trajectory of a mobile host, network resources are reserved near its present location and along its direction of travel. A less-ambiguous scheme can be found in the Profile-Based Next-Cell Prediction,[50] where resources are reserved only at the most-likely visiting cell, and further QoS commitments depend on the reservation process after the next handover. It is noticeable that the further ahead a scheme tries to predict the movement, the more likely it is to support the lifetime of a session. However, this is achieved at the expense of overall network utilization because of poor prediction accuracy.

3. Coarse-grained allocation: This scheme does not reserve resources on a per-user or per-cell basis, but works on a logical model called the Virtual Bottleneck Cell,[51] which treats a cluster of base stations as an aggregate virtual system. We believe that by controling the parameters and functions of a virtual bottleneck cell, the QoS agreements at each base station inside the cluster can be satisfied, even in environments with heterogeneous demands among base stations. However, it is not obvious how to decide

the boundary of a virtual bottleneck cell, so that it is large enough for users to stay for a sufficient duration but small enough to accurately reflect the characteristics of underlying base stations. Moreover, because of its design philosophy, it is difficult to integrate this aggregated admission control with flow-specific mobility protocols such as MRSVP or Advanced Reservation Signaling.

9.3.3.1.2 Discussion of Advance Reservation Issues

Advance reservation should make a compromise between the continuity of QoS support and the risk of overreserving resources in the mobile network. The coarse-grained allocation appears to be a scalable solution to this problem, but the feasibility of aggregated functions and the scope of a virtual bottleneck cell both require further investigations. The neighborhood-based allocation is the simplest scheme to implement. However, resources are likely to be oversubscribed because mobile users are seldom moving randomly in real life. By applying user mobility patterns, the history-based allocation scheme reserves resources in selective surrounding cells, and thereby attempts to minimize the probability of overreserving resources in the mobile network. This view has been supported by simulation results from various studies,[40,49,52] but its usefulness in real life cannot be fully verified unless the actual user mobility in wireless networks is better understood.[53]

9.3.3.2 Reactive Enhancements to Support Multimedia Delivery

The second class of measures that can be taken to increase QoS in a mobile environment is reactive measures. These are measures that react when the network characteristics fail to meet the requirements of the application. If the proactive measures work properly, there is no need for reactive measures because the application requirements would be met and therefore the two classes of measures are orthogonal.

The most commonly proposed reactive measure is to make the applications elastic through adaptation. For example, the adaptive multirate (AMR) codec in UMTS networks measures the available data rate due to packet loss and sets the encoding parameters accordingly. In this way, the voice quality can be degraded but still be continuous when the signal quality goes down. The proactive alternative, to rely only on resource reservation, means that resources will have to be overreserved in order to maintain a quality buffer for the codec or the voice output will be intermittent when the network fails to meet the codec demands.

The User Services Assistant (USA)[54] architecture takes this a step further. The architecture introduces a reactive QoS manager through which users can start and register application sessions and subsequently maintain them. USA allows users to make initial resource reservations, but rather than making hard reservations for maximum usage, minimum reservations are made and the decision of when to react to low quality is left to the user. Users indicate dissatisfaction to the manager, which then proposes adaptations to improve the quality of the session. The adaptations can then range from lowering encoding rate or increasing reservations to inserting transcoding functions or protocol translators in the data path.

Thus, USA integrates both proactive and reactive QoS management into the one architecture. Initially, the proactive measure of resource reservations takes place, and subsequently when a user indicates that the QoS level is too low, reactive measures take place. The steps can take place completely independently because the two measures are orthogonal and complementary.

9.4 CHALLENGES AND RECENT DEVELOPMENTS OF PERSONAL MOBILITY

Terminal mobility support mechanisms regard the terminal as the endpoint for communication and this has been the traditional view in the single-service telephony systems as well. Personal mobility shifts the focus from the terminal to the user. From this viewpoint, the terminal becomes a means for transferring the information from the network to the user and for enabling the user to interact with the network. When placing a phone call, the number is linked to the terminal and not the user. In the future, users will have identifiers that are independent of the terminal they use but can be resolved into the terminal address.

9.4.1 HETEROGENEITY

When personal mobility becomes a reality, the users' operational environment becomes dynamic and has to be taken into consideration for any communication. One example regarding the emerging mobile Internet infrastructure is the composition of vastly different network technologies with equally different characteristics. Most services on the Internet have been designed with certain assumptions about network characteristics such as a certain bandwidth, minimum delay, etc. Much of the content assumes capabilities of the end hosts such as screen size, color depth, codec availability, etc.

The mobile Internet is moving toward a situation where these assumptions often are broken due to the variations in terminals (PCs, laptops, PDAs, mobile phones, etc.) and the differentiation in access network characteristics. Today, users connect to the Internet through such different networks as Ethernet, cable modems, dial-up modems, ISDN, GSM, and satellite connections. To cope with this heterogeneity, recent proposals introduced proxy-based solutions that tailor the media to suit the overall characteristics of the environment.[55,56] Furthermore, TCP assumes that all lost packets are due to congestion in the network. In wireless networks, this does not hold true. Fluctuations in packet loss are, in these networks, more likely to be transient effects of the signal-to-noise ratio than dependent on congestion. Therefore, recent suggestions place performance enhancing proxies (PEPs) in the radio access networks.[57] These proxies implement a modified TCP solution or at least behave differently than standard end-to-end TCP.

This discussion illustrates the need for systems that can describe and categorize the user's operational environment so that applications can adapt to it. It is critical to know by what means and under which circumstances a user can be contacted before attempting it.

9.4.2 MOBILE AGENTS

One interesting aspect of the personal mobility systems presented earlier in this chapter is that all the systems that support personalization use mobile agent technologies in their design. This is not a coincidence as the characteristics of mobile agents provide many benefits in supporting personal mobility.

Mobile agent technology is regarded by many as the next step from the object-oriented paradigm, and is gaining popularity among software designers. Mobile agents are software agents that are not bounded to the system where they commence their execution. Once they have been created by a host, they can suspend their execution at any time, transport from one execution environment to another, and resume their execution.[58] This ability, in certain realizations, allows mobile agents to overcome network latency and reduce network traffic. In addition, mobile agents are autonomous; they have the ability to decide for themselves when and where to migrate. This characteristic allows mobile agents to operate asynchronously and independently of the process that created them, which can aid in making a system robust and fault tolerant. These two characteristics – autonomy and mobility – provide a good basis for designers to design any type of personal mobility framework.

One of the many benefits with mobile agents is that they are naturally heterogeneous. Because mobile agents are generally system and transport layer independent and are dependent only on their execution environment, they provide an optimal condition for seamless system integration.

According to the mobile agent list published in [MOBI02],[59] currently there are more than 70 different mobile agent systems available and this number is growing steadily. The contributions on the development of these systems come from both the research and commercial communities. As each system is designed with different philosophies and built for different purposes, each one has its own characteristics, such as migration and agent communication mechanisms. Despite their differences, many mobile agent systems have one thing in common: they use Java as the development and supporting programming language. The introduction of Java has helped solve many of the issues associated with mobile agents, such as performance, security, and agent migration.

Although mobile agents can provide many benefits to mobility systems design, the technology has not yet gained widespread commercial use. Issues such as standardization, security, and performance still need to be addressed before mobile agents can be used in a wider context. Despite these issues, the technology has proved to be a useful tool in solving the problems associated with the area of personal mobility.

9.4.3 INTEGRATED PRESENCE

As illustrated earlier, the current personal mobility systems provide support in two very distinctive areas. They support either contactability or personalization. To the best of our knowledge, currently there are no personal mobility systems that provide complete personal mobility support. Today's devices are no longer restricted to perform just one function. A computer terminal can act as a communicating device with voice over IP and PSTN gateways, and a communicating device has sufficient processing capability to perform operations that used to be restricted to desktop PCs. Thus, having a framework that supports only one aspect of personal mobility is insufficient.

It is likely that a user would like to keep all personal settings and at the same time be reached by others when migrating from one location to another. Thus, providing true personal mobility with the systems that have been made thus far will require the integration of methods of providing personal communications such as Mobile People Architecture with a scheme to support personalization of the user's operating environment, such as NetChaser. However, as each of these schemes was designed with objectives using different philosophies, combining them is at best cumbersome, and will lead to complex compatibility problems.

9.4.3.1 IPMoA

The Integrated Personal Mobility Architecture (IPMoA)[16] is a personal mobility system that attempts to address the above-mentioned issue by introducing an overlay network that caters to various personal mobility services through the use of mobile agent technology. The services the system supports include interpersonal communication (location support), customized Internet services, remote application execution, and file synchronization (personalization support).

9.5 CONCLUDING REMARKS

Currently, wireless Internet access does not differ much from wired access in terms of features. The majority of work so far has focused on rolling out standards for access technologies for basic connectivity. However, once the wireless Internet gains popularity the demand for additional services will become prevalent, and architectures will emerge from research laboratories to be implemented in the infrastructure.

In this chapter, we have tried to give an overview of the existing accepted solutions and list the most burning issues for the mobile Internet. Each of these issues will have to be taken into account when designing mobility solutions. There will undoubtedly emerge further issues in the future, as the mobile Internet matures and new types of services are introduced. One of the real challenges will be to devise solutions that cater to all the aspects and issues.

It is possible that there will be a number of coexisting solutions that are designed to address different scenarios and use cases. The Internet as a connecting infrastructure enables players as different as single applications developers and multinational operators to share the one infrastructure for data transfer. The demands put on solutions to support single applications and global user mobility are very different indeed. It is evident also that the trend of an increasingly heterogeneous Internet will continue in the future. The available access technologies will become even more diversified and terminals will become increasingly personalized and reflect user habits.

In this environment, mobility will take on different roles. For the traveling businessman, global connectivity with a traditional roaming agreement infrastructure will be necessary for all data services, as it is today for 2G cellular telephony. For a teenager, high bandwidth hot spots for networked gaming consoles, without the requirement of seamless mobility coupled with a separate cellular terminal for messaging, might be the right solution. The important issue is that mobility management will be one of the infrastructure components that will act as a service enabler. Therefore the primary goal of these components will be to impose as few restrictions on the services as possible. Only then will the mobile Internet reach its full potential.

References

1. Richardson, T. et al., Teleporting in an X window system environment, *IEEE Personal Communications Magazine,* 1(3), 6–12, 1994.
2. Perkins, C., Ed., "IP Mobility Support for IPv4," IETF RFC 2002, Oct. 1996.
3. Ford, P., Rekhter, Y., and Braun, H.-W., Improving the routing and addressing of IP, *IEEE Network,* May 1993.
4. Deering, S. and Hinden, R., "Internet Protocol, Version 6 (IPv6) Specification," IETF RFC 2460, Dec. 1998.
5. Montenegro, G., "Reverse Tunneling for Mobile IP," IETF RFC 2344, May 1998.
6. Perkins, C., Ed., "IP Mobility Support for IPv4," IETF RFC 3220, Jan. 2002.
7. "Principles of universal personal telecommunication (UPT)," ITU-T Recommendation F.850, Mar. 1993.
8. Rosenberg, J. et al., "SIP: Session Initiation Protocol," Internet Engineering Task Force (IETF) Internet draft, Feb. 2002.
9. Maniatis, P. et al., The Mobile People Architecture, *Mobile Computing and Communications Review,* July 1999.
10. Wang, H. et al., ICEBERG: An Internet-core network architecture for integrated communications, *IEEE Personal Communications,* Aug. 2000.
11. The Ninja Project, http://ninja.cs.berkeley.edu/.
12. Distefano, A. and Santoro, C., "NetChaser: Agent Support for Personal Mobility," *IEEE Internet Computing,* March-April 2000, 74–79.
13. Bellavista, P., Corradi, A., and Stefanelli, C., Mobile agent middleware for mobile computing, *IEEE Computer,* 34 (3), 73–81, 2001.
14. Thanh, D. et al., Using Mobile Agent Paradigm in Mobile Communications, Ericsson Conference Software Engineering, 1999.
15. Takashio, K., Soeda, G., and Tokuda, H., A Mobile Agent Framework for Follow-Me Applications in Ubiquitous Computing Environment, International Conference on Distributed Computing Systems Workshop, 2001.
16. Thai, B. et al., Integrated personal mobility architecture: a complete personal mobility solution, *MONET Journal on Personal Environment Mobility in Multi-Provider and Multi-Segment Networks,* submitted, 2002.
17. Perkins, C. and Johnson, D., "Route Optimization in Mobile IP," IETF Internet draft, Nov. 2000, work in progress.
18. Johnson, D. and Perkins, C., "Mobility Support in IPv6," IETF Internet draft, May 2002, work in progress.

19. Jain, R. et al., Mobile Internet Access and QoS Guarantees Using Mobile IP and RSVP with Location Registers, Proc. IEEE ICC'98, June 1998.

20. Wedlund, E. and Schulzrinne, H., Mobility Support using SIP, Proc. ACM WoWMo'99, Aug. 1999.

21. Landfeldt, B. et al., SLM, A Framework for Session Layer Mobility Management, Proc. IEEE ICCCN, Oct. 1999.

22. Maltz, D. and Bhagwat, P., MSOCKS: An Architecture for Transport Layer Mobility, Proc. IEEE INFOCOM'98, Mar. 1998.

23. Snoeren, A. and Balakrishnan, H., An End-to-End Approach to Host Mobility, ACM/IEEE Mobicom'00, Aug. 2000.

24. Gustafsson, E., Jonsson, A., and Perkins, C., "Mobile IP Regional Registration," IETF Internet draft, Mar. 2001, work in progress.

25. Foo, S. and Chua, K.C., Regional Aware Foreign Agent Scheme for Mobile-IP, Proc. MoMuC'99, Nov. 1999.

26. McCann, P. et al., "Transparent Hierarchical Mobility Agents (THEMA)," IETF Internet draft, Mar. 1999 (outdated).

27. Ramjee, R. et al., IP-based access network infrastructure for next-generation wireless data networks, *IEEE Personal Communications,* 7 (4), 2000.

28. Campbell, A. et al., Design, implementation, and evaluation of Cellular IP, *IEEE Personal Communications,* Aug. 2000.

29. Acharya, A. et al., Mobility support for IP over wireless ATM, *IEEE Personal Communications,* Apr. 1998.

30. Zhong, R. et al., "Integration of Mobile IP and MPLS," IETF Internet draft, June 2000 (outdated).

31. Terzis, A. et al., "RSVP operation over IP tunnels," IETF RFC 2746, Jan. 2000.

32. Bernet, Y. et al., "A Framework for Integrated Services Operation over DiffServ Networks," IETF RFC 2998, Nov. 2000.

33. Black, D., "Differentiated Services and Tunnels," IETF RFC 2983, Oct. 2000.

34. Fikouras, N. et al., Performance of TCP and UDP during Mobile IP Handoffs in Single-Agent Subnetworks, Proc. IEEE WCNC'99, Sep. 1999.

35. Fladenmuller, A. and De Silva, R., The effect of mobile IP handoffs on the performance of TCP, *ACM/Baltzer MONET,* 4 (2), 1999.

36. El Malki, K. and Soliman, H., "Fast Handoffs in Mobile IPv4," IETF Internet draft, Sep. 2000, work in progress.

37. Calhoun, P. et al., "Foreign Agent Assisted Hand-off," IETF Internet draft, Nov. 2000, work in progress.

38. Liu, G. and Maguire, G. Jr., A class of mobile motion prediction algorithms for wireless mobile computing and communications, *ACM/Baltzer MONET,* 1 (2), 1996.

39. Levine, D., Akyildiz, I., and Naghshineh, M., A resource estimation and call admission algorithm for wireless multimedia networks using the shadow cluster concept, *IEEE/ACM Trans. Networking,* 5 (1), Feb. 1997.

40. Oliveira, C., Kim, J., and Suda, T., An Adaptive Bandwidth Reservation Scheme for High-speed Multimedia Wireless Networks, *IEEE JSAC,* 16 (6), Aug. 1998.

41. Liu, T., Bahl, P., and Chlamtac, I., Mobility modeling, location tracking, and trajectory prediction in wireless ATM networks, *IEEE JAC,* 16 (6), Aug. 1998.

42. Talukdar, A., Badrinath, B., and Acharya, A., MRSVP: A resource reservation protocol for an integrated services network with mobile hosts, to appear in *ACM/WINET,* May 1999.

43. Pajares, A. et al., An Approach to Support Mobile QoS in an Integrated Services Packet Network, Proc. IQWiM Workshop, Apr. 1999.

44. Awduche, D. and Agu, E., Mobile Extensions to RSVP, Proc. IEEE ICCN'97, 1997.
45. Mahadevan, I. and Sivalingam, K., Architecture and experimental results for quality of service in mobile networks using RSVP and CBQ, *ACM/Baltzer Wireless Network,* 6 (3), July 2000.
46. Terzis, A., Srivastava, M., and Zhang, L., A Simple QoS Signaling Protocol for Mobile Hosts in the Integrated Services Internet, Proc. IEEE INFOCOM'99, Mar. 1999.
47. Terzis, A. et al., "RSVP operation over IP tunnels," IETF RFC 2746, Jan. 2000.
48. Acampora, A. and Naghshineh, M., An architecture and methodology for mobile-executed handoff in cellular ATM networks" *IEEE JACS,* 12 (8), Oct. 1994.
49. Levine, D., Akyildiz, I., and Naghshineh, M., A resource estimation and call admission algorithm for wireless multimedia networks using the shadow cluster concept, *IEEE/ACM Trans. Networking,* 5 (1), Feb. 1997.
50. Bharghavan, V. and Mysore, J., Profile Based Next-Cell Prediction in Indoor Wireless LANs, Proc. IEEE SICON'97, Apr. 1997.
51. Jain, R., Sadeghi, B., and Knightly, E., Toward Coarse-Grained Mobile QoS, Proc. ACM WoWMoM'99, Aug. 1999.
52. Ramanathan, P., Sivalingam, K., Agrawal, P., and Kishore, S., "Dynamic Resource Allocation Schemes During Handoff for Mobile Multimedia Wireless Networks," *IEEE/JSAC,* 17(7), July 1999, pp. 1270–1283.
53. Chan, J. et al., Integrating Mobility Prediction and Resource Pre-allocation into a Home-Proxy Based Wireless Internet Framework, Proc. IEEE ICON 2000, Sep. 2000.
54. Landfeldt, B., Seneviratne, A., and Diot, C., User Services Assistant: An End-to-End Reactive QoS Architecture, Proc. IWQOS98, Napa, California, 1998.
55. Ardon, S. et al., Mobile Aware Server Architecture: A DIstributed Proxy Architecture for Content Adaptation, INET2001, Stockholm, 2001.
56. Fox, A. et al., Adapting to network and client variation using active proxies: lessons and perspectives, *IEEE Personal Communications,* Aug. 1998.
57. "Performance Enhancing Proxies," http://search.ietf.org/internet-drafts/draft-ietf-pilc-pep-05.txt
58. Lange, D. and Oshima, M., *Programming and Deploying Java Mobile Agents with Aglets,* Addison-Wesley, Reading, MA, 1998.
59. The Mobile Agent List, http://mole.informatik.uni-stuttgart.de/mal/mal.html
60. Jung, E. et al., Mobile Agent Network for Support Personal Mobility, Proc. International Conference on Information Network (ICOIN), 1998, pp.131–136.
61. Di Stefano, A. and Santoro, C., NetChaser: Agent Support for Personal Mobility, *IEEE Internet Computing,* Mar.–Apr. 74–79, 2000.

10 Wireless Local Access to the Mobile Internet

José Antonio García-Macías and Leyla Toumi

CONTENTS

10.1 INTRODUCTION

The Internet has been around for more than three decades now. A key factor for its longevity is its flexibility to incorporate new technologies. However, this is not always a seamless process, as some of these new technologies break the basic assumptions under which the Internet works. For instance, the Internet was born at a time when all nodes in a network were fixed devices. Therefore, all the basic protocols were designed assuming that the end-points would stay fixed. Obviously, with the recent arrival of mobile networking devices (PDAs, laptops, 3G phones, etc.), these assumptions no longer hold.

We will discuss the problem of mobility in IP networks, with a special emphasis on the case of mobility within a restricted geographical span, also called micro-mobility.

TABLE 10.1
Activities of the Task Groups Working on the 802.11 Standard

Task Group	Activities
802.11	Initial standard, 2.4-GHz band, 2 Mbps
802.11a	High speed PHY layer in the 5-GHz band, up to 24 or 54 Mbps
802.11b	High speed PHY layer in the 2.4-GHz band, up to 11 Mbps
802.11d	New regulatory domains (countries)
802.11e	Medium access control (MAC) enhancements: Multimedia, QoS, enhanced security
802.11f	Interaccess point protocol for AP interoperability
802.11g	Higher data rate extension in the 2.4-GHz band, up to 22 Mbps
802.11h	Extensions for the 5-GHz band support in Europe

Before addressing the problem of local IP mobility, we will examine the technologies that allow local connectivity, so-called "last-meter" technologies such as 802.11 (WiFi), Bluetooth, and HiperLan.

10.2 LOCAL ACCESS TECHNOLOGIES

Among the different technologies available for wireless local networks, the most popular without a doubt is IEEE 802.11. Such popularity is evidenced by the number of products based on this standard that are commercially available. We will describe technologies used for access networks, with a particular emphasis on the 802.11 standard; also, we will discuss other technologies such as Bluetooth and HiperLAN.

It is also worth noting that some companies (e.g., Airify) have announced products to support multiple wireless standards using the same network interface; this way, the same device could be used to take advantage of WLAN technologies such as 802.11 or Bluetooth, or wide area wireless, such as GSM or GPRS. However, these types of products have yet to be commercially available.

10.2.1 THE 802.11 STANDARD

The Institute of Electrical and Electronic Engineers (IEEE) ratifed the original 802.11 specification in 1997 as the standard for wireless LANs (WLANs). That version of 802.11 provides for 1 and 2 Mbps data rates and a set of fundamental signaling methods and other services. Some disadvantages with the original 802.11 standard are the data rates that are too slow to support most general business requirements. Recognizing the critical need to support higher data transmission rates, the IEEE ratified the 802.11b standard for transmissions of up to 11 Mbps. With 802.11b (also known as WiFi), WLANs are able to achieve wireless performance and throughput comparable to wired 10-Mbps Ethernet. 802.11a offers speeds of up to 54 Mbps, but runs in the 5-GHz band, so products based on this standard are not compatible with those based on 802.11b.[1] Several task groups are working on further developments for the 802.11 standard, as shown in Table 10.1.

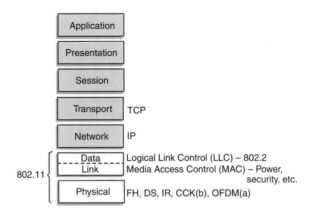

FIGURE 10.1 The 802.11 standard and the ISO model.

Like all 802.x standards, 802.11 focuses on the bottom two layers of the OSI Reference Model, the physical and the data link layers. In fact, the standard covers three physical layer implementations: direct-sequence (DS) spread spectrum, frequency hopping (FH) spread spectrum, and infrared (IR). A single medium access control (MAC) layer supports all three physical layer implementations, as shown in Figure 10.1. We will further discuss the two ISO layers that the 802.11 standard deals with.

10.2.2 802.11 ARCHITECTURE

Each computer (mobile, portable, or fixed) is referred to as a station in 802.11. Mobile stations access the LAN during movement. The 802.11 standard defines two modes: infrastructure mode and *ad hoc* mode. In infrastructure mode (Figure 10.2), the wireless network consists of at least one access point (AP) connected to the wired network infrastructure and a set of wireless end stations.

This configuration is called a basic service set (BSS). An extended service set (ESS) is a set of two or more basic service sets forming a single subnetwork. Two or more ESSs are interconnected using a distribution system (DS). In an extended service set, the entire network looks like an independent BSS to the logical link control (LLC) layer; this means that stations within the extended service set can communicate or even move between basic service sets transparently to the logical link control. The distribution system can be thought of as a backbone network that is responsible for MAC-level transport of MAC service data units (MSDUs). The distribution system, as specified by 802.11, is implementation independent; therefore, the distribution system could be a wired 802.3 Ethernet LAN, an 802.4 token bus LAN, an 802.5 token ring LAN, a fiber distributed data interface (FDDI) metropolitan area network (MAN), or another 802.11 wireless medium. Note that while the distribution system could physically be the same transmission medium as the basic service set, they are logically different because the distribution system is solely used as a transport backbone to transfer packets between different basic service

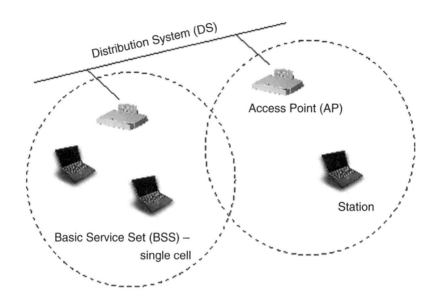

FIGURE 10.2 Infrastructure mode.

sets in the extended service set. An extended service set can provide gateway access for wireless users into a wired network such as the Internet. This is accomplished via a device known as a portal. The portal is a logical entity that specifies the integration point on the distribution system where the 802.11 network integrates with a non-802.11 network. If the network is an 802.x, the portal incorporates functions that are analogous to a bridge, i.e., it provides range extension and the translation between different frame formats.

The *ad hoc* mode (also called peer-to-peer mode or an independent basic service set, or IBSS) is simply a set of 802.11 stations that communicate directly with one another without using an access point or any connection to a wired network (Figure 10.3). In *ad hoc* networks, there is no base and no one gives permission to talk; these networks are spontaneous and can be set up rapidly, but are limited both temporally and spatially.

10.2.3 THE PHYSICAL LAYER

The three physical layers originally defined in the 802.11 standard included two spread spectrum radio techniques and a diffuse infrared specification. The radio-based standards operate within the 2.4-GHz ISM (industrial, scientific, and medical) band. These frequency bands are recognized by international regulatory agencies, such as the FCC (United States), ETSI (Europe) and the MKK (Japan) for unlicensed radio operations. As such, 802.11-based products do not require user licensing or special training. Spread spectrum techniques, in addition to satisfying regulatory requirements, boost throughput and allow many unrelated products to share the spectrum without explicit cooperation and with minimal interference.

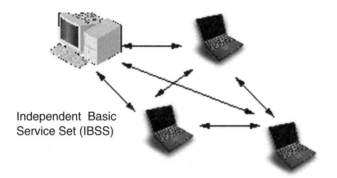

FIGURE 10.3 *Ad hoc* mode.

The original 802.11 wireless standard defines data rates of 1 and 2 Mbps via radio waves using frequency hopping (FH) spread spectrum or direct sequence (DS) spread spectrum. It is important to note that these are fundamentally different transmission mechanisms and will not interoperate with each other. Direct sequence has a more-robust modulation and a larger coverage range than FH, even when FH uses twice the transmitter power output level. Frequency hopping gives a large number of hop frequencies, but the adjacent channel interference behavior limits the number of independently operating collocated systems. Hop time and a smaller packet size introduce more transmission time overhead into FH, which affects the maximum throughput. Although FH is less robust, it gives a more-graceful degradation in throughput and connectivity.

Under poor channel and interference conditions, FH will continue to work over a few hop channels a little longer than over the other hop channels.

Direct sequence, however, still gives reliable links for a distance at which very few FH hop channels still work. For collocated networks (access points), DS gives a higher potential throughput with fewer access points than FH, which has more access points. The smaller number of access points used by DS lowers the infrastructure cost.

10.2.4 THE DATA LINK LAYER

The data link layer within 802.11 consists of two sublayers: logical link control (LLC) and media access control (MAC). 802.11 uses the same 802.2 LLC and 48-bit addressing as other 802.x LANs, allowing for very simple bridging from wireless to wired networks, but the MAC is unique to WLANs.

Of particular interest in the specification is the support for two fundamentally different MAC schemes to transport asynchronous and time-bounded services. The first scheme, distributed coordination function (DCF), is similar to traditional legacy packet networks. The DCF is designed for asynchronous data transport, where all users with data to transmit have an equally fair chance of accessing the network. The point coordination function (PCF) is the second MAC scheme. The PCF is based on polling that is controlled by an access point.

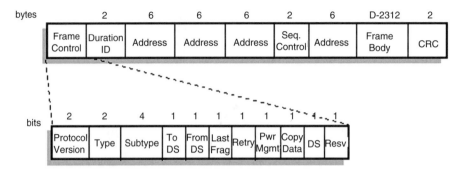

FIGURE 10.4 Standard 802.11 frame format.

The basic access method, DCF, is drawn from the family of Carrier Sense Multiple Access with Collision Avoidance (CSMA/CA) protocols. The collision detection (CD) mechanism as used in the CSMA/CD protocol of Ethernet cannot be used under 802.11 due to the near/far problem: to detect a collision, a station must be able to transmit and listen at the same time, but in radio systems the transmission drowns out the ability of the station to hear a collision. So, 802.11 uses CSMA/CA, under which collisions are avoided by using explicit packet acknowledgment (ACK) to confirm that the data packet arrived intact.

802.11 supports three different types of frames: management, control, and data. The management frames are used for station association and disassociation with the access point, timing and synchronization, and authentication and deauthentication. Control frames are used for handshaking during a contention period (CP), for positive acknowledgment during the CP, and to end the contention-free period (CFP). Data frames are used for the transmission of data during the CP and CFP, and can be combined with polling and acknowledgments during the CFP. Figure 10.4 shows the standard 802.11 frame format.

10.2.5 OTHER RELATED STANDARDS

There are other WLAN technologies available besides 802.11, and we will review some of the most prominent ones, namely Bluetooth and HiperLAN. It is relevant to point out that up to now the market for WLANs has been dominated by products based on the 802.11 standard. There are starting to appear some products based on Bluetooth but they have been very deceiving and many important equipment and software manufacturers have decided not to support this standard,[2,3] at least temporarily. Although some early prototypes for HiperLAN 2 have been demonstrated,[4] there are no commercial products available yet.

10.2.5.1 HiperLAN

Between 1990 and 1992, the European Telecommunications Standards Institute (ETSI) noticed the trend toward faster and better wireless networks and started the development of standards for this type of network. Within this framework, the

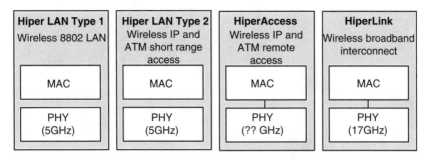

FIGURE 10.5 Overview of HiperLAN standards.

Broadband Radio Access Networks (BRAN) Project 3 of ETSI is working on a standard called High Performance Radio Local Area Network (HiperLAN). This project quickly separated into four different HiperLAN types:

1. HiperLAN 1 is a standard for ad hoc networking operating in the 5.2-GHz band with a spectrum of 100 MHz and speeds of up to 19 Mbps. It offers one-to-one communications as well as one-to-many broadcasts. Using the CSMA/CA technique for resolving contention, the scheme shares available radio capacity between active users who attempt to transmit data during an overlapping time span. Although HiperLAN 1 provides a means of transporting time-bounded services, it does not control nor guarantee QoS on the wireless link. This is what motivated ETSI to develop a new generation of standards that support asynchronous data and time-critical services bounded by specific time delays.
2. HiperLAN 2 specifies a radio-access network that can be used with a variety of core networks (e.g., IP, ATM, UMTS). HiperLAN 2 operates in the 5.2-GHz band with 100 MHz spectrum, but at speeds of up to 54 Mbps.[5]
3. HiperAccess is the next step from HiperLAN 2, providing outdoor wireless access. It gives up to 5 km coverage between wireless access points and wireless termination points and is therefore intended for stationary and semistationary applications. The original operating frequencies were in the 5-GHz band, but this is currently under discussion.
4. HiperLink is the standard meant to provide interconnecting services for high data rate sources, such as networks (e.g., HiperLANs). Therefore, HiperLink provides point-to-point interconnections at very high data rates of up to 155 Mbps over distances up to 150 meters. The operating frequency is in the 17-GHz band with 200 MHz spectrum at the moment.

The standard for HiperLAN 1 was finalized in 1996, although amendments were made to it in 1998. HiperLAN types 2 through 4 were designed to support only ATM networks, but at the moment HiperLAN 2 supports access to IP and UMTS networks. The names for types 3 and 4 were changed to HiperAccess and HiperLink, respectively. Figure 10.5 gives an overview of the different HiperLAN standards.

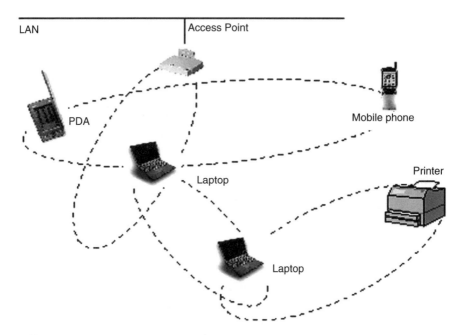

FIGURE 10.6 A Bluetooth scatternet of four piconets.

10.2.5.2 Bluetooth

Bluetooth is a protocol intended to wirelessly connect cellular phones, laptops, handheld computers, digital cameras, printers, and other devices.[6] It operates over short distances of up to 10 meters, basically being a wireless replacement for data cables and infrared connections. There are currently some discussions underway to extend its range to 100 meters by increasing the transmit power to 100 mW. Although Bluetooth was initially developed by Ericsson in the late 1990s, it is currently led by the Bluetooth SIG 4, including members such as Nokia, IBM, Toshiba, Intel, 3Com, Motorola, Lucent Technologies, and Microsoft. It is, then, not a technology backed by an standards body, but instead backed by an industry consortium.

The Bluetooth system supports point-to-point or point-to-multipoint connections. In point-to-multipoint, the channel is shared among several Bluetooth units. Two or more units sharing the same channel form a piconet. There is one master unit and up to seven active slave units in a piconet. These devices can be in any of the following states: active, park, hold, and sniff. Multiple piconets with overlapping coverage areas form a scatternet (Figure 10.6).

The Bluetooth system consists of a radio unit, a link control unit, and a support unit for link management and host terminal interface functions.

The radio operates in the 2.4-GHz ISM band. Depending on the class of the device, a Bluetooth radio can transmit up to 100 mW (20 dBm) to a minimum of 1 mW (0 dBm) of power. It uses frequency hopping for low interference and fading, and a TDD (time-division duplex) scheme for full-duplex transmission and transmits using GFSK (Gaussian frequency shift keying) modulation.[7]

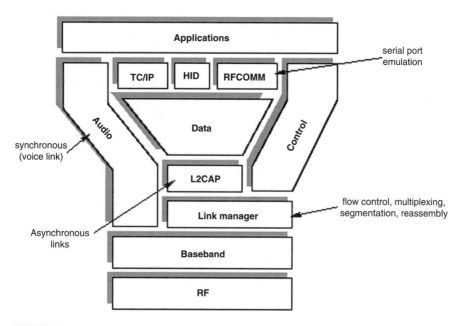

FIGURE 10.7 The Bluetooth protocol stack.

The Bluetooth protocol uses a combination of circuit and packet switching. The channel is slotted and slots can be reserved for synchronous packets. The protocol stack can support an asynchronous connectionless link (ACL) for data and up to three simultaneous synchronous connection-oriented (SCO) links for voice or a combination of asynchronous data and synchronous voice (DV packet type). Each voice channel supports a 64 kbps synchronous channel in each direction. The asynchronous channel can support a maximum of 723.2 kbps uplink and 57.6 kbps downlink (or vice versa) or 433.9 kbps symmetric links. The stack (shown in Figure 10.7) primarily contains a physical level protocol (baseband) and a link level protocol (LMP) with an adaptation layer (L2CAP) for upper layer protocols to interact with lower layer ones.

It should be clear, given its distance coverage, bandwidth, and other characteristics, that Bluetooth does not really fit within the profile for supporting WLANs as their promoters portend through intense marketing campaigns. Bluetooth fits more within the profile of technologies used for wireless personal area networks (WPANs), as those studied by the 802.15 Working Group.

10.2.6 WLAN Interoperability

Given their ease of installation, dropping prices, and increasingly higher speeds, WLANs are gradually replacing many wired LANs as the networks of choice for typical activities such as Internet access. There are currently coexistence problems between some of the technologies we mention here, namely between 802.11 and Bluetooth. A source of problems is the fact that Bluetooth has been designed to transmit blindly, whenever its timing dictates, as if there was no possibility that a collocated system might be using the same frequency (as 802.11 does). This has earned it the reputation of a "bad neighbor" in the 2.4-GHz band. There are also other common sources of interference in this unregulated band, including microwave ovens and newer generations of cordless phones. Historically, microwave ovens are by far the most-significant source of interference in residential and office environments, but with the impending avalanche of new communications devices with embedded Bluetooth radios, serious questions have been raised about their interference on wireless LANs. The 802.15 WPAN Task Group 2 is developing recommended practices and mechanisms to facilitate coexistence between WLANs (such as 802.11) and WPANs (such as Bluetooth).

10.3 MOBILITY AND THE INTERNET PROTOCOLS

The Internet was born in an era when no mobile networking equipment was available. Therefore, all the basic protocols were designed under the tacit assumption that the end-points of a communication would stay fixed all along a session. With the arrival of modern communications equipment that allows these end-points to change their position, new protocols for handling mobility have been proposed.

Mobile IP allows mobility of devices, potentially around the world; this is why the type of mobility support it provides is sometimes referred to as global mobility. However, as we will see, mobility within a limited geographical area (called micromobility), has different characteristics and requirements that pose the need for specialized support.

10.3.1 THE PROBLEM OF IP-BASED MOBILITY

Although networking-enabled mobile devices are becoming more common everyday, most networking protocols — including the TCP/IP protocol suite — have been designed assuming that hosts are always attached to the network at a single physical location. Therefore, host mobility is seen as a rarely occurring fact that can be handled manually. Consider for instance the following scenario: a business executive is usually connected to the network in the office, but occasionally needs to use a laptop computer for meetings; the meeting facilities may be elsewhere in the building or perhaps in a different building or city. If the executive's desk and the meeting room have direct access to the same IP subnet, then the mobility process is trivial. In situations where this is not the case, the only solution is for the user to acquire a new IP address from the appropriate local authority. Then, several configuration files on the moving machine, on various name servers, and on other machines that use the original IP address to identify the moving machine need to be modified. Thus, moving the computer from one place to another involves a slow, error-prone, manual procedure that a typical user does not have the skills or the inclination to deal with. Moreover, even if the process is successfully performed, the mobile host will lose its former identity and will usually need rebooting.

The situation is that, given TCP/IP's early design assumptions that end systems are stationary, if during an active connection one end system moves, then the whole connection breaks, obviously disrupting all networking services layered on top of

TCP/IP. Evidence has been given[8] that in order to retain transport layer connections, a mobile host's address must be preserved regardless of its point of attachment to the network. The problem with a transport layer protocol such as TCP is that a TCP connection is identified by a 4-tuple:

```
<src IP address, src TCP port, dest IP address, dest
  TCP port>
```

So, if neither host moves, all elements of the tuple remain fixed and the TCP connection can be preserved. However, if either end of the connection moves, the following problem will take place:

- If the mobile host acquires a new IP address, then its associated TCP connection identifier also changes. This causes all TCP connections involving the mobile host to break.
- If the mobile retains its address, then the routing system cannot forward packets to its new location.

These problems come from the very design of IP which, in addition to fragmentation and reassembly, is responsible for "providing the functions necessary to deliver a package of bits (an Internet datagram) from a source to a destination over an interconnected system of networks."[9] So, this definition designates responsibility to IP for routing datagrams to and from mobile hosts transparently to higher layers. The problem is that IP addresses serve a dual purpose, as they are not only used by higher layers to identify source and destination hosts, but also by their division into network and host parts which contain location information. Therefore, in its role as an identifier, an IP address must be constant during mobility to avoid affecting higher layers.

Research studies on IP mobility have suggested that mobility is essentially an address translation problem and is best resolved at the network layer.[8] As Figure 10.8 shows, a mobile host *MH* can move away from its home network and attach to the Internet through a foreign network. While away, *MH* obtains a forwarding address derived from the address space of the foreign network. However, if another host *S* tries to send packets to *MH*, it will do so using *MH*'s home address. The problem is resolved by the use of an address translation agent (ATA) at the home network, and a forwarding agent (FA) at the foreign network. These agents perform functions *f* and *g*, respectively, which are defined as follows:

- *f*: home address → forwarding address
- *g*: forwarding address → home address

This way, when *S* sends packets to *MH*, they first pass through ATA. This agent performs mapping f to send the packets to the address that *MH* acquired in the foreign network. At the foreign network, *FA* intercepts all packets containing *MH*'s forwarding address. It then proceeds to apply the function g to map from this forwarding address to *MH*'s original home address and effectively forward the packets.

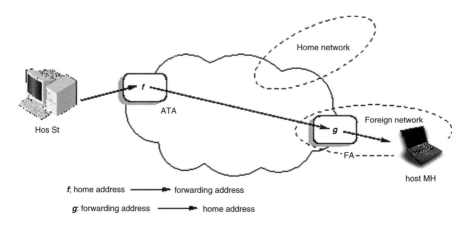

f, home address ─────▶ forwarding address

g, forwarding address ─────▶ home address

FIGURE 10.8 Mobility as an address translation problem.

10.3.2 MOBILE IP

In order to react to the new challenges posed to the Internet architecture by the arrival of mobile networking devices, the IETF created the Mobile IP Working Group. The basic Mobile IP standard[10] specifies a mobility management architecture for the Internet. In principle, both local-area and wide-area mobility across wired and wireless networks can be handled, although certain inefficiencies have been detected. Later, we will see extensions to Mobile IP proposed to overcome such inefficiencies.

Figure 10.9 shows the basic operation of Mobile IP. A mobile node is normally attached to its home network using a static home address. When the mobile node moves to a foreign network, it makes its presence known by registering with a foreign agent (FA). The mobile node then communicates with a home agent (HA) in its home network, giving it the care-of address (COA), which identifies the foreign agent's location. Typically, routers in a network will implement the roles of home and foreign agents. When IP datagrams are exchanged over a connection between the mobile node A and a correspondent host B, the following operations occur:[11]

1. Host *B* transmits an IP datagram destined for mobile node *A*, with *A*'s home address in the IP header. The IP datagram is routed to *A*'s home network.
2. At the home network, the incoming IP datagram is intercepted by the home agent. The home agent encapsulates the entire datagram inside a new IP datagram, which has *A*'s care-of address in the header, and retransmits the datagram. The use of an outer IP datagram with a different destination IP address is known as tunneling.
3. The foreign agent strips off the outer IP header, encapsulates the original IP datagram in a MAC-level PDU (for example, an Ethernet frame), and delivers the original datagram to *A* across the foreign network.

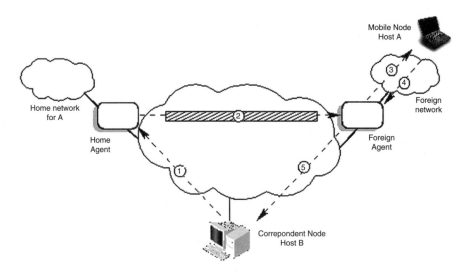

FIGURE 10.9 Basic mobile IP scenario.

4. When *A* sends IP traffic to *B,* it uses *B*'s IP address. In our example, this is a fixed address; i.e., *B* is not a mobile node. Each IP datagram is sent by *A* to a router on the foreign network for routing to *B.*
5. The IP datagram from *A* to *B* travels directly across the Internet to *B,* using *B*'s IP address.

10.3.4 MOBILE IP PROBLEMS

There are currently several outstanding problems facing Mobile IP, posing technical as well as practical obstacles for its deployment.[12] One of the most notable problems is due to routing inefficiencies. In the basic Mobile IP protocol, IP packets destined to a mobile node (MN) that is outside its home network are routed through the home agent. However, packets from the MN to the corresponding nodes are routed directly. This is known as triangle routing (see Figure 10.10).

This method may be inefficient when the correspondent host and the MN are in the same network, but not in the same home network of the MN. In such a case, the messages will experience unnecessary delay because they have to be routed first to the HA that resides in the home network. In order to alleviate this, a technique known as route optimization has been proposed.[13] However, implementing it requires changes in the correspondent nodes that will take a long time to deploy in IPv4.

Some other problems are related to performance and scaling issues. Studies have shown that Mobile IP can suffer from unacceptably long handoff latencies when the mobile host is far from its home network.[14] Scalability can be a problem as the number of mobile hosts grow, but in this case the network is the bottleneck, as mobility agents (i.e., HAs, FAs) can easily service at least a few hundred hosts. Suggestions have been made that using a hierarchical model to manage mobility

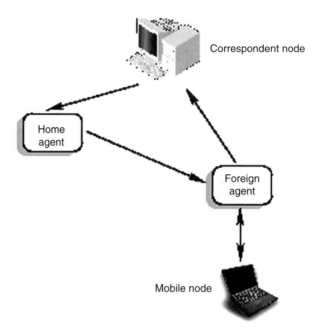

Correspondent node

Home
agent

Foreign
agent

Mobile node

FIGURE 10.10 Triangular routing.

could reduce or eliminate these performance and scaling problems.[15,16] Security is
also a particular area of attention in Mobile IP.

A lot of the problems of Mobile IP are related to the lack of features for
streamlining mobility support in IPv4.[17] Some of these problems may be solved by
IPv6. While Mobile IP was originally designed for IPv4, IPv6[18] incorporates features
that support mobility much more easily; several mechanisms that had to be specified
separately now come integrated with IPv6. Some of these IPv6 features include
stateless address autoconfiguration[19] and neighbor discovery.[20] IPv6 also attempts
to drastically simplify the process of renumbering, which may be critical to the
future of routability of the Internet.[21] Security is a required feature for all IPv6 nodes.

10.3.5 MICRO-MOBILITY

As several studies indicate,[22,23] users' mobility patterns are highly localized. For
instance, business professionals may spend a considerable amount of time away
from their desks, but once away, most of their mobility will take place within the
same building. While the mobile user is at the foreign administrative domain, there
is no need to expose motion within that domain to the home agent or to correspondent
hosts in other domains. Therefore, mobility management within an administrative
domain should be separate from global mobility management.

In principle, Mobile IP can handle both global and local mobility. However, it
requires that the mobile's home network is notified of every change in location.
Moreover, route optimization extensions[13] further require that every new location is
registered with hosts that are actively communicating with the mobile node. All

these location updates incur communications latency and also add traffic to the wide-area portion of the internetwork. Therefore, Mobile IP does not extend well to large numbers of portable devices moving frequently between small cells. It also has been demonstrated that, when used for micro-mobility support, Mobile IP incurs disruption to user traffic during handoff, and high control overhead due to frequent notifications to the home agent.[15] Another type of protocol, a micro-mobility protocol,[24] is then needed for local environments where mobile hosts change their point of attachment to the network so frequently that the basic Mobile IP tunneling mechanism introduces network overhead in terms of increased delay, packet loss, and signaling.

Acknowledging the fact that Mobile IP may not be the universal end-all solution for mobility on the Internet, its performance and scalability challenges have been under discussion. Within this context, the Mobile IP working group has recently started discussing the subject of micro-mobility protocols. There are several attributes that micro-mobility protocols aim toward:

- Minimum (or zero) packet loss: Fast handoff techniques have been developed to achieve this, and they may reduce latency or delay.
- Reduced signaling: Techniques for locating mobile hosts, known as paging, have been proposed in order to reduce signaling. Reduced registration is also an outcome of these techniques.

HAWAII and Cellular IP are two prominent proposals for micro-mobility management and we give a brief presentation of both:

- HAWAII[25,26] (Handoff-Aware Wireless Access Internet Infrastructure) is an alternative for providing domain-based mobility (i.e., micro-mobility). Under this approach, Mobile IP is used as the basis for mobility management in wide-area wireless networks, but new methods for managing mobility within an administrative domain are developed. One point worth highlighting is that mobile hosts retain their network address while moving within a domain; this way, the Home Agent (HA) — if using Mobile IP — and any corresponding hosts are not aware that the host has performed intradomain mobility. Dividing the network into hierarchies, loosely modeling the autonomous system hierarchy used in the Internet, is part of the HAWAII approach. Indeed, the gateway into each domain is called the domain root router, and each host is assumed to have an IP address and a home domain. As already stated, hosts retain their address while moving within a domain, so when packets destined to a mobile host arrive at the domain root router, they are forwarded over specially established paths to reach the mobile host. However, if the mobile host moves to a foreign domain, traditional Mobile IP mechanisms are used.
- Cellular IP[27,28] aims to integrate cellular technology principles with the IP networking paradigm; this poses difficult challenges, as there are fundamental architectural differences between cellular and IP networks. A Cellular IP node constitutes the universal component of a Cellular IP

network, because it serves as a wireless access point but at the same time routes IP packets and integrates cellular control functionality traditionally found in mobile switching centers (MSC) and base station controllers (BSC). Cellular IP nodes are modified IP nodes where standard routing is replaced by Cellular IP's own routing and location-management functions. A Cellular IP network is connected to the Internet via a gateway router. Mobility between gateways (i.e., Cellular IP access networks) is managed by Mobile IP, while mobility within access networks is handled by Cellular IP. Mobile hosts attached to the network use the IP address of the gateway as their Mobile IP care-of address.

Another important aspect that has received little attention in the design of micro-mobility protocols is that of quality of service (QoS). Triangular routing, address translation, and complex interaction between agents make Mobile IP unsuitable for QoS support in local environments.[29–31]

10.4 PERSPECTIVES AND CONCLUSIONS

Among local wireless access technologies, WLANs have a predominant place in the market, as they are increasingly replacing wired LANs as the method of choice for accessing the Internet. By far, the most popular WLAN technology is currently 802.11 (particularly the 802.11b variation, also named Wi-Fi). Wireless technologies allow hosts to freely roam between cells, but the Internet's core protocols were not designed with mobility in mind. Even though Mobile IP has been proposed as a solution to handle IP mobility, it is not very suitable for the case of micro-mobility (i.e., mobility within a very limited geographical span). Thus, IP micro-mobility protocols have been proposed.

Recent research has addressed the problem of providing QoS guarantees in micro-mobility environments. Some have proposed RSVP-like signaling protocols to make resources reservations,[32] while others have taken the differentiated services approach (proposed by the IETF), where no hard QoS guarantees are provided, but only statistical guarantees.[33] Also, work in progress within the IETF's SeaMoby Working Group is currently addressing problems related to QoS in mobile environments, although not exclusively for the case of micro-mobility.

References

1. Nobel, C., Making 802.11 standards work together, *eWeek,* July 19, 2000.
2. Staff, N., Psion backtracks on consumer plans, http://news.cnet.com, July 12, 2001.
3. Orlowski, A., Microsoft turns the drill on Bluetooth, August 1, 2001, available at http://www.theregister.co.uk.
4. Ericsson, Ericsson demonstrates HiperLAN 2 prototypes, Press release, December 11, 2000, available at http://www.ericsson.com/press/20001211–0067.html.
5. Khun-Jush, J. et al., HiperLAN type 2 for broadband wireless communication, *Ericsson Review,* 2, 108, 2000.

6. Haarsten, J., Bluetooth – the universal radio interface for ad hoc, wireless connectivity, *Ericsson Review,* 3, 110, 1998.
7. Haarsten, J., The Bluetooth radio system, *IEEE Personal Communications Magazine,* 7, 28, 2000.
8. Bhagwat, P., Perkins, C., and Tripathi, S., Network layer mobility: an architecture and survey, *IEEE Personal Communications Magazine,* 3, 54, 1996.
9. DARPA, DARPA Internet Program Protocol Specification, Internet RFC 791, 1981.
10. Perkins, C., Mobile IP specification, Internet RFC 2002, 1996.
11. Stallings, W., Mobile IP, *The Internet Protocol Journal,* 4, 2, 2001.
12. Chesire, S. and Baker, M., Internet mobility 4x4, in *ACM SIGCOMM Computer Communications Review,* 318, 1994.
13. Johnson, D. and Perkins, C., Route optimization in mobile IP, IETF Mobile-IP draft, July 1995.
14. Mukkamalla S. and Raman, B., Latency and scaling issues in mobile IP, Iceberg Project technical report, University of California, Berkeley, 2001.
15. Caceres, R. and Padmanabhan, V., Fast and scalable handoffs for wireless internet-works, in *ACM Mobicom 96,* 1996.
16. Soliman, H. et al., Hierarchical MIPv6 mobility management (HMIPv6), Internet draft draft-ietf-mobileip-hmipv6–04.txt, work in progress, July 2001.
17. Perkins, C., Mobile networking through mobile IP, *IEEE Internet Computing,* 2 (1), 1998.
18. Deering, S. and Hinden, R., Internet Protocol version 6 (IPv6), Internet RFC 1883, 1995.
19. Thomson, S. and Narten, T., IPv6 stateless address autoconfiguration, Internet RFC 1971, 1996.
20. Narten, T., Nordmark, E., and Simpson, W., Neighbor discovery for IP version 6 (IPv6), Internet RFC 1970, 1996.
21. Castineyra, I., Chiappa, J., and Steenstrup, M., The Nimrod routing architecture, Internet RFC 1992, 1996.
22. Kirby, G., Locating the user, *Communications International,* 1995.
23. Toh, C., The design and implementation of a hybrid handover protocol for multimedia wireless LANs, in Proc. 1st International Conference on Mobile Computing and Networking, 1995.
24. Campbell, A. and Gomez-Castellanos, J., IP micro-mobility protocols, *ACM Sigmobile Mobile Computer and Communications Review,* 2001.
25. Ramjee, R. et al., IP micro-mobility support using HAWAII, Internet draft draft-ietf-mobileip-hawaii-01.txt, work in progress, July 1999.
26. Ramjee, R. et al., HAWAII: a domain-based approach for supporting mobility in wide-area wireless networks, in IEEE International Conference on Network Protocols, 1999.
27. Valko, A., Cellular IP — a new approach to Internet host mobility, *ACM Computer Communication Review,* 1999.
28. Campbell, A. et al., An overview of cellular IP, in IEEE Wireless Communications and Networks Conference, WCNC, 1999.
29. Chan, J. et al., The challenges of provisioning real-time services in wireless Internet, *Telecommunications Journal of Australia,* 2000.
30. Helal, A. et al., Towards integrating wireless LANs with wireless WANs using mobile IP, in IEEE Wireless Communications and Networks Conference, WCNC, 2000.
31. Mukkamalla, S. and Raman, B., Latency and scaling issues in mobile IP, ICEBERG Project technical report, University of California, Berkeley, 2001.

32. Legrand, G., Qualité de Service dans les Environnements Internet Mobile, Ph.D. thesis, Université Pierre et Marie Curie, Paris VII, July 2001.

33. García-Macías, J.A. et al., Quality of service and mobility for the wireless Internet, in ACM/IEEE Mobicom 2001, Workshop on Mobile Internet (WMI), Rome, Italy, July 2001.

11 Location Prediction Algorithms for Mobile Wireless Systems

Christine Cheng, Ravi Jain, and Eric van den Berg

CONTENTS

ABSTRACT

Predicting the location of a mobile wireless user is an inherently interesting and challenging problem. Location prediction has received increasing interest over the past decade, driven by applications in location management, call admission control, smooth handoffs, and resource reservation for improved quality of service. It is likely that location prediction will receive even more interest in the future, especially given the increased availability and importance of location estimation hardware and applications.

In this chapter, we present an overview of location prediction in mobile wireless systems. We do not attempt to provide a comprehensive survey of all techniques and applications, but offer instead a description of several types of algorithms used for location prediction. We classify them broadly into two types of approaches: (1) domain-independent algorithms that take results from Markov analysis or text compression algorithms and apply them to prediction, and (2) domain-specific algorithms that consider the geometry of user motion as well as the semantics of the symbols in the user's movement history. We briefly mention other algorithms using Bayesian or neural network approaches, and end with some concluding remarks.

11.1 INTRODUCTION

Predicting the location of a user or a user's mobile device is an inherently interesting problem and one that presents many open research challenges. The explosion in mobile wireless technologies and applications over the past decade has sparked renewed interest in location prediction techniques. The advent of new access technologies such as wireless local area network (WLAN) and third-generation (3G) systems, location based services, and pervasive computing and communications indicate that location prediction will become even more important in the future.

There are two classes of applications that can benefit from accurate prediction of a user's location:

1. End-user applications, where the object is to predict location so that a human user can prepare or react accordingly. An example of an end-user application is one that predicts the location of a moving vehicle for road traffic optimization or for catching thieves if the vehicle is stolen.
2. System-enhancement applications, where location prediction can be used to enhance system performance, availability, or other metrics. An example of a system-enhancement application is one that predicts the location of a moving vehicle where a passenger is using a cell phone so as to reserve resources in adjoining cells and provide a smooth handoff.

Location can be specified in an absolute coordinate system, e.g., latitude/longitude, or in symbolic coordinates (e.g., cell ID). In some cases, both may be available. For example, a facilities administrator in an office building is likely to have a detailed map of the room layout, showing both absolute locations (in meters from some fixed origin) as well as symbolic locations (room numbers).

While in principle the same basic prediction techniques can be used for both end-user and system-enhancement applications, the constraints and metrics differ. For example, in end-user applications it may be important to know the user's geographical location, while for a system-enhancement application knowing parameters required for signaling (e.g., cell ID or paging area) is more relevant. In this chapter, we have assumed that system-enhancement applications are the target. We assume that time is discretized and a user's location is given in symbolic coordinates. The task of the location prediction algorithm is to provide the user's (symbolic) location at the next time step or, if possible, the path of the user (a sequence of locations) over several time steps. Note that the user's predicted location at the next time step may be the same as the current location.

Location prediction has been implicitly or explicitly utilized in many areas of mobile and wireless system design. For example, consider the problem of locating a cellular phone user in order to deliver a call to that user or, more specifically, to determine in which set of cells, called a location area (or registration area or paging area), the user is currently located. Broadly speaking, the strategies employed in cellular systems essentially consist of having the mobile device report its location area to a set of databases that are queried when an incoming call arrives for the user.[1,2] Analysis showed that these strategies placed a heavy burden on the SS7 signaling network used in the PSTN wired backbone, in particular on the Home Location Register (HLR) database in the user's home network. Early work on reducing this signaling impact used the following simple idea: the caller's switch recorded (cached) the location area at which the called party was found when the switch last queried the HLR database.[3] For the new call, it attempted to locate the user at that location area first, and only queried the HLR if the user was no longer found there. Essentially, the switch using this caching strategy employed a simple location prediction algorithm in order to reduce the overall signalling load in the system. As we discuss later in this chapter, this can be regarded as a type of order-1 Markov predictor where the next term in the sequence is assumed to be identical to the present term. In this example, as in other applications, in abstract terms the location prediction algorithm is worthwhile if, over the entire population of users:

$$pS > A + (1-p)F \qquad (11.1)$$

where p is the probability of successful prediction, S is the benefit of success, A is the cost of running the prediction algorithm itself, and F is the cost of failure. Of course, this general relation has to be made specific and evaluated for any particular application, architecture, and prediction algorithm.

We briefly mention types of location and mobility prediction that we do not consider in this chapter. Efforts on location management and prediction for other types of mobile objects, including software objects such as agents,[4,5] are outside the scope of this chapter. We also do not cover efforts on predicting the amount of time that the wireless link that a mobile host is using will stay usable (e.g., Su and coworkers[6]), or on predicting the link quality and availability.[7]

In Section 11.2, we begin our discussion with some definitions and preliminaries. In Section 11.3, we describe location prediction algorithms that do not explicitly take advantage of the specifics of the mobile wireless environment. These algorithms generally are based on order-k Markov prediction or on the prediction capabilities inherent in text compression algorithms. In Section 11.4, we describe algorithms that have been designed for location prediction in mobile wireless environments and explicitly take advantage of their characteristics. Location prediction techniques have been developed or suggested for many domain-specific applications, including location management (e.g., see references 8 through 10, and references therein), smooth handoffs (e.g., see references 10 through 13, and references therein), resource reservations (e.g., see references 14 through 19, and references therein), call admission control (e.g., see

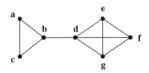

Time	9:00	9:04	9:18	(:20	9:31	9:43	9:56
Cross	a	b	d	c	f	d	b

FIGURE 11.1 Example of a Cell Boundary Graph and Movement History

references 20 through 25, and references therein) and adaptive resource management (e.g., see references 11 and 26, and references therein). We do not attempt to provide a comprehensive survey of all these domain-specific techniques; instead we briefly present some domain-specific algorithms that suggest slightly different approaches to the prediction problem.

All the algorithms we discuss essentially compare the sequence of recent movements the user has made to the sequence of locations $\mathcal{H}$ representing users' (or this particular user's) stored movement history. One way that the domain-independent algorithms discussed in Section 11.3 differ from the domain-specific algorithms discussed in Section 11.4 is in how the stored history $\mathcal{H}$ is partitioned into substrings for the purposes of this comparison. The order-k Markov predictors (Section 11.3.1) essentially compare the most recent k movements of the user with every length k substring in $\mathcal{H}$. The LZ-based predictors (Section 11.3.2) partition $\mathcal{H}$ based on techniques used in text compression algorithms. In contrast, the domain-specific algorithms partition the history based on the semantics of the location prediction domain, such as considering a location as a substring delimiter if the user was stationary there a significant amount of time or if the location is at the boundary of the geographical service area.

11.2 PRELIMINARIES

11.2.1 MOVEMENT HISTORY

We will assume that the user's location is given in symbolic coordinates, and that the system has a record of the user's past movements based on its location updates. The user's movement history is thus represented as a sequence of symbols from an alphabet $\mathcal{A}$, which is finite. The information contained in the record depends on the system's update scheme. How this information is interpreted affects also the results of prediction algorithms.

For example, in movement-based update schemes, updates occur every time the user has crossed M cell boundaries.[27,28] If $M = 1$, then a record can look like the table in Figure 11.1, which has the details of all the user's cell crossings of the map on the left from 9 a.m. to 10 a.m.

In this case, each symbol of the sequence is an ordered pair (t, v), where t is the time of update (and is discretized) and v is the user's new location. Depending on the purpose of a prediction algorithm, this sequence may be transformed to a new one so that the size of $\mathcal{A}$ is smaller.* For instance, if a prediction algorithm's objective is to predict the user's next cell, it may just consider the sequence *abdefdb*. In this case, $\mathcal{A} = V$, the set of all the cell IDs.

For a time-based update scheme such as described in Rose,[29] updates occur every T time units. If we set $T = 5$ minutes, then the sequence generated from Figure 11.1 would be *abbbeeeffdddb*. Notice that while this sequence is able to capture the duration of residence of the user at a cell, it misses the cell crossings that took place between updates. It will, nonetheless, be useful for a prediction algorithm that seeks to predict the user's location in the next T time units.

In Bhattacharya and Das,[30] the authors suggested generating a movement history that reflects both cell crossings and durations of residence of the user at each cell, while keeping $\mathcal{A} = V$. Such a history is generated when a user updates every T time units and every M cell crossings. If $T = 5$ and $M = 1$, the sequence that reflects the movement in Figure 11.1 would be *abbbbdeeefffddddb*.

Hence, there are different ways of representing a user's movement history as a sequence from a finite alphabet. It is imperative that the choice of sequence be matched to the purpose(s) of the prediction algorithm.

In the following discussion, we will assume that the appropriate movement history has been chosen. A history of length n is denoted as a sequence

$$\mathcal{H}_n = \langle X_1 = a_1, \ldots, X_n = a_n \rangle$$

where each X_i is a random variable and each $a_i \in \mathcal{A}$. For brevity we will sometimes denote a sequence as $a_1 a_2 \ldots a_n$. The notation $P(X_i = a_i)$ denotes the probability that X_i takes the value a_i and $\hat{P}(X_i)$ denotes an estimate of $P(X_i)$.

11.2.2 APPROACH

We will discuss various prediction algorithms which use different approaches to predict the next term of the user's itinerary, i.e., sequence L. When possible, we discuss also how these methods can be extended to predict not just the next term but the future terms of the sequence as well.

Prediction algorithms usually consist of two steps: (1) to assign conditional probabilities to the elements of $\mathcal{A}$ given the user's movement history $\mathcal{H}$, and (2) to use these values to predict the next term in the sequence.

We note that there are some applications where a single guess for the next term may be too restrictive. For example, to satisfy QoS requirements, Chan and coworkers[31] proposed an algorithm that outputs a subset of $\mathcal{A}$ so that the probability that the next term of the movement history is in this set is above some threshold.

11.3 DOMAIN-INDEPENDENT ALGORITHMS

We discuss two families of domain-independent algorithms that have been used as the core of techniques for location prediction in mobile wireless systems.

---

* Intuitively, the smaller $|\mathcal{A}|$ is, the better because there will be fewer choices for a prediction.

11.3.1 THE ORDER-K MARKOV PREDICTOR

The order-k Markov predictor assumes that the next term of the movement history depends only on the most recent k terms. Moreover, the next term is independent of time, i.e., if the user's history consists of $\mathcal{H}_n = \{X_1 = a_1, ..., X_n = a_n\}$, then for all $a \in \mathcal{A}$,

$$P(X_{n+1} = a \mid \mathcal{H}_n) = P(X_{n+1} = a \mid X_{n-k+1} = a_{n-k+1}, ..., X_n = a_n)$$
$$= P(X_{i+k+1} = a \mid X_{i+1} = a_{n-k+1}, ..., X_{i+k} = a_n), \forall i \in \mathbb{N}$$

The current state of the predictor is assumed to be $< a_{n-k+1}, a_{n-k+2}, ..., a_n>$.

If the movement history was truly generated by an order-k Markov source, then there would be a transition probability matrix M that encodes these probability values. Both the rows and columns of M are indexed by length-k strings from $\mathcal{A}^k$ so that $P(X_{n+1} = a \mid \mathcal{H}_n) = M(s, s')$, where s' and s are the strings $a_{n-k+1}a_{n-k+2}...a_n$ and $a_{n-k+2}a_{n-k+3}...a_n a$, respectively. In this case, knowing M would immediately provide the probability for each possible next term of $\mathcal{H}_n$.

Unfortunately, even if we assume the movement history is an order-k Markov chain for some k we do not know M. Here is how the order-k Markov predictor estimates the entries of M. Let $N(t, s)$ denote the number of times the substring t occurs in the string s. Then, for each $a \in \mathcal{A}$,

$$\hat{P}(X_{n+1} = a \mid \mathcal{H}_n) = \frac{N(a_{n-k+1}...a_n a, \mathcal{H}_n)}{N(a_{n-k+1}...a_n, \mathcal{H}_n)} \tag{11.2}$$

If r predictions are allowed for X_{n+1} then the predictor chooses the r symbols in $\mathcal{A}$ with the highest probability estimates. In other words, the predictor always chooses the r symbols which most frequently followed the string $a_{n-k+1}...a_n$ in $\mathcal{H}_n$.

Vitter and Krishnan[32] suggested using the above predictor in the context of prefetching Web pages. Chan and coworkers[31] considered five prediction algorithms, three of which can be expressed as order-k predictors. (We will briefly describe the other two in a later section.) Two of them, the location based and direction-based prediction algorithms, are equivalent to the order-1 and order-2 Markov predictors, respectively, when $\mathcal{A}$ is the set of all cell IDs. The time-based prediction algorithm is an order-2 Markov predictor when $\mathcal{A}$ is the set of all time-cell ID pairs.

We emphasize that the above prediction scheme can be used even if the movement history is not generated by an order-k Markov source. If the assumption about the movement history is true, however, the predictor has the following nice property. Consider $\mathcal{F}$, the family of prediction algorithms that make their decisions based only on the user's history, including those that have full knowledge of the matrix M. Suppose each predictor in $\mathcal{F}$ is used sequentially so that a guess is made for each X_i. We say that a predictor has made an error at step i if its guess $\hat{X}_i$ does not equal X_i. Let $\hat{\pi}(\mathcal{H}_n) = \sum_{i=1}^{n} I(\hat{X}_i \neq X_i)/n$, where I is the indicator function, denote the average error rate for the order-k Markov predictor. Let $\mathcal{F}(\mathcal{H}_n)$ be the best possible average error rate achieved by any predictor in $\mathcal{F}$. Vitter and Krishnan[32] showed that as $n \to \infty$, $\hat{\pi}(\mathcal{H}_n) \to \pi_{\mathcal{F}}(\mathcal{H}_n)$, i.e., the average error rate of the order-k predictor is the best possible as, $n \to \infty$, or that the average error rate is asymptotically optimal. This result holds for any given value of k.

We observe that the algorithm can (naively) be generalized for predicting location beyond the next cell, i.e., predicting the user's path, as follows. If M is known, $P(X_t = a \mid \mathcal{H}_n)$ for any $t > n + 1$ and each $a \in \mathcal{A}$ can be determined exactly from $M^{(t-n)}$. The process to estimate $M^{(t-n)}$ is to first construct $\hat{M}$, the estimate for M, and then raising it to the $(t - n)$th power. Then the value(s) of X_t can be predicted using the same procedure as for X_{n+1}. However, any errors in the estimate of M will accumulate as prediction is attempted for further steps in the future.

11.3.2 THE LZ-BASED PREDICTORS

LZ-based predictors are based on a popular incremental parsing algorithm by Ziv and Lempel[33] used for text compression. Some of the reasons this approach was considered were (1) most good text compressors are good predictors[32] and (2) LZ-based predictors are like the order-k Markov predictor except that k is a variable allowed to grow to infinity.[30] We first describe the Lempel-Ziv parsing algorithm.

11.3.2.1 The LZ Parsing Algorithm

Let γ be the empty string. Given an input string s, the LZ parsing algorithm partitions the string into distinct substrings $s_0, s_1, \ldots, s_m$ such that $s_0 = \gamma$, for all j ≥ 1, substring s_j without its last character is equal to some s_i, $0 \leq i < j$, and $s_0, s_1, \ldots, s_m = s$. Observe that the partitioning is done sequentially, i.e., after determining each s_i, the algorithm only considers the remainder of the input string. For example, $\mathcal{H}_n = abbbbdeeefffd\text{-}dddb$ is parsed as γ, a, b, bb, bd, e, ee, f, ff, d, dd, db.

Associated with the algorithm is a tree, which we call the LZ tree, that is grown dynamically to represent the substrings. The nodes of the tree represent the substrings where node s_i is an ancestor of node s_j if and only if s_i is a prefix of s_j. Typically, statistics are stored at each node to keep track of information such as the number of times the corresponding substring has been seen as a prefix of $s_0, s_1, \ldots, s_m$ or the sequence of symbols that has followed the substring. The tree associated with this example is shown in Figure 11.2.

Suppose s_{m+1} is the newest substring parsed. The process of adding the node corresponding to the s_{m+1} in the LZ tree is equivalent to tracing a path starting from the root of the tree through the nodes that correspond to the prefixes of s_{m+1} until a leaf is reached. The node for s_{m+1} is then added to this leaf. Path tracing resumes at the root.

11.3.2.2 Applying LZ to Prediction

Different predictors based on the LZ parsing algorithm have been suggested in the past.[19,30,32,34,35] We describe some of these here and then discuss how they differ.

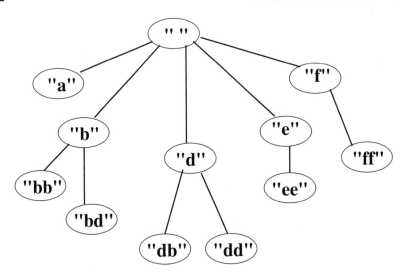

FIGURE 11.2 An example LZ parsing tree.

Suppose $\mathcal{H}_n$ has been parsed into $s_0, s_1, \ldots, s_m$. If the node associated with s_m is a leaf of the LZ tree, then LZ-based predictors usually assume that each element in $\mathcal{A}$ is equally likely to follow s_m. That is, $\hat{P}(X_{n+1} = a | \mathcal{H}_n) = 1/|\mathcal{A}|$. Otherwise, LZ-based predictors estimate $P(X_{n+1} = a \mid \mathcal{H}_n)$ based on the symbols that have followed s_m in the past when $\mathcal{H}_n$ was parsed.

1. Vitter and Krishnan[32] considered the case when the generator of $\mathcal{H}_n$ is a finite-state Markov source, which produces sequences where the next symbol is dependent on only its current state. (We note that a finite-state Markov source is more general than the order-k Markov source in that the states do not have to correspond to strings of a fixed length from $\mathcal{A}$.) They suggested using the following probability estimates: for each $a \in \mathcal{A}$, let

$$\hat{P}(X_{n+1} = a | \mathcal{H}_n) = \frac{N^{LZ}(s_m a, \mathcal{H}_n)}{N^{LZ}(s_m, \mathcal{H}_n)} \qquad (11.3)$$

where $N^{LZ}(s', s)$ denotes the number of times s' occurs as a prefix among the substrings $s_0, \ldots, s_m$, which were obtained by parsing s using the LZ algorithm.

It is worthwhile comparing Equation 11.2 with Equation 11.3. While the former considers how often the string of interest occurs in the entire input string, (i.e., in our application, the history $\mathcal{H}_n$), the latter considers how often it occurs in the partitions s_i created by LZ. Thus, in the example of Figure 11.2, while bbb occurs in $\mathcal{H}_n$, it does not occur in any s_i.

If r predictions are allowed for X_{n+1}, then the predictor chooses the r symbols in $\mathcal{A}$ that have the highest probability estimates. Once again, Vitter and Krishnan showed that this predictor's average error rate is asymptotically optimal when used sequentially.

2. Feder and coworkers[35] designed a predictor for *arbitrary* binary sequences, i.e., sequences where $A = \{0,1\}$. The following are their probability estimates:

$$\hat{P}(X_{n+1} = 0 \mid \mathcal{H}_n) = \frac{(N^{LZ}(s_m 0) + 1)}{(N^{LZ}(s_m 0) + N^{LZ}(s_m 1) + 2)}$$

(11.4)

$$= 1 - \hat{P}(X_{n+1} = 1 \mid \mathcal{H}_n).$$

For convenience let $a = \hat{P}(X_{n+1} = 0 \mid \mathcal{H}_n)$. Then the predictor guesses that the next term is 0 with probability ϕ/α, where for some chosen $\varepsilon > 0$

$$\phi(\alpha) = \begin{cases} 0 & 0 \le \alpha < \dfrac{1}{2} - \varepsilon \\ \dfrac{1}{2\varepsilon}\left[\alpha - \dfrac{1}{2}\right] + \dfrac{1}{2} & \dfrac{1}{2} - \varepsilon \le \alpha \le \dfrac{1}{2} + \varepsilon \\ 1 & \dfrac{1}{2} + \varepsilon < \alpha \le 1. \end{cases}$$

Essentially, the predictor outputs 0 if α is above $1/2 + \varepsilon$, 1 if α is below $1/2 + \varepsilon$, and otherwise outputs 0 or 1 probabilistically.

Example: Let $s_m = 00$ and suppose $N^{LZ}(000) = 11$ and $N^{LZ}(001) = 9$. Thus, $\hat{P}(X_{n+1} = a \mid \mathcal{H}_n) = 12/22 = 0.545 = 1 - \hat{P}(X_{n+1} = 1 \mid \mathcal{H}_n)$. If $\varepsilon = 0.01$, then the predictor would guess 0 for X_{n+1}; if $\varepsilon = 0.25$, then the predictor would guess 0 for X_{n+1} with probability of 13/22 and 1 with probability 9/22.

Compare the output of Vitter and Krishnan[32] with Feder and coworkers' algorithm[35] for this example assuming a single prediction is desired ($r = 1$). The former simply calculates the probability estimates $\hat{P}(X_{n+1} = 0 \mid \mathcal{H}_n) = 12/22$ and $\hat{P}(X_{n+1} = 1 \mid \mathcal{H}_n) = 10/22$, so that the prediction is 0. The latter provides predictions with certainty only if the probability estimates for 0 and 1 are not too close (i.e., not within 2ε of each other).

If the predictor is used sequentially, then Feder and coworkers[35] showed that its asymptotic error rate is the best possible among predictors with finite memory.

3. Krishnan and Vitter generalized Feder and coworkers' procedure and result to arbitrary sequences generated from a bigger alphabet;[34] i.e., $|A| \ge 2$. Their scheme for computing $\hat{P}(X_{n+1} = a \mid \mathcal{H}_n)$ for each $a \in A$ is not only based on how frequently a followed s_m, but also on the order in which the symbols followed s_m. Specifically, they assigned probability estimates in the following manner. Suppose after the first occurrence of s_m, the next occurrence is $s_m h_1$ for some symbol h_1, the following occurrence is $s_m h_2$, etc. Consider all the symbols h_i that have followed s_m (after its first occurrence), and create the sequence $h = h_1 h_2 h \dots h_r$. Let $h(i, j)$ denote the

subsequence $h_i h_{i+1} h \ldots h_j$. Let $q = \lceil \sqrt[4]{t} \, \rceil$ and $h' = h(4^{q-1} + 2, t)$. Then for each $a \in \mathcal{A}$,

$$\hat{P}(X_{n+1} = a \mid \mathcal{H}_n) = \frac{(N(a,h'))^{2^q}}{\displaystyle\sum_{a \in A} (N(a,h'))^{2^q}}. \qquad (11.5)$$

If r predictions are allowed for X_{n+1}, then the predictor uses these probability estimates to choose without replacement r symbols from $\mathcal{A}$.

Example: Suppose 9 symbols from $\mathcal{A} = \{0,1,2\}$ followed s_m and the sequence of the symbols is $h = 210011102$. Thus, $q = 2$. The relevant subsequence h for predicting X_{n+1} is 1102. The frequency of 0, 1, and 2 in the subsequence are 1, 2, and 1, respectively, so their probability estimates are 1/18, 16/18, and 1/18, respectively. The predictor will pick r of these symbols without replacement using these probabilities

4. Bhattacharya and Das[30] proposed a heuristic modification to the construction of the LZ tree, as well as a way of using the modified tree to predict the most-likely cells that the user will reside in so as to minimize paging costs to locate the user. The resulting algorithm is called *LeZi-Update*. Although their application (similar to that of Yu and Leung[19]) lies in the mobile wireless environment, the core prediction algorithm itself is not specific to this domain. For this reason, and for ease of exposition, we include it in this section.

As pointed out earlier, not every substring in $\mathcal{H}_n$ forms a leaf s_i in the LZ parsing tree. In particular, substrings that cross boundaries of the s_i, $0 < i \leq m$, are missed. Further, previous LZ-based predictors take into account only the occurrence statistics for the prefixes of the leaves s_i. To overcome this, the following modification is proposed. When a leaf s_i is created, all the proper suffixes of s_i are considered (i.e., all the suffixes not including s_i itself.) If an interior node representing a suffix does not exist, it is created, and the occurrence frequency for every prefix of every suffix is incremented.

Example: Suppose the current leaf is $s_m = bde$ and the string de is one that crosses boundaries of existing s_i for $1 \leq i < m$ (see Figure 11.2). Thus de has not occurred as a prefix or a suffix of any s_i, $0 < i < m$. The set of proper suffixes of s_m is $S_m = \{\gamma, e, de\}$ and because there is no interior node for de, it is created. Then the occurrence frequency is incremented for the root labeled γ, the first-level children b and d, and the new interior node de.

We observe that this heuristic only discovers substrings that lie within a leaf string. Also, at this point it would be possible to use the modified LZ tree and apply one of the existing prediction heuristics, e.g., use Equation 11.3 and the Vitter-Krishnan method.

However, in Bhattacharya and Das[30] a further heuristic is proposed to use the modified LZ tree for determining the most-likely locations of the user. This second heuristic is based on the prediction by partial match (PPM) algorithm for text compression.[36] (The PPM algorithm essentially attempts to "blend" the predictions of several order-k Markov predictors, for $k = 1, 2, 3, ...$; we do not describe it here in detail.) Given a leaf string s_m, the set of proper suffixes S_m is found. Observe that each element of S_m is an interior node in the LZ tree. Then, for each suffix, the heuristic considers the subtree rooted at the suffix and finds all the paths in this subtree originating from the root. (Thus these paths would be of length l for $l = 1, 2,...t$, where t is the height of the suffix in the LZ tree.) The PPM algorithm is then applied. PPM first computes the predicted probability of each path in the entire set of paths and then uses these probabilities to compute the most-probable symbol(s), which is the predicted location of the user.

5. Yu and Leung[19] use LZ prediction methods for call admission control and bandwidth reservation. Their mobility prediction approach is novel in that it predicts both the location and handoff times of the users. Assume time is discretized into slots of a fixed duration. The movement history $\mathcal{H}_n$ of a user is recorded as a sequence of ordered pairs $(S,l_1),(T_2,l_2),...,(T_n,l_n)$, where S is the time when the call was initiated at cell l_1 and $S + T_i$ is when the i-th handoff occurred to cell l_i for $i \geq 2$. In other words, T_i is the relative time (in time slots) that has elapsed since the beginning of the call when the i-th handoff was made.

Similar to LeZi-Update, if the user is currently at cell l and time $S + T$, the predictor uses the LZ tree to determine the possible paths the user might take and then computes the probabilities of these paths. Unlike LeZi-Update, the computation is easier and is not based on PPM. The algorithm estimates the probabilities $P_{i,j}(T_k)$, the probability that a mobile in cell i will visit cell j at timeslot $S + T_k$, by adding up the probabilities of the paths in the LZ tree that are rooted at the current time-cell pair and contain the ordered pair (j, T_k).

11.3.3 OTHER APPROACHES

Chan et al.[31] suggest a different approach for location prediction based on using an order-2 Markov predictor with Bayes' rule. The idea is to first predict the general direction of movement and then use that to predict the next location. For the order-2 predictor, the last two terms of the user's itinerary, $L = <L_1, L_2>$ are used. First, the most-likely location m steps away from the current location, i.e., L_{2+m}, is predicted based on the user's past history. Then the next location L_3 is predicted using Bayes' rule and the reference point L_{m+2} by choosing the location B_x with the highest probability as follows.

$$P(L_1 L_2 B_x \mid L_{2+m}) = \frac{P(L_{2+m} \mid L_1 L_2 B_x)P(L_1 L_2 B_x)}{\sum_{j-1}^{n} P(L_1 L_2 B_j)P(L_{2+m} \mid L_1 L_2 B_j)} \qquad (11.6)$$

11.4 DOMAIN-SPECIFIC HEURISTICS

In this section, we discuss several location prediction algorithms that have been proposed for specific application domains.

11.4.1 MOBILE MOTION PREDICTION (MMP)

Liu and Maguire[11] present a location prediction algorithm that can be used for improving mobility management in a cellular network. The movement of a user is modeled as a process $\{M(a,t) : a \in \mathcal{A}, t \in T\}$, where $\mathcal{A}$ is the set of possible locations (called *states*) and T is an index set indicating time. It is assumed that the user's movement is composed of a regular movement process $\{S(a,t)\}$ and a random movement process $\{X(a,t)\}$.

A location is called a stationary state if the user resides there longer than some threshold time interval, and a transitional state otherwise. A location at the geographical boundary of the service area is called a boundary state. For convenience we call the stationary and boundary states marker states. Two types of movement patterns are then defined. A movement circle (MC) is a sequence of locations that begins and ends with the same location and contains at least one marker state. A movement track (MT) is a sequence of locations that begins and ends with a marker state. It is possible for an MC to be an MT and vice versa. It is assumed that the regular movement process $\{S(a,t)\}$ consists only of the MC process $\{MC(a,t)\}$ and the MT process $\{MT(a,t)\}$. The random movement process is further assumed to be a pure (i.e., order-1) Markov process.

The mobile motion prediction (MMP) algorithm consists of a regularity detection algorithm (RDA) that builds up a database of MC and MT seen for each user over time, and a motion prediction algorithm (MPA) that uses this database. Although the details of these algorithms are not clearly specified, it appears that MPA operates as follows (for convenience we describe the algorithm using MTs, although the process for MCs is similar). Suppose the most-recent $k - 1$ locations of the mobile's history are the sequence $L = l_1 l_2 \ldots l_{k-1}$, i.e., L is the suffix of $\mathcal{H}$ of length $k - 1$. Suppose there exists an MT stored in the database, $C = c_0 \ldots c_n$, where c_0 and c_n are marker states. Using a matching algorithm (described later), suppose L matches C; we call C a candidate MT. If the current location of the mobile l_k equals that predicted by C, then C continues as the candidate MT and MPA uses it for prediction (as described later). Otherwise, MPA uses the matching algorithm on the sequence $L = l_i l_{i+1} \ldots l_{k-1} l_k$, where l_i is the most recent marker state in L, to find a new MT candidate D.

The matching algorithm uses three matching heuristics. The first is called state matching and computes a state matching index μ indicating the degree of similarity in the locations of the mobile's actual itinerary compared to the candidate MT. Using the notation above, for the itinerary L, let m, $0 < m < k$, be the number of locations that appear in both L and C. Then $\mu = m/(k - 1)$, and higher values indicate a better match. The second heuristic is time matching and computes an index η indicating the degree of similarity in the residence times of the mobile in each location for the mobile's itinerary compared to the candidate MT. Let r_i be the time the mobile spends at each location l_i in L, and similarly r_i be the residence time for location c_i in C. Then

$$\eta = \frac{\sum_{i=1}^{k-2} |r_i - s_i|}{\sum_{i=1}^{k-2} |r_i + s_i|} \tag{11.7}$$

and lower values indicate a better match. The third heuristic is frequency matching and computes an index Φ comparing F' and F, where F' is how often the mobile's itinerary appears in a given time period, and F is how often the candidate MT appears over the time period in the database. (Unfortunately, only approximation equations and an example are given for F' and F, so this heuristic is quite unclear.) Then $\Phi = |(F'/F) - 1|$ and lower values indicate a better match. The matching algorithm applies the three matching heuristics in sequence, so that the final prediction is dependent on μ, η, and Φ.

Note that in MMP any itinerary that cannot be classified based on the stored MT and MC is assumed to be a random movement. The MMP algorithm is not clearly specified and lacks a theoretical foundation but does contain interesting ideas in terms of classifying location types (stationary and boundary states), as well as movement patterns (MC, MT) and different matching heuristics applied in sequence. Because it was one of the first attempts at location prediction for mobility management, it is often referenced.

11.4.2 SEGMENT MATCHING

Chan et al.[31] use a simplification of the Liu and Maguire algorithm, which they call the segment criterion algorithm. Like Liu and Maguire's stationary states, they define stationary cells based on the residence time of the user in the cell. They then partition the individual user's history into segments, where a segment is a sequence of cells that starts with a stationary cell and ends with the same or different stationary cell. Thus a segment is similar to an MT in Liu and Maguire except that it applies only to stationary cells; there is no concept of boundary cells.

The prediction algorithm begins constructing a segment as the user moves, i.e., the user's itinerary after k moves is $L = l_1 l_2 \ldots l_k$, where l_1 is a stationary cell. L is compared with the user's stored segments. A match is found if $l_i = c_i$, $1 \leq i \leq k \leq n$, for some stored candidate segment $C = c_1 c_2 \ldots c_n$. In that case, the prediction is the cell c_{k+1}. If there are multiple candidates, then the prediction is the most frequently occurring cell in position $k + 1$ among the candidate segments.

Chan et al.[31] use two heuristics for overcoming the limitations of relying on the individual user's history. The first heuristic attempts to compensate for sudden changes in movement behavior as follows. The last ten predictions are compared with the user's actual itinerary; a higher weight is assigned to the latest movement of the user if six of the predictions were incorrect, and this weight is decreased gradually if predictions come inside a preset criterion of success. (The weight, the way in which it is decreased, and the criterion are not specified.) The second heuristic attempts to compensate for users who do not have a movement history, and uses the

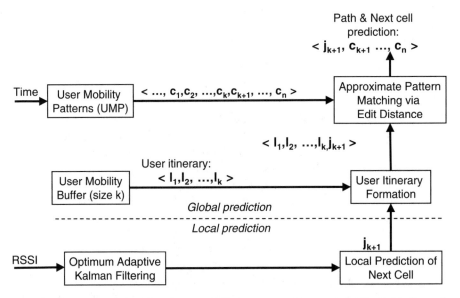

FIGURE 11.3 Hierarchical location prediction process. (*Source:* Liu, T., Bahl, P., and Chlamtac, I., Mobility modeling, location tracking, and trajectory prediction in wireless ATM networks, *IEEE J. Sel. Areas Commun.*, 16 (6), 922–936, 1998.)

aggregate history over all users instead. These heuristics are used also for Chan et al.'s Markov prediction schemes (see Section 11.3.1), as well as the probabilistic scheme (Section 11.3.3).

11.4.3 HIERARCHICAL LOCATION PREDICTION (HLP)

Liu et al.[10] have developed a two-level prediction scheme intended for use in mobility management in a wireless ATM environment, but with wider applicability. The lower level uses a local mobility model (LMM), which is a stochastic model for intracell movements, while the top level uses a deterministic model (the global mobility model, or GMM) dealing with intercell movements. The two-level scheme is depicted in Figure 11.3 and summarized below.

The local prediction algorithm is intended only for predicting the next cell that the user will visit, while the global prediction can predict the future path. The local prediction algorithm uses consecutive radio signal strength indication (RSSI) measurements and applies a modified Kalman filtering algorithm to estimate the dynamic state of a moving user, where the dynamic state consists of the position, velocity, and acceleration of the user. When the user is "close" to a cell boundary, (i.e., in an area called a correlation area defined precisely using the geometry of hexagonal cells), the estimated dynamic state is used to determine cell-crossing probability for each neighboring cell, and the cell with the highest crossing probability is output as the predicted next cell. This prediction is used as input to the global prediction algorithm.

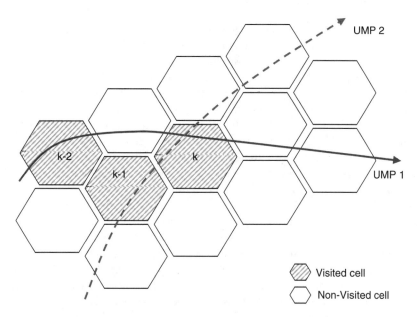

FIGURE 11.4 Benefit of local prediction for selecting a candidate UMP. (*Source:* Liu, T., Bahl, P., and Chlamtac, I., Mobility modeling, location tracking, and trajectory prediction in wireless ATM networks, *IEEE J. Sel. Areas Commun.,* 16 (6), 922–936, 1998.)

As in the Liu and Maguire MMP algorithm, the global prediction algorithm relies on a number of user mobility patterns (UMP) recorded for each user. The user's itinerary so far, along with the next cell predicted by the local prediction algorithm, is compared to these stored UMPs and an edit distance is generated, which is based on the smallest number of cell insertion, cell deletion, and cell ID modification operations required to make the itinerary identical to a UMP. If the edit distance is less than a threshold value, the UMP with the smallest edit distance is found using a dynamic programming method; this UMP is assumed to be the candidate UMP and to indicate the general direction of user movement. The remaining portion of the candidate UMP is output as the predicted path for the user.

Liu et al.[10] show by simulation that their scheme has a better prediction accuracy than MMP for mobility patterns with a moderate or high degree of randomness.

It is also worth noting that unlike the MMP algorithm the accuracy of next cell prediction using the local prediction algorithm is based purely on RSSI measurements and is independent of the long-term movement patterns of the mobile. This local prediction can be used to improve the path prediction as depicted in Figure 11.4. Next cell prediction can help choose between two candidate UMPs when the user's itinerary (shown by the shaded cells) is equidistant in terms of edit distance from them.

11.4.4 OTHER APPROACHES

One approach we have not considered so far is to take the help of the user for
prediction. Obviously, as little burden as possible should be placed on the user, but
one could envision a situation where the user is only prompted for current destination
(which could be collected, for example, via a voice prompt) and this is used (possibly
along with history information) to do cell and path prediction. Another variation
could be that at the start of the day, the user is prompted for a list of the day's likely
destinations (or is approximately inferred from her calendar), so that interaction is
minimized further. Madi et al.[37] have developed schemes for prompting for user
destination input in this way, although no prediction is carried out as such.

Biesterfeld et al.[38] propose using neural networks for location prediction. They
have considered both feedback and feed-forward networks with a variety of learning
algorithms. Their preliminary results indicated, somewhat nonintuitively, that feed-
forward networks delivered better results than feedback networks.

Das and Sen[39] consider how to use location prediction to assign cells to location
areas so as to minimize mobility management cost (i.e., the combined paging and
location update cost), where the location areas are arranged hierarchically. The
assignment is dynamic and is calculated periodically, every τ seconds. The user's
movement history is assumed only to consist of a set, $L_c = \{(l_i, f_i) : 0 \leq i \leq c\}$,
recording the frequency f_i with which each cell l_i is visited during the previous
period, where c is the number of distinct cells visited.

Then the probability of the user visiting a given cell l_i is calculated simply as
the relative frequency with which the user has visited that cell in the previous period,

i.e., $f_i/\sum_j f_j$. (Similarly, frequencies for visiting areas where two or more cells

overlap are recorded, and the probability of visiting the overlapping area calculated.)
We observe that this scheme is similar to order-0 Markov prediction, as described
in Bhattacharya and Das.[30]

These cell visit probabilities are used to assign cells to a most-probable location
area (MPLA). However, it is possible that the user has left this area and moved to
an adjoining area, called the future probable location area (FPLA). If the user is not
found in the MPLA, the FPLA is paged. The cells belonging to the FPLA are
determined based on the current mean velocity, the last cell visited, and (optionally)
the direction of future movement. Given the cells in the FPLA, the probability that
the user will visit a particular cell is estimated by a heuristic that takes into account

the total number of cell crossings $\sum_j f_j$, the frequency $\max_i f_i$, and c, the number

of distinct cells visited.

11.5 CONCLUSIONS

In this chapter, we have provided an overview of different approaches to predicting
the location of users in a mobile wireless system. This chapter is not intended to be
a comprehensive survey, and in particular we have only summarized a few of the
approaches being used in domain-specific algorithms for location prediction.

We see two general ideas pursued in the literature: domain-independent algorithms that take results from Markov analysis or text compression algorithms and apply them to prediction, and domain-specific algorithms that consider the geometry of user motion as well as the semantics of the symbols in the movement history. Domain-independent algorithms tend to have well-founded theoretical principles on which they are based and can make analytical statements about their prediction accuracy. However, in some cases, these statements refer to the asymptotic optimality of their accuracy, i.e., that as the input history approaches infinite length, no similar prediction algorithm can do any better. While satisfying from a theoretical point of view, it is unclear how relevant these results are in practice. On the other hand, some domain-specific algorithms offer heuristics that appear intuitively appealing but have no explicit theoretical analysis to support them. Clearly, a better bridge between engineering intuition and theoretical analysis would be helpful.

One of the major barriers to practical advancement in this area is the lack of publicly available empirical data to guide future research. Most studies have used artificial mobility models; relatively few, e.g., Chan and coworkers,[31] have collected empirical data for the domain of interest (cellular handoffs) and used them for validation. We compare the situation to the early work done on caching disk pages in computer systems. A large variety of cache replacement policies, many of which were intuitively plausible, were proposed. It was only empirical data from page fault traces that enabled the conclusion that the Least Recently Used (LRU) algorithm offered the best compromise between simplicity and effectiveness in most cases. Large-scale statistical data for the domains of interest is sorely needed to help provide benchmarks and directions for future research.

ACKNOWLEDGMENTS

We thank Prof. John Kieffer for interesting and helpful discussions, as well as Dr. Xiaoning He for comments on a draft of this chapter.

References

1. Jain, R., Lin, Y.-B., and Mohan, S., Location strategies for personal communications services, *Mobile Communications Handbook,* 2nd ed., Gibson, J., Ed., CRC Press, Boca Raton, FL, 1999.
2. Akyildiz, I.F. et al., Mobility management for next generation wireless systems, *Proc. IEEE,* 87(8), 1347–1385, 1999.
3. Jain, R. et al., A caching strategy to reduce network impacts of PCS, *IEEE J. Selected Areas Commun.,* 12(8), 1434–1444, 1994.
4. Wolfson, O. et al., Cost and imprecision in modeling the position of moving objects, Proc. IEEE Intl. Conf. Data Eng. (ICDE), Feb. 1998.
5. Pitoura, E. and Samaras, G., Locating objects in mobile computing, *IEEE Trans. Knowledge Database Eng.,* 13 (4), 571–692, 2001.
6. Su, W., Lee, S., and Gerla, M., Mobility prediction and routing in ad hoc wireless networks, *Intl. J. Net. Mgmt., (*11), 3–30, 2001.

7. Jiang, S., He, D., and Rao, J., A prediction-based link availability estimation algorithm for mobile ad hoc networks, *Proc. IEEE InfoCom,* 2001.

8. Krishna, P., Vaidya, N., and Pradhan, D., Static and adaptive location management in mobile wireless networks, *Computer Commun.,* 19 (4), 321–334, 1996.

9. Shivakumar, N., Jannink, J., and Widom, J., Per-user profile replication in mobile environments: algorithms, analysis, and simulation results, *ACM/Baltzer Mobile Networks Appl. (MONET),* 2 (2), 129–140, 1997.

10. Liu, T., Bahl, P., and Chlamtac, I., Mobility modeling, location tracking, and trajectory prediction in wireless ATM networks, *IEEE J. Selected Areas Commun.,* 16 (6), 922–936, 1998.

11. Liu, G. and Maguire, G. Jr., A class of mobile motion prediction algorithms for wireless mobile computing and communications, *ACM/Baltzer Mobile Networks Appl. (MONET),* 1 (2), 113–121, 1996.

12. Chan, J. et al., A framework for mobile wireless networks with an adaptive QoS capability, Proc. Mobile Mult. Comm. (MoMuC), Oct. 1998, pp. 131–137.

13. Erbas, F. et al., A regular path recognition method and prediction of user movements in wireless networks, IEEE Vehic. Tech. Conf. (VTC), Oct. 2001.

14. Levine, D., Akyildiz, I., and Naghshineh, M., A resource estimation and call admission algorithm for wireless multimedia networks using the shadow cluster concept, *IEEE/ACM Trans. Networking,* 2, 1–15, 1997.

15. Bharghavan, V. and Jayanth, M. Profile-based next-cell prediction in indoor wireless LAN, Proc. IEEE SICON, Singapore, Apr. 1997.

16. Riera, M. and Aspas, J., Variable channel reservation mechanism for wireless networks with mixed types of mobility platforms, Proc. IEEE Vehic. Tech. Conf. (VTC), May 1998, pp. 1259–1263.

17. Oliveira, C., Kim, J., and Suda, T., An adaptive bandwidth reservation scheme for high-speed multimedia wireless networks, *IEEE J. Selected Areas Commun.,* 16 (6), 858–874, 1998.

18. Chua, K.C. and Choo, S.Y., Probabilistic channel reservation scheme for mobile pico/microcellular networks, *IEEE Commun. Lett.,* 2 (7), 195–196, 1998.

19. Yu, F. and Leung, V., Mobility-based predictive call admission control and bandwidth reservation in wireless cellular networks, *Computer Networks,* 38, 577–589, 2002.

20. Posner, C. and Guerin, R., Traffic policies in cellular radio that minimize blocking of handoff calls, Proc. 11th Int. Teletraffic Cong., Kyoto, Japan, 1985.

21. Ramjee, R., Nagarajan, R., and Towsley, D., On optimal call admission control in cellular networks, Proc. IEEE Infocom, San Francisco, 1996.

22. Naghshineh, M. and Schwartz, M., Distributed call admission control in mobile/wireless networks, *IEEE J. Selected Areas Commun.,* 14, 711–717, 1996.

23. Chao C. and Chen, W. Connection admission control for mobile multiple-class personal communications networks, *IEEE J. Selected Areas Commun.,* 15, 1618–1626, 1997.

24. Luo, X., Thng, I., and Zhuang, W., A dynamic channel pre-reservation scheme for handoffs with GoS guarantee in mobile networks, Proc. IEEE ICC, Vancouver, Canada, 1999.

25. Zhang, T. et al., Local predictive reservation for handoff in multimedia wireless IP networks, *IEEE J. Selected Areas Commun.,* 19, 1931–1941, 2001.

26. Bharghavan, V. et al., The TIMELY adaptive resource management architecture, *IEEE Pers. Commun.,* 20–31, 1998.

27. Akyildiz, I., Ho, J., and Lin, Y., Movement based location update and selective paging for PCS networks, *IEEE/ACM Trans. Networking,* 4 (4), 629–638, 1996.

28. Bar-Noy, A., Kessler, I., and Sidi, M., Mobile users: to update or not to update?, *ACM/Baltzer J. Wireless Networks,* 1 (2), 175–195, 1995.
29. Rose, C., Minimizing the average cost of paging and registration: a timer-based method, *Wireless Networks,* 2 (2), 109–116, 1996.
30. Bhattacharya, A. and Das, S.K., LeZi-update: an information-theoretic framework for personal mobility tracking in PCS networks, *ACM/Kluwer Wireless Networks,* 8 (2-3), 121–135, 2002.
31. Chan, J., Zhou, S., and Seneviratne, A., A QoS adaptive mobility prediction scheme for wireless networks, Proc. IEEE Globecom, Sydney, Australia, Nov. 1998.
32. Vitter, J. and Krishnan, P., Optimal prefetching via data compression, *J. ACM,* 43 (5), 771–793, 1996.
33. Ziv, J. and Lempel, A., Compression of individual sequences via variable-rate coding, *IEEE Trans. Inf. Theory,* 24 (5), 530–536, 1978.
34. Krishnan, P. and Vitter, J., Optimal prediction for prefetching in the worst case, *SIAM J. Computing,* 27 (6), 1617–1636, 1998.
35. Feder, M., Merhav, N., and Gutman, M., Universal prediction of individual sequences, *IEEE Trans. Inf. Theory,* 38, 1258–1270, 1992.
36. Bell, T.C., Cleary, J.G., and Witten, I.H., *Text Compression,* Prentice Hall, New York, 1990.
37. Madi, M., Graham, P., and Barker, K., Mobile computing: predictive connection management with user input, Technical report, Department of. Computer Science, University of Manitoba, Aug. 1996.
38. Biesterfeld, J., Ennigrou, E., and Jobmann, K., Location prediction in mobile networks with neural networks, Proc. Intl. Workshop Appl. Neural Networks to Telecom., June 1997, pp. 207–214.
39. Das, S.K. and Sen, S.K., Adaptive location prediction strategies based on a hierarchical network model in a cellular mobile environment, *Computer J.,* 42 (6), 473–486, 1999.

12 Handoff and Rerouting in Cellular Data Networks

Gopal Racherla and Sridhar Radhakrishnan

CONTENTS

12.1 INTRODUCTION

Cellular mobile data networks consist of wireless mobile hosts (MH), static hosts (SH), and an underlying wired network consisting of base stations (BS) and intermediate routers. Each base station has a geographical area of coverage called a cell. Hosts communicate with each other using the base stations and the underlying wired network. Figure 12.1 shows the architecture of a cellular data network. The figure shows the fixed cellular backbone consisting of base stations and a group of mobile hosts that can move from one cell to another. When a mobile host moves from one cell to another, it registers with the base station of the new cell. If there is an ongoing communication session between two hosts and one of the hosts moves out of its present cell, the session is interrupted. In order for the session to be restarted, a handoff or handover needs to take place in the network. Handoff is the process of transferring the control and responsibility for maintaining communication connectivity from one base station to another. Handoff is used by the mobile network to provide the mobile hosts with seamless access to network services and the freedom of mobility beyond the cell coverage of a base station. Rerouting is the process of setting up a new route (path) between the hosts after the handoff has occurred. Handoff[1-6] in mobile and wireless networks has been an active topic of research and development for the past several years. The rerouting problem also has been studied extensively in cellular, mobile, and wireless networks, including wireless ATM,[1,2,7-11] picocellular networks,[12] cellular networks,[3] wireless LANs,[5,6,13] and connectionless networks.[14-18]

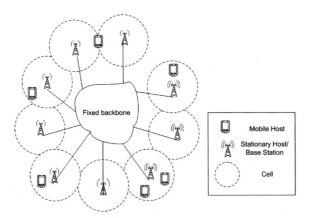

FIGURE 12.1 Architecture of a cellular data network.

Communication in a mobile data network can be either between two static hosts (static–static), a static host and a mobile host (static–mobile), or two mobile hosts (mobile–mobile). Static–static communication and its related routing algorithms have been studied extensively in the literature. Static–mobile communication and the consequent handoff and rerouting also have been studied in detail.[7,8,12,13,18] However, mobile–mobile rerouting has not been explored much in the literature. There are only a few[4,12,18] suggested schemes for mobile–mobile data communication and rerouting in mobile data networks. However, these mobile–mobile rerouting schemes are suboptimal. In addition, these schemes do not look at different rerouting strategies. Racherla and coworkers[4] have proposed a scheme for performing optimal rerouting in mobile–mobile networks.

In this chapter, we survey, classify, analyze, and evaluate several known rerouting (static–mobile and mobile–mobile) techniques for connection-oriented cellular data networks. We study connection-oriented networks as they provide performance guarantees needed for delivery of multimedia data to mobile hosts. We classify the various rerouting schemes in four major categories and do a survey of related work in detail. We use a set of rerouting metrics in order to compare and classify various static–mobile and mobile–mobile rerouting schemes. We discuss the characteristics and performance metrics used for the comparison of static–mobile and mobile–mobile rerouting in more detail later in the chapter.

The rest of the chapter is organized as follows. In this section, we continue to explore the nuances of the rerouting process in more detail. We study the characteristics of rerouting and use them for comparison and classification of rerouting schemes proposed in the literature. We classify various rerouting schemes in four categories. In Section 12.2, we analyze and evaluate the various rerouting classes. Each class is explained in detail, including the protocol used for rerouting, the advantages and disadvantages of the class, and implementation examples and variations of the rerouting class. In Section 12.3, we evaluate the rerouting schemes using several performance metrics that are calculated using analytical cost modeling. We study the issues involved in mobile–mobile rerouting, including potential problems and solution ideas for alleviating these problems, as well as all the known

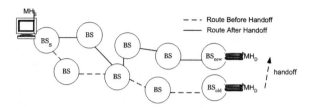

FIGURE 12.2 Rerouting process.

solutions for rerouting in mobile–mobile connections, in Section 12.4. In Section 12.5, we compare these schemes using several performance metrics including the total rerouting distance, the cumulative connection path length and the number of connections as the mobile hosts move. Finally, in Section 12.6, we present our conclusions and the plan for future work.

12.1.1 CLASSIFICATION OF REROUTING SCHEMES

Rerouting in cellular mobile environments occurs as result of a handoff. When a mobile host moves from one cell to another, a handoff is said to have taken place. In order to maintain communication connectivity, packets have to be rerouted to and from the MH. This process of reestablishment of a route (connection) is called as rerouting. Figure 12.2 depicts the rerouting process. Initially, the source mobile host (MH$_S$) is in a session with the destination mobile host (MH$_D$). MH$_S$ is in the cell being administered by source base station (BS$_S$). After some time, MH$_D$ moves from the cell of BS$_{old}$ to BS$_{new}$ after performing a handoff while MH$_S$ is stationary. The old route (between BS$_S$ and BS$_{old}$) and the new route (between BS$_S$ and BS$_{new}$) may be the same, partially the same or completely different. Because of overlaps in the cell coverage of adjacent cells, MH$_D$ may get a radio "hint" before it enters its cell. Using the radio hint, MH$_D$ can request BS$_{old}$ to inform BS$_{new}$ to set up the required connections in advance. This mechanism of using radio hints is known as radio hint processing.

We classify the rerouting strategies broadly as full rerouting, partial rerouting, tree-based rerouting, and cell forwarding rerouting. Each of the schemes can be either with or without a radio hint.[3–5] Full rerouting involves establishing a new routing path from BS$_{new}$ to BS$_S$. Full rerouting schemes are slow and inefficient and hence perform poorly. Examples of such schemes include full reestablishment without hints and full reestablishment with hint rerouting.[3] Partial rerouting involves finding the crossover point of the route between BS$_S$ and BS$_{old}$ and the route between BS$_S$ and BS$_{new}$ with a view to increasing route reuse. Examples of these schemes include incremental reestablishment without hints and incremental reestablishment with hint rerouting[3,5] and Nearest Common Neighbor Routing (NCNR).[8] Tree rerouting involves routing using a tree-based structure for communication. The base stations in the network form the nodes of the tree. The tree has a specially designated base station that acts as the root of the tree. Some implementations may have multiple trees that form the communication structure. This scheme typically uses multicasting for communication. Tree rerouting can be either (a) to a group (tree-group rerouting) as described in multicast reestablishment rerouting (with and without hint)[3] and the

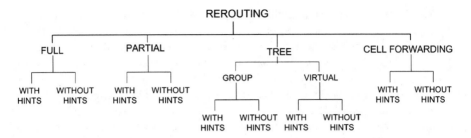

FIGURE 12.3 Classification of rerouting schemes.

picocellular network architecture rerouting,[12] or (b) from a single source to a single destination using a virtual tree (tree-virtual rerouting), where only one branch of the tree is active at a time. Examples of this scheme include the virtual tree scheme[1] and the SRMC scheme.[19] Also, tree-group rerouting can have either a static[3,19] or a dynamic group[12] to communicate with. Static tree-group rerouting involves a group consisting of members that do not change over time, while in the case of a dynamic tree-group rerouting the membership of a group may change. Cell forwarding rerouting involves designating a specialized base station to forward data packets to the MH_D when it moves from a "home" area. Such schemes include the ones described in the adaptive routing scheme[20] and the BAHAMA scheme.[21] Figure 12.3 shows the classification scheme.

12.1.2 RELATED WORK

In this section, we briefly describe related work in the area of comparative analysis of handovers and rerouting.

Toh explained how handovers in multicast connections can be achieved irrespective of the kind of multicast tree (source-based, server-based, or core-based) used in a wireless ATM environment.[6] Toh has proposed solutions to handle handover and rerouting for both multicast and unicast connections without categorizing rerouting strategies. However, his scheme also does not consider pure cell-forwarding schemes. The rerouting algorithm used in Toh's approach is partial rerouting using a crossover discovery algorithm. Toh demonstrates how this handoff and rerouting scheme can be used for both unicast and multicast connections using either distributed or centralized connection management. Toh's work considers partial rerouting with static–mobile connections.

Ramanathan and Steenstrup[22] have made an extensive survey of routing techniques for cellular, satellite, and packet radio networks. They categorize different types of handoffs in cellular telecommunication networks. These include mobile-controlled handoff (the MH chooses its new BS based on the relative signal strength), network controlled handoff (the MH's current BS decides the occurrence of a handoff based on the signal strength from the mobile host), mobile-assisted handoff (the MH's current BS requests the MH for information on the signal strengths from several nearby base stations and then decides, in consultation with the mobile switching center, when a handoff has occurred), and soft handoff (the MH may be affiliated to multiple base stations with approximately equal signal quality).

Bush[2] has classified handoff schemes for mobile ATM networks. The classification is specific to handovers (and not rerouting) for connection-oriented networks. These include pivotal connection handoff (a specific base station is chosen as a pivot to perform handoff), IP mobility-based handoff (handoffs require IP packet forwarding using mechanisms such as loose source routing), and handoff tree (a preestablished virtual circuit tree is used to automatically detect handoffs in a wireless ATM environment).

Cohen and Segall[23] describe a scheme for connection management and rerouting in standard (not wireless/mobile) ATM networks. Their rerouting is a Network Node Interface protocol, which can be invoked when an intermediate link or node in the virtual path fails. The protocol reroutes all the affected nodes to an alternate virtual path.

Ramjee et al.[24] have performed experimental performance evaluations of five types of rerouting protocols for wireless ATM networks. Their rerouting protocols are primarily based on crossover switch-based rerouting using mobile-directed handoff. In order to perform rerouting in an ATM environment, the ATM switch at the crossover point must change the appropriate entry in the translation table. This involves dismantling the old entry (break) and installing the new entry (make). Their evaluation includes the following rerouting schemes: make–break (make followed by break), break–make (break followed by make), chaining (cell forwarding from the old base station to the new base station), make–break with chaining, and break–make with chaining.

Song and coworkers[25] have defined five kinds of rerouting schemes for connection-oriented networks:

1. *Connection-extension rerouting:* This rerouting is the same as cell forwarding rerouting.
2. *Destination-based rerouting:* In this scheme, the rerouting point is predetermined at the connection time. This is similar to connection-extension rerouting except the rerouting base station is a predetermined base station and not necessarily the old base station.
3. *Branch-point-traversal-based rerouting:* This rerouting is the same as partial rerouting.
4. *Multicast-join-based rerouting:* This rerouting is the same as tree-group rerouting.
5. *Virtual-tree-based rerouting:* This rerouting is the same as tree-virtual rerouting.

Mishra and Srivastava[10] have classified rerouting schemes in an ATM environment as:

- *Extension:* This rerouting is the same as cell forwarding rerouting.
- *Extension with loop removal:* This rerouting is a specialized case of cell forwarding rerouting with removal of any possible path loops caused by chaining.
- *Total rebuild:* This rerouting is the same as full rerouting.
- *Partial rebuild:* This rerouting is the same as partial rerouting. There are two variants in this scheme (fixed or dynamically chosen crossover point).
- *Multicast to neighbors:* This rerouting is the same as tree-group rerouting.

Naylon et al.[9] have provided classification of rerouting in the wireless ATM LANs as follows:

- *Virtual connection tree:* This rerouting is the same as tree-virtual rerouting.
- *Path rerouting:* This rerouting is the same as partial rerouting.
- *Path extension scheme:* This rerouting is the same as cell forwarding rerouting.

From the related work described here, we see that our classification encompasses all the proposed rerouting schemes we have described. In this sense, our work can be viewed as a generalization of previous rerouting classifications. As we shall we in the subsequent sections of the chapter, our contribution in this work is fourfold. First, we provide a comprehensive framework for comparing rerouting strategies for connection-oriented cellular data networks. We subsume also the classification proposed by various authors in the literature. Second, we analyze and evaluate the performance of the rerouting schemes using a large array of metrics, including the ones proposed by other researchers. We propose and evaluate specialized metrics that are applicable to each class of rerouting schemes. Third, we abstract the common handshaking signals from all the rerouting schemes (with and without hints) to avoid repetition and provide also a framework to compare these common handshaking signals. Finally, we compare and contrast the performance of rerouting in mobile–mobile connections. This last contribution is the first such attempt, to the best of our knowledge.

12.2 ANALYSIS OF REROUTING SCHEMES

In this section, we describe for each class of rerouting the basic protocol and its different implementations, advantages, and disadvantages. Self-descriptive figures are provided to aid in the explanation.

In the descriptions of the rerouting schemes, we make the following assumptions. There is coverage overlap between cells. An MH communicates with only one BS at a time. Each MH can measure the radio signal strength of the base stations in order to know about its entry into a new cell. Handoff and rerouting is initiated by the destination mobile host (MH_D). The source mobile host (MH_S) is assumed to be stationary during the rerouting process.

12.2.1 COMMON HANDSHAKING SIGNALS
FOR REROUTING SCHEMES

Many handshaking signals during the rerouting process are common to all the rerouting schemes. We have abstracted these signals in order to avoid repetition in all the rerouting schemes. Using the framework that we describe here, we can selectively compare the rerouting schemes, with or without these handshaking signals. The handshaking protocol is assumed to be completed before the actual rerouting protocol. We describe these signals for rerouting schemes, with and without hints, separately. In case of schemes without hints, the handshaking protocol is the

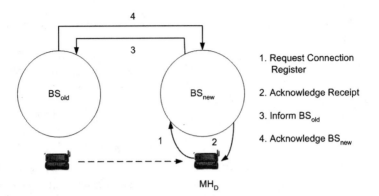

FIGURE 12.4 Rerouting handshaking without hints.

same for all the rerouting schemes. However, the handshaking protocol varies for schemes that use hints. We now describe these handshaking schemes briefly.

12.2.1.1   Without Hints

The handshaking protocol for all rerouting schemes without hints is shown in Figure 12.4. The messages are:

1. MH_D enters the new cell and requests a connection with BS_{new} after identifying itself. MH_D informs BS_{new} the identity of BS_{old} and its present connections.
2. BS_{new} acknowledges the MH_D request. MH_D can continue its transmissions.
3. BS_{new} requests BS_{old} to forward MH_D's data to BS_{new}. The message requests also that BS_{new} be allowed to be the "anchor" for MH_D's transmission.
4. BS_{old} acknowledges and grants permission to BS_{new}. After this, BS_{new} begins forwarding MH_D's transmitted data to MH_S through BS_{old}.

12.2.1.2   With Hints

The handshaking protocol for rerouting with hints is dependent on the rerouting scheme. Only the partial rerouting scheme has a different handshaking protocol, as shown in Figure 12.5. It should be noted that BS is a special base station that has different functionality for each of the rerouting schemes. In the schemes other than partial rerouting, the handshaking protocol is as follows:

1. MH_D requests BS_{old} to send a list of active connections to BS_{new}.
2. BS_{old} sends the list to BS_{new}.
3. BS_{new} establishes the connections with BS.
4. BS acknowledges the establishing of the connections.

In case of partial rerouting, BS_{old} invokes the crossover discovery algorithm and BS_{new} establishes the connections with BS. BS then acknowledges the establishment of the connections.

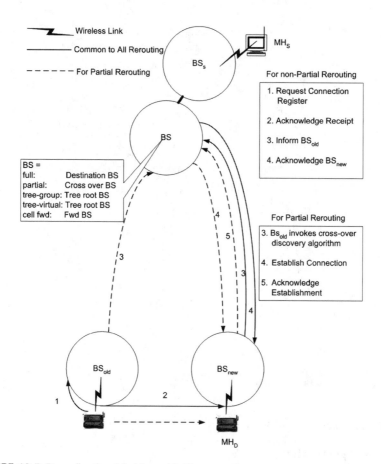

FIGURE 12.5 Rerouting handshaking with hints.

12.2.2 FULL REROUTING

Full rerouting can occur with or without hints. We describe later the generic full rerouting schemes without hints. Detailed protocol descriptions of all rerouting schemes (with and without hints) are explained in Seshan[3] and Racherla and coworkers.[4] Figures 12.6 and 12.7 describe the protocol of a generic full rerouting without hints and full rerouting with hints, respectively.

12.2.2.1 Implementations

An implementation of full rerouting, namely, full reestablishment schemes with and without hints, is described by Seshan.[3] The generic full rerouting explained prviously is the same as Seshan's implementation. The source (a video server in the implementation) is assumed to be fixed, while the destination is the MH_D. The mobile host moves from the old base station BS_{old} to the new base station BS_{new}. The rerouting involves finding a new route from BS_{new} to BS_S.

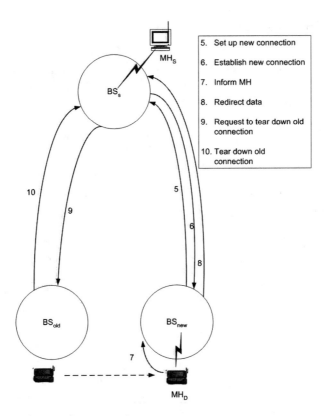

FIGURE 12.6 Full rerouting without hints.

12.2.2.2 Special Metrics

With respect to full rerouting, we look at two special metrics, namely, old connection teardown time and new connection setup time.

1. *Old connection tear down time (T_{tear}):* It is the total time required for BS_{dest} to inform BS_{old} to tear down the old connection and for BS_{old} to comply.
2. *New connection setup time (T_{tear}):* It is the total time elapsed from the time MH_D informs BS_{old} to send a list of active connections to the time when BS_{dest} confirms the establishment of the connections.

12.2.3 PARTIAL REROUTING

Partial rerouting tries to use as much of the old route as possible in the new route. The heart of partial rerouting is the crossover discovery algorithm.[5,6] The algorithm aims to find the base station (called the crossover point) that belongs to both the old and the new route so that the overlap between the old and the new path is maximized. We assume that BS_{cross} is the base station at the crossover point. Figure 12.8 shows the protocol of a generic partial rerouting without hints.

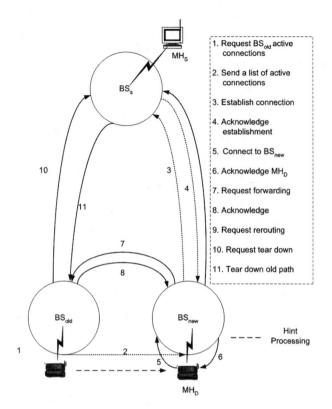

FIGURE 12.7 Full rerouting with hints.

12.2.3.1 Implementations

There are many variations of implementing partial rerouting. Seshan[3] gives an implementation called incremental reestablishment rerouting with and without hints. The protocol of Seshan's implementation is the same as the generic protocol described previously. However, the protocol does not specify the crossover discovery mechanism.

Toh[6] gives an implementation of partial rerouting, also called incremental reestablishment with and without hints. Toh's implementation is tailored for wireless LANs (local area networks). Toh's strategy for partial rerouting without hints, unlike the generic partial rerouting without hints described in Figure 12.8, assumes that the wireless link between the MH and BS_{old} has failed, resulting in the unavailability of hints. In Toh's incremental reestablishment rerouting without hints protocol, BS_{old} does not acknowledge the MH (message 2 in the generic protocol). Also, messages 3 and 4 are absent, and there is cell loss as BS_{new} does not request BS_{old} to forward cells. In addition, there are no explicit messages to tear down the old connection (messages 10 and 11). In case of Toh's incremental reestablishment rerouting with hints, BS_{new} invokes the crossover discovery algorithm (message 3) instead of BS_{old} as described in the generic partial rerouting with hints scheme.[4] Also, messages 8

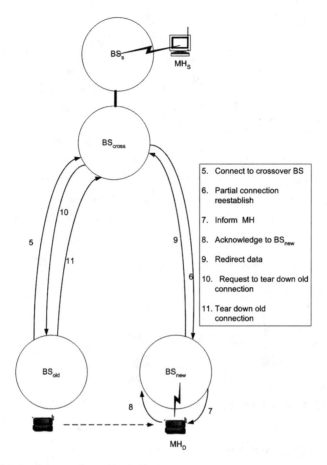

FIGURE 12.8 Partial rerouting without hints.

and 9 intended for cell forwarding from BS_{old} to BS_{new} are absent, resulting is cell loss. Toh describes many algorithms for discovery of the crossover switch. Akyol and Cox's[7,8] strategy, called NCNR rerouting, performs partial rerouting by choosing the crossover discovery as the nearest common neighbor of the BS_{old} and BS_{new}. Their scheme checks to see if there is a direct link between BS_{new} and BS_{old} and whether the traffic is time dependent (e.g., voice, video) or throughput dependent (e.g., data). It should be noted that the performance of the partial rerouting is strongly dependent on the performance of the crossover discovery algorithm.

12.2.3.2 Special Metrics

With respect to partial rerouting, we study several performance metrics, including old connection teardown time, new connection setup time, the time required to invoke crossover discovery algorithm, the time required to actually discover the crossover switch, and the efficiency of path reuse.

1. *Old connection teardown time (T_{tear}):* It is the time required for BS_{cross} to inform BS_{old} to tear down the old connection and for BS_{old} to comply.
2. *New connection setup time (T_{new}):* It is the total time elapsed since MH_D requests a connection with BS_{new} till the confirmation of the establishment of the new connection.
3. *Time required for invoking the crossover discovery algorithm (T_{cross}):* The total time in the rerouting process until the crossover switch discovery algorithm has been invoked.
4. *Time to discover the crossover switch ($T_{discover}$):* The total time for finding the crossover point after invoking the crossover discovery algorithm. This term depends on the algorithm used.
5. *Partial reuse efficiency (η_{part}):* The fraction of the new path that has been reused.

12.2.4 TREE REROUTING

Tree routing involves setting up and using a tree with base stations as nodes to communicate to a group of base stations. The tree can be either static or dynamic.

12.2.4.1 Tree-Group Rerouting

Figure 12.9 describes the protocol of a generic tree-group rerouting without hints. In this case, there is a dynamic multicast tree built that is used to multicast data to a group of base stations. The base station BS_{root} is the root of the multicast tree. The messages used to perform rerouting are described in the figure.

12.2.4.2 Tree-Virtual Rerouting

Figure 12.10 describes the protocol of a tree-virtual rerouting without hints. Let BS_{root} be the root of the virtual tree that connects a group of base stations. There are no tree-virtual rerouting algorithms with hints described in the literature. However, it is easy to conceive such a class. The most important aspect of this class of rerouting is that one path (from the root of the tree to the leaves of the tree) is active at a time unlike the tree-group rerouting. We assume that the virtual tree is established statically and is in place before the commencement of the rerouting process.

12.2.4.3 Implementations

Acampora[1] describes a rerouting scheme using virtual connection trees in a wireless ATM environment. A virtual connection tree is a collection of base stations and wired switching nodes and the connecting links. Each virtual connection tree, which is statically built prior to rerouting, has a fixed root node and the leaves of the tree are base stations. For each mobile connection, the tree provides a set of virtual connection numbers in each direction, each associated with a path from a leaf and root. When a mobile host that is already in the tree wants to handoff to another base station (in the same tree), it starts to transmit ATM cells with the connection number assigned for use between itself and the new base station. The cells flow along the

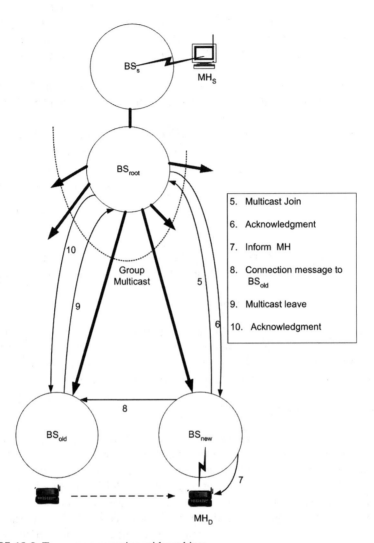

FIGURE 12.9 Tree-group rerouting without hints.

fixed path between the new base station and the root. Appropriate translation tables are maintained at the nodes of the tree to understand and implement handoff and the consequent rerouting. Seshan[3] describes a scheme for tree-group rerouting called multicasting reestablishment. The protocol of the generic tree-group rerouting described earlier is based on Seshan's proposed scheme, which includes strategies for multicasting rerouting with and without hints. However, Seshan's scheme does not address the issue of how the members of the groups are decided. Ghai and Singh[12] describe a tree-group rerouting scheme based on a dynamic grouping of base stations and the mobility characteristics of the MH. The scheme is for a

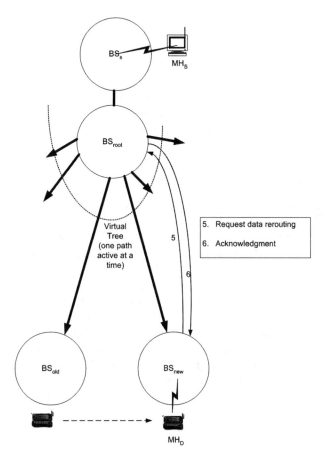

FIGURE 12.10 Tree-virtual rerouting without hints.

picocellular network with a three-tier hierarchy of cells. The scheme does not take advantage of hints to aid in rerouting.

12.2.4.4 Special Metrics

12.2.4.4.1 Tree-Virtual Rerouting

- *Virtual tree setup time* ($T_{vtree-setup}$): It is the time required for setting up the virtual tree, which includes the time to choose the root, broadcast the information to all the nodes of the tree, and for all the nodes to join the tree.
- *Virtual tree teardown time* ($T_{vtree-tear}$): It is the time required to tear down the virtual tree.
- *Number of nodes in virtual tree* (N_{vtree}): N_{vtree} is determined statically and remains fixed.

12.2.4.4.2   Tree-Group Rerouting

- *Multicast tree setup time ($T_{mcast-setup}$):* The time required to set up the multicast tree.
- *Multicast tree teardown time ($T_{mcast-tear}$):* The time required to tear down the multicast tree.
- *Number of nodes in the multicast tree (N_{mcast}):* N_{mcast} depends on the algorithm in question. In static algorithms, this value is fixed and remains constant. In algorithms that dynamically decide the number on-the-fly, the number changes dynamically.
- *Multicast join/leave time ($T_{mcast-jn}$, $T_{mcast-lv}$):* The time required for nodes of the multicast tree to join or leave the tree.

12.2.5   CELL FORWARDING REROUTING

Cell forwarding rerouting involves using a specialized base station (BS_{fwd}) as an "anchor." Cell forwarding is used in many rerouting schemes implicitly. Full rerouting and partial rerouting can be considered specialized cell forwarding rerouting schemes. The anchor in all these cases is BS_{old}. Figure 12.11 describes the protocol of a generic cell forwarding rerouting without hints. There is no known cell forwarding scheme described in the literature, although such a scheme can be easily conceived. In the remainder of the discussion, when we refer to cell forwarding, we assume that the anchor is the old base station; thus, in this case data is simply forwarded from the previous destination base station. Therefore, we simply do not need any signaling other than the regular handshake without hints. In other words, steps 5 through 10 do not exist; this way, we avoid the overhead of setting up the new path and tearing down the old path.

12.2.5.1   Implementations

Cell forwarding and its variations are used in almost all rerouting schemes. In most cases, the old base station, BS_{old}, acts as the anchor. Yuan[20] describes a scheme using a specialized "anchor" to perform the cell forwarding.

12.2.5.2   Special metrics

In the case of cell forwarding, we study the following special metrics:

- *Old connection teardown time (T_{tear}):* This is true only if there is a specialized anchor used for forwarding data.
- *New connection setup time (T_{new}):* This is true only if there is a specialized anchor used for forwarding data.
- *Time to establish a path between BS_{old} and BS_{fwd} ($T_{cellfwd-path}$):* It is the time required to set up a path between BS_{old} and BS_{fwd}.

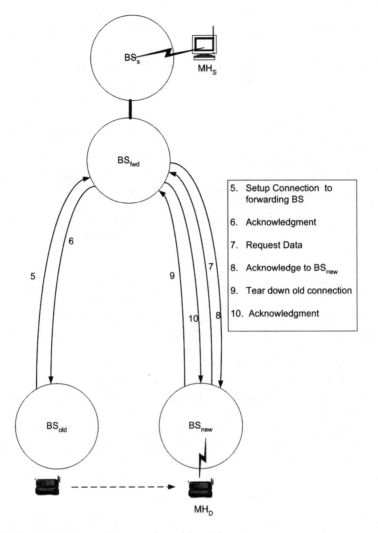

FIGURE 12.11 Cell forwarding rerouting without hints.

12.3 PERFORMANCE EVALUATION OF REROUTING SCHEMES

In this section, we use analytical cost modeling to calculate the performance metrics. Our cost model heavily borrows from the work done by Seshan[3] and Toh.[5,6] The cost modeling involves calculating the time required for each step in the protocol of the rerouting scheme. These individual times are used to calculate the proposed metrics.

We use the network parameters described in the literature[3–6] for our analytical cost models: bandwidth of the wireless link = 2 Mbps; bandwidth of the wired

TABLE 12.1
Advantages and Disadvantages of Various Rerouting Schemes

Rerouting	Advantages	Disadvantages
Full	Simple, easy to implement	Naive, inefficient, slow, prone to data loss
Partial	Maximizes resource utilization	Prone to data loss, onus of crossover discovery
Tree-virtual	Fastest, efficient, low data loss, works well if enough resources present	Multiple connections, static membership, membership difficult to decide
Tree-group	Fast, efficient, low data loss, dynamic membership	Multiple connections, resource intensive, wasted bandwidth
Cell forwarding	Simple, easy to implement	Requires special anchor node, requires a lot of buffering, inefficient, slow, prone to data loss

backbone network = 155 Mbps; latency of the wireless link, including data link and network layer processing = 2 ms; latency of the wired backbone, including data link and network layer processing = 500 μs; protocol processing time for control messages = 0.5 ms; protocol processing time for admission control = 2 ms ; maximum size of a control packet = 50 B; maximum size of a data packet = 1 kb; wireless channel acquisition time for a MH from a BS = 5 ms.

We measure the performance of the metrics by changing the number of hops in the appropriate paths, depending on the rerouting scheme. For cost modeling, we assume a perfect delivery of messages and that maximum throughput for a connection is the throughput of the bandwidth of the wireless link. Also, the calculations used in the model are on a per-channel basis. Detailed cost modeling and performance metric calculations can be found in Racherla and coworkers.[4]

12.3.1 COMPARISON OF REROUTING SCHEMES

We first look at the advantages and the disadvantages of each of the rerouting schemes. In order to compare the rerouting schemes, we built cost models for several metrics for each rerouting scheme using the system parameters and the length of the routes. Then, we compare the schemes using the metrics. To this end, we first consider metrics that are not dependent on the path length (in terms of number of hops) of the old/new connection. Next, we consider the metrics that are dependent on the length of the old/new connection. In this case, we vary the path length to determine its effect on these metrics.

12.3.1.1 Advantages and Disadvantages of the Rerouting Schemes

The advantages and disadvantages of the various rerouting schemes are described in Table 12.1.

TABLE 12.2
Comparison of Rerouting Schemes

Scheme	Type	$M_h{}^a$	$M_r{}^b$	Nodes	UC/BW[c]
Full (no hint)	Full	2	8	2 + D[d]	1/1
Full (hint)	Full	6	5	2 + D	1/1
Incremental (no hint)	Partial	2	9	3 + D	1/1
Incremental (hint)	Partial	7	5	3 + D	1/1
Multicast (no hint)	Tree-group	2	8	N[e] + D	N/N
Multicast (hint)	Tree-group	6	3	N + D	N/N
SRMC	Tree-virtual	3	9	4 + D	N/1
BAHAMA	Cell forwarding	2	5	N + D	1/1
Virtual Tree	Tree-virtual	1	1	3 + D	1/1
Adaptive	Cell forwarding	2	4	3 + D	1/1
Incremental[f] (no hint)	Partial	4	7	3 + D	1/1
Incremental[f] (hint)	Partial	3	4	2	1/1
NCNR (direct link)	Partial	7	2	4 + D	1/1
NCNR (link)	Partial	9	4	N + D	1/1
Picocellular[g] (same subnet)	Tree-group	2	0	N + D	N/N
Picocellular[g] (different subnet)	Tree-group	4	2	N + D	N/N

Note: All the schemes can handle time-dependent and throughput-dependent data.

[a] M_h: Number for messages to perform handoff.
[b] M_r: Number for messages to perform rerouting.
[c] UC/BW: Number of user connections/bandwidth.
[d] D: Topology dependent.
[e] N: Number of leaves in tree.
[f] Tailored for LANs. Other schemes are tailored for WANs.
[g] Tailored for picocells/micocells. Other schemes are tailored for microcells/macrocells.

12.3.1.2 Metrics not Dependent on the Connection Length

These metrics are fixed and give a complexity of the rerouting protocol in question. We base this comparison on the work done by Akyol and Cox.[7,8] These include the number of messages exchanged during handoff and rerouting, the number of nodes and user connections involved for rerouting, the user bandwidth allocated, whether the scheme is tailored for WAN (wide area networks), local area networks, or ATM-based networks. The comparison for the rerouting metrics is shown in Table 12.2.

12.3.1.2.1 Number of Messages Exchanged during Handoff

The partial rerouting scheme with hints requires the maximum number of messages for handoff while full rerouting without hints, partial rerouting without hints, tree-group rerouting without hints, tree-virtual rerouting, and cell forwarding rerouting require the least.

12.3.1.2.2 Number of Messages Exchanged during Rerouting

Cell forwarding rerouting requires the maximum number of messages exchanged during rerouting while tree-group with hints requires the minimum number.

12.3.1.2.3 Number of Nodes Involved in Handoff and Rerouting

The number of the nodes required depends on the network topology. However, in terms of the minimum requirements, the full rerouting schemes are the best, as they require two nodes (BS_{old} and BS_{new}) in addition to BS_{dest}, while the other schemes require at least three nodes in addition to BS_{dest}.

12.3.1.2.4 Number of User Connections Established for Rerouting

The tree-group rerouting schemes can handle N_{mcast} connections, while the others handle only one connection.

12.3.1.2.5 User Bandwidth Allocated for Handoff and Rerouting

If we consider the bandwidth allocated for handoff and rerouting for the full rerouting without hints as unity, then all the schemes have the same bandwidth requirements with the exception of the tree-group rerouting schemes. The tree-group rerouting requires a bandwidth that is N times the unit bandwidth requirements.

12.3.1.3 Metrics Dependent on the Connection Length

We now study the metrics that depend on the connection path length. Figures 12.12(a–d) show the effect of old connection path length on the service disruption time, total rerouting time, buffering requirements at the mobile host, buffering requirements at the base station on the uplink, and buffering requirements at the base station on the downlink, respectively. In these figures, we vary the old connection path length from 1 hop to 10 hops. We assume that the new connection path length is of the same length as the old connection and the control path length is twice the size of the connection path length (based on the assumptions in Seshan[3] and Gopal and co-workers).[4] We now discuss the performance of the various rerouting schemes in detail.

12.3.1.3.1 Service Disruption Time

Figure 12.12(a) depicts the effect of old connection path length on the service disruption time. From the figure, we see that:

1. The minimum service disruption time is for the tree-group with hints rerouting scheme. It does not vary with the number of hops in the path.
2. For all schemes, except the tree-group with hints, the service disruption time is dependent on the control path length between BS_{old} and BS_{new}. This is because, in the case of tree-group with hints scheme, there is no need to forward the data from BS_{old} to BS_{new} as the data is being multicast to both BS_{old} and BS_{new}.

(a)

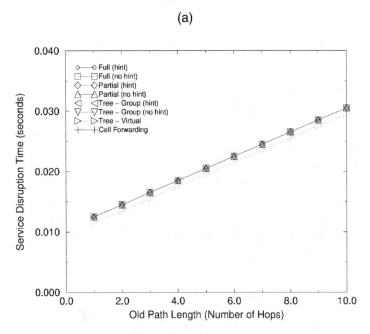

FIGURE 12.12 Performance metrics dependent on path length for rerouting: (a) service disruption time; (b) buffering at mobile host; (c) buffering at base station (uplink); (d) buffering at base station (downlink); (e) total rerouting completion time.

(b)

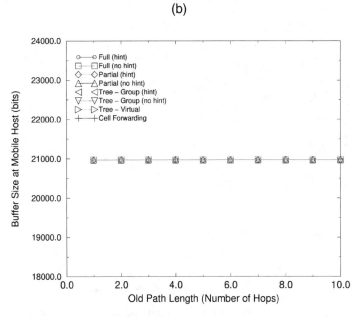

FIGURE 12.12 (continued)

(c)

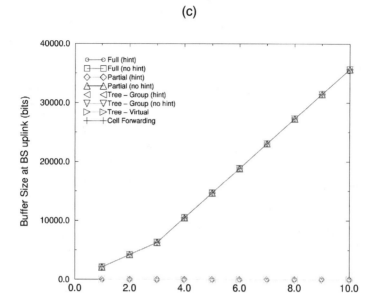

FIGURE 12.12 (continued)

(d)

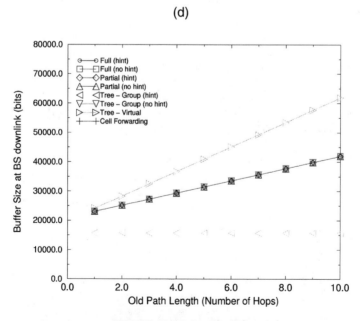

FIGURE 12.12 (continued)

(e)

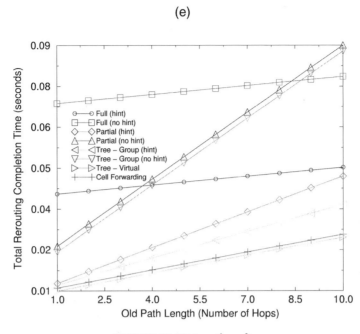

FIGURE 12.12 (continued)

3. The service disruption time for the full without hints, full with hints, partial without hints, partial with hints, tree-group without hints, and cell forwarding rerouting schemes is the same because all of these schemes depend on forwarding of downlink data from BS_{old}. The service disruption time for tree-virtual rerouting is not dependent on the length of control path, as it does not rely on downlink data forwarding. However, it is dependent on the length of the new path. In case of the tree-virtual rerouting, when the mobile host moves to a new base station, the root automatically recognizes that a handoff has taken place.

12.3.1.3.2   Buffering Required at the Mobile Host

Figure 12.12(b) depicts the effect of old connection path length on the buffering requirements at the mobile host. From the figure, we see that in each rerouting scheme, the buffering at the MH is only used to buffer data for the two specific messages in their respective protocol. The first is the registration message that MH sends to the BS_{new} and the second is for the acknowledgment that it receives in response to the first message; the cost of the messages is constant for the given parameters for all the schemes. All the schemes require the same amount of buffering at the MH.

12.3.1.3.3   Buffering Required in Base Station for the Uplink

Figure 12.12(c) depicts the effect of old connection path length on the buffering requirements at the base station for the uplink. From the figure, we see that:

1. The buffering requirements for uplink data for all the schemes without hints, including tree-virtual and cell forwarding, are the same.
2. There are no buffering requirements for uplink data for all the schemes with hints if the length of new and old forwarding path is the same. In general, buffering is dependent on the difference of the old and new path lengths.
3. Except for the case of tree-virtual rerouting, the buffering requirement for uplink data is proportional to the forwarding path length. In the case of tree-virtual rerouting, the buffering requirement for uplink data is proportional to the new path length between BS_{src} and BS_{new}.

12.3.1.3.4   Buffering Required in Base Station for the Downlink

Figure 12.12(d) depicts the effect of the old connection path length on the buffering requirements at the base station for the downlink. From the figure, we see that:

1. This metric is very closely dependent on the time required for forwarding downlink data from BS_{old} to BS_{new}.
2. Because all schemes with the exception of the tree-group with hints and tree-virtual scheme require data forwarding of downlink data, the buffering required in the base station for downlink for them is larger than the requirements for the tree-group with hints scheme.
3. The buffering requirements are the maximum for the tree-virtual rerouting as it does not rely on downlink data forwarding on the control path between BS_{old} to BS_{new}. BS_{new} has to buffer all the data on the downlink until it gets an acknowledgment from the server (source) that it has accepted the request to reroute data to BS_{new}.
4. There is a fixed amount of downlink buffering for the tree-group with hints scheme. This is used to buffer the data during the handoff period alone while the others require downlink buffering for the handoff and the time period until BS_{new} requests forwarding of data.

12.3.1.3.5   Rerouting Completion Time

Figure 12.12(e) depicts the effect of old connection path length on the rerouting completion time. From the figure, we see that:

1. The minimum rerouting time is for the tree-virtual rerouting scheme. The metric varies with the length of the new path (and hence with the number of hops in the old path). However, there is no need to either forward downlink data or form new connections or delete connections as the virtual tree is fixed.
2. The rerouting completion time for cell forwarding is comparable to the rerouting time for tree-virtual rerouting. In case of cell forwarding, there is no need to form new connections or delete old connections as the connection path is simply extended. The problem with cell forwarding is in the case of multiple handoffs, the length of the downlink forwarding path can become very large, giving rise to inefficiency. For tree-group with hints, rerouting

TABLE 12.3
Special Metrics for Rerouting Schemes

Rerouting Scheme Special Metrics	Without Hints	With Hints
Full		
• Old path teardown time	12.36 ms	12.36 ms
• New path setup time	23.82 ms	23.76 ms
Partial		
• Old path teardown time	12.36 ms	12.36 ms
• New path setup time	16.18 ms	15.55 ms
• Crossover discovery time	1.06 ms	1.06 ms
• Partial reuse efficiency	0.5	0.5
Tree-group		
• Tree teardown time	12.36 ms	12.36 ms
• Tree setup time	16.43 ms	16.43 ms
Tree-virtual		
• Old path teardown time	12.36 ms	12.36 ms
• Tree setup time	16.43 ms	16.43 ms
Cell forwarding		
• Old path teardown time	12.36 ms	12.36 ms
• New path setup time	16.18 ms	16.18 ms

the completion time is dependent on old path length and performs fairly well because it does not require any downlink data forwarding.

3. Full rerouting and partial rerouting perform worse than the other rerouting schemes. When the old path is small, full rerouting performs worse than partial rerouting. However, as the old (and the new) path length increases, partial rerouting performs worse than full rerouting because of the additional onus of computing the crossover point.

4. The completion time for the rerouting schemes with hints is less than their counterparts, which do not take advantage of hints. The completion time for rerouting schemes without hints is dependent on the length of the new path. In case of rerouting schemes with hints, the completion time is dependent and closely related to time needed for forwarding downlink data and to tear down old connections.

12.3.1.3.6 Special Metrics

Table 12.3 shows the values of some of the special metrics (described earlier) for the different rerouting schemes. More analysis of these results can be found in Racherla and coworkers.[4] We see from the previous discussions that:

1. The tree-group with hints and tree-virtual schemes perform the best among the rerouting schemes. However, they use a lot more resources for preestablishing the tree and maintaining it. This overhead may be worthwhile

for time-critical applications that require low rerouting completion time and service disruption.

2. In order to minimize service disruption, tree-group rerouting with hints is the best choice. If the rerouting completion time is the criterion considered, tree-virtual and cell forwarding rerouting perform the best, as they do not require forming new paths and tearing down old paths.

3. It is certainly advantageous to use rerouting with hints. This readily improves service disruption time (for tree-group rerouting), total rerouting completion time, and uplink buffering requirements at the base station.

4. Needless to say, the topology of the network is important. In essence, the lengths of the new and old paths are of vital importance.

5. Most of the rerouting schemes behave somewhat similarly because the underlying mechanisms use downlink data forwarding.

6. The full rerouting and the partial rerouting schemes perform the worst, as they involve forming new paths, tearing down old paths, and downlink data forwarding.

12.4 MOBILE–MOBILE REROUTING IN CONNECTION-ORIENTED NETWORKS

Static–static communication has been studied extensively and the routing algorithms for this type of communication are well known. The problem with the rerouting schemes for connection-oriented mobile networks described in the literature is the assumption that only one of the parties communicating in a session can be a mobile host (typically, the destination) and the other is stationary. Only Racherla and coworkers,[4] Ghai and Singh,[12] Biswas,[18] and cellular telecommunications standard IS-41(c)[26,27] suggest schemes for mobile-host-to-mobile-host communication. Biswas'[18] strategy uses an already preestablished route between two stationary hosts that house the mobile agents in charge of the communication. The only rerouting involves both the source and destination mobile hosts establishing paths to these stationary hosts. The disadvantage of this scheme is that if the mobile hosts keep moving, the path used for communication may be inefficient, as the scheme does not dynamically update the paths based on the location of the mobile hosts. The scheme proposed by Ghai and Singh suffers from the same problem. In addition, Ghai and Singh's scheme uses only dynamic multicast rerouting. The complex setup architecture uses a three-tier hierarchy (mobile host, base station, and supervisory host). In case of cellular telecommunications using the EIA/TIA IS-41(c) standard, cell forwarding rerouting is used continuously for multiple handoffs, as the mobile hosts move away from the original source and the destination base stations. The obvious disadvantage is that the new cell forwarding routes used are not optimal. We shall see this in more detail later as we compare all these schemes. In addition, these schemes do not look at different rerouting strategies known for static–mobile connections. Racherla and coworkers[4] have proposed a generic framework for mobile–mobile rerouting allowing unlimited movement by both the source and destination mobile hosts, while alleviating the problem of nonoptimal paths. Also, this framework for mobile–mobile rerouting is not tied to the type of rerouting (full, partial, cell, tree-based). These rerouting schemes basically concentrate on deciding the endpoints on the connections and can, in theory, use any type of rerouting. In this chapter, we compare the performance of these rerouting algorithms. Our comparison is based on calculating the total rerouting distance, the cumulative connection path length, and the amount of resources used as the mobiles move. As seen earlier, the metrics that determine efficiency of rerouting schemes in static–mobile connections are dependent on connection length. For example, the total rerouting path length gives a good estimate of metrics such as throughput, the total rerouting time, the service disruption time, and buffering requirements at the base stations.

We know of only these four techniques to perform rerouting in mobile–mobile connections. In this section, we discuss these techniques in more detail and look at their drawbacks and at the technique suggested by Racherla to alleviate them.

12.4.1 Problems in Mobile–Mobile Rerouting

Rerouting techniques that work for static–mobile connections do not work for mobile–mobile connections, as they cause some problems. We discuss the problems in this section.

12.4.1.1 Inefficiency

The original protocol assumes that only one end of a session is mobile. So, if MH_S moves (and assumes that MH_D is fixed), it establishes the new path between BS_{SN} and BS_{DO}, whereas if MH_D moves also (and assumes that MH_S is fixed), it establishes the new path between BS_{DN} and BS_{SO} instead of the correct path being established between BS_{SN} and BS_{DN}. So, the rerouting and path establishment would be inefficient.

12.4.1.2 Lack of Coordination

There is no mechanism to coordinate between the source and destination base stations. The bottom line is that the algorithm suffers from the classic problems of asynchronous messages and the lack of synchronization.

12.4.2 Techniques for Mobile–Mobile Rerouting

We now discuss the known techniques for rerouting in mobile–mobile connections. Specifically, we will look into the details of the schemes proposed by Racherla and coworkers,[4] Ghai and Singh,[12] Biswas,[18] and the EIA/TIA IS-41(c) cellular telecommunications standard. We look also at an approach for extending Biswas' work for mobile–mobile rerouting using core-based trees.

12.4.2.1 Biswas' Strategy: Mobile Representative and Segment-Based Rerouting

Biswas[18] proposes the use of software mobile agents called mobile representatives that handle the connection management operations of the mobile host. The representatives reside on one of the intermediate routers. Each mobile host has a corresponding mobile representative. It is assumed that the source mobile host MH_S (resp. the destination mobile host MH_D) is in the cell corresponding to the base station BS_S (resp. BS_D) and its mobile representative is MR_S (resp. MR_D). Thus, the initial route in the source mobile host–destination mobile host connection is $MH_S – BS_S – MR_S – MR_D – BS_D – MH_D$. The crux of Biswas' rerouting strategy is that the path connecting the corresponding source and the destination mobile representative is the same during the lifetime of the connection except the portion of the connection between the mobile host and the mobile representative that changes with handoff.

12.4.2.1.1   Problem with Biswas' Strategy
The disadvantage of this scheme is that if the mobile hosts keep moving, the path used for communication may be inefficient, as the scheme does not dynamically update the paths based on the location of the mobile hosts.

12.4.2.2   CBT (Core-Based Tree) Strategy: Extending Biswas' Work

We propose an extension of Biswas' approach by using a core-based tree connecting the source mobile representative to a group of destination mobile representatives instead of a simple path connecting the mobile representatives. The advantage with this scheme is that the total rerouting distance can be reduced significantly compared to Biswas' strategy. We will see in detail how this can be accomplished when we analyze the performance of this strategy.

12.4.2.2.1   Problem with the CBT Strategy
The only problem with the CBT is the excess use of resources as the core-based tree has to be built *a priori* for the purpose of rerouting.

12.4.2.3   Ghai and Singh's Strategy: Two-Level Picocellular Rerouting

Ghai and Singh[12] present a picocellular-based architecture wherein the number of handoffs and therefore handoff overhead within the network increases as cell size decreases. Their proposal describes a network architecture that supports a method for reducing the handoff overhead and the buffer space requirements using multicast groups and mobile trajectory prediction. Base stations (referred in this proposal as mobility support stations or MSSs) do not have any intelligence, but rely on a centralized authority called a supervisory host. The supervisory host calculates the mobile's likely trajectory, and forms a multicast group for MSSs that the mobile is likely to handoff to in the near future. All packets are multicast to this group. MSSs do not currently host the mobile host buffer packets in anticipation of a handoff. Such buffered packets are tossed out when the MSSs receive an update of the mobile's acknowledged sequence numbers. A connection-oriented network architecture is described also in this proposal. Virtual circuits are set up between the endpoints for communication. The communication is optimized for the case where two mobiles use the base stations and supervisory host(s). Multiple supervisory hosts are needed if the communicating mobile hosts are under the supervision of different supervisory hosts.

12.4.2.3.1 Problem with Ghai and Singh's Strategy

The scheme proposed by Ghai and Singh suffers from several drawbacks. The architecture is fixed, rigid, and requires a tree or tree-like topology. In addition, it requires an extra level of hierarchy (supervisory hosts), compared to the other schemes that we discussed earlier. This hierarchy results in longer paths. In addition, this scheme performs rerouting using a tree-group mechanism and does not allow for the other rerouting strategies.

12.4.2.4 EIA/TIA IS-41(c) Rerouting

The EIA/TIA IS-41(c) Protocol for cellular communications is designed to deal with handoffs and the subsequent forwarding of connections as the mobile hosts move. This is done by cell forwarding rerouting (called chaining in the scheme). Chaining in this case is done using switches as opposed to base stations used in all the other schemes described earlier. Because the connections are through switches, the links to old base stations are automatically removed during switching. These protocols are meant for circuit-switched networks and not for packet-switched networks. Data loss and data ordering is not an important concern in this case.

12.4.2.4.1 Problem with EIA/TIA IS-41(c) Rerouting

We cannot compare this scheme with the other schemes described earlier as this scheme is for circuit-switched networks and switches are used instead of base stations. However, because the rerouting involves cell forwarding, the rerouting is nonoptimal and inefficient.

12.4.2.5 Racherla's Framework for Mobile–Mobile Rerouting

In order to avoid the above-mentioned problems, Racherla and coworkers[4] proposed a source-initiated distributed rerouting algorithm. In this algorithm, the source base station is responsible for initiating the rerouting. The destination base station informs the source base station of the movement of MH_D. The establishment and removal of routes is initiated by the appropriate source base stations. The crux of this proposed framework is that incorrect requests for new path establishment are rejected, unlike the other proposed rerouting algorithms described earlier. In addition, Racherla's framework for rerouting in mobile–mobile communications is independent of the type of rerouting scheme (full, partial, tree-based or cell forwarding). Each base station maintains data structures in its local memory to keep track of connections, movements, and identities of mobile hosts MH_S and MH_D. This data structure allows base stations to decide whether or not to accept the creation or termination of a route during the rerouting procedure. The framework assumes that there are no lost messages and that the messages are delivered in a first-in, first-out manner on a channel. A mobile host can move any number of times. However, it eventually stays in a cell long enough for the rerouting to be completed. Each MH receives a radio hint informing it of its movement into another cell, which can be used to set up connections *a priori*. The framework assumes that there is initially a connection between the base stations corresponding to MH_S and MH_D. This proposed scheme has the following advantages over the other schemes described earlier:

- It uses optimal routes and avoids creating unnecessary routes.
- It gives the flexibility to use either full, partial, tree, or cell forwarding routing.
- It is independent of the network architecture and network topology.
- Packet loss is kept to a minimum using radio hints effectively.
- Processing at the mobile hosts is kept at a minimum.

12.4.3 COMPARISON OF REROUTING SCHEMES FOR MOBILE–MOBILE CONNECTIONS

We compare the previously described schemes based on the rerouting distances. We use the extended cross-bar network and analysis as suggested by Song and coworkers[25] for calculating the rerouting distance. The extended cross-bar architecture allows for simplified calculations. The analysis can be extended to other types of networks. In the architecture, the nodes represent the base stations. The horizontal and vertical links are of length 1, while the slanted links are $\sqrt{2}$ in length. We use $RD_{S,D}$ to represent the reroute distance between nodes S and D. We use $TRD(a, d)$ to represent the total rerouting distance between BS_{SN} and BS_{DN}, where the initial path length between the source and destination base station is a and the movement distance of the MH from its initial position is d. We denote the performance metrics for each of the rerouting schemes by using the name of the scheme as a subscript.

We assume that:

- The MH_S and MH_D are moving in a straight line in the same direction, and the top and the bottom of the network grid respectively.
- The mobile hosts both moved hops.
- MH_S moves from cell of its old base station BS_{SO} to a new old base station BS_{SN}.
- MH_D moves from cell of its old base station BS_{DO} to a new old base station BS_{DN}.
- There is a connection path between the old base stations for the mobile hosts to start with. The minimum length of this route is a, and is shown as the straight line connecting BS_{SO} and BS_{DO}.
- The mobile hosts stay at the new base station long enough for the rerouting to be completed.
- For simplicity of analysis, we assume that the mobile hosts start moving at the same time and continue moving at the same speed.

We calculate the following metrics for comparing the performance for all of the mobile–mobile rerouting schemes. We consider each case separately for calculating the metrics.

- *Total rerouting distance (TRD(a, d)):* This is the rerouting path length connecting the new source and destination base stations after the mobile hosts have moved d hops from their initial positions and the initial path length between their corresponding base stations is a hops. This metric gives a good estimate of the system throughput for the connection between the source and the destination mobile hosts.
- *Cumulative connection path length (CCPL):* This is the cumulative total of the lengths of new reroute paths formed and torn down as the mobile hosts move. In order to calculate the CCPL for the rerouting scheme, we calculate the cumulative length of new paths formed (CLNP) and the cumulative length of old paths torn down (CLOP). So, CCPL is the sum of CLNP and CLOP. This metric is an indicator of the system disruption, the cumulative rerouting time, and the buffering requirements for the connection between the source and the destination mobile hosts.
- *Number of connections (NC):* We calculate the amount of resources reserved of each connection pair in terms of the maximum number of connections that can exist at any time during the entire rerouting process. This metric gives a reflection of the resources used by the system.

12.5 PERFORMANCE OF MOBILE–MOBILE REROUTING

We now analyze the performance of the mobile–mobile rerouting algorithms. It should be noted for all the subsequent analysis below that for Biswas' scheme, the initial path length between the source and destination base stations is equal to $a = b + 2c$ hops, where b is the length of the fixed path between the source and destination mobile representatives and c is the length to the path between the initial source (resp. destination) base station and the source (resp. destination) mobile representative. In the case of the CBT strategy, the initial path length from the source (resp. destination) base station to the source (resp. initial destination) mobile representative is equal to c hops and the path length between the initial pair of source and destination mobile representative is b hops. Similarly, in the case of Ghai and Singh's scheme, b is the path length between the old (resp. new) source and destination supervisory hosts and c is the length of the path between the initial source (resp. destination) base station and the source (resp. destination) supervisory host. Figures 12.13 through 12.17 are graphical representations of these mobile–mobile rerouting schemes. In these figures, we specify the values of the parameters a, b, and c. For all the metrics that we discuss later, we plot values for rerouting distance for d varying from 1 to 25 for the rerouting schemes. Biswas' scheme, the CBT scheme, Ghai and Singh's worst and best schemes, the IS-41(c) cellular scheme, and Racherla's scheme are represented by the legends Biswas, CBT, Ghai (worst case), Ghai (best case), Cell, and Ours (Racherla), respectively.

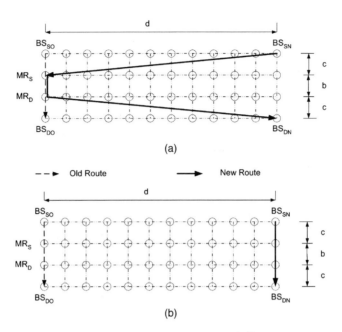

FIGURE 12.13 Ghai and Singh's scheme: (a) worst case; (b) best case.

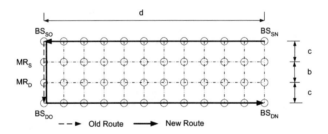

FIGURE 12.14 Core-based tree scheme.

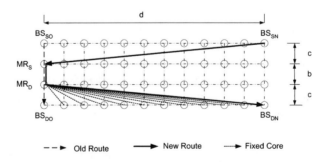

FIGURE 12.15 IS-41(c) scheme.

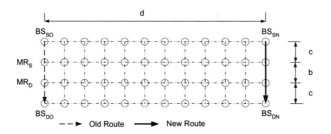

FIGURE 12.16 Racherla's scheme.

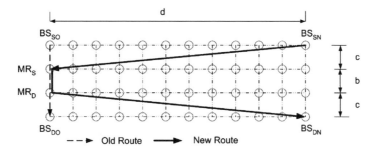

FIGURE 12.17 Biswas' scheme.

12.5.1 TOTAL REROUTING DISTANCE

Figure 12.18 shows the total rerouting distance as a function of mobile host movement distance (d) for the five rerouting schemes described previously. The figure contains four plots for various values of the initial path length a (a = 3, 27, 39, 63). We make the following observations:

- Racherla's proposed scheme and Ghai's (best case) scheme perform the best because these schemes set up the optimal route. IS-41(c) performs the worst with increasing values of d because of its naive approach of cell forwarding continuously. The rerouting distance for Racherla's scheme and Ghai (best case) is optimal, and for the given topology is a constant and does not vary with increasing values of d. For all the other schemes, the rerouting distance increases with increasing values of d.
- For Biswas' scheme, it can be seen that total rerouting length depends on the length of the initial fixed path. If this initial fixed path $b = a$ (that is, $c = 0$), then it is the exact same scheme as cell forwarding rerouting. Biswas' strategy is a specialized case of the CBT strategy wherein a core-based tree is replaced by a simple path. It is better to use Biswas' strategy than the CBT strategy in order to reduce the total rerouting distance, if $c < b$. In case, $b > c$, it is better to use the CBT strategy.
- For Ghai's (worst case), the CBT and Biswas' scheme, the rerouting distance increases linearly with the increasing values of d. However, the rerouting distance improves when $c = d$. This is intuitive because when $c = d$, the new route can be done along the slanted links.

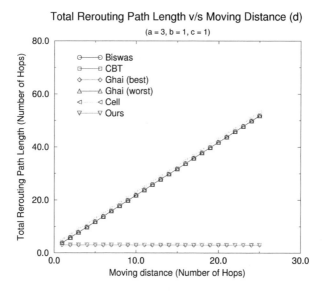

FIGURE 12.18 Effect of the moving distance on the total rerouting distance for various rerouting schemes.

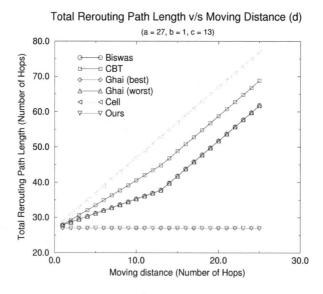

FIGURE 12.18 (continued)

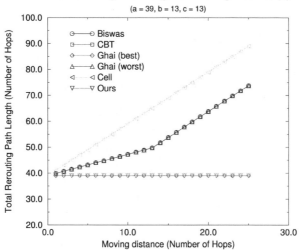

FIGURE 12.18 (continued)

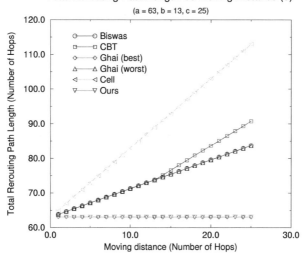

FIGURE 12.18 (continued)

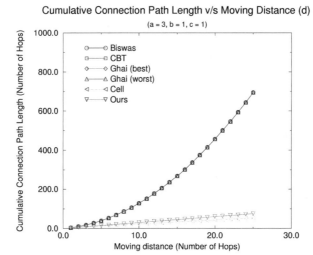

FIGURE 12.19 Effect of the moving distance on the cumulative connection path length for various rerouting schemes.

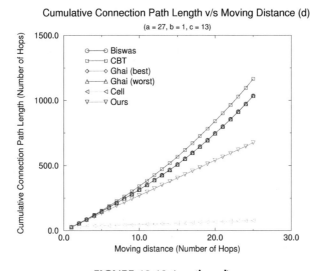

FIGURE 12.19 (continued)

12.5.2 CUMULATIVE CONNECTION PATH LENGTH

Figure 12.19 shows the cumulative connection path length as a function of mobile host movement distance (d) for the rerouting schemes. From the figure, we make the following observations:

Cumulative Connection Path Length v/s Moving Distance (d)

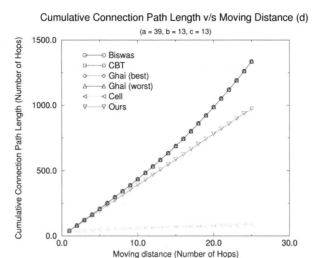

FIGURE 12.19 (continued)

Cumulative Connection Path Length v/s Moving Distance (d)

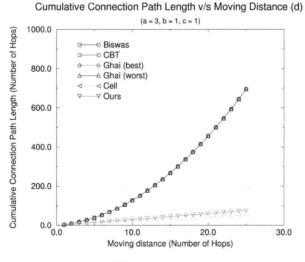

FIGURE 12.19 (continued)

- The IS-41(c) scheme does the best because there is no need to remove any of the old paths and every move of the mobile translates to an increase of a unit length increase in the cumulative path.
- Racherla's strategy performs the next best as the cumulative path length is increased by a hops every time the mobile moves. This is because a new optimal path of length a hops needs to be established.

- In case of all the other schemes, the cumulative path length is much higher and grows exponentially. The CCPL for d moves in each of these cases is a sum of all the total rerouting distance values for all moves 1 through d.
- We see again that it is better to use Biswas' strategy than the CBT strategy in order to reduce the cumulative path length, if $c < b$. In case, $b > c$, it is better to use the CBT strategy.

12.5.3 NUMBER OF CONNECTIONS

In Figure 12.20, we show the cumulative connection path length as a function of mobile host movement distance (d) for the rerouting schemes. The figure contains two parts showing the plots for various values of the initial path length a ($a = 3$, 39). From this figure, we observe that the number of connections is minimum and constant (equal to 2) in the case of Racherla's strategy. Biswas' strategy requires also a constant number of connections, namely, 5. Also, the number of connections for the other schemes is dependent on the moving distance.

Among the remaining schemes, the CBT scheme requires the least number of connections. Ghai (best case) and Ghai (worst case) and IS-41(c) require the same number of connections. From the observations of the performance of all these rerouting schemes with respect to the metrics, we infer that Racherla's scheme performs the best as far as the total rerouting distance and the number of connections are concerned. Ghai (best case) performs as well as our scheme in terms of the total rerouting distance. However, it requires many more connections to be established and incurs more cumulative cost in establishing connections. Also in the worst case, the routes for Ghai and Singh's scheme are suboptimal. In addition, Ghai and Singh's strategy requires specialized supervisory hosts and a tree-based network for optimal performance. IS-41(c) performs the best with respect to the cumulative cost incurred in establishing connections. However, IS-41(c) requires a high number of connections and its total rerouting distance is very high. CBT is a generalized case of Biswas' scheme. In general, the CBT scheme requires more resources than Biswas' scheme and its performance advantage in terms of total rerouting distance over Biswas' scheme is highly dependent on the underlying topology and the details of the core-based tree.

12.6 CONCLUSION

In this chapter, we examined various existing static–mobile rerouting schemes for connection-oriented cellular mobile computing environments and have proposed a generic framework to classify them. We presented an exhaustive survey of various rerouting techniques and rerouting classification schemes proposed in the literature. We proposed generic protocols also for each of the rerouting classes. In addition, we studied each class by looking at its variations, which have been proposed in the literature. We used a variety of performance metrics to compare and contrast these rerouting schemes. These metrics include the ones common to all the rerouting schemes and the ones that are specific to each class. We surveyed and compared

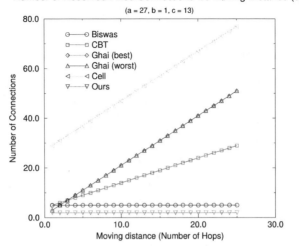

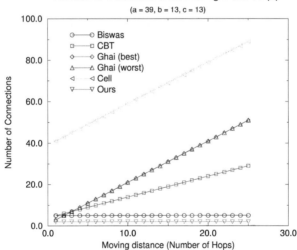

FIGURE 12.20 Effect of the moving distance on the number of connections established or torn down for various rerouting schemes.

rerouting techniques for mobile–mobile connections. Our future work in this area includes increasing the array of performance metrics for comparison, studying the effect of traffic patterns and traffic distribution on the performance of rerouting schemes, and developing simulation models for the implementation of more-efficient rerouting schemes based on the lessons learned in this chapter.

References

1. Acampora, A., An architecture and methodology for mobile-executed handoff in cellular ATM networks, *IEEE J. Selected Areas Commun.*, 12 (8), 1365–1375, 1994.
2. Bush, S., "Handoff mechanism for mobile ATM systems," Internet draft, 1995.
3. Seshan, S., Low latency handoff for cellular data networks, Ph.D. diss., University of California, Berkeley, 1995.
4. Racherla, G., Radhakrishnan, S., and Sekharan, C.N., A framework for evaluation of rerouting in connection-oriented mobile networks, Technical report, School of Computer Science, University of Oklahoma, Norman, 1998.
5. Toh, C-K., The design and implementation of a hybrid handover protocol for multimedia wireless LANs, Proc. ACM Mobicom, Berkeley, California, November 1995.
6. Toh, C-K., A unifying methodology for handovers of heterogeneous connections in wireless ATM networks, in *Computer Communication Review*, 1997, pp. 12–30.
7. Akyol, B. and Cox, D., Handling mobility in a wireless ATM network, Proc. IEEE InfoCom 96, 3, 8, 1405–1413, March 1996.
8. Akyol, B. and Cox, D., Rerouting for handoff in a wireless ATM network, *IEEE Personal Commun.*, 3 (5), 26–33, 1996.
9. Naylon, J. et al., Low-latency handover in a wireless ATM LAN, *IEEE J. Selected Areas Commun.*, 16 (6), 909–921, 1998.
10. Srivastava, M. and Mishra, P.P., Effect of connection rerouting on application performance in mobile networks, *IEEE Trans. Comput.*, 47 (4), 371–390, 1998.
11. Keeton, K. et al., Providing connection-oriented network services to mobile hosts, Proc. USENIX Symposium on Mobile and Location Independent Computing, Cambridge, Massachusetts, August 1993, pp. 83–102.
12. Ghai, R. and Singh, S., An architecture and communication protocol for picocellular networks, *IEEE Personal Commun.*, Third Quarter, 36–47, 1994.
13. Toh, C-K., Performance evaluation of crossover switch discovery algorithms for wireless ATM LANs, Proc. IEEE InfoCom, San Francisco, March 1996, pp. 1380–1387.
14. Rekhter, Y. and Perkins, C., Optimal routing for mobile hosts using IPs loose source route option, IETF RFC, 1992.
15. Teraoka, F. et al., VIP: a protocol providing host mobility, *Commun. ACM,* 37 (8), 67–75, 1994.
16. Mobile IP Working Group, Routing support for IP mobile hosts, Internet draft, 1993.
17. Wada, H. et al., Mobile computing environment based on Internet packet forwarding, Proc. USENIX Tech. Conf., San Diego, California, Winter 1993, pp. 503–517.
18. Biswas, S.K., Handling real-time traffic in mobile networks, Ph.D. diss., University of Cambridge, 1994.
19. Yuan, R., Biswas, S.K., and Raychaudhuri, D., Mobility support in a wireless ATM network, Proc. 5th Workshop on Third Generation Wireless Information Networks, Kluwer, Dordrecht, 1995, pp. 335–45.
20. Eng, K.Y. et al., A wireless broadband ad-hoc ATM local area network, *ACM J. Wireless Networks (WINET)*, 1 (2), 161–174, 1995.
21. Ramanathan, S. and Streenstrup, M., A survey of routing techniques for mobile communication networks, *ACM/Baltzer Mobile Networks Appl.*, Special issue on Routing in Mobile Communications Networks, 1 (2), 89–104, 1996.
22. Cohen, R. and Segall, A,. Connection management and rerouting in ATM networks, Internet draft, 1998.

23. Ramjee, R., La Porta, T.F., Kurose, J., and Towsley, D., Performance evaluation of connection rerouting schemes for ATM-based wireless networks, *IEEE/ACM Trans. Networking,* 6 (3), 249–261, 1998.

24. Song, M., Choi, Y., and Kim, C., Connection-information-based connection rerouting for connection oriented mobile communication networks, *Distributed Systems Engineering,* 5, 47–65. 1998.

25. EIA/TIA, IS-41(c), Cellular radio telecommunications intersystem operations, 1995.

26. Racherla, G., Radhakrishnan, S., and Sekharan, C.N., A distributed rerouting algorithm for mobile–mobile connections in connection-oriented mobile networks, Proc. 7th Int. Conf. Comput. Commun. Networks (ICCCN), September 1998, pp. 40–44.

27. Wu, O.T.W. and Leung, V.C.M., B-ISDN architectures and protocols to support wireless personal communications internetworking, Proc. PIMRC'95, Toronto, Canada, September 1995.

13 Wireless Communications Using Bluetooth

Oge Marques and Nitish Barman

CONTENTS

0-8493-1502-6/03/$0.00+$1.50
© 2003 by CRC Press LLC

13.1 INTRODUCTION

Bluetooth is a wireless communications standard that allows compliant devices to exchange information with each other. The technology makes use of the globally available, unlicensed ISM (Industrial, Scientific, and Medical) band. Although it was initially developed as a cable-replacement technology, it has grown into a standard that is designed to support an open-ended list of applications (including multimedia applications). As a short-range, low-power technology with data rates of up to 720 kbps, it is ideally suited for use in establishing *ad hoc* personal area networks (PANs).

The Bluetooth specification emerged from a study undertaken by Ericsson Mobile Communications in 1994 to find alternatives to using cables to facilitate communications between mobile phones and accessories. As this study grew in scope, other companies joined Ericsson's efforts to utilize radio links as cable replacements. In 1998 these companies – Ericsson, Intel, IBM, Toshiba, and Nokia – formally founded the Bluetooth Special Interest Group (SIG). In July 1999 this core group of promoters published version 1.0 of the Bluetooth specification.[1] Shortly after the specification was published, the group of core promoters was enlarged further with the addition of four more companies: Microsoft, Agere Systems (a Lucent Technologies spin-off), 3COM, and Motorola.

In addition to the core promoters group, many hundreds of companies have joined the SIG as Bluetooth adopter companies. In fact, any incorporated company can join the SIG as an adopter company by signing the Bluetooth SIG membership agreement (available on the Bluetooth Web site[1]). Joining the SIG entitles an adopter company to a free license to build Bluetooth-based products, as well as the right to use the Bluetooth brand.

The list of adopter companies continues to grow in part because there is no cost associated with intellectual property rights, but primarily because there are so many potential applications and usage models for Bluetooth:

1. Cordless desktop: In this cable-replacement usage model, all (or most) of the peripheral devices (e.g., mouse, keyboard, printer, speakers, etc.) are connected to the PC cordlessly.
2. Ultimate headset: This usage model would allow one headset to be used with myriad devices, including telephones, portable computers, stereos, etc.
3. Automatic synchronization: This usage model makes use of the hidden computing paradigm, which focuses on applications in which devices automatically carry out certain tasks on behalf of the user without user intervention or awareness. Consider the following scenario: A user attends a business meeting, exchanges contact information with other attendees, and stores this information on a PDA. Upon returning to the user's office, the PDA automatically establishes a Bluetooth link with the user's desktop PC and the information stored on the PDA is automatically uploaded to the PC. All this happens without the user's conscious involvement.

There are many other usage models that have been proposed by the Bluetooth SIG as well as other contributors. With increasing numbers of applications being developed around Bluetooth, it is highly likely that the technology will be well accepted by the market.

The remainder of this chapter is organized as follows. Section 13.2 provides a broad overview of the Bluetooth protocol stack and introduces some concepts and terminology. Section 13.3 gives a more-detailed explanation of the key features of the various layers of the Bluetooth protocol stack. Section 13.4 introduces the second volume of the Bluetooth specification, the Profile specification. It serves as a basic introduction to the concept of Bluetooth profiles, and gives a brief description of the fundamental profiles. Section 13.5 discusses security and power management issues, two items that are critical for the acceptance of Bluetooth in the consumer market. Finally, Section 13.6 concludes this chapter with some anecdotal evidence of the interest evinced in Bluetooth application development.

13.2 OVERVIEW

The Bluetooth specification comprises two parts:

1. The core specification,[2] which defines the Bluetooth protocol stack and the requirements for testing and qualification of Bluetooth-based products
2. The profiles specification,[3] which defines usage models that provide detailed information about how to use the Bluetooth protocol for various types of applications

The Bluetooth core protocol stack consists of the following five layers:

1. Radio specifies the requirements for radio transmission – including frequency, modulation, and power characteristics – for a Bluetooth transceiver.
2. Baseband defines physical and logical channels and link types (voice or data); specifies various packet formats, transmit and receive timing, channel control, and the mechanism for frequency hopping (hop selection) and device addressing.
3. Link Manager Protocol (LMP) defines the procedures for link set up and ongoing link management.
4. Logical Link Control and Adaptation Protocol (L2CAP) is responsible for adapting upper-layer protocols to the baseband layer.
5. Service Discovery Protocol (SDP) – allows a Bluetooth device to query other Bluetooth devices for device information, services provided, and the characteristics of those services.

In addition to the core protocol stack, the Bluetooth specification also defines other layers that facilitate telephony, cable replacement, and testing and qualification. The most important layers in this group are the Host Controller Interface (HCI) and RFCOMM, which simply provides a serial interface that is akin to the EIA-232 (formerly RS-232) serial interface. HCI provides a standard interface between the host controller and the Bluetooth module, as well as a means of testing/qualifying a Bluetooth device.

The Bluetooth protocol stack is discussed in greater detail in Section 13.3. In the remainder of this section we will introduce key concepts and terminology that are helpful for an understanding of Bluetooth technology.

13.2.1 MASTERS AND SLAVES

All active Bluetooth devices must operate either as a master or a slave. A master device is a device that initiates communication with another Bluetooth device. The master device governs the communications link and traffic between itself and the slave devices associated with it. A slave device is the device that responds to the master device. Slave devices are required to synchronize their transmit/receive timing with that of the masters. In addition, transmissions by slave devices are governed by the master device (i.e., the master device dictates *when* a slave device may transmit). Specifically, a slave may only begin its transmissions in a time slot immediately following the time slot in which it was addressed by the master, or in a time slot explicitly reserved for use by the slave device.

13.2.2 FREQUENCY HOPPING SPREAD SPECTRUM (FHSS) AND TIME-DIVISION DUPLEXING (TDD)

Bluetooth employs a frequency hopping technique to ensure secure and robust communication. The hopping scheme is such that the Bluetooth device hops from one 1-MHz channel to another in a pseudorandom manner – altogether there are 79 such 1-MHz channels defined by the SIG. Typically, each hop lasts for 625 μs. At the end of the 625-μs time slot, the device hops to a different 1-MHz channel. The technology ensures full-duplex communication by utilizing a TDD scheme. In the simplest terms this means that a Bluetooth device must alternate between transmitting and receiving (or at least listening for) data from one time slot to the next. The Bluetooth specification facilitates this TDD scheme by (1) using numbered time slots (the slots are numbered according to the clock of the master device), and (2) mandating that master devices begin their transmissions in even-numbered time slots and slave devices begin their transmissions in odd-numbered time slots (Figure 13.1). Both these techniques will be further discussed in Section 13.3.

13.2.3 PICONETS AND SCATTERNETS

Bluetooth technology furthers the concept of *ad hoc* networking where devices within range of each other can dynamically form a localized network for an indefinite duration. *Ad hoc* Bluetooth networks in which all member devices share the same (FHSS) channel are referred to as piconets.[4] A piconet can consist of up to eight devices: a single master and up to seven slave devices. The frequency hopping sequence of the piconet is determined by the Bluetooth device address (BD_ADDR) of the master device. At the time that the piconet is established, the master device's address and clock are communicated to all slave devices on the piconet. The slaves

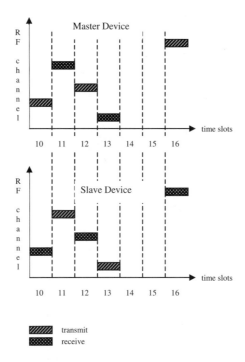

FIGURE 13.1 Frequency hopping and time-division duplexing. During even-numbered slots, the master device transmits and the slave device receives. During odd-numbered slots, the slave device transmits and the master device receives.

use this information to synchronize their hop sequence as well as their clocks with that of the master's.

Scatternets (Figure 13.2) are created when a device becomes an active member of more than one piconet. Essentially, the adjoining device shares its time slots among the different piconets. Scatternets are not limited to a combination of only two piconets; multiple piconets may be linked together to form a scatternet. However, as the number of piconets increases, the total throughput of the scatternet decreases. Note that the piconets that make up a scatternet still retain their own FH sequences and remain distinct entities.

Scatternets offer two advantages over traditional collocated *ad hoc* networks. First, because each individual piconet in the scatternet retains its own FH sequence, the bandwidth available to the devices on any one piconet is not degraded by the presence of the other piconets (although the probability of collisions may increase). Second, devices on different piconets are able to "borrow" services from each other if they are on the same scatternet.

13.3 PROTOCOL STACK

The five layers of the Bluetooth core protocol are divided into two logical parts (Figure 13.3). The lower three layers (Radio, Baseband, and LMP) comprise the

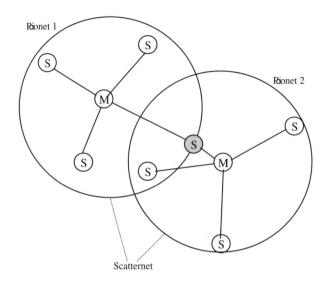

(a)

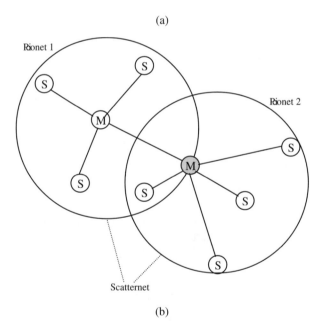

(b)

FIGURE 13.2 Scatternet examples: (a) piconet 1 and piconet 2 share a common slave device; (b) a device acts as slave in piconet 1 and master in piconet 2.

Bluetooth module, whereas the upper layers (L2CAP and SDP) make up the host. The interface between these two logical groupings is called the Host Controller Interface (HCI). Note that the HCI is itself a distinct layer of the Bluetooth protocol stack. The reason for separating the layers into two groups stems from implementation issues. In

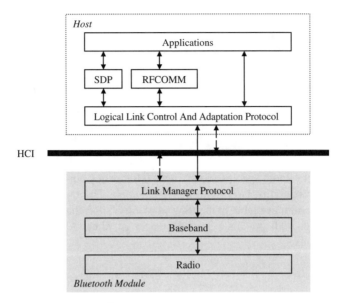

FIGURE 13.3 Bluetooth protocol stack.

many Bluetooth systems the lower layers reside on separate hardware than the upper layers of the protocol. A good example of this is a Bluetooth-based wireless LAN card. Because the PC to which the LAN card connects typically has enough spare resources to implement the upper layers in software, the LAN card itself only contains the lower layers, thus reducing hardware complexity and user cost. In such a scenario the HCI provides a well-defined, standard interface between the host (the PC) and the Bluetooth module (the LAN card).

13.3.1 THE RADIO LAYER

Bluetooth devices operate in the 2.4-GHz ISM band. In North America, Europe, and most of the rest of the world, a bandwidth of 83.5 MHz is available in the ISM band. In France, however, the ISM band is much narrower. The Bluetooth SIG has actively lobbied the French regulatory bodies to release the full ISM band; France is expected to comply by 2003. Although the Bluetooth specification defines a separate RF specification for the narrower (French) band, we will not consider it here.

The ISM band is littered with millions of RF emitters ranging from random noise generators, such as sodium vapor street lamps, to well-defined, short-range applications, such as remote entry devices for automobiles. Microwave ovens are particularly noisy and cause significant interference. Clearly, the ISM band is not a very reliable medium. Bluetooth overcomes the hurdles presented by the polluted environment of the ISM band by employing techniques that ensure the robustness of transmitted data. Specifically, Bluetooth makes use of frequency hopping (along with fairly short data packets) and adaptive power control techniques in order to ensure the integrity of transmitted data.

TABLE 13.1
Bluetooth Radio Power Classes

Power Class	Range of Transmission (meters)	Maximum Power Output (mW/dBm)
1	3	1/0
2	10	2.5/4
3	100	100/20

The Bluetooth Radio specification defines a frequency hopping spread spectrum radio transceiver operating over multiple RF channels. The specification divides the 83.5-MHz bandwidth available in the ISM band into 79 RF channels of 1 MHz each, a 2-MHz lower guard band, and a 3.5-MHz upper guard band. The channel frequencies are defined as

$$2402 + k \text{ MHz}$$

where $k = 0, 1, \ldots, 78$.

In order to maximize the available channel bandwidth, the data rate for the RF channels is set at 1 Mbps. Gaussian frequency shift keying (GFSK) is used as the modulation scheme, with a binary 1 defined as a positive frequency deviation from the carrier frequency and a binary 0 defined as a negative frequency deviation.

In addition to specifying the modulation scheme, the Radio specification defines also three power classes (Table 13.1) for Bluetooth devices. Each power class has associated with it a nominal transmission range and a maximum power output. The nominal power output for all three classes is 1 mW (0 dBm).

13.3.2 THE BASEBAND LAYER

The baseband layer is by far the most-complex layer defined in the Bluetooth specification. It encompasses a wide range of topics, not all of which can be discussed in this short introduction to the Bluetooth technology. What follows is a brief overview of the key aspects of the baseband layer.

13.3.2.1 Device Addressing

The radio transceiver on each Bluetooth device is assigned a unique 48-bit Bluetooth device address (BD_ADDR) at the time of manufacture. Portions of this address are used to generate three types of access codes: the device access code (DAC), the channel access code (CAC), and the inquiry access code (IAC). The DAC and the IAC are used for paging and inquiry procedures. The CAC is specified for the entire piconet and is derived from the device address of the master device. It is used as the preamble of all packets exchanged on that piconet.

With regard to slave devices on a piconet, there are two other addresses of interest. The first is the active member address (AM_ADDR). This is a 3-bit address that is assigned to an active slave device on the piconet. Because the piconet may have seven active slaves at any given time, a slave's AM_ADDR uniquely identifies it within its piconet. The second address of interest is the parked member address (PM_ADDR). This is an 8-bit address assigned to slave devices that are parked, i.e., devices that are not active on the piconet but remain synchronized to the piconet's master. An active slave device loses its AM_ADDR as soon as it is parked. Similarly, a parked device that becomes active immediately relinquishes its PM_ADDR and takes on an AM_ADDR. Note that the reactivated device may or may not be assigned the same AM_ADDR that it was assigned before it was parked.

13.3.2.2 Frequency Hopping

In order to ensure secure and robust data transmission, Bluetooth technology utilizes a frequency hopping spread spectrum technique. Under this scheme, Bluetooth devices in a piconet hop from one RF channel to another in a pseudorandom sequence. Typically, a hop occurs once at the beginning of every time slot. Because each time slot has a 625-μs duration, the nominal hopping frequency is 1600 hops per second. All devices that are part of the same piconet hop in synchrony. The hopping sequence is based on the master's BD_ADDR and consequently is unique for each piconet.

Because there can be multiple piconets operating in close proximity to each other, it is important to minimize collisions between the devices in different piconets. The specification addresses this issue in two ways. First, the pseudorandom hop selection algorithm is designed to generate the maximum distance between adjacent hop channels. Second, the duration of the time slots is kept very small (625 μs). Time slots of such short duration ensure that even if a collision occurs it will not last long. A collision is unlikely to recur during the next time slot because the piconet will have hopped to a different RF channel.

Strictly speaking, the baseband specification defines a different hopping sequence for different types of operations. The earlier description is that of the channel hopping sequence, which utilizes all 79 RF channels and defines a unique physical channel for the piconet. However, the other hopping sequences (associated with device discovery procedures such as paging and inquiry) utilize only 32 RF channels.

One final note: It was previously stated that the nominal hop rate is 1600 hops per second. This is true only for packets (discussed later) that occupy a single time slot. For packets that occupy three (or five) time slots, the same RF channel is used for transmission during all three (or five) time slots.

13.3.2.3 Link Types (ACL and SCO)

The baseband specification defines two types of links between Bluetooth devices.

1. Synchronous Connection Oriented (SCO) link: This is a bidirectional (64 kbps each way), point-to-point link between a master and a single slave device. The master maintains the SCO link by using reserved time slots

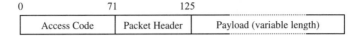

FIGURE 13.4 Packet structure of a typical Bluetooth packet.

at regular intervals. The slots are reserved as consecutive pairs: one for transmissions from the master to the slave and the other for transmissions in the opposite direction. The master can support up to three SCO links simultaneously. A slave device can support up to three simultaneous SCO links with one master or — if it is the adjoining device on a scatternet — it can support up to two simultaneous SCO links with two different masters. SCO links are used for exchanging time-bound user information (e.g., voice). Packets sent on these links are never retransmitted.

2. Asynchronous Connectionless (ACL) link: This is a point-to-multipoint link between the master and all the slaves on a piconet. An ACL link can utilize any packet that is not reserved for an SCO link. The master can exchange packets with any slave on the piconet on a per-slot basis, including slaves with which it has an established SCO link. Only one ACL link can exist between any master–slave pair. ACL links are used for the exchange of control information and user data, therefore packet retransmission is applied to ACL packets received in error.

13.3.2.4 Packet Definitions

The baseband layer defines the format and structure of the various packet types used in the Bluetooth system. The length of a packet can range from 68 bits (shortened access code) to a maximum of 3071 bits. A typical packet (Figure 13.4), however, consists of a 72-bit access code, a 54-bit packet header, and a variable length (0 to 2745 bits) payload.

Every packet must begin with an access code (see Section 3.2.1). The access code is used for synchronization and identification at either the device level (DAC/IAC) or the piconet level (CAC).

The packet header contains link control information, so it is important to ensure its integrity. This is accomplished by applying a 1/3-rate FEC scheme to the entire header. The packet header consists of six different fields:

1. AM_ADDR (3-bit field): This field contains the active member address of the slave device that transmitted the packet (or, if the master sent the packet, the AM_ADDR of the slave device to which the packet was addressed).
2. TYPE (4-bit field) : This field identifies the type of packet, as explained later in this section.
3. FLOW (1-bit field): This field is used for providing flow control information on an ACL link. A value of 0 (STOP) indicates a request to the sender to stop sending information, whereas a value of 1 (GO) indicates that the receiver is once again ready to receive additional packets. If no packet is received from the receiver, a GO is implicitly assumed.

4. ARQ (1-bit field): This field is used for acknowledging payload data accompanied by a CRC (cyclic redundancy check). If the payload was received without error, a positive acknowledgment (ACK) is sent; otherwise, a negative acknowledgment (NACK) is sent. An ACK corresponds to a binary 1 and a NACK corresponds to a binary 0. Note that NACK is the default value for this field and is implicitly assumed.

5. SEQN (1-bit field): This field is used for a very simple sequential numbering scheme for transmitted packets. The SEQN bit is inverted for each new transmitted packet containing a payload accompanied by a CRC. This field is required for filtering out retransmissions at the destination.

6. HEC (8-bit field) : This field, the header error check, is used for ascertaining the integrity of the header. If the header is received in error, the entire packet is ignored.

The payload portion of the packet is used to carry either voice or data information. Typically, packets transmitted on SCO links carry voice information – with the exception of the DV packet type, which carries both voice and data – and packets transmitted on ACL links carry data. For packets carrying voice information, the payload length is fixed at 240 bits (except DV packets). Packets carrying data, on the other hand, have variable length payloads that are segmented into three parts: a payload header, payload body, and a CRC.

The payload header is either one or two bytes long. It specifies the logical channel on which the payload is to be carried, provides flow control information for the specified channel, and indicates the length of the payload body. The only difference between the one-byte and two-byte headers is that the two-byte headers use more bits to specify the length of the payload body. The payload body simply contains the user host data and the 16-bit CRC field is used for performing error checking on the payload body.

13.3.2.4.1   Packet Types

The packets defined for use in the Bluetooth system (Table 13.2) are identified by the 4-bit TYPE field in the packet header. As is evident from the length of the TYPE field, 16 possible packet types can be defined altogether. Bluetooth packet types can be categorized in four broad groups: control packets, single-slot packets, three-slot packets, and five-slot packets. Single-slot packets require only one time slot for complete transmission. Similarly, three-slot and five-slot packets require three and five time slots, respectively, for complete transmission. The single-slot, three-slot, and five-slot packets are defined differently for the two different link types (SCO and ACL), while the control packets are common to both links.

The four control packets defined in the specification are:

1. NULL: This packet type is used to return acknowledgments and flow control information to the source.

2. POLL: This packet is used by the master to poll slave devices in a piconet. Slave devices must respond to this packet even if they have no information to send.

TABLE 13.2
Bluetooth Packet Types

Packet Group	Type Code	Packet Name on SCO Link	Packet Name on ACL Link	Number of Slots
	0000	NULL	NULL	1
Control group	0001	POLL	POLL	1
	0010	FHS	FHS	1
	0011	DM1	DM1	1
	0100	Undefined	DH1	1
	0101	HV1	Undefined	1
Single-slot group	0110	HV2	Undefined	1
	0111	HV3	Undefined	1
	1000	DV	Undefined	1
	1001	Undefined	AUX1	1
	1010	Undefined	DM3	3
Three-slot group	1011	Undefined	DH3	3
	1100	Undefined	Undefined	3
	1101	Undefined	Undefined	3
Five-slot group	1110	Undefined	DM5	5
	1111	Undefined	DH5	5

3. FHS: The frequency hop selection packet is used to identify the frequency hop sequence before a piconet is established or when an existing piconet changes to a new piconet as a result of a master–slave role switch. This control packet contains the BD_ADDR and clock of the sender (recall that the master's device address and clock are used to generate a pseudo-random frequency hop sequence). In addition, the packet contains also the AM_ADDR to be assigned to the recipient along with other information required in establishing a piconet.

4. DM1: The DM1 (data–medium rate) packet is used for supporting control information on any link. This is necessary, for example, when synchronous information (on an SCO link) must be interrupted in order to supply control information. In an ACL link, this packet can be used to carry user data. The packet's payload can contain up to 18 bytes of information. The payload data is followed by a 16-bit CRC. Both the data and the CRC are encoded with a 2/3-rate forward error correction (FEC) scheme.

The remaining 12 packet types are defined differently for each link. Some packet types have not yet been defined or have been set aside for future use.

13.3.2.4.2   SCO Packets

All packets defined specifically for the SCO link are single-slot packets. They all have a 240-bit payload and do not include a CRC. The first three SCO packets are designed to carry high-quality voice (HV) information (64 kbps) and are defined as follows:

- HV1: This packet contains 10 bytes of data and is protected by 1/3-rate FEC. Because one packet can carry about 1.25 ms of speech, HV1 packets need to be transmitted once every two time slots.
- HV2: This packet contains 20 bytes of data and is protected by 2/3-rate FEC. Because one packet can carry about 2.5 ms of speech, HV2 packets only need to be transmitted once every four time slots.
- HV3: This packet contains 30 bytes of data and is not encoded with an error correction scheme. Because one packet can carry about 3.75 ms of speech, HV3 packets need to be transmitted only once every six time slots.

The last packet type defined for the SCO link is the DV (combined data and voice) packet. This packet contains a 10-byte voice field and a data field containing up to 10 bytes of data. The data field is encoded with 2/3-rate FEC and is protected by a 16-bit CRC. It is interesting to note that the voice and data information contained in this packet are treated completely differently. The voice information – which is synchronous – is never retransmitted, whereas the data information *is* retransmitted if it is found to be in error. So, for example, if the data field of a packet is received in error, the packet is retransmitted (i.e., the data field contains the same information), but the voice field contains new (different) information.

13.3.2.4.3 ACL Packets

Six different packet types have been defined specifically for the ACL link. All the ACL packets are designed to carry data information and are distinguished from one another by two basic criteria: (1) whether they are encoded with a FEC scheme, and (2) how many time slots they require for complete transmission. The packet types protected with an FEC scheme are referred to as medium data rate (DM) packets, whereas the unprotected packets are referred to as high data rate (DH) packets.

There are two DM packet types defined: DM3 and DM5. Both these packet types are similar to the DM1 control packet in that they are protected with a 2/3-rate FEC scheme. The difference is that unlike DM1, which requires one time slot for complete transmission, DM3 and DM5 require three time slots and five time slots, respectively. Additionally, as one might expect, DM5 packets can carry more data than DM3 packets, which in turn can carry more data than DM1.

Three different DH packet types are defined for the ACL link: DH1, DH3 and DH5. All these packets can carry more data than their DM counterparts because the DH packets are not encoded with an error correction scheme. As with the DM packet types, DH3 and DH5 carry more information than DH1.

The sixth packet type defined for the ACL link is the AUX1 (auxiliary) packet type. This packet type is very similar to DH1 except it does not contain a CRC code.

The remaining packet types have not been defined as yet and have been set aside to accommodate packet types that may be required in the future.

13.3.2.5 Logical Channels

Bluetooth specification defines five logical channels for carrying either link control and management information or user data.

The two channels that carry link control information are the link control (LC) channel and the link manager (LM) channel. The LC logical channel is mapped to the header of every packet (except the ID packet) exchanged on a Bluetooth link. This channel carries low level link information such as flow control, ARQ, etc. The LM channel is typically carried on protected DM packets and it can utilize either link type (SCO or ACL). A packet used for the LM channel is identified by an L_CH value of 11 in the packet's header. The LM channel carries LMP traffic and therefore takes priority over user data channels.

Three logical channels are defined for conveying user data: the user asynchronous (UA) channel, the user isochronous (UI) channel, and the user synchronous (US) channel. The UA and UI channels are normally mapped to the payload of packets transmitted over an ACL link, but they can be mapped also to the data field of the DV packet on an SCO link. The US channel carries synchronous user data (e.g., voice communication) and is mapped to the payload field of packets transmitted over an SCO link.

13.3.2.6 Channel Control

Channel control deals with the establishment of a piconet and adding/removing devices to/from a piconet. The specification defines two primary states for a Bluetooth device: standby and connected. The standby state is the default, low-power state. A device is in the connected state when it is a member of a piconet, i.e., when a device is synchronized to the hop sequence of a piconet.

Devices do not transition directly from the standby state to the connected state, and vice versa. In fact, the specification describes several substates that a device must transition through in order to move from one primary state to the other. These substates are associated with inquiry and paging procedures that are required for slave devices to join or leave the piconet. Figure 13.5 depicts the state transitions involved in going from the standby state to the connected state and back. The observant reader will note that there is a direct transition from the connected state back to the standby state. This direct transition occurs only in case of a hard reset. Normally, devices must methodically detach from the piconet, which requires transitioning through different substates.

The Bluetooth system makes provisions for connected devices to be put into one of three low-power modes while remaining synchronized to the hop sequence of the piconet. These modes (discussed further in Section 3.3) can be considered as substates of the connected state (Figure 13.5).

13.3.2.7 Error Checking and Correction

Bluetooth provides a variety of error checking and error correction mechanisms. Bluetooth packets can be checked for errors at three levels. When a packet is received, the receiver immediately checks its channel access code (recall that the CAC is the preamble for all packets). If the CAC embedded in the packet does not match the CAC of the piconet, then either the packet was received in error, or the packet was destined for another piconet. In either case, the packet is discarded. If the CAC

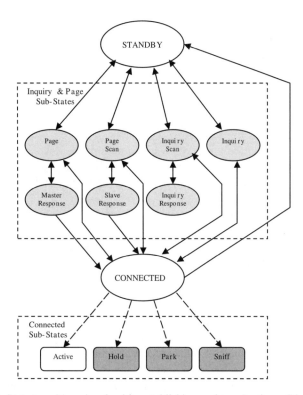

FIGURE 13.5 State transitions involved in establishing and terminating a Bluetooth link.

matches the CAC for the piconet, the next error checking mechanism is employed. This involves checking the header error check. If the HEC indicates that the header data has irrecoverable errors, the packet is again discarded. If the packet passes this second error-checking test, the final error checking mechanism involves checking the CRC to check the integrity of the packet payload.

Bluetooth also makes use of three different error correction schemes. The first two schemes are 1/3-rate FEC and 2/3-rate FEC. Using these two schemes, it may be possible to recover from bit errors without having to retransmit the packet. However, if the packet payload cannot be recovered despite the use of these two correction schemes or if no FEC was used, then an ARQ scheme is used to request retransmission of the packet. For more information on either FEC scheme, please refer to the Bluetooth specification.[2]

13.3.2.8 Security

Security is of paramount importance in any wireless (RF) communication. It is even more important in the case of Bluetooth, which aims at becoming a ubiquitous, *de facto* standard for wireless communications. With this in mind, the specification provides for security mechanisms in more than one layer of the protocol stack. At the baseband (lower link level), the facility for link security between two devices relies on two different procedures:

1. Authentication: Procedure used to verify the identity claimed by a Bluetooth device
2. Encryption: Procedure used to encode user data so that it is unintelligible until decoded using a specific key

The security algorithms defined in the baseband specification utilize the following four parameters:

1. BD_ADDR: Publicly known 48-bit address of the device
2. Authentication key: 128-bit secret key preconfigured with the device
3. Privacy key: Variable length (4 to 128 bits) secret key that is preconfigured also with the device
4. Random number: 128-bit number used for device authentication

In addition to these key responsibilities, the baseband layer addresses other issues, including transmit/receive timing, data whitening, and encoding of audio information for transmission over Bluetooth links.

13.3.3 THE LMP LAYER

The Link Manager Protocol is responsible for establishing and maintaining ACL and SCO links, and establishing, managing, and terminating the connections between different devices. LMP provides a mechanism for the link managers (LMs) on separate devices to exchange messages containing link management information with each other. These messages are sent as LMP protocol data units (PDUs) and are carried in the payload of single-slot packets (DM1 or DV) transmitted on the LM logical channel.

Because LMP PDUs are used for link management, they have been assigned a very high priority. Consequently, LMP PDUs may be transmitted during slots reserved for an SCO link. The PDU contains three fields:

1. A 1-bit transaction ID field, which specifies whether the link management transaction was initiated by the master (0) or the slave (1) device
2. A 7-bit OpCode field, which identifies the PDU and provides information about the PDUs payload
3. A payload field, which contains the actual information necessary to manage the link

Many PDUs have been defined for the different transactions defined in the specification. Two of the most general response PDUs are LMP_Accepted and LMP_Not_Accepted. Although many link management transactions have been defined in the specification, not all of them are mandatory. However, the LM on every Bluetooth device must be able to recognize and respond to any of the specified transactions. In case the LM on the receiving device does not support a particular transaction, it must still respond to the requesting device's LM with an LMP_Not_Accepted PDU.

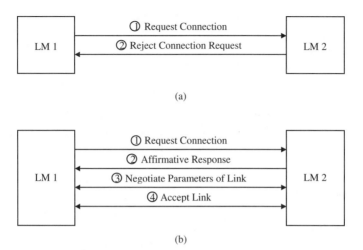

FIGURE 13.6 LM connection request transactions: In both cases, the transaction is initiated by LM 1 (on device 1). (a) LM 2 rejects the request for establishing a connection, and the transaction terminates; (b) LM 2 accepts the request, LM 1 and LM 2 negotiate the link parameters, the negotiation process is completed, and the connection is established.

In the remainder of this section we will briefly mention some of the more-important link management transactions defined in the specification. These transactions are grouped into three broad categories: link control, power management, and security management.

Link management transactions used for link control include:

- Connection establishment: Before two devices can exchange L2CAP information, they must first establish a communications link (Figure 13.6). The procedure for establishing a link is as follows: the LM of the device requesting the connection sends an LMP_Host_Connection_Req PDU to the receiving device. The LM of the receiving device may then either accept or reject the request. If the request is accepted, the LMs of the two devices further negotiate the parameters of the link to be set up. When this negotiation is completed, the LMs of both devices send an LMP_Setup_Complete PDU to each other and the link is established.

- Link detachment: When a device wishes to terminate its link to another device, it issues an LMP_Detach PDU. The receiving device cannot reject the detach request and the link between the two devices is immediately terminated.

- Clock and timing information: LMP PDUs in clock-and-timing-related transactions are used by devices to obtain clock offsets and slot offsets from each other. Support for many of these PDUs is mandatory because link timing cannot be established or negotiated without them.

- Master–slave role switch: This transaction plays an important role in the establishment of a scatternet.[5] The device that wants to switch roles initiates the process by sending an LMP_Switch_Req PDU. If the request is accepted, the role switch occurs after the two devices exchange some timing information.

- Information exchange: Transactions of this type allow devices to request information from each other. As an example, a (local) device may want to inquire another (remote) device about the optional link manager features supported by the receiving device. In that case, the local device sends an LMP_Features_Req PDU to the remote device. In addition to requesting a list of features supported by the remote device, the PDU contains information about the optional features supported by the local device. When the remote device receives the request, it sends an LMP_Features_Res PDU back to the local device. The LMP_Features_Res PDU contains information about the optional features supported by the remote device. At the end of the transaction, both devices are aware of the optional services supported by the other.

Power management transactions allow Bluetooth devices the flexibility to conserve power when they are not actively exchanging information. The transactions in this category include:

- Sniff mode: While an active slave must monitor every even-numbered time slot for transmissions from the master, a slave device in sniff mode has to monitor far fewer slots. Note that slots reserved for an SCO link are not affected by sniff mode. A master may force a slave device into sniff mode by sending an LMP_Sniff PDU to the slave. Alternatively, either the master or the slave may request that the slave device be placed in sniff mode by issuing an LMP_Sniff_Req PDU.
- Hold mode: In an established piconet, there may be times when a slave device will not be addressed for a significant duration. In such instances, it is highly desirable to place the slave device in sleep-like mode for the duration of time that it will not be addressed, in order to conserve power. In hold mode, the ACL link between the master and slave is temporarily suspended (i.e., there is no ACL traffic between the master and slave). The duration for which the link is suspended is called the hold time and is stipulated at the time the link is suspended. Note that any SCO links between the master and slave device are unaffected. As with sniff mode, the master may either force a slave into hold mode (LMP_Hold PDU) or either the master or the slave may request that the slave be placed in hold mode (LMP_Hold_Req).
- Park mode: In the sniff and hold modes, the slave device is considered an active member of the piconet (i.e., it is still one of the seven active slave devices on the piconet). In park mode, however, the slave device is no longer active on the piconet; however, it still remains synchronized to the piconet. The advantage is that if the device needs to rejoin the piconet, it can do so very quickly. Interested readers are referred to the LMP specification for further information.

Security is addressed in more than one layer of the Bluetooth protocol stack. In the LMP layer, the key security management transactions are device authentication and link encryption; while the former is a mandatory feature of all Bluetooth devices, the latter is optional.

- Device authentication: This transaction is based on a challenge response scheme. The transaction begins by one device (the verifier) sending an LMP_Au_Rand PDU to another device (the claimant). The PDU contains a 128-bit random number. The claimant uses this number as the input to an encryption function. The output of this function is then transmitted back to the verifier. If the value received by the verifier matches the value it expects, the claimant is authenticated. At the end of this transaction, the verifier and claimant may swap roles in order to authenticate the link in the opposite direction. The authentication transaction is far more complex than the simple procedure outlined here. A much more detailed explanation can be found in the Bluetooth LMP specification.[2]
- Link encryption: Link encryption is often necessary to protect the privacy of the data transmitted over a Bluetooth link. This transaction is initiated when a device issues an LMP_Encryption_Mode_Req PDU. The encryption mode determines whether the encryption is to be applied to a point-to-point link only or to broadcast packets as well. If the request is accepted, the devices negotiate the size of the encryption key to be used by exchanging LMP_Encryption_Key_Size_Req PDUs. Once the encryption key size has been determined, the devices begin the encryption process by issuing an LMP_Start_Encryption PDU.

13.3.4 THE L2CAP LAYER

The L2CAP layer serves to insulate higher-layer protocols (including non-Bluetooth-specific protocols such as TCP/IP, PPP, etc.) from the lower-layer Bluetooth transport protocols. In addition, it supports protocol multiplexing with respect to higher-layer protocols. Another key feature of the L2CAP layer is that it facilitates segmentation and reassembly of higher-layer packets. Note that L2CAP itself does not segment (or reassemble) packets. It simply provides packet length information to the higher-layer protocols, which allows those protocols to segment (and reassemble) the packets submitted to (and received from) the L2CAP layer. L2CAP supports quality-of-service (QoS) features by allowing the exchange of QoS information between the L2CAP layers on different devices.

The L2CAP layer messages are sent in packets known as L2CAP PDUs. These PDUs are carried on ACL packets transmitted on the user asynchronous logical channel. Because these PDUs may be carried on multislot packets, the L_CH field in the payload header of the ACL packets is used to indicate whether the current packet is the start of the L2CAP PDU (denoted by an L_CH value of 10) or a continuation of the L2CAP PDU (denoted by an L_CH value of 01). The L2CAP layer does not itself guarantee the reliable transmission of L2CAP PDUs. It assumes that the packet retransmission facility provided by the baseband layer is sufficient for ensuring a reliable communication channel.

TABLE 13.3
L2CAP CID Definitions

CID hex	Description
0000	NULL identifier
0001	Signaling channel
0002	Connectionless reception channel
0003-003F	Reserved
0040-FFFF	Dynamically allocated

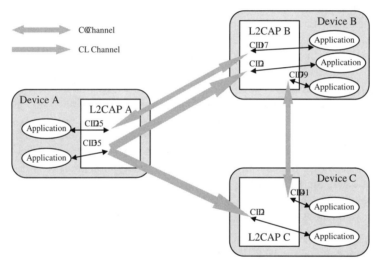

FIGURE 13.7 L2CAP channels between three different devices. Device A maintains two connectionless (CL) channels, one each with Device B and Device C. Devices A and B share a bidirectional connection-oriented (CO) channel also, as do Device B and Device C. Note that all the channels terminate at endpoints in the L2CAP entities of the different devices. Each of these endpoints is assigned a CID by its L2CAP entity. The endpoints in each of the devices are uniquely associated with some recipient application.

All communication between the L2CAP layers on different devices occurs over logical links referred to as channels (Figure 13.7). These channels are identified by the two endpoints (one on the first device, the other on the second device) of the link. Each endpoint is assigned a unique 16-bit channel identifier (CID). The CIDs are assigned locally, i.e., the endpoint local to a particular device is assigned its CID by the L2CAP layer of that same device. These endpoints are in turn uniquely associated with some recipient (e.g., higher-layer protocol) to which the payload of the L2CAP PDU is delivered. The CIDs are administered locally and the scheme for assigning CIDs is left up to the implementer. The only exception is that some CIDs have been reserved (Table 13.3) for specific channels.

The Bluetooth specification defines three types of L2CAP channels:

1. Connection-oriented (CO) channel: This is a persistent channel that supports bidirectional communication and requires the exchange of signaling information before it can be established.
2. Connectionless (CL) channel: CL channels are unidirectional, are not persistent, and are typically used for broadcast messages. If a device wants to respond to a L2CAP transmission on a CL channel, it must send its response on another channel. Also, CL channels allow the L2CAP layer to provide protocol multiplexing (refer to the specification for additional details).
3. Signaling channel: This is very similar to the CO channel except that its CID (and other channel parameters) are fixed, and therefore no signaling information is exchanged in order to establish the channel. Indeed, the signaling channel is itself used for exchanging signaling information.

13.3.4.1 L2CAP Channel Management

In order for a CO channel to be established or terminated, signaling information has to be exchanged between the local and remote devices. This signaling information is exchanged by means of a request-and-response transaction mechanism. The signaling commands used in this transaction are L2CAP_Connection_Req, L2CAP_Connection_Res, L2CAP_Configuration_Req, L2CAP_Configuration_Res, L2CAP_Disconnection_Req, and L2CAP_Disconnection_Res. There are additional signaling commands defined in the specification, but they will not be discussed here.

The procedure for establishing a CO channel is as follows: the local device (i.e., the device that wants to establish the CO channel) sends an L2CAP_Connection_Req command to the remote device. The command contains the source CID (i.e., the CID of the endpoint on the local device) and other signaling information. If the remote device accepts the connection request, it sends an L2CAP_Connection_Res command back to the local device. The response command contains the destination CID (i.e., the CID of the endpoint on the remote device), as well as other signaling information. Once the CO channel has been established, it has to be configured. The local and remote devices negotiate channel configuration by exchanging L2CAP_Configuration_Req and L2CAP_Configuration_Res commands. If the two devices cannot agree on configuration parameters, the CO channel is either terminated or the default configuration parameters (implementation dependent) are used. The most important of the configuration parameters negotiated between the two devices are the QoS parameters.

13.3.5 THE SDP LAYER

Bluetooth technology was developed to support mobility and facilitate the formation of *ad hoc* networks. This scenario is qualitatively different from "static" networks in which member devices do not constantly leave/join the network. In an *ad hoc* network, devices are free to join or leave the network at whim. This presents a complex challenge, because a device leaving a network may deprive the network of some service that only that device provides. On the other hand, a device that joins a network (piconet) may provide a service that was previously not available on any other device in that network. So how does any device on the network know what services are available on the network at any given time? This is precisely the issue addressed by the SDP.

SDP utilizes the concept of a client/server model. Clients are devices that are searching for services, whereas servers are devices that provide services. These roles are entirely interchangeable, i.e., any device can be either a client or a server, depending on whether it is using a service provided by another device or providing a service to another device. SDP requires that all devices (in their role as servers) maintain a service registry. The service registry is a collection of service records that provide information (service attributes) about the services provided by the device.

SDP specifies two types of service attributes: (1) universal service attributes, which could apply to any class of service (e.g., printing, telephony, LAN access, etc.); and (2) service-specific attributes, which are meaningful only in the context of a particular service class. It is important to note that the universal service attributes are not necessarily mandatory. In fact, the specification mandates only two such attributes: the service class attribute, which identifies the class or type of service, and the service record handle, which is a pointer to the service record of that service.[6]

Service discovery is accomplished by means of SDP transactions. SDP transaction information is communicated using SDP PDUs. In order for service discovery to work, it is necessary that the services are represented in a standard format, one that both the client and the server can use to refer to the service. SDP uses universally unique identifiers (UUIDs) to represent the services. Typically, two transactions are required in order for a client to obtain service information from the server: a service search transaction, and a service attribute transaction. The first transaction is initiated by the client when it sends an SDP_Service_Search_Req PDU to the server. Upon receiving this request, the server responds to the client with an SDP_Service_Search_Res PDU. The response PDU returns a list of service record handles that match the service requested by the client. Of course, if the server does not support the requested service, it still responds to the client, but the list of service record handles is empty. Following the response from the server, the client begins the second transaction by sending an SDP_Service_Attribute_Req PDU to the server. The parameters of this PDU are the service record handle (for the service of interest) and the attribute IDs of the attributes desired by the client. When the server receives the SDP_Service_Attribute_Req PDU from the client, it responds with an SDP_Service_Attribute_Res PDU that contains a list of attributes (attribute ID and the corresponding value) retrieved from the service record specified by the service record handle supplied by the client. At the end of the second transaction, the client should have enough information to connect to the service. It is important to note that SDP does not actually provide the client with the requested service. If the client wants to connect to the service, it utilizes some other protocol to do so.

There is a third type of transaction defined in the SDP specification, but it is simply a composite of the first two transactions, and is not discussed here. Interested readers may refer to the specification itself for additional details.

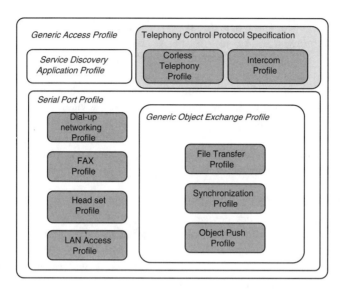

FIGURE 13.8 All thirteen profiles and their inheritance relationships are depicted. Each profile inherits from the profile that encloses it. The four fundamental profiles (GAP, SDAP, SPP, and GOEP) are not shaded.

13.4 BLUETOOTH PROFILES SPECIFICATION

As previously mentioned, the Bluetooth specification includes not just the core protocol specification, but also a second volume referred to as the Profiles specification.[2] The Bluetooth SIG made a conscious decision to make this second volume a part of the specification. This makes sense considering that one of the key features of Bluetooth technology is the emphasis on universal interoperability. Bluetooth devices from different manufacturers are expected to work seamlessly with each other. In order to facilitate this seamless interoperability, the Bluetooth specification provides usage models for the many different types of foreseen applications built around Bluetooth technology. These usage models are termed profiles in the specification. It should be noted that this part of the specification does not limit implementers to the profiles defined therein; it simply provides guidelines for using Bluetooth technology for different application types. Indeed, as more applications are devised for Bluetooth, it is inevitable that new usage models will be added to the profiles specification.

The Bluetooth profiles specification defines 13 different profiles (Figure 13.8), which are logically divided into two types. The first are fundamental profiles, which are essentially building-block profiles for the other type of profiles, the usage profiles. All usage profiles inherit from at least one of the fundamental profiles. The four fundamental profiles are (briefly) discussed in the following sections.

13.4.1 GAP

The generic access profile (GAP) is the fundamental Bluetooth profile. All other profiles stem from GAP. GAP defines the key features necessary for Bluetooth devices to successfully establish a baseband link. The features defined in GAP are:

- Conformance: Every Bluetooth device that does not support some other profile must at least conform to GAP. Essentially, this means that GAP specifies certain features that must be implemented in all Bluetooth devices.
- Discovery procedures: The minimum set of procedures required for a Bluetooth device to discover another Bluetooth device.
- Security procedures: Procedures required for using the different security levels.
- Link management facilities: Facilities that ensure that Bluetooth devices can connect to each other.

In addition to these features, GAP defines also the mandatory and optional modes of operation for a Bluetooth device. Please refer to the specification for further information. Finally, GAP defines a standard set of terminology that is to be used with respect to user interfaces developed for Bluetooth devices. Defining standardized terminology ensures that users of the technology will recognize Bluetooth functionality across different user interface designs.[5]

13.4.2 SDAP

The service discovery application profile describes, in general terms, how applications that use the SDP should be created, and how they should behave. Fundamentally, SDAP specifies which services an SDAP-based application should provide to its users. These services are defined as "service primitive abstractions," and four such primitives are defined:

1. ServiceBrowse: Used by a local device when conducting a general search for services available on a set of remote devices
2. ServiceSearch: Used by a local device when searching for a specific service type on a set of remote devices
3. EnumerateRemDev: Used by a local device to search for remote devices in its vicinity
4. TerminatePrimitive: Used to terminate the operations initiated by the other three primitives

13.4.3 SPP

The serial port profile is concerned with providing serial port emulation to two devices that want to utilize a Bluetooth link for serial communication. As can be expected, SPP uses the RFCOMM protocol for providing serial port emulation. The key feature of the SPP is that it outlines the procedures necessary for using the RFCOMM protocol to establish a serial link between two devices. The overarching goal of the SPP is to ensure transparency, i.e., an application using the emulated serial link should not be able to distinguish it from a physical serial link. By defining the SPP profile, the SIG has assured that legacy applications (that make use of serial communication links) will not have to be modified in order to use Bluetooth.

13.4.4 GOEP

The generic object exchange profile is based on the object push/pull model, as defined in Infrared Data Association's (IrDA) OBEX layer. GOEP distinguishes between a client device and a server device. The client device is the device that initiates the object exchange operation by requesting the OBEX service from the server. The client either pushes a data object onto or pulls a data object off of the server device. The server device is the device that provides the client device with this push/pull object exchange service. Note that there is no correlation between master/slave (in the context of Bluetooth) on the one hand, and client/server in the context of OBEX on the other. In the simplest terms, GOEP defines the primitives that allow objects to be exchanged between a client and server. The two most important of these primitives are object push and object pull. An interesting side note regarding GOEP is that it was originally developed to provide Bluetooth devices with a synchronization capability, but during the course of development, it grew into the concept of IrDA interoperability.[6]

The remaining nine profiles are usage profiles. As stated earlier, these profiles inherit features from at least one of the four fundamental profiles. Although we will not discuss the usage profiles, they are listed below for reference:

1. Cordless telephony
2. Intercom
3. Headset
4. Fax
5. Dial-up networking
6. LAN access
7. Object push
8. Synchronization
9. File transfer (FTP)

13.5 ADDITIONAL CONSIDERATIONS

13.5.1 POWER MANAGEMENT

Bluetooth technology is motivated by the desire to provide a universal interface to battery-driven portable devices. As such, the technology makes several provisions for effective power management in order to conserve power. Some of the key power management features are implemented at the micro level. The first such feature is identified with respect to frequency hopping. The mechanism for frequency hopping is defined such that no dummy data has to be exchanged in order to keep the master and slave devices synchronized to each other. This obviates the need for the transceivers on both the master and slave devices to periodically wake up and transmit dummy packets. The second feature is inherent in the packet format. Because Bluetooth packets begin with the access code of the piconet for which they are intended, a device that is listening on the piconet's hop frequency during its receive slot can quickly ascertain whether the packet carried on the current hop frequency is intended for its piconet. If the device determines that the packet was not intended for its piconet, the device's receiver can go to sleep for the remainder of the time slot. Moreover, the packet header contains information about the length of the payload. So, if the payload is very small, the device does not have to keep its receiver on for the entire duration of the time slot. It can put its receiver to sleep as soon as the entire payload has been received. Aside from the micro-level power management features discussed here, Bluetooth provides support for macro-level power management in that it allows devices to be put in any one of three power saving modes: hold, sniff, and park (see Section 13.3.3).

13.5.2 SECURITY

Because Bluetooth is an RF-based wireless technology, data exchanged between Bluetooth devices can be easily intercepted. Clearly, there is a need to protect personal and private data from would-be eavesdroppers. The Bluetooth SIG has made a conscious effort to provide various security mechanisms at many different levels of the specification. To begin with, the fact that the technology employs a frequency hopping spread spectrum technique to establish a channel for communication itself provides a certain degree of security. Consider that without knowing the hop sequence of the piconet, the eavesdropper will not know which 1-MHz channel to listen to during the next 625-μs time slot. But there are other intentional security mechanisms provided in the Bluetooth specification.

At a macro level, the specifications provide three security modes for a Bluetooth device. Devices in security mode 1 never initiate any security procedures. Additionally, devices in this mode are not required to support device authentication. In security mode 2, devices do not need to initiate security procedures until an L2CAP channel is established. Once the L2CAP link is established, the device can decide which security procedures to enforce. Security mode 3 is the most stringent of the three security modes. In this mode, security procedures are initiated at the link level, i.e., before any link is established between devices.

There are two key security procedures defined in the specification: device authentication and link encryption (see Section 13.3). In addition to the security mechanisms provided by Bluetooth at the lowest levels, more-advanced security mechanisms can be employed at higher layers.[4,7,8]

13.6 CONCLUDING REMARKS

Although the Bluetooth profiles specification defines usage models for several foreseen applications of Bluetooth technology, these are by no means the only applications that can be built around Bluetooth. In fact, new applications are constantly being defined. One source for innovative new applications is the Computer Society International Design Competition.[9] In 2001 and 2002, this international competition, sponsored by the IEEE Computer Society, focused on Bluetooth technology. The competition required student teams to build applications around Bluetooth technology. Some of the applications envisioned by the student teams have prompted members of the SIG to consider new profiles (e.g., a profile designed specifically for medical devices). All this development activity bodes well for the acceptance of Bluetooth technology by the consumer market.

In this chapter we have endeavored to present a salient, albeit cursory overview of Bluetooth technology. It must be borne in mind, however, that Bluetooth is a feature-rich and well-defined wireless communications standard. Indeed, the Bluetooth specification spans more than 1500 pages. Interested readers are encouraged to refer to the specifications,[2,3] as well as several publications that discuss the technology in much greater detail.[5,10]

ACKNOWLEDGMENT

Nitish Barman gratefully acknowledges the support and encouragement provided by his family (especially his brother) and the generous assistance (proofreading and feedback) provided by his good friends and colleagues, Ms. Shefat Sharif and Mr. Scott Bowser.

References

1. The Official Bluetooth Web site, http://www.bluetooth.com.
2. Specification of the Bluetooth System – Core online, available at http://www.bluetooth.com.
3. Specification of the Bluetooth System – Profiles online, available at http://www.bluetooth.com.
4. Haartsen, J.C., Bluetooth – ad hoc networking in an uncoordinated environment, *Proc. IEEE International Conference on Acoustics, Speech, and Signal Processing*, 4, 2029–2032, 2001.
5. Miller, B.A. and Bisdikian, C., *Bluetooth Revealed: The Insider's Guide to an Open Specification for Global Wireless Communications*, Prentice-Hall, Upper Saddle River, NJ, 2001.
6. Stallings, W., *Wireless Communications and Networks*, Prentice-Hall, Upper Saddle River, NJ, 2002.
7. Haartsen, J. C., The Bluetooth radio system, *IEEE Personal Commun.*, 7 (1), 28–36, 2000.
8. Haartsen, J. C. and Mattisson, S., Bluetooth – a new low-power radio interface providing short-range connectivity, *Proc. IEEE*, 88 (10), 1651–1661, 2000.
9. CSIDC Web site, http://www.computer.org/csidc.
10. Bray, J. and Sturman, C.F., *Bluetooth 1.1: Connect Without Cables*, 2nd ed., Prentice-Hall, Upper Saddle River, NJ, 2001.

14 Multiantenna Technology for High-Speed Wireless Internet Access

Angel Lozano

CONTENTS

14.1 INTRODUCTION

With the explosive growth of both the wireless industry and the Internet, it is inevitable that demand for seamless mobile wireless access to the Internet explodes as well. Limited Internet access, at very low speeds, is already available as an enhancement to some second-generation (2G) cellular systems. Third-generation (3G) mobile wireless systems will bring true packet access at significantly higher speeds.[1]

Traditional wireless technologies, however, are not particularly well suited to meet the extremely demanding requirements of providing the very high data rates and low cost associated with wired Internet access and the ubiquity, mobility, and portability characteristics of cellular systems. Some fundamental barriers, associated with the nature of the radio channel as well as with limited bandwidth availability at the frequencies of interest, stand in the way. As a result, the cost-per-bit in wireless is much higher than in the wired world, wherein an entire generation of Internet users has grown accustomed to accessing huge volumes of information at very high speed and negligible cost.

14.2 FUNDAMENTAL LIMITS TO MOBILE DATA ACCESS

14.2.1 CAPACITY AND BANDWIDTH EFFICIENCY

Ever since the dawn of the Information Age, capacity has been the principal metric used to assess the value of a communication system.[2,3] Irrespective of whether it is applied to an individual radio link, to a cell, or even to an entire system, the capacity signifies the largest volume of data throughput that can be communicated with arbitrary reliability.

Because capacity grows linearly with the amount of spectrum utilized, the most immediate way in which capacity can be enlarged is by allocating additional bandwidth. However, radio spectrum is a scarce and very expensive resource at the frequencies of interest, where propagation conditions are favorable.* Hence, it is imperative that the available bandwidth is utilized as efficiently as possible. Consequently, bandwidth efficiency — defined as the capacity per unit bandwidth — has become a key figure of merit.

Besides bandwidth, the capacity also is a function of the received signal power or, more specifically, of the signal-to-interference-and-noise ratio (SINR) at the receiver.

However, unlike with bandwidth, the capacity only scales logarithmically with the SINR and thus trying to enhance the capacity by simply transmitting more power is extremely costly. Furthermore, it is futile in the context of a dense interference-limited cellular system, wherein an increase in everybody's transmit power scales up both the desired signals as well as their mutual interference, yielding no net benefit. Therefore, power increases are useless once a system has become limited in essence by its own interference. Furthermore, because mature systems designed for high capacity tend to be interference-limited,[4] it is power itself — in the form of interference — that ultimately limits their performance.

In order to improve bandwidth efficiency, multiple access methods — originally rather conservative in their design — have evolved toward much more sophisticated schemes. In the context of frequency division multiple access (FDMA) and time division multiple access (TDMA), this evolutionary path has led to advanced forms of dynamic channel assignment, as well as the incorporation of frequency hopping.

---

* Efforts to exploit the larger bandwidths available at frequencies above 10 to 20 GHz are under way, but radio propagation and equipment cost pose serious challenges and thus the realm of portable and mobile systems appear to be, for now, confined to the range around 1 to 5 GHz.

In the context of code division multiple access (CDMA), it has led to a variety of multiuser detection and interference cancellation techniques.[5,6] In all cases, the objective is to attain the highest possible degree of bandwidth utilization through aggressive frequency reuse.

14.2.2 SPACE: THE FINAL FRONTIER

As a key ingredient in the design of more spectrally efficient systems, space has become, in recent years, the last frontier. Nonetheless, the use of the spatial dimension in wireless is hardly new. In fact, one could argue that the entire concept of frequency reuse on which cellular systems are based constitutes a simple way of exploiting the spatial dimension. Cell sectorization, a widespread procedure that reduces interference, can be regarded also as a form of spatial processing. These basic concepts can be taken to the limit, and the area capacity can be increased almost indefinitely,* by shrinking the cells and deploying additional base stations.[4] However, the cost and difficulty of deploying the vast infrastructure required to provide ubiquitous coverage using only microcells has proved prohibitive in the past; it remains to be seen whether that will change with the advent of wireless data. In light of these developments, the use of the spatial dimension is now geared mostly toward maximizing the system capacity on a per-base-station basis.** Here, base station antenna arrays are the enabling tool for a wide range of spatial processing techniques.[7] Because capacity grows roughly linearly with the number of sectors per cell, the most immediate use for such arrays is an increase in the number of sectors. This idea can be refined by making such sectors adaptive using beam-steering and beam-forming techniques devised to enhance desired signals and mitigate interference. All such schemes, however, are fundamentally limited by the multipath nature of the radio channel: sectors and beams are only effective as long as they are sufficiently broad with respect to the angular dispersion or spread introduced by the channel.

Any attempts to create excessively narrow sectors or beams will result in distorted patterns and unforeseen interference. This fundamental barrier, however, can be overcome by incorporating a second antenna array at the terminal.

14.2.3 PUSHING THE LIMITS WITH MULTIANTENNA TECHNOLOGY

Until recently, the deployment of antenna arrays in mobile systems had been contemplated exclusively at base station sites because of size and cost considerations. One of the principal role of those arrays was to provide spatial diversity against signal fading.[4,8] Such fading, arising from multipath propagation caused by scattering, had always been regarded as an impairment that had to be mitigated. However, recent advances in information theory have shown that, with the simultaneous use of antenna arrays at both base station and terminal, multipath interference cannot

---

* Up to the point where the propagation exponent becomes too small for effective distance decay and frequency reuse.
** The use of microcells is still actively considered for dense urban areas, hot spots, indoor environments, etc., but often as a complementary overlay to macrocells.

only be mitigated, but actually exploited to establish multiple parallel channels that operate simultaneously and in the same frequency band.[9-11] Based on this fundamental idea, an entire new class of multiple-transmit multiple-receive (MTMR) communications architectures has emerged.* A critical feature of these MTMR architectures is that the total radiated power is held constant irrespective of the number of transmit antennas. Extraordinary levels of bandwidth efficiency can thus be achieved without any increase in the amount of interference caused to other users.

Imagine a number of single-antenna user terminals collocated into an MTMR terminal that handled their multiple signals simultaneously. Intuitively, this would require the base station to be able to resolve the individual antennas within the terminal array, which in turn would require synthesizing an impossibly narrow beam. The novelty in MTMR communication, however, is that the scattering environment around the terminal is used as an aperture through which those antennas become effectively resolvable.

Notice that, by reusing the same frequency band at each antenna, very large increases in throughput are achieved without increasing the user bandwidth. Hence, in many respects, these MTMR schemes can be regarded as the ultimate step in the quest for ever-tighter levels of frequency reuse, for here every individual user is reusing its bandwidth multiple times.

14.3 MODELS

With n_T transmit and n_R receive antennas, a baseband discrete-time model for the multiantenna channel with frequency-flat fading** is

$$\mathbf{y} = \sqrt{g}\mathbf{H}\mathbf{x} + \mathbf{n} \tag{14.1}$$

where $\mathbf{x}$ is the n_T-dimensional vector representing the transmit signal and $\mathbf{y}$ is the n_R-dimensional received vector. The vector $\mathbf{n}$, containing both thermal noise and interference, is modeled as Gaussian with zero-mean independent components and power σ^2 per receive antenna.*** The channel, in turn, is represented by the $(n_R \times n_T)$ random matrix $\sqrt{g}\mathbf{H}$ containing the transfer coefficients from each transmit to each receive antenna. For convenience, we choose to factor out the scalar $\sqrt{g}$ so as to yield a normalized channel $\mathbf{H}$, the second-order moment of whose entries is unity.

We define also the ratio

$$\beta = \frac{n_T}{n_R} \tag{14.2}$$

---

* These communication architectures are also referred to as multiple-input multiple-output (MIMO).
** The analysis and results to follow can be extended to the more general case of frequency-selective fading.
*** While thermal noise is inherently white, interference tends to be spatially colored, and thus its components are not necessarily independent. Nonetheless, for the sake of simplicity the entire vector $\mathbf{n}$ can be modeled as white to yield a lower bound on the bandwidth efficiency.

The transmit power is constrained to some value P and thus

$$E[\|\mathbf{x}\|^2] = P \tag{14.3}$$

While power control proved to be an essential ingredient in telephony systems, where source rate variability was minimal, in mobile data systems rate adaptation becomes not only an attractive complement, but even a full alternative to power control.[12] Hence, we restrict ourselves to the case where the total transmit power is held constant while the data rate is being adapted.

The thermal noise power per receive antenna is $\sigma^2 = N_0 BF$, where N_0 is the one-sided noise spectral density, B is the signal bandwidth, and F is the receiver noise figure. We set the noise figure to an optimistic value of $F = 3$ dB and use the noise spectral density corresponding to a standard temperature of 300 K. In line with the 3G framework, the available bandwidth is set to $B = 5$ MHz.

Within the channel,[4] different levels of randomness exist:

- The large-scale randomness associated with distance decay, shadow fading, etc., determines the average conditions at every location. Its impact is absorbed into the path gain g, which has a coherence distance of tens or even hundreds of wavelengths, and thus can be regarded as deterministic within a local area. The path gain has a range-dependent component, which we model using the well-established COST231 model,[13] and a shadow fading component, which is taken to be log-normally distributed with an 8-dB standard deviation.[4] In suburban environments at 2 GHz, the average path gain is given by*

$$E[g] = 4 \cdot 10^{-14} d^{-3.5} G$$
$$= -134 - 35 \log_{10} d + 10 \log_{10} G \quad \text{dB} \tag{14.4}$$

   d where the expectation is over the shadow fading, the range in km and G the total (combined transmit and receive) antenna gain.
- The small-scale randomness caused by multipath propagation can be modeled as a locally stationary random process. This level of randomness, contained within $\mathbf{H}$, has a coherence distance that is on the order of a wavelength. Because the small-scale fading encountered in wireless systems tends to be Rayleigh in distribution, the entries of $\mathbf{H}$ can be realistically modeled as zero-mean complex Gaussian random variables.** With that, the characterization of $\mathbf{H}$ entails simply determining the correlation between its entries. A typical propagation scenario, portrayed in

---

* The base station and terminal heights are set to 35 and 2 m, respectively. The path gain can be adjusted for other types of environment and frequency bands.
** Channels that are non-Gaussian and behave abnormally may in theory occur.[40,41]

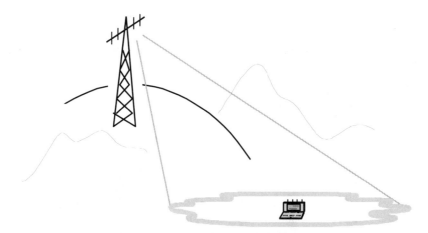

FIGURE 14.1 Propagation scenario with local scattering around the terminal spanning a certain azimuth angle spread at the base station.

Figure 14.1, contains an area of local scattering around each terminal with little or no local scattering around the elevated base stations.* As a consequence of this local scattering, antennas mounted on a terminal can be presumed, to first order, to be mutually uncorrelated even when the physical separation between antennas is as small as a fraction of wavelength. From the perspective of a base station, the angular distribution of power that gets scattered to every terminal is much narrower, characterized by its root-mean-square width or angular spread. Typical values for the angular spread at a base station are in the range of 1 to 10 degrees, depending on the environment and range.[14] With such narrow spreads, ensuring low levels of correlation between those antennas may require larger physical separation (typically a few wavelengths) or the use of orthogonal polarizations,[15] but to first order we can again model them as uncorrelated.

Most wireless systems are equipped with pilots that are needed for synchronization, identification, and a number of other purposes, and which may be used also to obtain an estimate of the channel. Therefore, accurate information about **H** can be gathered by the receiver.[16] Consequently, throughout the chapter we focus on those scenarios wherein **H** is known to the receiver.** The transmitter, however, is presumed unaware of the state of the channel for otherwise a heavy burden would be placed on the system in terms of fast feedback requirements.

At the receiver, the SINR is given by

---

* Local scattering around the base stations would only reinforce the model.
** If the channel changes so rapidly that it cannot be properly estimated, a different class of multiantenna techniques based on differential encoding can be applied.[42–44] Although inferior in potential to the coherent techniques discussed in the chapter, these schemes could be relevant to certain services and applications such as high-speed trains, etc.

$$\text{SINR} = \frac{E[\|\mathbf{Hx}\|^2]}{E[\|\mathbf{n}\|^2]}$$

$$= g\frac{P}{\sigma^2} \tag{14.5}$$

We shall concentrate mostly on the downlink, which has the most stringent demands for Internet access, but occasional references to the uplink will be made as well. The analysis of both links is quite similar, with the exception of much tighter transmit power constraints for the uplink, which originates at the terminal. In terms of system structure, a cellular layout with fairly large hexagonal cells is assumed, with every cell partitioned into three equal-sized sectors.

14.4 SINGLE-USER THROUGHPUT

14.4.1 Single-User Bandwidth Efficiency

We first consider an isolated single-user link limited only by thermal noise. Within the context of a real system, this would correspond to an extreme case wherein the entire system bandwidth is allocated to an individual user. Furthermore, it would require that no other users are active anywhere in the system or that their interference is perfectly suppressed. Clearly, these are unrealistic conditions, and thus the single-user analysis provides simply an upper bound, only a fraction of which is attainable. In addition, this analysis determines what cell sizes can be supported.

With a single transmit and a single receive antenna, the normalized channel is not a matrix but rather a scalar H and the single-user bandwidth efficiency can be expressed as

$$C = E\left[\log_2\left(1 + \text{SINR}|H|^2\right)\right] \tag{14.6}$$

with expectation over the distribution of H. Implicit in this expectation is the use of interleaving and coding over the small-scale fading fluctuations.[5] Shadow fading, however, cannot be similarly averaged out without imposing an unacceptable degree of latency. Thus, with respect to the large-scale variations, we prefer to resort to the idea of outage bandwidth efficiency, which is the value of C supported with certain (high) probability.

When multiple antennas are used at the transmitter or receiver, the bandwidth efficiency can be generalized[9] to

$$C = E\left[\log_2\left(\mathbf{I} + \tfrac{\text{SINR}}{n_\text{T}}\mathbf{HH}^H\right)\right] \tag{14.7}$$

with $\mathbf{I}$ the identity matrix and with $|H|^2$ replaced by $\mathbf{HH},^H$ where $(\cdot)^H$ indicates the Hermitian transpose of a matrix. Although a closed-form solution for Equation 14.7

can be obtained, the corresponding expression is rather involved.[10] More insightful expressions can be obtained by making the number of antennas large and, remarkably, such asymptotic expressions can be scaled to yield a very tight approximation to the capacity for any number of antennas.[17] Therefore, throughout the rest of this section we shall evaluate these limiting behaviors to gain some insight while illustrating the exact behavior numerically.

14.4.2 TRANSMIT DIVERSITY

A basic downlink strategy based on the deployment of base station arrays only, which has already been incorporated into the 3G roadmap, is that of transmit diversity. In this case, the base station is equipped with n_T uncorrelated antennas, while the terminal is equipped with a single antenna. Thus, the normalized channel **H** adopts the form of a vector and the single-user bandwidth efficiency becomes

$$C = E\left[\log_2\left(1 + \text{SINR}\,\frac{\|\mathbf{H}\|^2}{n_T}\right)\right] \qquad (14.8)$$

From the law of large numbers, the term $\|\mathbf{H}\|^2/n_T$ converges to unity as the number of transmit antennas grows,[10] and thus Equation 14.8 converges to

$$C = \log_2(1 + \text{SINR}) \qquad (14.9)$$

showing no dependence on the number of antennas. Hence, the bandwidth efficiency saturates rapidly. The single-user throughput achievable with $B = 5$ MHz as a function of the range is plotted in Figure 14.2 parameterized by the number of transmit antennas. As certified by this figure, there is little advantage in increasing the number of base antennas beyond $n_T \approx 3$ to 4, because of the increasingly diminishing returns. The limiting throughput achieved with an infinite number of antennas, corresponding to the bandwidth efficiency in Equation 14.9, is also shown.

14.4.3 RECEIVE DIVERSITY

The same structure that provides transmit diversity to the downlink enables, by reciprocity, receive diversity for the uplink. The uplink channel **H** is still a vector, the exact transpose of its downlink counterpart, and the corresponding efficiency is given by

$$C = E\left[\log_2\left(1 + \text{SINR}\|\mathbf{H}\|^2\right)\right] \qquad (14.10)$$

Again, as n_R grows the law of large numbers can be invoked to yield, in the limit

$$C = \log_2\left(1 + n_R\,\text{SINR}\right) \qquad (14.11)$$

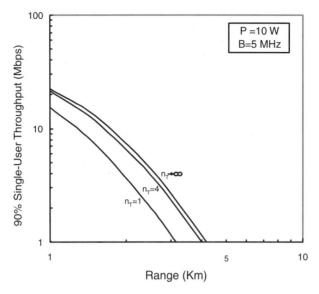

FIGURE 14.2 Single-user throughput (Mbps) supported in 90 percent of locations vs. range (km), with transmit diversity at the base station and a single omnidirectional antenna at the terminal. n_T is the number of 15-dB uncorrelated antennas at the base. Transmit power P = 10 W; bandwidth B = 5 MHz.

displaying the well-known logarithmic improvement with the number of receive antennas, improvement that is a direct consequence of a progressively higher SINR, as more power is being captured by the additional antennas. Notice that this is in sharp contrast with the transmit diversity case, where the total transmit power is constrained, and thus does not grow with the number of antennas.

Hence, the uplink efficiency does not saturate, but it grows at an increasingly slower pace.

14.4.4 MULTIPLE-TRANSMIT MULTIPLE-RECEIVE ARCHITECTURES

We now turn our attention to MTMR architectures. In this case, the terminal must be equipped with its own array of antennas. For the sake of concreteness, we consider a symmetric scenario with $n = n_T = n_R$, but similar analysis can be performed for $n_T \neq n_R$. As n is driven to infinity, the bandwidth efficiency converges[18,19] to

$$C = n \left[2 \log_2 \left(\frac{1 + \sqrt{1 + 4\ \text{SINR}}}{2} \right) - \frac{\log_2 e}{4\ \text{SINR}} \left(\sqrt{1 + 4\ \text{SINR}} - 1 \right)^2 \right] \quad (14.12)$$

indicating that the bandwidth efficiency grows linearly with the number of (uncorrelated) antennas, which is a key result that contrasts with conventional diversity systems, using multiple antennas at either transmitter or receiver exclusively, wherein the growth is at best logarithmic. The bandwidth efficiency becomes particularly revealing at high SINR, wherein Equation 14.12 particularizes[9] to

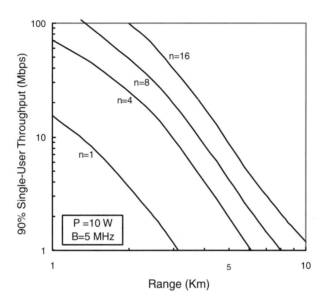

FIGURE 14.3 Single-user throughput (Mbps) supported in 90 percent of locations vs. range (km) with MTMR technology. n is the number of 15-dB antennas at the base station, as well as the number of omnidirectional antennas at the terminal. Transmit power P = 10 W; bandwidth B = 5 MHz.

$$C \approx n \, \log_2\left(\frac{\text{SINR}}{e}\right) \qquad\qquad (14.13)$$

The attainable throughput in $B = 5$ MHz as a function of the range is depicted in Figure 14.3, parameterized by the number of transmit and receive antennas. Notice the extraordinary growth in achievable throughput unleashed by the additional signaling dimensions provided by the combined use of multiple transmit and receive antennas. With only $n = 8$ antennas at both transmitter and receiver, the single-user throughput can be increased by an order of magnitude. Furthermore, the growth does not saturate as long as additional uncorrelated antennas can be incorporated.

14.5 SYSTEM THROUGHPUT

In this section, we extend our analysis in order to reevaluate the throughput achievable in much more realistic conditions. To that extent, we incorporate an entire cellular system into the study.

Most emerging data-centric systems feature time-multiplexed downlink channels, certainly those evolving from TDMA, but also those evolving from CDMA.[20,21] Hence, same-cell users are mutually orthogonal, and thus the interference arises exclusively from other cells. Accordingly, we consider a time-multiplexed multicell

TABLE 14.1
System Parameters

Multiplexing	Time division
Sectors per cell	3
Base station antennas	120° perfect sectorization
Terminal antennas	Omnidirectional
Frequency reuse	Universal
Propagation exponent	3.5
Log-normal shadowing	8 dB standard deviation
Small-scale fading	Rayleigh
	Independent per antenna
Transmit power	Fixed
Thermal noise	Negligible

system layout with base stations placed on a hexagonal grid. Users are uniformly distributed and connected to the sector from which they receive the strongest signal. To further mimic actual 3G data systems, rate adaptation with no power control is assumed. Altogether, the results presented in this section can be considered as upper bounds for a 5-MHz data-oriented 3G system.

The results correspond to Monte-Carlo simulations conducted on a 19-cell universe: a central cell, wherein statistics are collected, surrounded by two rings of interfering cells. The cell size is scaled to ensure that the system is basically interference-limited, and thus thermal noise can be neglected. The simulation parameters are summarized for convenience in Table 14.1.

Figure 14.4 displays cumulative distributions of system throughput (in Mbps per sector) over all locations with multiple transmit antennas only, as well as with multiple transmit and receive antennas. These curves can be interpreted also as peak single-user throughputs, i.e., single-user throughputs (in Mbps) when the entire capacity of every sector is allocated to an individual user. With only multiple transmit antennas, the benefit appears only significant in the lower tail of the distribution, corresponding to users in the most detrimental locations. The improvements in average and peak system capacities are negligible. Moreover, the gains saturate rapidly as additional transmit antennas are added. The combined use of multiple transmit and receive antennas, on the other hand, dramatically shifts the curves offering multiple-fold improvements in throughput at all levels. Notice that, without multiple terminal antennas, the peak single-user throughput that can be supported in 90 percent of the system locations is only on the order of 500 kbps with no transmit diversity and just over 1 Mbps with diversity. Moreover, these figures correspond to absolute upper bounds. With modulation excess bandwidth, training overhead, imperfect channel estimation, realistic coding schemes, and other impairments, only a fraction of these bounds can be actually realized. Without an antenna array mounted on the terminal, user rates on the order of several Mbps can only be supported within a restricted portion of the coverage area and when no other users compete for bandwidth within the same sector.

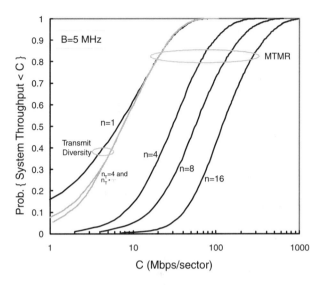

FIGURE 14.4 Cumulative distributions of system throughput (Mbps per sector) with multiple transmit antennas only, as well as with multiple transmit and receive antennas. n is the number of antennas. System bandwidth B = 5 MHz.

14.6 IMPLEMENTATION:
REALIZING THE MTMR POTENTIAL

In order to realize the bandwidth efficiency potential promised by the use of MTMR technology in multipath environments, a number of practical approaches have been proposed in recent years. These approaches can be naturally grouped in two distinct categories:

1. Space–time coding schemes, wherein the signals radiated from the various transmit antennas are jointly encoded and must, therefore, be jointly decoded.[22-26] These schemes tend to be more robust, but the joint decoding process required for good performance suffers rapid increases in complexity as the number of antennas grows. Additionally, new (vector) coding formats may have to be devised. It appears, however, that both these shortcomings may have remedies. Recent results appear to indicate that conventional (scalar) codes may be used to build good vector codes,[27,28] while at the same time some reduced-complexity decoding strategies are emerging.[29]

2. An alternative approach is that of layered architectures, wherein each transmit antenna radiates a separately encoded signal. At the receiver, these signals can be successively decoded and their interference canceled.[30,31] The decoding complexity of these architectures increases more gracefully with the number of antennas. Furthermore, they make direct use of existing scalar coding formats. As an added benefit, layered architectures may offer interesting synergies with upper layers on the data

communication protocol.[32] These incentives, however, come at the expense of reduced robustness because now each signal must be decoded without support from the others, which are conveying independent data. Furthermore, errors in the detection of each of the signals may propagate through the interference cancellation process and adversely impact the detection of other signals. Chief among these layered architectures is the original Bell Labs layered space–time (BLAST) scheme proposed by Foschini and co-workers[30,31] and later refined by other authors. Extensions of the BLAST concept to frequency-selective environments have also been put forth.[34] Also, because the detection problem in a layered architecture bears close resemblance to the more-general problem of multiuser detection, the reader is referred also to the abundant literature on this topic.[6]

Needless to say, a number of hurdles must be overcome before these new concepts can be widely implemented. First of all, it is necessary to assess the antenna arrangement and spacings that are required, as well as the multipath richness of the environments of interest. In that respect, very encouraging experimental data — both indoor and outdoor — has been surfacing.[35–39] Second, the historical opposition to installing multiple antennas on a terminal must be conquered. It is to be expected that terminals requiring increasingly higher throughputs will tend to be naturally larger in size and, as a result, they will offer additional room for multiple, closely spaced antennas.

14.7 SUMMARY

Traditional wireless technologies are not very well suited to meet the demanding requirements of providing very high throughputs with the ubiquity, mobility, and portability characteristics of cellular systems. Given the scarcity and exorbitant cost of radio spectrum, such throughputs dictate the need for extremely high bandwidth efficiencies, which cannot be achieved with classical schemes in systems that are inherently self-interfering. Consequently, increased processing across the spatial dimension appears as the only means of enabling the types of throughputs that are needed for ubiquitous wireless Internet and exciting multimedia services. Whereas the most natural way of utilizing the space dimension may be to deploy additional base stations in order to allow for more frequent spectral reuse with smaller cells, economical and environmental considerations require that performance be enhanced on a per-base-station basis; that, in turn, calls for the use of multiantenna technology. While the deployment of base station antenna arrays is becoming universal, it is really the simultaneous deployment of base station and terminal arrays that unleashes vast increases in throughput by opening up multiple signaling dimensions.

Throughout the chapter, we have quantified the benefits of using such multiantenna technology, in the context of emerging mobile data systems, as a function of the number of available antennas. Although absolute throughput levels are very sensitive to the specifics of the propagation environment, the improvement factors are not. Hence, it is the relative improvement rather than the absolute numbers themselves that is relevant.

REFERENCES

1. Ojanpera, T. and Prasad, R., An overview of third-generation wireless personal communications: a European perspective, *IEEE Personal Commun.,* 5 (6), 59–65, 1998.
2. Shannon, C.E., A mathematical theory of communications, *Bell Syst. Tech. J.,* 27, 379–423 and 623–656, 1948.
3. Cover, T.M. and Thomas, J.A., *Elements of Information Theory,* John Wiley & Sons, New York, 1990.
4. Cox, D.C., Universal digital portable radio communications, *Proc. IEEE,* 75 (4), 436–477, 1987.
5. Biglieri, E., Proakis, J., and Shamai, S., Fading channels: information-theoretic and communication aspects, *IEEE Trans. Information Theory,* 44 (6), 2619–2692, 1998.
6. Verdú, S., *Multiuser Detection,* Cambridge University Press, New York, 1998.
7. Winters, J.H., Smart antennas for wireless systems, *IEEE Personal Commun.,* 5, 23–27, 1998.
8. Jakes, W.C., *Microwave Mobile Communications,* IEEE Press, New York, 1974.
9. Foschini, G.J. and Gans, M.J., On the limits of wireless communications in a fading environment when using multiple antennas, *Wireless Personal Communications,* 1998, pp. 315–335.
10. Telatar, I.E., Capacity of multiantenna Gaussian channels, *Eur. Trans. Telecommun.,* 10, Nov. 1999, pp. 585–595.
11. Raleigh, G. and Cioffi, J.M., Spatio-temporal coding for wireless communications, *IEEE Trans. Commun.,* 46 (3), 357–366, 1998.
12. Goldsmith, A.J. and Varaiya, P., Capacity of fading channels with channel side information, *IEEE Trans. Information Theory,* 1985–1992, Nov. 1997.
13. European Corporation in the Field of Scientific and Technical Research EURO-COST 231, Urban Transmission Loss Models for Mobile Radio in the 900 and 1800 MHz Bands, Revision 2, The Hague, Sept. 1991.
14. Chu, T.-S. and Greenstein, L.J., A semiempirical representation of antenna diversity gain at cellular and PCS base stations, *IEEE Trans. Commun.,* 45–46, June 1997.
15. Gesbert, D. et al., MIMO Wireless Channels: Capacity and Performance Prediction, Proc. IEEE GLOBECOM'00, San Francisco, Dec. 2000.
16. Marzetta, T.L., BLAST Training: Estimating Channel Characteristics for High Capacity Space-Time Wireless, Proc. 37th Annual Allerton Conference on Communication, Control, and Computing, Monticello, Illinois, Sept. 1999.
17. Lozano, A. and Tulino, A.M., Capacity of multiple-transmit multiple-receive antenna architectures, *IEEE Trans. Inf. Theory,* 48(12), 3117–3128, Dec. 2002.
18. Verdú, S. and Shamai, S., Spectral efficiency of CDMA with random spreading, *IEEE Trans. Inf. Theory,* 45, 622–640, 1999.
19. Rapajic, P. and Popescu, D., Information capacity of a random signature multiple-input multiple-output channel, *IEEE Trans. Commun.,* 48 (8), 1245–1248, 2000.
20. Bender, P. et al., CDMA/HDR: a bandwidth-efficient high-speed wireless data service for nomadic users, *IEEE Commun.,* 38 (7), 70–77, 2000.
21. 3G TR 25.950, UTRA High Speed Downlink Packet Access, Third Generation Partnership Project, Technical Specification Group Radio Access Network, March 2001.
22. Guey, J.-C. et al., Signal design for transmitter diversity wireless communication systems over Rayleigh fading channels, Proc. IEEE Vehicular Technology Conference (VTC'96), Atlanta, 1996, pp. 136–140.

23. Tarokh, V., Seshadri, N., and Calderbank, A.R., Space–time codes for high data rate wireless communications: performance criterion and code construction, *IEEE Trans. Inf. Theory,* 44, 744–765, 1998.

24. Hammons, A.R. Jr. and El Gamal, H., On the theory of space–time codes for PSK modulation, *IEEE Trans. Inf. Theory,* 46 (2), 524–542, 2000.

25. Tao, M. and Cheng, R.S., Improved design criteria and new trellis codes for space–time coded modulation in slow flat-fading channels, *IEEE Commun. Lett.,* 5 (7), 313–315, 2001.

26. Byun, M.-K. and Lee, B.G., New Bounds of Pairwise Error Probability for Space–Time Codes in Rayleigh Fading Channels, Proc. Wireless Communication and Networking Conference (WCNC'02), March 2002, 89–93.

27. Hochwald, B.M. and ten Brink, S., Achieving Near-Capacity on a Multiple-Antenna Channel, Proc. Allerton Conference on Communication, Control, and Computing, Oct. 2001, 815–824.

28. Biglieri, E., Tulino, A.M., and Taricco, G., Performance of space–time codes for a large number of antennas, *IEEE Trans. Inf. Theory,* 48 (7), 1794–1803, 2002.

29. Vikalo, H. and Hassibi, B., Maximum-likelihood sequence detection of multiple antenna systems over dispersive channels via sphere decoding, *EURASIP J. Appl. Signal Process.,* Special issue on space–time coding and its applications, Part II, 2002.

30. Foschini, G.J., Layered space-time architecture for wireless communications in a fading environment when using multielement antennas, *Bell Labs Tech. J.,* 41–59, 1996.

31. Foschini, G.J. et al., Simplified processing for high spectral efficiency wireless communication employing multielement arrays, *J. Selected Areas Commun.,* 17 (11), 1841–1852, 1999.

32. Zheng, H., Lozano, A., and Haleem, M., Multiple ARQ Processes for MIMO Systems, 13th IEEE International Symposium on Personal, Indoor and Mobile Radio Communications (PIMRC'2000), Lisbon, Portugal, Sept. 15–18, 2002.

33. Ariyavisitakul, S.L., Turbo space–time processing to improve wireless channel capacity, *IEEE Trans. Commun.,* 48 (8), 1347–1358, 2000.

34. Lozano, A. and Papadias, C.B., Layered space–time receivers for frequency-selective wireless channels, *IEEE Trans. Commun.,* 50 (1), 65–73, 2002.

35. Kermoal, J. P. et al., Experimental Investigation of Multipath Richness for Multielement Transmit-and-Receive Antenna Arrays, Proc. IEEE Vehicular Technology Conference (VTC'00 Spring), Tokyo, May 2000.

36. Martin, C.C., Winters, J. H., and Sollenberger, N.R., Multiple-Input Multiple-Output (MIMO) Radio Channel Measurements, Proc. IEEE Vehicular Technology Conference (VTC'00 Fall), Boston, Sept. 2000.

37. Xu, H. et al., Experimental verification of MTMR system capacity in a controlled propagation environment, *IEEE Electron. Lett.,* July 2001.

38. Ling, J. et al., Multiple transmit multiple receive (MTMR) capacity survey in Manhattan, *IEEE Electron. Lett.,* 37 (16), 1041–1042, 2001.

39. Erceg, V. et al., Capacity Obtained from Multiple-Input Multiple-Output Channel Measurements in Fixed Wireless Environments at 2.5 GHz, Int. Conf. on Communications (ICC'02), New York, Apr. 2002.

40. Chizhik, D. et al., Keyholes, Correlations and capacities of multielement transmit-and-receive antennas, *IEEE Trans. Wireless Commun.,* 2 (1), 361–368, 2002.

41. Dietrich, C.B. Jr. et al., Spatial, polarization, and pattern diversity for wireless handheld terminals, *IEEE Trans. Antennas Propagation,* 49 (9), 1271–1281, 2001.

42. Marzetta, T.L. and Hochwald, B.H., Capacity of a mobile multiple-antenna communication link in Rayleigh flat fading, *IEEE Trans. Inf. Theory,* 45 (1), 139–157, 1999.

43. Hochwald, B.H. and Marzetta, T.L., Unitary space–time modulation for multiple-antenna communications in Rayleigh flat fading, *IEEE Trans. Inf. Theory,* 46, 543–564, Mar. 2000.

44. Hassibi, B., Cayley Codes for Multiple-Antenna Differential Modulation, Proc. Asilomar Conference on Signals, Systems, and Computers, Pacific Grove, California, Vol. 1, Nov. 2001.

15 Location Management in Mobile Wireless Networks

*Amitava Mukherjee, Debashis Saha,
and Sanjay Jha*

CONTENTS

ABSTRACT

Location management schemes are essentially based on users' mobility and incoming call rate characteristics. The network mobility process has to face strong antag-
onism between its two basic procedures: location update (or registration) and paging.
The location update procedure allows the system to keep location knowledge more
or less accurately in order to find the user in case of an incoming call, for example.
Location registration also is used to bring the user's service profile near its location
and allows the network to rapidly provide the user with services. The paging process
achieved by the system consists of sending paging messages in all cells where the
mobile terminal could be located. A detailed description of the means and techniques
for user location management in present cellular networks is addressed.

      A network must retain information about the locations of endpoints in the
network in order to route traffic to the correct destinations. Location tracking (also
referred to as mobility tracking or mobility management) is the set of mechanisms
by which location information is updated in response to endpoint mobility. In
location tracking, it is important to differentiate between the identifier of an endpoint
(i.e., what the endpoint is called) and its address (i.e., where the endpoint is located).
Mechanisms for location tracking provide a time-varying mapping between the
identifier and the address of each endpoint. Most location tracking mechanisms may
be perceived as updating and querying a distributed database (the location database)
of endpoint identifier-to-address mappings. In this context, location tracking has two
components: (1) determining when and how a change in a location database entry
should be initiated; and (2) organizing and maintaining the location database. In
cellular networks, endpoint mobility within a cell is transparent to the network, and
hence location tracking is only required when an endpoint moves from one cell to
another. Location tracking typically consists of two operations: (1) updating (or
registration), the process by which a mobile endpoint initiates a change in the location
database according to its new location; and (2) finding (or paging), the process by which
the network initiates a query for an endpoint's location (which also may result in an
update to the location database). Most location tracking techniques use a combination
of updating and finding in an effort to select the best trade-off between update overhead
and delay incurred in finding. Specifically, updates are not usually sent every time an
endpoint enters a new cell, but rather are sent according to a predefined strategy so that
the finding operation can be restricted to a specific area. There is also a trade-off,
analyzed formally between the update and paging costs.

      Location management methods are most adapted and widely used in current
cellular networks, e.g., GSM, IS-54, IS-95, etc. The location management methods
are broadly classified into two groups. The first group includes all methods based

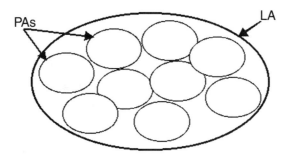

FIGURE 15.1 Number of PAs within an LA.

on algorithms and network architecture, mainly on the processing capabilities of the system. The second group contains the methods based on learning processes, which require the collection of statistics on subscribers' mobility behavior, for instance.

15.1 PAGING

Paging involves messages sent over the radio informing the mobile user that an incoming call is pending. When the mobile station replies, the exact base station to which it is attached will be known to the network, and the call setup can proceed. The network knows the position of the mobile station only at the location area level. Because radio spectrum is scarce, these messages must be kept to a minimum by paging a minimum of cells. The trade-off, as mentioned previously, is that in order to minimize the number of cells that must be paged, location updates must be more frequent. It should be taken into account that because of the unpredictable nature of radio communications, paging messages may not arrive at the mobile with the first attempt, and there is usually some number of repetitions. Because the arrival of paging messages cannot be predicted, a mobile station must listen to the paging channel continuously or almost continuously, as explained in GSM.

For location management purposes, cells are usually grouped together into location areas (LAs) and paging areas (PAs) (see Figure 15.1). A location area is a set of cells, normally (but not necessarily) contiguous, over which a mobile station may roam without needing any further location updates. In effect, a location area is the smallest geographical scale at which the location of the mobile station is known. A paging area is the set of cells over which a paging message is sent to inform a user of an incoming call. In most operational systems, location area and paging area are identical, or paging areas are a subset of location area. For this reason, any grouping of cells for location management purposes is usually called a location area.

15.1.1 Blanket Paging

Two major steps are involved in call delivery. These are (1) determining the serving VLR (visitor location register) of the called MT (mobile terminal) and (2) locating the visiting cell of the called MT. Locating the serving VLR of the MT involves the following database lookup procedures:[1]

1. The calling MT sends a call initiation signal to the serving MSC (mobile switching center) of the MT through a nearby base station.
2. The MSC determines the address of the HLR (home location register) of the called MT by global title translation (GTT) and sends a location request message to the HLR.
3. The HLR determines the serving VR of the called MT and sends a route request message to the MSC serving the MT.
4. The MSC allocates a temporary identifier called temporary local directory number (TLDN) to the MT and sends a reply to the HLR together with the TLDN.
5. The HLR forwards this information to the MSC of the calling MT.
6. The calling MSC requests a call setup to the called MSC through the CCS 7 network.

The procedure described here allows the network to set up a connection from the calling MT to the serving MSC of the called MT.

Because each MSC is associated with a location area, a mechanism is therefore necessary to determine the cell location of the called MT. In current cellular networks, this is achieved by a paging procedure so that polling signals are broadcast to all the cells within the residing LA of the called MT over a forward control channel. On receiving the polling signal, the MT sends a reply over a backward control channel, which allows the MSC to determine its current residing cell. This is called the blanket-paging method. In a selective paging scheme, instead of polling all the cells in an LA, a few cells are polled at a time. The cluster of cells polled at the same time constitutes the paging area. Here, a factor called granularity, K, is defined as the ratio of the number of cells in the PA to the number of cells in the LA. K denotes the fineness in the polling scheme. In a purely sequential polling scheme, $K = 1/S_j$, whereas the granularity factor is 1 in case of blanket polling and S_j is the number of cells in the j-th LA.

15.1.2 DIFFERENT PAGING PROCEDURES

The work reported in Rose[2] developed methods for balancing call registration and paging. The probability distribution on the user location as a function of time is either known or can be calculated, the lower bounds on the average cost of paging are used in conjunction with a Poisson incoming-call arrival model to formulate the paging/registration optimization problem in terms of time-out parameters.

In another work by Rose and Yates,[3] efficient paging procedures are used to minimize the amount of bandwidth expended in locating a mobile unit. Given the probability distribution on user location, they have shown that optimal paging strategy, which minimizes the expected number of locations polled, is to query each location sequentially in order of decreasing probability. Because sequential search over many locations may impose unacceptable polling delay, they considered optimal paging subject to delay constraint.

Akyildiz and Ho[4] proposed a mobile user location mechanism that incorporates a distance-based location update scheme and a paging mechanism that satisfied predefined delay requirements. Akyildiz and coworkers[5] have introduced a mobility tracking mechanism that combines a movement-based location update policy with a selective paging scheme. This selective paging scheme decreases the location tracking cost under a small increase in the allowable paging delay.

Bar-Noy and Kessler[6] explored tracking strategies for mobile users in personal communications networks, which are based on topology of cells. The notion of topology-based strategies was introduced in a general form in this work. In particular, the known paging areas, overlapping paging areas, reporting centers, and distance-based strategies were covered by this notion.

Lyberopoulos et al.[7] proposed a method that aims at the reduction of signaling overhead on the radio link, produced by the paging procedure. The key idea is the application of a multiple-step paging strategy, which operates as follows: At the instance of a call terminating to a mobile user that roams within a certain location area, paging is initially performed in a portion of the location area (the paging area) that the so-called paging-related information indicates. Upon a no-paging response, the mobile user is paged in the complementary portion of the location area; this phase can be completed in more than one (paging) step. Several "intelligent" paging strategies were defined in this work. In Wang and coworkers,[8] various paging schemes were presented for locating mobile users in wireless networks. Paging costs and delay bounds are considered because paging costs are associated with bandwidth utilization and delay bounds influence call setup time. To reduce the paging costs, three paging schemes (reverse, semireverse, and uniform) were introduced to provide a simple way of partitioning the service areas and decreased the paging costs based on each mobile terminal's location probability distribution.

The several paging strategies mainly based on blanket paging were applied to reduce the paging costs, as well as update costs associated with constraints. The strategies briefly discussed here were widely used and few of them are applied in industry. In spite of having widespread use of those paging strategies, some disadvantages have been discovered. In the next section, we discuss new paging schemes to overcome the disadvantages in the different blanket-paging schemes.

15.2 INTELLIGENT PAGING SCHEME

The movement of MTs is modeled according to some ergodic, stochastic process.[9] To provide a ubiquitous communications link, irrespective of the location of MTs, the BSs (base stations) provide continuous coverage during the call, as well as in the idle state. When an incoming call comes to an MT, which roams within a certain LA, paging is initially performed within a portion of LA, which is a subset of the actual LA. This portion of the LA, which is a set of base stations of paging (BSPs), is termed a paging area (PA). Intelligent paging is a multistep paging strategy,[10] which aims at determining the proper PA within which the called MT currently roams. In order to quantitatively evaluate the average cost of paging, time-varying probability distributions on MTs are required. These distributions may be derived from the specific motion models, approximated via empirical data or even provided by the MTs in the form of partial itinerary at the time of last contact. It is assumed that

1. The probability density function of the speed of MTs is known.
2. The process of movement of MTs is isotropic, Brownian motion[11] with drift. In a one-dimensional version of Brownian motion, an MT moves by one step Δx to the right with some probability p, and to the left with probability q, and stays there with probability (1-p-q) for each time step Δt. Given the MT starts at time t = 0 for position x = 0, the Gaussian pdf on the location of an MT is given by

$$P_{X(t)}(x(t)) = (\pi Dt)^{-0.5} e^{-k(x-vt)*(x-vt)/Dt} \tag{15.1}$$

where $v = (p - q) * (\Delta x/\Delta t)$ is the drift velocity, and $D = 2[(1 - p)p + (1 - q)q + 2pq] (\Delta x)^2/\Delta t$ is the diffusion constant, both functions of the relative values of time and space steps. Drift is defined as mean velocity in a given direction and is used to model directed traffic such as vehicles along a highway.
3. Time has elapsed since the last known location.
4. The paging process described here is rapid enough to the rate of motion of MT (i.e., MT, to be found, does not change its location during the paging process).

The algorithm of the intelligent paging process on arrival of a PR is

```
While PR is attached {
   while MT is not busy {
      if current traffic load exceeds threshold traffic load {
         initialize the incremental counter i = 0;
         select the proper PA;
         page within the selected PA;
         if reply against PR received
            then stop;
         else {
            while (i < maximum value of incremental counter i)
               do {
                  page within another PA;
                     increase the incremental counter i = i +1;
               }
         }
      }
      else
         apply blanket paging;
   }
}
```

The intelligent paging strategy maps the cells inside the location area comprising S cells into a probability line at the time of arrival of the incoming call. This mapping depends on factors such as mobility of the MT, its speed profile, the incoming call statistics, and the state of the MT at that instant. This procedure is called attachment. If it is detached, the paging requests (PRs) are cancelled. If it is busy, a relation between the MT and the network already exists and therefore paging is not required. If it is free, the network proceeds for paging upon receipt of a PR (see above algorithm). In an intelligent paging scheme, the network determines the probability of occupancy of the called MT in different cells in an LA. These probabilities are arranged in descending order. The order in which cells are to be polled depends on the ordered set of probability occupancy vectors. In each paging cycle, the MSC serves PRs stored in its buffer, independently of each other. It is done by assigning a BS to each of the n requests in the buffer according to an assignment policy. The j-th PR may be sent to the i-th BS, in order to be paged in the corresponding cell. There are many ways to generate such assignments. Two methods will be presented subsequently in this chapter as part of the proposed intelligent paging scheme. If the buffer size is n and there are k PAs (denoted by $A_1, A_2, .., A_k$), then we can write

$$A_1 \cup A_2 \cup ..... \cup A_k = S$$

$$A_i \cup A_j = \phi$$

This means that the PAs are mutually exclusive and collectively exhaustive. Paging and channel allocation packets from a BS to MTs are multiplexed stream in a forward signaling channel. Paging rate represents the average number of paging packets, which arrive at a base station during unit time. Paging signals are sent to the BSs via landlines and are broadcast over the forward signaling channel. As each attached MT in the location area constantly monitors the paging channels to check whether it is paged or not, the distributor in the MSC which is a part of MM allocates the distribution of PRs to the BSs for each paging cycle based on the information collected over the previous paging cycle. As soon as an MT is found, the corresponding PR is purged from the buffer and a new PR replaces it. The function of the distributor in the MM is to map the PRs to the PAs:

$$g : (PR_1, PR_2, ..., PR_n) \rightarrow (A_1, A_2, ...,A_k)$$

Paging and channel allocation packets from a BS to MTs are multiplexed to stream in a forward signaling channel. Paging rate represents the average number of paging packets, which arrives at a base station during unit time. Paging signals are sent to the BSs via landlines and are broadcast over the forward signaling channel. As each attached MT in the location area constantly monitors the paging channels to check whether it is paged or not, the distributor in the MSC which is a part of MM allocates the distribution of PRs to the BSs for each paging cycle based on the information collected over previous paging cycle. As soon as an MT is found, the corresponding PR is purged from the buffer and a new PR replaces it.

Depending on the nature of polling, there may be two types of search, sequential and parallel-o-sequential. In purely sequential polling, one cell is polled in a paging cycle. Sometimes, due to delay constraint, instead of polling one cell at a time, we poll a cluster of cells in an LA. This is called parallel-o-sequential intelligent paging (PSIP), which is a special case of sequential intelligent polling (SIP).

The network first examines whether a multiple-step paging strategy should be applied or not. The decision is based on the current traffic load. Normally, when this load exceeds a threshold value, a multiple-step paging strategy is employed (Figure 15.1). In the very first phase, the network decides whether checking the current status of the MT needs paging. The network then examines whether the appropriate type of paging is blanket paging or multiple step paging. The granularity factor (K) shows fineness in polling. In general, the granularity factor is defined as

$$K = \text{(number of cells to be polled in a cycle)/(number of cells in an LA)}$$

The maximum value of granularity factor is 1, when all cells in an LA are polled in one polling cycle. The granularity factor in SIP is

$$K^{SIP} = 1/\text{(number of cells in an LA)}$$

The granularity factor in PSIP is

$$K^{PSIP} = \text{(number of cells in the cluster)/(number of cells in an LA)}$$

We assumed a perfect paging mechanism where an MT will always respond to a paging signal meant for it, provided it receives the PR. However, situations may leave an MT undetected even though the distributor in the MSC is able to select the corresponding BS and initiate PR for it. Such a situation will arise when there are more PRs assigned to a BS by the distributor than the number of paging channels available in a cycle. As there are only l paging channels per BS, the PRs in excess of l will be considered blocked. These excess PRs will be attempted for sending to select BSs in subsequent paging cycles. So the called MT may be inside the area of an overloaded cell. But the PR for it might be blocked in a paging cycle. So, the distributor must keep track of the number of times a search has been attempted for the PR.

The application of intelligent paging includes the event of paging failures due to wrong predictions of the locations of the called MT. In such cases, another step or more than one step will be required (i.e., the called MT will be paged in other PAs). Continuous unsuccessful paging attempts may lead to unacceptable network performance in terms of paging delay. Moreover, the paging cost will increase with each unsuccessful attempt to locate the called MT. In such cases, the network does not preclude the option of single-step paging at certain intermediate point of search.

15.2.1 SEQUENTIAL INTELLIGENT PAGING

In a sequential intelligent paging scheme, one cell is polled at a time and the process continues until such time the called MT is found or time out occurs, whichever is earlier. The selection of the cell to be polled sequentially depends on the value of occupancy probability vector, which is based on the stochastic modeling delineating the movement of the MT. In SIP, the PRs are stored in a buffer of MSC, and each

PR is sent to that BS where there is maximum probability of finding the called MT. When the paging is unsuccessful during a polling cycle, the MT is paged sequentially in other cells of the LA that have not been polled so far. This phase is completed in one or more paging step(s). The sequential paging algorithm is

STEP 1:   When an incoming call arrives, calculate the occupancy proba-
            bility vector [P] of an MT for the cells in the LA based on the
            probability density function, which characterizes the motion of
            the MT;
STEP 2:   Sort the elements of [P] in descending order;
STEP 3.0:   FLAG = False;
            i = 1;
STEP 3.1:   Poll the i-th cell for $I \in S$;
STEP 3.2:   If the MT is found
            FLAG = True;
            Go to ENDSTEP;
STEP 4.0:
            If time out occurs
              Go to ENDSTEP;
            Else
              i = i + 1;
              Go to STEP 3.1;
            Endif
ENDSTEP: If FLAG = True
            Declare "Polling is Successful";
          Else
            Declare "Polling is Unsuccessful";
          Endif

As extra processing is required to be done at the MSC, an inherent delay will be associated with this process, i.e., before the PR is sent to the appropriate BS. This delay includes the determination of the probabilities in different cells, sorting of these probabilities in descending order, and polling the cells sequentially depending on those values. This delay will be added to the call setup process. The amount of this delay will be $\sim [O(S) + O(S \log S) + O(S/2)]$.

15.2.2 Parallel-o-Sequential Intelligent Paging

Parallel-o-sequential intelligent paging is a special case of SIP where K > 1. Instead of polling a single cell in each cycle, here we partition the LAs into several PAs and poll those PAs sequentially comprising more than one cell. The benefit that accrues out of PSIP is significant improvement in expected discovery rate of called MTs and the overwhelming reduction in paging cost and signaling load. The number of steps in which the paging process should be completed depends on the allowed delay during paging. The application of PSIP also includes the event of paging failures due to unsuccessful predictions of location of called MT. In such cases, multiple steps are required and the called MT is paged in another portion of the LA. To obviate the deterioration of the network performance in such a situation and minimize the number of paging steps, the network should guarantee formation of PAs such that the P_{SFP} is high (typical value > 90 percent). So, the PA should consist of those cells where the sum of probabilities of finding the called MT is greater than or equal to the typical value chosen for P_{SFP}. The parallel-o-sequential paging algorithm is

STEP 1:   When an incoming call arrives, find out the current state of the called MT;

STEP 2:   If MT is detached
               PR is cancelled;
               Go to ENDSTEP;
   Else
               If MT is busy (location is known)
               Go to ENDSTEP;
   Else
               Find granularity factor K;
   Endif

STEP 3:   If granularity factor is 1
               Poll all the cells;
               Go to ENDSTEP;
   Else
               Find out [P], the occupancy probability vector of the MT for the cells in the LA, based on the probability density function, which characterizes the motion of the MT;
   Endif

STEP 4:   Sort the elements of [P] in descending order;
               Set all the cells as "unmarked";

STEP 5.0:   Select a proper PA consisting of "unmarked" cells for which Sp_i > PSFP;
               FLAG = False;

STEP 5.1:   Poll i-th cluster and label the cells in i-th cluster as "marked"

STEP 5.2:   If the MT is found
               FLAG = True;
               Go to ENDSTEP;

STEP 6.0:
               If time out occurs
               Go to ENDSTEP;
   Else
               Go to STEP 5.0;
   Endif

ENDSTEP: If FLAG = True
               Declare : "Polling is Successful";
   Else
               Declare "Polling is Unsuccessful";
   Endif

15.2.3 COMPARISON OF PAGING COSTS

In the conventional or the blanket paging, upon arrival of an incoming call the paging message is broadcast from all the BSs in the LA, which means all the cells in the LA are polled at a time for locating the called MT, i.e., each MT is paged S times before the called MT is discovered. The polling cost per cycle is

$$C_p^{conv} = S\ A_{cell}\ \rho\mu\ T_p B_p \qquad (15.2)$$

The SIP strategy described here aims at the significant reduction in load of paging signaling on the radio link by paging a cell sequentially. PRs arrive according to a Poisson process at the buffer of the MSC. The distributor issues the PRs to appropriate BSs. These PRs are queued at the location and serviced on FCFS at the average rate ζ. The result may be a success or a failure. The results of completed polls are fed back to the controller in the BS for further appropriate action. As pointed out earlier, a called MT may not be found during a paging cycle. Either the number of paging channels may be insufficient to accommodate the PR in a particular cycle or the search for the called MT in the cell results in a failure. In both cases, the polling process goes through more than one cycle. So, the paging cost per polling cycle in this scheme is

$$C_p^{SIP} = K^{SIP}\ S\ A_{cell}\ \rho\mu\ (1 + z)T_p B_p \qquad (15.3)$$

The variable z accounts for the unsuccessful PRs from the previous cycle due to either of two reasons: z depends on the success rate, time-out duration, and the number of paging channels available per BS. In GSM, assuming that a sufficient channel for paging is there, z becomes zero. In the best case, i.e., when the called MT is found during the first polling cycle of SIP, z also is zero. In the worst case, all the cells in the LA are to be polled before the MT is found. Then the polling cost just exceeds that of GSM. Moreover, the delay is maximum, i.e., S units. There may be a situation when the polling cost in the SIP scheme exceeds the cost in blanket polling significantly. If the MT resides in a cell with a low occupancy of probability and returns to one of the cells, which is polled already after the polling cycle, the called MT will not be found even after polling all the cells in the LA. Such an incidence is likely when the number of cells is more, a few cells have the same probability of occupancy, and the MT is very mobile. In this case, the call is blocked or the cells are polled sequentially once again to find out the MT. So, K has an inverse effect on z. The granularity factor K is generally chosen more than once to avoid such a scenario. The paging cost per polling cycle in PSIP is

$$C_p^{PSIP} = K^{PSIP}\ S\ A_{cell}\ \rho\mu\ (1 + z)T_p B_p \qquad (15.4)$$

The variable z also accounts for the unsuccessful PRs from the previous cycle. As mentioned earlier, K is chosen such that $P_{SFP} > 0.9$. The optimum value of K varies from case to case.

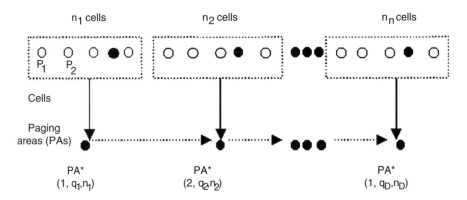

FIGURE 15.2 Partition of location area in paging areas.

15.3 OTHER PAGING SCHEMES

Assume that each LA consists of the same number N of cells in the system.[8] The worst-case paging delay is considered as delay bound, D, in terms of polling cycle. When D is equal to 1, the system should find the called MT in one polling cycle, requiring all cells within the LA to be polled simultaneously. The paging cost, C, which is the number of cells polled to find the called MT, is equal to N. In this case, the average paging delay is at its lowest, which is one polling cycle, and the average paging cost is at its highest, C = N. On the other hand, when D is equal to N, the system will poll one cell in each polling cycle and search all cells one by one. Thus, the worst case occurs when the called MT is found in the last polling cycle, which means the paging delay would be at its maximum and equal to N polling cycles.[12] However, the average paging cost may be minimized if the cells are searched in decreasing order of location probabilities.[3]

Consider the partition of PAs given that $1 \leq D \leq N$, which requires grouping cells within an LA into the smaller PAs under delay bound D. Suppose, at a given time, the initial state **P** is defined as $\mathbf{P} = [p_1 p_2 \ldots, p_j \ldots, p_N]$, where p_j is the location probability of the j-th cell to be searched in decreasing order of probability. Thus the time effect is reflected in the location probability distribution. We use triplets $PA^*_P(i, q_i, n_i)$ to denote the PAs under the paging scheme P in which i is the sequence number of the PA, q_i is the location probability that the called MT can be found within the i-th PA, and n_i is the number of cells contained in this PA. In Figure 15.2, an LA is divided into D PAs because the delay bound is assumed to be D. Thus, the worst-case delay is guaranteed to be D polling cycles. The system searches the PAs one after another until the called MT is found. Three paging schemes are discussed in this section.

15.3.1 REVERSE PAGING

This scheme is designed for a situation where the called MT is most probably to be found in a few cells. Consider the first (D − 1) highest probability cells as the first (D − 1) PAs to be searched. Each of these (D − 1) PAs consists of only one cell.

It is then lumped with the remaining $(N - D + 1)$ lower probability cells to be the last PA, i.e., the D-th PA. The newly formed PAs become $PA^*_r (1, p_1, 1)$, $PA^*_r (2, p_2, 1)$, ..., $PA^*_r (D - 1, p_{D-1}, 1)$, $PA^*_r (D, q_D, N - D + 1)$, where r denotes the reverse paging scheme.

15.3.2 SEMIREVERSE PAGING

Because the average paging cost can be minimized by searching cells in decreasing order of location probability if a delay bound D is not applied,[3] intuitively the paging cost can be reduced by searching the PAs in decreasing order of probability. Under a semireverse paging scheme, a set of PAs is created in a nonincreasing order of location probabilities. Combine first the two cells with the lowest location probabilities into one PA, and then reorder all PAs in nonincreasing order of location probabilities. Keep combining the two lowest probabilities PAs and reorder them until the total number of PAs is equal to D. If two PAs have the same probability, the PA with fewer cells has higher priority, i.e., its sequence number is smaller. The semireverse paging scheme guarantees that the location probability of each PA is in a nonincreasing order. However, the cell with lower probability may be searched before the cell with higher probability because the initial sequence of the cells is reordered during the semireverse paging procedure.

15.3.3 UNIFORM PAGING

Under this scheme, the LA is partitioned into a series of PAs in such a way that all PAs consist of approximately the same number of cells. The uniform paging procedure is as follows:

- Calculate the number of cells in each PA as $n_0 = \lfloor N/D \rfloor$, where $N = n_0 D + k$.
- Determine a series of PAs as $PA^*_u (1, p_1, 1)$, $PA^*_u (2, p_2, 1)$, ..., $PA^*_u (D, q_D, n_D)$. Note that there are n_0 cells in each of the first $(D - k)$ PAs and there are $n_0 + 1$ cells in each of the remaining PAs. This means $n_1 = n_2 = ... = n_{D-k} = n_0$, and $n_{D-k+1} = ... = n_D = n_0 + 1$. For example, the first PA consists of n_0 cells and the last PA, i.e., D-th PA, consists of $n_0 + 1$ cells.
- The network polls one PA after another sequentially until the called MT is found.

15.4 INTERSYSTEM PAGING

In a multitier wireless service area consisting of dissimilar systems, it is desirable to consider some factors that will influence the radio connections of the mobile terminals (MTs) roaming between different systems.[13] Consider there are two systems, Y and W, in the microcell tier, that may use different protocols such as DCS1800 and PCS1900. Each hexagon represents a location area (LA) within a stand-alone system and each LA is composed of a cluster of microcells. The terminals are required to update their location information with the system whenever they enter a new LA; therefore, the system knows the residing LA of a terminal all the time. In the macrocell tier there are also two systems, X and Z, in which different protocols (e.g., GSM and IS-41) are applied. For macrocell systems, one LA can be one macrocell. It is possible that systems X and W, although in different tiers, may employ similar protocols such as IS-95, GSM, or any other protocol. There are two types of roaming: intra- or intersystem. Intrasystem roaming refers to an MT's movement between the LAs within a system such as Y and Z. Intersystem roaming refers to the MTs that move between different systems. For example, mobile users may travel from a macrocell system within an IS-41 network to a region that uses GSM standard.

For intersystem location update, a boundary region called boundary location area (BLA) exists at the boundary between two systems in different tiers.[13] In addition to the concept of BLA, a boundary location register (BLR) is embedded in the BIU. A BLR is a database cache to maintain the roaming information of MTs moving between different systems. The roaming information is captured when the MT requests a location registration in the BLA. The BLRs enable the intersystem paging to be implemented within the appropriate system that an MT is currently residing in, thus reducing the paging costs. Therefore, the BLR and the BIU are accessible to the two adjacent systems and are colocated to handle the intersystem roaming of MTs. On the contrary, the VLR and the MSC provide roaming information within a system and deal with the intrasystem roaming of MTs. Besides, there is only one BLR and one BIU between a pair of neighboring systems, but there may be many VLRs and MSCs within a stand-alone system.

When a call connection request arrives at X, the call will be routed to the last registered LA of the called MT. Given that the last registered LA within X is adjacent to Y, the system needs to perform the following steps to locate the MT:

- Send a query signal to the BLR between X and Y to obtain the MT's location information. This step is used to ascertain whether the MT has crossed the boundary.
- If the MT has already moved to Y, only the LA in Y needs to be searched. Otherwise, the last registered LA within X will be searched. Within network X or Y, one or multiple polling messages are sent to the cells in the LA according to a specific paging scheme.

As a result, only one system (X or Y) is searched in the paging process for intersystem roaming terminals. This approach will significantly reduce the signaling cost caused by intersystem paging. In particular, it is very suitable for the high-traffic environment because it omits searching in two adjacent systems. Moreover, because the BLR is an additional level of cache database, it will not affect the original database architecture. Another advantage of the BLR is that it reduces the zigzag effect caused by intersystem roaming. For example, when an MT is moving back and forth on the boundary, it only needs to update the information in the BLR instead of contacting the HLRs. If the new BLR concept is not used, the intersystem paging can still take place. The system will search X first, if the called MT cannot be found, then Y will be searched. This method increases the paging cost as well as the paging delay, thus degrading the system performance.

15.5 IP MICROMOBILITY AND PAGING

Recent research[14] in Mobile IP has proposed that IP should take support from the underlying wireless network architecture to achieve good performance for handover and paging protocols. Recent IETF work defines requirements for layer 2 (the data link layer of the OSI model) to support optimized layer 3 (the network layer of the OSI model) handover and paging protocols. Layer 2 can send notification to layer 3 that a certain event has happened or is about to happen. The notification is sent using a trigger. Kempf et al.[15] discuss various ways of implementing triggers. A trigger may be implemented using system calls. The operating system may allow an application thread to register callback for a layer 2 trigger, using system calls of an application-programming interface (API). A system call returns when that particular event is fired in layer 2. Each trigger is defined by three parameters:

1. The event that causes the trigger to fire
2. The entity that receives the trigger
3. The parameters delivered with the trigger

Triggers were defined to aid low-latency hand-over in Mobile IP.[16] Another set of triggers was defined in Gurivireddy and coworkers[17] to aid IP paging protocols. CDMA, for example, works in conjunction with Mobile IP to support mobility in IP hosts. IP paging is a protocol used to determine the location of a dormant (a mode that conserves battery by not performing frequent updates) MN. Paging triggers were defined in Gurivireddy and coworkers[17] to aid movement of a MN in multiple IP subnets in the same layer 2 paging area. Paging triggers aid the MN to enter dormant mode in a graceful manner and make best use of paging provided by underlying wireless architecture.

15.6 LOCATION UPDATE

In the previous section, the tracking of an MT has been discussed when an incoming call is to be delivered. As the MTs move around the network service area, the data stored in these databases may no longer be accurate. To ensure that calls can be delivered successfully, a mechanism is needed to update the databases with up-to-date location information. This is called location update (LU) or registration. Several location updating methods are based on LA structuring. Two automatic location area management schemes are very much in use:[18]

1. Periodic location updating. This method, although the simplest, has the inherent drawback of having excessive resource consumption, which at times is unnecessary.
2. Location updating on LA crossing. A network must retain information about the locations of endpoints in the network, in order to route traffic to the correct destinations. Location tracking (also referred to as mobility tracking or mobility management) is the set of mechanisms by which

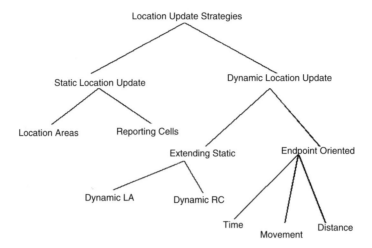

FIGURE 15.3 Classifications of location update strategies.

location information is updated in response to endpoint mobility. In location tracking, it is important to differentiate between the identifier of an
endpoint (i.e., what the endpoint is called) and its address (i.e., where the
endpoint is located). Mechanisms for location tracking provide a time-
varying mapping between the identifier and the address of each endpoint.

Most location-tracking mechanisms may be perceived as updating and querying
a distributed database (the location database) of endpoint identifier-to-address mappings. In this context, location tracking has two components: (1) determining when
and how a change in a location database entry should be initiated, and (2) organizing
and maintaining the location database. In cellular networks, endpoint mobility within
a cell is transparent to the network, and hence location tracking is only required
when an endpoint moves from one cell to another. Location tracking typically
consists of two operations: (1) updating (or registration), the process by which a
mobile endpoint initiates a change in the location database according to its new
location; and (2) finding (or paging), the process by which the network initiates a
query for an endpoint's location (which may result also in an update to the location
database). Most location-tracking techniques use a combination of updating and
finding in an effort to select the best trade-off between update overhead and delay
incurred in finding. Specifically, updates are not usually sent every time an endpoint
enters a new cell, but rather are sent according to a predefined strategy such that the
finding operation can be restricted to a specific area. There is also a trade-off,
analyzed formally in Madhow and coworkers,[19] between the update and paging costs.
Figure 15.3 illustrates a classification of possible update strategies.[18]

15.6.1 LOCATION UPDATE STATIC STRATEGIES

In a static update strategy, there is a predetermined set of cells at which location
updates may be generated. Whatever the nature of mobility of an endpoint, location

updates may only be generated when (but not necessarily every time) the endpoint enters one of these cells. Two approaches to static updating are as follows.

1. Location areas (also referred to as paging or registration areas).[20] In this approach, the service area is partitioned into groups of cells with each group as a location area. An endpoint's position is updated if and only if the endpoint changes location areas. When an endpoint needs to be located, paging is done over the most-recent location area visited by the endpoint. Location tracking in many 2G cellular systems, including GSM[21] and IS-41,[22] is based on location areas.[23] Several strategies for location area planning in a city environment have been evaluated,[10] including strategies that take into account geographical criteria (such as population distribution and highway topology) and user mobility characteristics.

2. Reporting cells (or reporting centers).[24] In this approach, a subset of the cells is designated as the only one from which an endpoint's location may be updated. When an endpoint needs to be located, a search is conducted in the vicinity of the reporting cell from which the most-recent update was generated. In Bar-Noy and Kessler,[24] the problem of which cells should be designated as reporting cells so as to optimize a cost function is addressed for various cell topologies.

The principal drawback to static update strategies is that they do not accurately account for user mobility and frequency of incoming calls. For example, although a mobile endpoint may remain within a small area, it may cause frequent location updates if that area happens to contain a reporting cell.

15.6.2 LOCATION UPDATE DYNAMIC STRATEGIES

In a dynamic update strategy, an endpoint determines when an update should be generated, based on its movement. Thus, an update may be generated in any cell. A natural approach to dynamic strategies is to extend the static strategies to incorporate call and mobility patterns. The dynamic location area strategy proposed in Xie and coworkers[25] dynamically determines the size of an endpoint's location area according to the endpoint's incoming call arrival rate and mobility. Analytical results presented in Xie and coworkers[25] indicate that this strategy is an improvement over static strategies when call arrival rates are dependent on user or time. The dynamic reporting centers strategy proposed in Birk and Nachman[26] uses easily-obtainable information to customize the choice of the next set of reporting cells at the time of each location update. In particular, the strategy uses information recorded at the time of the endpoint's last location update, including the direction of motion, to construct an asymmetric distance-based cell boundary and to optimize the cell search order. In Bar-Noy and coworkers,[27] three dynamic strategies are described in which an endpoint generates a location update: (1) every T seconds (time-based), (2) after every M cell crossings (movement based), or (3) whenever the distance covered (in terms of number of cells) exceeds D (distance based). Distance-based strategies are inherently the most difficult to implement because the mobile endpoints need information about the topology of the cellular network. It was shown in Bar-Noy and coworkers,[27] however, that for memoryless movement patterns on a ring topology, distance-based updating outperforms both time-based and movement-based updating. In Madhow and coworkers,[19] a set of dynamic programming equations is derived and used to determine an optimal updating policy for each endpoint, and this optimal policy is in fact distance-based. Strategies that minimize location-tracking costs under specified delay constraints (i.e., the time required to locate an endpoint) also have been proposed. In Rose and Yates,[3] a paging procedure is described that minimizes the mean number of locations polled with a constraint on polling delay, given a probability distribution for endpoint locations. A distance-based update scheme and a complementary paging scheme that guarantee a predefined maximum delay on locating an endpoint are described in Akyildiz and Ho.[4] This scheme uses an iterative algorithm to determine the optimal update distance D that results in minimum cost within the delay bound.

In organizing the location database, one seeks to minimize both the latency and the overhead, in terms of the amount of storage and the number of messages required, in accessing location information. These are, in general, counteracting optimization criteria. Most solutions to the location database organization problem select a point, which is a three-way trade-off between overhead, latency, and simplicity. The simplest approach to location database organization is to store all endpoint identifier-to-address mappings in a single central place. For large numbers of reasonably mobile endpoints, however, this approach becomes infeasible in terms of database access time and storage space and also represents a single-point-of-failure.

The next logical step in location database organization is to partition the network into a number of smaller pieces and place a portion of the location database in each piece. Such a distributed approach is well suited to systems where each subscriber is registered in a particular area or home. With this organization, the location database in an area contains the locations of all endpoints whose home is that area. When the endpoint moves out of its home area, it updates its home location database to reflect the new location. The home location register (HLR) and visitor location register (VLR) schemes of emerging wireless cellular networks[23] are examples of this approach, as are the Mobile IP scheme[28] for the Internet and the GSM-based General Radio Packet Switching (GPRS) network for data transport over cellular networks. Studies[29,30] have shown that with predicted levels of mobile users, signaling traffic may exceed acceptable levels. Thus, researchers have considered augmenting this basic scheme to increase its efficiency under certain circumstances. For instance, in Jain et al.,[31] per-user caching is used to reuse location information about a called user for subsequent calls to that user, and is particularly beneficial for users with high call-to-mobility ratios (i.e., the frequency of incoming calls is much larger than the frequency of location updates). In Ho and Akyildiz,[32] "local anchoring" is used to reduce the message overhead by reporting location changes to a nearby VLR instead of to the HLR, thus increasing the location tracking efficiency when the call-to-mobility ratio is low and the update cost is high. As with most large organizational problems, a hierarchical approach provides the most general and scalable solution. By hierarchically organizing the location database, one can exploit the fact that many movements are local. Specifically, by confining location update propagation to the lowest level (in the hierarchy) containing the moving endpoint, costs can be made proportional to the distance moved. Several papers address this basic theme. In Awerbuch and Peleg,[33] a hierarchy of regional directories is prescribed where each regional directory is based on a decomposition of the network into regions. Here, the purpose of the i-th level regional directory is to enable tracking of any user residing within a distance of 2^i. This strategy guarantees overheads that are poly-logarithmic in the size and diameter of the network. In Anantharam et al.,[34] the location database is organized so as to minimize the total rate of accesses and updates. This approach takes into account estimates of mobility and calling rates between cells and a budget on access and update rates at each database site. In Badrinath and coworkers,[35] location database organization takes into account the user profile of an endpoint (i.e., the predefined pattern of movement for the endpoint). Partitions of the location database are obtained by grouping the locations among which the endpoint moves frequently and by separating those to which the endpoint relocates infrequently. Each partition is further partitioned in a recursive fashion, along the same lines, to obtain a location database hierarchy.

In the above strategies, the emphasis is on reducing update overhead, but it is equally important to reduce database access latency. One strategy for doing so is replication, where identical copies of the database are kept in various parts of the network so that an endpoint location may be obtained using a low-latency query to a nearby server. The problem here is to decide where to store the replications. This is similar to the classical database allocation[36] and file allocation[37] problems, in which databases or files are replicated at sites based on query–update or read–write access patterns. In Shivakumar and Widom,[38] the best zones for replication are chosen per endpoint location entry, using a minimum-cost maximum-flow algorithm to decide where to replicate the database, based on the calling and mobility patterns for that endpoint.

15.7 LOCATION MANAGEMENT

Location management schemes are essentially based on users' mobility and incoming call rate characteristics. It is a two-stage process that enables a network to discover the current attachment point of the mobile user for call delivery. The first stage is location registration or location update. In this stage the mobile terminal periodically notifies the network of it new access point, allowing the network to authenticate the user and revise the user's location profile. The second stage is call delivery. Here, the network is queried for the user location profile and the current position of the mobile host is found. Current techniques for location management involve database architecture design and the transmission of signaling messages between various components of a signaling network. Other issues include security, dynamic database updates, querying delays, terminal paging methods, and paging delays.

There are two standards for location management currently available: Electronic/Telecommunications Industry Associations (EIA/TIA) Interim Standard 41(IS-41)[39] and the Global System for Mobile Communications (GSM) mobile application part (MAP).[23] The IS-41 scheme is commonly used in North America for Advanced Mobile Phone System (AMPS), IS-54, IS-136, and personal access communications system (PACS) networks, while the GSM MAP is mostly used in Europe for GSM and digital cellular service at 1800 MHz (DCS-1800) networks. Both standards are based on a two-level data hierarchy. Location registration procedures update the location databases (HLR and VLRs) and authenticate the MT when up-to-date location information of an MT is available. Call delivery procedures locate the MT based on the information available at the HLR and VLRs when a call for an MT is initiated. The IS-41 and GSM MAP location management strategies are very similar. While GSM MAP is designed to facilitate personal mobility and to enable user selection of network provider, there are a lot of commonalities between the two standards.[1,21] The location management scheme may be categorized in several ways.

15.7.1 WITHOUT LOCATION MANAGEMENT

This level 0 method is no location management is realized,[40] the system does not track the mobiles. A search for a called user must therefore be done over the complete radio coverage area and within a limited time. This method is usually referred to as the flooding algorithm.[41] It is used in paging systems because of the lack of an uplink channel allowing a mobile to inform the network of its whereabouts. It is used also in small private mobile networks because of their small coverage areas and user populations. The main advantage of not locating the mobile terminals is obvious simplicity; in particular, there is no need to implement special databases. Unfortunately, it does not fit large networks dealing with high numbers of users and high incoming call rates.

15.7.2 MANUAL REGISTRATION IN LOCATION MANAGEMENT

This level 1 method[40] is relatively simple to manage because it requires only the management of an indicator, which stores the current location of the user. The mobile is also relatively simple; its task is just limited to scanning the channels to detect paging messages. This method is currently used in telepoint cordless systems (such as CT2, Cordless Telephone 2). The user has to register when moving to a new island of CT2 beacons. To page a user, the network first transmits messages through the user's registered beacon and, if the mobile does not answer, extends the paging to neighboring beacons. The main drawback of this method is the constraint for a user to register with each move.

15.7.3 AUTOMATIC LOCATION MANAGEMENT
USING LOCATION AREA

Presently, this level 2 location method[40] most widely implemented in first- and second-generation cellular systems (NMT, GSM, IS-95, etc.) makes use of location areas (LAs) (Figure 15.1). In these wide-area radio networks, location management is done automatically. Location areas allow the system to track the mobiles during their roaming in the network(s): subscriber location is known if the system knows the LA in which the subscriber is located. When the system must establish a communication with the mobile (typically, to route an incoming call), the paging only occurs in the current user LA. Thus, resource consumption is limited to this LA; paging messages are only transmitted in the cells of this particular LA. Implementing LA-based methods requires the use of databases. Generally, a home database and several visitor databases are included in the network architecture.

15.7.4 MEMORYLESS-BASED LOCATION MANAGEMENT METHODS

All methods are included based on algorithms and network architecture, mainly on the processing capabilities of the system.

15.7.4.1 Database Architecture

LA partitioning, and thus mobility management cost, partly relies on the system architecture (e.g., database locations). Thus, designing an appropriate database organization can reduce signaling traffic. The various database architectures are proposed with this aim.[1,42–44] An architecture where a unique centralized database is used is well suited to small and medium networks, typically based on a star topology. The second one is a distributed database architecture, which uses several independent databases according to geographical proximity or service providers. It is best suited to large networks, including subnetworks managed by different operators and service providers. The GSM worldwide network, defined as the network made up of all interconnected GSM networks in the world, can be such an example of a large network. The third case is the hybrid database architecture that combines the centralized and distributed database architectures. In this case, a central database (HLR-like) is used to store all user information. Other smaller databases (VLR-like) are distributed all over the network. These VLR databases store portions of HLR user records. A single GSM network is an example of such architecture.

15.7.4.2 Optimizing Fixed Network Architecture

In 2G cellular networks and 3G systems, the intelligent network (IN) manages signaling.[45] Appropriately organizing mobility functions and entities can help reduce the signaling burden at the network side. The main advantage of these propositions is that they allow one to reduce the network mobility costs independent of the radio interface and LA organization.

15.7.4.3 Combining Location Areas and Paging Areas

In current systems, an LA is defined as both an area in which to locate a user and an area in which to page him. LA size optimization is therefore achieved by taking into account two antagonistic procedures, locating and paging. Based on this observation, several proposals have defined location management procedures, which make use of LAs and paging areas (PAs) of different sizes.[46] One method often considered consists of splitting an LA into several PAs. An MS registers only once, i.e., when it enters the LA. It does not register when moving between the different PAs of the same LA. For an incoming call, paging messages will be broadcast in the PAs according to a sequence determined by different strategies. For example, the first PA of the sequence can be the one where the MS was last detected by the network. The drawback of this method is the possible delay increase due to large LAs.

15.7.4.4 Multilayer LAs

In present location management methods, LU traffic is mainly concentrated in the cells of the LA border. Based on this observation and to overcome this problem, Okasaka et al. have introduced the multiplayer concept.[47] In this method, each MS is assigned to a given group, and each group is assigned one or several layers of LAs. This location updating method, although it may help reduce channel congestion, does not help reduce the overall signaling load generated by LUs.

15.7.5 MEMORY-BASED LOCATION MANAGEMENT METHODS

The design of memory-based location management methods has been motivated by the fact that systems do a lot of repetitive actions, which can be avoided if predicted. This is particularly the case for LUs. Indeed, present cellular systems achieve every day, at the same peak hours, almost the same LU processing. Systems act as memoryless processes.

15.7.5.1 Dynamic LA and PA Size

The size of LAs is optimized according to mean parameter values, which, in practical situations, vary over a wide range during the day and from one user to another. Based on this observation, it is proposed to manage user location by defining multilevel LAs in a hierarchical cellular structure.[48] At each level, the LA size is different, and a cell belongs to different LAs of different sizes. According to past and present MS mobility behavior, the scheme dynamically changes the hierarchical level of the LA to which the MS registers. LU savings can thus be obtained.

An opposite approach considers that instead of defining LA sizes *a priori,* these can be adjusted dynamically for every user according to the incoming call rate (a) and LU rate (u_k), for instance. In Xie and coworkers,[25] a mobility cost function denoted $C(k, a, u_k)$ is minimized so that k is permanently adjusted. Each user is therefore related to a unique LA for which size k is adjusted according to the particular mobility and incoming call rate characteristics. Adapting the LA size to each user's parameter values may be difficult to manage in practical situations. This led to the definition of a method where the LA sizes are dynamically adjusted for the whole population, not per user.[49]

15.7.5.2 Individual User Patterns

Observing that users show repetitive mobility patterns, the alternative strategy (AS) is defined.[50,51] Its main goal is to reduce the traffic related to mobility management — and thus reduce the LUs — by taking advantage of users' highly predictable patterns. In AS, the system handles a profile recording the most probable mobility patterns of each user. The profile of the user can be provided and updated manually by the subscriber himself or determined automatically by monitoring the subscriber's movements over a period of time. For an individual user, each period of time corresponds to a set of location areas, k. When the user receives a call, the system pages him sequentially over the LA *ai s* until getting an acknowledgment from the mobile. When the subscriber moves away from the recorded zone $\{a_1,...,a_k\}$, the terminal processes a voluntary registration by pointing out its new LA to the network. The main savings allowed by this method are due to the nontriggered LUs when the user keeps moving inside his profile LAs. So, the more predictable the user's mobility, the lower the mobility management cost. The main advantage of this method relies on the reduction of LUs when a mobile goes back and forth between two LAs.

15.7.6 LOCATION MANAGEMENT IN 3G-AND-BEYOND SYSTEMS

The next generation in mobility management will enable different mobile networks to interoperate to ensure terminal and personal mobility and global portability of network services. However, in order to ensure global mobility, the deployment and integration of both wire and wireless components are necessary. The focuses are given on issues related to mobility management in a future mobile communications system, in a scenario where different access networks are integrated into an IP core network by exploiting the principles of Mobile IP. Mobile IP,[52] the current standard for IP-based mobility management, needs to be enhanced to meet the needs of future fourth-generation (4G) cellular environments. In particular, the absence of a location management hierarchy leads to concerns about signaling scalability and handoff latency, especially for a future infrastructure that must provide global mobility support to potentially billions of mobile nodes and accommodate the stringent performance bounds associated with real-time multimedia traffic. In this chapter, the discussion is confined to Mobile IP to describe the aspects of location management in 3G and beyond.

The 4G cellular network will be used to develop a framework for truly ubiquitous IP-based access by mobile users, with special emphasis on the ability to use a wide variety of wireless and wired access technologies to access the common information infrastructure. While the 3G initiatives are almost exclusively directed at defining wide area packet-based cellular technologies, the 4G vision embraces additional local area access technologies, such as IEEE 802.11-based wireless local area networks (WLANs) and Bluetooth-based wireless personal area networks (WPANs). The development of mobile terminals with multiple physical or software-defined interfaces is expected to allow users to seamlessly switch between different access technologies, often with overlapping areas of coverage and dramatically different cell sizes.

Consider one example[53] of this multitechnology vision at work in a corporate campus located in an urban environment. While conventional wide area cellular coverage is available in all outdoor locations, the corporation offers 802.11-based access also in public indoor locations such as the cafeteria and parking lots, as well as Bluetooth-based access to the Internet in every individual office. As mobile users drive to work, their ongoing Voice over IP (VoIP) calls are seamlessly switched, first from the wide area cellular to the WLAN infrastructure, and subsequently from the

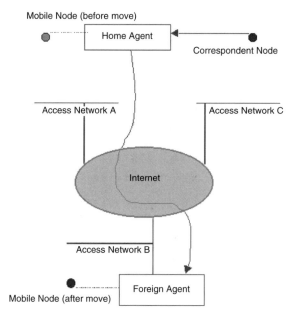

FIGURE 15.4 Mobile IP architecture.

802.11 access point (AP) to the Bluetooth AP located in their individual cubicles or offices. Because a domain can comprise multiple access technologies, mobility management protocols should be capable of handling vertical handoffs (i.e., handoffs between heterogeneous technologies).

Due to the different types of architecture envisaged in the multiaccess system, three levels of location management procedures can be envisaged[54]

1. Internet (interdomain) network location management: Identifies the point of access to the Internet network
2. Intrasegment location management: Executed by segment-specific procedures when the terminal moves within the same access network
3. Intersegment location management: Executed by system-specific entities when the terminal moves from one access network to another

In Mobile IP (Figure 15.4),[52] each mobile node is assigned a pair of addresses. The first address is used for identification, known as the home IP address, which is defined in the address space of the home subnetwork. The second address is used to determine the current position of the node and is known as the care-of address (CoA), which is defined in the address space of the visited/foreign subnetwork. The continuous tracking of the subscriber's CoA allows the Internet to provide subscribers with roaming services. The location of the subscriber is stored in a database, known as a binding table, in the home agent (HA) and in the corresponding node (CN). By using the binding table, it is possible to route the IP packets toward the Internet point of access to which the subscriber is connected.

The terminal can be seen from the Internet perspective as a mobile terminal (MT). Once the MT selects an access segment, the access point to the Internet network is automatically defined. The MT is therefore identified by a home address of the home subnetwork and by a CoA of the access segment. In the target system, location management in the Internet network is based on the main features of Mobile IP. Nevertheless, a major difference can be identified between the use of Mobile IP in fixed and mobile networks. In the fixed Internet network, IP packets are routed directly to the mobile node, whereas in the integrated system considered in this chapter, packets will be routed up to the appropriate edge router. Once a packet leaves the edge router and reaches the access network, the routing toward the final destination will be performed according to the mechanisms adopted by each access segment (intrasegment mobility). When the MMT decides to change access segments, its CoA will be changed. Therefore, the new CoA has to be stored in the corresponding binding tables. Because these binding tables can be seen as a type of location management database, this binding update also can be seen as a form of location update on the Internet.

Intersegment location management is used to store information on the access segments at a particular time. The information is then used to perform system registration, location update, and handover procedures. Using certain parameters, including the condition of the radio coverage and QoS perceived by the user, the MT continuously executes procedures with the objective of selecting the most suitable access segment. Any modifications to these parameters could therefore lead to a change of access segment. This implies also a change in the point of access to the Internet network. Therefore, in order to route these packets correctly it is necessary to have information on the active access segment, particularly information concerning the edge router that is connected to the node of the access segment. Thus, from the Internet point of view, no additional procedure or database is required because the information is implicitly contained in the CoA assigned to the MT.

15.8 LOCATION AREA PLANNING

Location area (LA) planning in minimum cost plays an important role in cellular networks because of the trade-off caused by paging and registration signaling. The upper bound on the size of an LA is the service area of a mobile switching center (MSC). In that extreme case, the cost of paging is at its maximum, but no registration is needed. On the other hand, if each cell is an LA, the paging cost is minimal, but the registration cost is the largest. In general, the most important component of these costs is the load on the signaling resources. Between the extremes lie one or more partitions of the MSC service area that minimizes the total cost of paging and registration. In this section, a few approaches are discussed that address LA planning-related issues.

15.8.1 TWO-STEP APPROACH

This approach[55] deals with the planning of LAs in a personal communication services network (PCSN) to be overlaid on an existing wired network. Given the average speed of mobile terminals, the number of MSCs, their locations, call handling capacity of each MSC, handoff cost between adjacent cells, and call arrival rate, an important consideration in a PCSN is to identify the cells in every LA to be connected to the corresponding MSC in a cost-effective manner. While planning a location area, a two-step approach is presented, namely, optimization of total system recurring cost (subproblem I), and optimization of hybrid cost (subproblem II). The planning first determines the optimum number of cells in an LA from subproblem I. Then it finds out the exact LAs by assigning cells to the switches, while optimizing the hybrid cost which comprises the handoff cost and the cable cost, in subproblem II. The decomposition of the problem provides a practical way for designing LAs. As this approach toward LA planning takes into account both cost and network planning factors, this unique combination is of great interest to PCSN designers. It develops an optimum network-planning method for a wide range of call-to-mobility ratio (CMR) that minimizes the total system recurring cost, while still ensuring a good system performance. Approximate optimal results, with respect to cell-to-switch assignment, are achievable with a reasonable computational effort that supports the engineered plan of an existing PCSN.

In order to design a feasible PCSN, constraints such as traffic-handling capacity of MSCs and costs related to paging, registration, and cabling should be considered. Utilizing the available information of MTs and the network in a suitable manner, it is possible to devise a technique for planning of location areas in a PCSN that optimizes both system-recurring cost and hybrid cost.

This design is not restricted to any particular assumption on the mobility pattern of MTs or the mobility model either. Because the optimum LA size decreases significantly with the increase in CMR, as the corresponding hike in system cost is very high, design parameters at BS and MSCs cannot be specified until cell allocation is completed. Finally, channel assignment, which can further improve system performance in terms of QoS and improved carrier interference ratio, can only be determined once the architecture of the PCSN has been obtained.

15.8.2 LA PLANNING AND SIGNALING REQUIREMENTS

In 2G mobile systems,[10] LA planning does not generate significant problems because the number of generated LUs (as well as the amount of paging signaling) remains relatively low because of the low number of users. For 3G mobile telecommunications systems, several alternative location-tracking techniques have been used (e.g., based on the use of reporting centers or a dynamic location area management protocol). Nevertheless, in UMTS an approach similar to 2G location finding has been used, i.e., the system area is divided into LAs and a called user is located in two steps (determine the LA within which the user roams and perform paging within this LA). The main issue concerning location area planning in 3G mobile telecommunications systems is the amount of location finding related signaling load (paging signaling, location updating, and distributed database queries).

Because the size and shape of an LA affects the signaling requirements due to paging and LU, it is obvious that LA planning should minimize both, if possible. In order to provide a clear view of the relation between the LA planning and the above-mentioned parameters, we consider two extreme LA planning approaches:

1. The system area equals an LA. Whenever an MT is called, it is paged over the whole system coverage area, while no LUs are performed due to MU movements. In this case, paging signaling load can be enormous, especially during rush hour.
2. The cell area equals an LA. In this case, the location of an MT is determined with accuracy of a single cell area. The need for paging here is minimal; paging does not locate the MT, it just alerts the terminal for the incoming call. However, the number of LUs is expected to be enormous due to the small cell size and user mobility. A brief description of the LA planning methods under consideration follows:
   - LA planning based on heuristic algorithms: This is a method to approximate the optimum LA planning as a set of cells. According to the example heuristic algorithm used in this chapter, cells are randomly selected to form LAs.
   - LA planning based on area zones and highway topology: Area zones are defined according to geographical criteria (e.g., city center), and the approach considers the population distribution and the way that people move via city highways so as to determine the proper LA configuration.
   - LA planning based on overlapping LA borders: This method can be considered as an attempt to improve the previous one by means of reducing the number of generated LUs. In this case, LAs have overlapping borders so as to avoid LUs due to MU movements around the LA borders.
   - Time-dependent LA configuration: According to this scenario the network alters the LA configuration based on either some predefined timetable or monitoring of the number of LUs and the number of paging messages. The LA configuration here is selected so as to fit the time variable mobility and traffic conditions.
   - LA planning based on MU grouping: This method considers the mobility behavior of each individual MU so as to minimize the number of LUs generated due to daily MU movements. To apply this method, MUs are grouped based on their mobility behavior (e.g., high mobility MUs) and different LA configuration is determined for each group.
   - LA planning using simulated annealing.[56] This research focusing on LA management in wireless cellular networks has minimized the total paging and registration cost. This chapter finds an optimal method for determining the location areas. To that end, an appropriate objective function is defined with the addition of paging and registration costs. For that purpose, the available network information to formulate a realistic optimization problem is used. In reality, the load (i.e., cost) of paging and registration to the network varies from cell to cell. An algorithm based on simulated annealing for the solution of the resulting problem is used.

References

1. Akyildiz, I.F. and Ho, J.S.M., On location management for personal communications networks, *IEEE Communications Magazine,* Sept. 1996, pp. 138–145.
2. Rose, C., Minimizing the average cost of paging and registration: a timer-based method, *ACM J. Wireless Networks,* 109–116, Feb. 1996.
3. Rose, C. and Yates, R., Minimizing the average cost of paging under delay constraints, *Wireless Networks,* 1, 211–219, 1995.
4. Akyildiz, I.F. and Ho, J.S.M., A mobile user location update and paging mechanism under delay constraints, Proc. of ACM SIGCOMM, Cambridge, Massachusetts, 1995, pp. 244–255.
5. Akyildiz, I.F., Ho, J.S.M., and Lin, Y.-B., Movement-based location update and selective paging for PCS networks, 1996.
6. Bar-Noy, A. and Kessler, I., Mobile users: to update or not to update? Proc. INFO-COM '94, June 1994, pp. 570–576.
7. Lyberopoulos, G.L., Markoulidakis, J.G., Polymeros, D.V., Tsirkas, D.F., and Sykas, E.D., Intelligent Paging Strategies for Third Generation Mobile Telecommunication Systems,
8. Wang, W., Akyildiz, I.F., and Stüber, G.L., Effective paging schemes with delay bounds as QoS constraints in wireless systems, *Wireless Networks,* 7, 455–466, 2001.
9. Bhattacharjee, P.S., Saha, D., and Mukherjee, A., Paging strategies for future personal communication services networks, Proc. 6th Int. Conf. on High Performance Computing (HiPC'99), Calcutta, India, Dec. 1999.
10. Markoulidakis, G.L. et al., Evaluation in LA planning in future mobile telecommunication systems, *Wireless Networks,* 1995.
11. Papoulis, A., *Probability, Random Variable and Stochastic Processes,* 3rd ed., McGraw-Hill, New York.
12. Wang, W., Akyildiz, I.F., and Stüber, G.L., Reducing the paging costs under delay bounds for PCS networks, Proc. of IEEE WCNC'2000, Sept. 2000.
13. Akyildiz, I.F. and Wang, W., A dynamic location management scheme for next generation multi-tier PCS systems, *IEEE Trans. Wireless Commun.,* Jan. 2002.
14. Sarikaya, B. and Gurivireddy, S., Evaluation of CDMA2000 support for IP micromobility handover and paging protocols, *IEEE Communications Magazine,* May 2002, pp. 146–149.
15. Kempf, J. et al., Requirements for layer 2 protocols to support optimized handover for IP mobility, IETF draft, July 2001.
16. Kempf, J. et al., Bidirectional edge tunnel handover for IPv6, IETF draft, Sept. 2001.
17. Gurivireddy, S., Sarikaya, B., and Krywaniuk, A., Layer-2 aided mobility independent dormant host alerting protocol, IETF draft, Sept. 2001
18. Ramanathan S. and Steenstrup, M., A Survey of Routing Techniques for Mobile Communications Networks, 1996.
19. Madhow, U., Honig, M.L., and Steiglitz, K., Optimization of wireless resources for personal communications mobility tracking, *IEEE/ACM Trans. Networking,* 3 (6), 698–706, 1995.
20. Ketchum, J.W., Routing in cellular mobile radio communication networks, in *Routing in Communication Networks,* Steenstrup, M., Ed., Prentice-Hall, Englewood Cliffs, NJ, 1995.
21. Mouly, M. and Pautet, M.B., *The GSM system for mobile communications,* M. Mouly, Palaiseu, France, 1992.

22. Telecommunications Industry Association, Cellular radio telecommunication inter-system operation, TIA/EIA IS-41B, 1991.
23. Mohan, S. and Jain, R., Two user location strategies for personal communications services, *IEEE Personal Commun.*, First Quarter, 42–50, 1994.
24. Bar-Noy, A. and Kessler, I., Tracking mobile users in wireless communications networks, *IEEE Trans. Infor. Theory,* 39 (6), 1877–1886, 1993.
25. Xie, H., Tabbane, S., and Goodman, D.J., Dynamic location area management and performance analysis, Proc. 43rd IEEE Vehicular Tech. Conf., 1993, pp. 536–539.
26. Birk Y. and Nachman, Y., Using direction and elapsed-time information to reduce the wireless cost of locating mobile units in cellular networks, *Wireless Networks,* 1 (4), 403–412, 1995.
27. Bar-Noy, A., Kessler, I., and Sidi, M., Mobile users: to update or not to update?, *Wireless Networks,* 1 (2), 175–186, 1995.
28. IETF Mobile-IP Working Group, IPv4 Mobility Support, Working draft, 1995.
29. Meier-Hellstern, K. and Alonso, E., The use of SS7 and GSM to support high density personal communications, Proc. ICC, 1992, pp. 1698–1702.
30. Lo, V.N., Wolf, R.S., and Bernhardt, R.C., Expected network database transaction volume to support personal communications services, 1st International Conference Universal Personal Communications Services, Dallas, 1992.
31. Jain, R. et al., A caching strategy to reduce network impacts of PCS, *IEEE J. Selected Areas Commun.,* 12 (8), 1434–1444, 1994.
32. Ho, J.S.M. and Akyildiz, I.F., Local anchor scheme for reducing location tracking costs in PCNs, Proc. ACM MOBICOM, Berkeley, California, 1995, pp. 181–194.
33. Awerbuch, B. and Peleg, D., Concurrent online tracking of mobile users, Proc. ACM SIGCOMM, Zurich, Switzerland, 1991, pp. 221–234.
34. Anantharam, V. et al., Optimization of a database hierarchy for mobility tracking in a personal communications network, *Perform. Eval.,* 20, 287–300, 1994.
35. Badrinath, B.R., Imielinski, T., and Virmani, A., Locating strategies for personal communication networks, Proc. Workshop on Networking of Personal Communications Applications, 1992.
36. Ozsu, M.T. and Valduriez, P., *Principles of Distributed Systems,* Prentice-Hall, Englewood Cliffs, NJ, 1991.
37. Dowdy, L.W. and Foster, D.V., Comparative models of the file allocation problem, *ACM Comput. Surv.,* 14 (2), 287–313, 1982.
38. Shivakumar, N. and Widom, J., User profile replication for faster location lookup in mobile environments, Proc. ACM MOBICOM, Berkeley, California, 1995, pp. 161–169.
39. EIA/TIA, Cellular Radio-Telecommunications Intersystem Operations, Tech. rep. IS-41 Revision C, 1995.
40. Tabbane, S., Location management methods for third-generation mobile systems, *IEEE Communications Magazine,* Aug. 1997, pp. 72–84.
41. Lee, W.C.Y., *Mobile Cellular Telecommunications Systems,* McGraw-Hill, New York, 1989.
42. Wang,, D.C.C., A survey of number mobility techniques for PCS, Proc. IEEE Int. Conference on Personal Communications, Tokyo, Nov. 6–10, 1995.
43. Wang, D.C.C., A survey of number mobility techniques for PCS, Proc. IEEE ICC, 1994.
44. Tabbane, S., Database architectures and location strategies for mobility management in mobile radio systems, Proc. Workshop Multiaccess, Mobility and Teletraffic for Personal Communications, Paris, France, May 1996.

45. Jabbari, B., Intelligent network concepts in mobile communications, *IEEE Communications Magazine,* Feb. 1992.

46. Plassmann, D., Location management for MBS, Proc. IEEE VTC, Stockholm, June 8–10, 1994, pp. 649–653.

47. Okasaka S. et al., A new location updating method for digital cellular systems, Proc. IEEE VTC '91, Saint Louis, Missouri, May 1991.

48. Hu, L.-R. and Rappaport, S.S., An adaptive location management scheme for global personal communications, Proc. IEEE Int. Conf. Personal Communications, Tokyo, Nov. 6–10, 1995.

49. RACE II deliverable, Location areas, paging areas and the network architecture, R2066/PTTNL/MF1/DS/P/001/b1, Apr. 1992.

50. Tabbane, S., Comparison between the alternative location strategy (AS) and the classical location strategy (CS), WINLAB Tech. Rep. 37, Aug. 1992.

51. Tabbane, S., An alternative strategy for location tracking, *IEEE JSAC,* 13 (5), 1995.

52. Perkins, C., IP Mobility Support for IPv4, revised, IETF, draft-ietf-mobileip-rfc2002-bis-08.txt, Sept. 2001.

53. Misra, A. et al., IDMP-based fast handoffs and paging in IP-based 4G mobile networks, *IEEE Communications Magazine,* March 2002, pp 138–145.

54. Chan, P.M. et al., Mobility management incorporating fuzzy logic for a heterogeneous IP environment, *IEEE Communications Magazine,* Dec. 2001, pp. 42–51.

55. Bhattacharjee, P.S. et al., A practical approach for location area planning in a personal communication services network, Proc. MMT '98, 1998.

56. Demirkol, I. et al., Location area planning in cellular networks using simulated annealing, Proc. INFOCOM, 2001.

16 Mobile *Ad Hoc* Networks: Principles and Practices

Sridhar Radhakrishnan, Gopal Racherla, and David Furuno

CONTENTS

0-8493-1502-6/03/$0.00+$1.50

16.1 INTRODUCTION

An *ad hoc* network[1,2] is characterized by a collection of hosts that form a network "on-the-fly." These hosts typically communicate with each other using wireless channels; they will communicate with each other also using other hosts as intermediate hops in the communication path, if necessary. Thus, an *ad hoc* network is a multihop wireless network wherein each host acts also as a router. A true *ad hoc* network does not have an existing infrastructure to begin with; however, most real-life *ad hoc* networks only contain subnetworks that may be truly *ad hoc*. Mobile *ad hoc* networks (MANETs)[1–4] are *ad hoc* networks wherein the wireless hosts have the ability to move. Mobility of hosts in MANETs has a profound impact on the topology of the network and its performance. An *ad hoc* network can be modeled as a graph whose nodes represent the hosts, and an edge exists between a pair of nodes if the corresponding hosts are in communication range of each other. Such a graph represents the topology of the *ad hoc* network, and in the case of a MANET the topology will constant change due to the mobility of the nodes. The complexity of maintaining communication increases with the increase in the rate of change of the network topology. The protocols that allow communication on the Internet tolerate very small and slow changes in the network topology as routers and hosts are added and removed. However, if applied to MANETs, these protocols would fail as the rate of change of topology is much higher. The aim of researchers working in the area of MANETs has been to develop network protocols that adapt to the fast and unpredictable changes in the network topology. In the rest of the chapter, we use the terms *ad hoc* network, wireless *ad hoc* network, and MANET synonymously; also the terms node and host in a MANET are used interchangeably.

MANETs are characterized by:[1]

- *Dynamic network topology:* As the nodes move arbitrarily, the network topology changes randomly and suddenly. This can result in broken and isolated subnetworks. MANETs need to be resilient and self-healing.
- *Bandwidth-constrained, variable capacity, possibly asymmetric links:* Wireless links are error-prone and have significantly lower capacity than wired links, and hence network congestion is more pronounced. Guaranteed quality of service (QoS) is difficult to accomplish and maintain in MANETs.
- *Power-constrained operation:* Some or all nodes rely on batteries for energy.
- *Wireless vulnerabilities and limited physical security:* MANETs are generally more prone to information and physical security threats than wired networks.

A MANET can be formed using a wireless radio network between a collection of hosts such as individuals assembled in a lecture hall, pedestrians on a street, soldiers in a battlefield, relief workers in a disaster area, or a fleet of ships and aircraft. The collection of hosts in a MANET need not be homogeneous; for example, the collection in a battlefield environment might include tanks, all-terrain vehicles, transport vehicles, infantry, and aircraft. A MANET need not be isolated or self-contained; one or more nodes in the collection of hosts could be connected to a wide area network (WAN), such as the Internet. The nodes that communicate directly with the nonmobile nodes of the WAN move traffic in and out of the MANET.

The rate of change of topology of a MANET is dependent on the characteristics of its nodes, as well as the environment in which it is used.[5,6] For example, in the case of a group of business delegates gathering in a conference room, the topology is fairy stable after an initial setup of multiple new connections to the MANET. When the delegates leave the conference room, the MANET experiences a large number of disconnections. In a battlefield environment, where soldiers are constantly moving in different formations, the topology changes are characterized by increased numbers of link additions and deletions. Thus the rate of change of topology of the MANET is a function of the number of hosts that join and leave the network (node addition/deletion), as well as the number of connections that are added and removed (edge addition/deletion) as the nodes move in and out of the network.

In Figure 16.1, the possible stages of MANET applications, as a function of the rate of connections and disconnections, are illustrated. Applications' stages (or phases) can be static (low rate of disconnections, low rate of connections), high connection (low rate of disconnections, high rate of connections), high disconnection (high rate of disconnections, low rate of connections), and chaotic (high rate of disconnections, high rate of connections). A single application can be categorized under any of these schemes, depending on the phase in the life cycle of the application. For example, consider a battlefield environment wherein initially the soldiers (each carrying a wireless PDA for communication) are briefed at the field headquarters, the connectivity between users is high as they are in close physical proximity, and the rate of disconnections is low as the soldiers remain there until the briefing is over. This phase of the activity is classified as a high connection phase. When the same set of soldiers move in small groups on the battlefield, the rate of connection is small as they are not in radio range with all the other soldiers, and the rate of disconnection is low because the soldiers are moving together as a group. This phase of the activity can be classified as static. When the military exercise is over and the soldiers leave the network, the application is in a high disconnection phase. When reinforcements arrive to relieve a group of soldiers, the network is in a chaotic phase as the rate of disconnections and connections is high.

For a MANET to function well, every node in the network must perform its routing duties efficiently. In addition, there must be a high level of cooperation among nodes that form the *ad hoc* network. As mentioned earlier, the nodes of the network can be heterogeneous in terms of their communication and computation capacities. MANET protocols must be designed and implemented with all of the

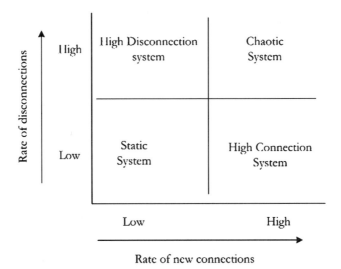

FIGURE 16.1 Stages of MANET applications based on rate of connections and disconnections.

MANET characteristics and issues discussed here taken into account. Network protocol issues that are important for successful deployment of an efficient MANET are discussed in this chapter. In order to better understand, we present a real-life application that requires the use of a MANET. Section 16.2 describes a MANET application involving mobile robots. Section 16.3 highlights issues that must be taken into consideration by the application, transport, network, data link, and physical layers of the protocol stack. When discussing the issues related to each of the layers, we summarize the state-of-art solutions proposed in the literature. Section 16.4 discusses the various technologies and standards that contribute to real-life implementation of MANETs. Section 16.5 summarizes the chapter and presents our conclusions.

16.2 A WIRELESS *AD HOC* NETWORK APPLICATION

Consider a team of robots, with communication and computation capabilities, that have been placed in an unexplored facility thought to contain a contaminator. The goal of the robot team is to map the entire terrain and to locate the contaminator. All robots can move around the facility freely, avoiding obstacles in their way; the exception is a single fixed anchor robot that communicates with the external world. Messages are sent between the controller and the anchor robot and such communication is established using the wireless WAN (see Figure 16.2).* The robots form a network dynamically and communicate with each other using the wireless channels

---

* One might imagine the anchor robot at the entrance of a tunnel, while the other robots are inside the tunnel. The anchor robot has the ability to communicate with the WAN.

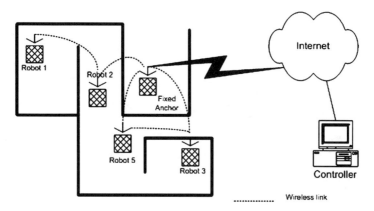

FIGURE 16.2 Mobile robots application.

without the aid of a fixed infrastructure. Two robots can exchange messages directly if they are in communication range of each other. The robot team must complete its task quickly. Due to limited battery life, the robots must be very prudent in power usage. We assume that an intelligent coordination protocol exists that avoids inspecting previously inspected areas. The robots exchange command-and-control messages apart from data messages that contain the information captured about the terrain. The control messages are sent from the controller to the anchor robot. These messages eventually are delivered from the anchor robot to all the mobile robots. The robots send video images to the anchor robot upon receiving commands from it. All messages are sent reliably, except the video images. This scenario application is very typical of the MANET environment. The issues that the protocols must address as a result in this scenario are:

- Network partition is caused due to the host mobility and the associated packet losses.
- Mobility may cause changes in the communication path. This may cause out-of-order delivery of packets.
- The transmission range of the hosts is limited and this results in the need for extensive cooperation among nodes in the MANET for proper delivery of messages.
- The broadcast nature of the wireless medium allows itself to be vulnerable to snooping and other security risks. It is subject also to frequent packet distortions due to collisions.
- Batteries carried by the mobile hosts have a limited life.

Figure 16.3 illustrates how the various layers of the OSI protocol stack must operate in order to successfully complete a communication session. Each of these layers is explained in detail in the sections that follow. For the sake of simplicity, the session layer and the presentation layer are assumed to be merged with the application layer.

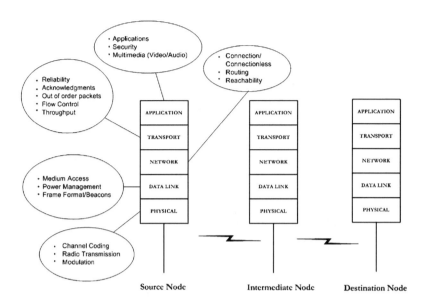

FIGURE 16.3 Issues to be addressed by each layer of the protocol stack.

16.3   ISSUES FOR PROTOCOL LAYERS IN MANETS

In this section, we look at the various issues for the OSI reference model layers for MANETs and briefly describe the functionality of the layers as they apply to MANETs (see Figure 16.3 for an overview). We pay special attention to the application, transport, network, data link, and physical layers. The OSI reference model contains presentation and session layers, which are not described in great detail in this chapter for the sake of simplicity. A more detailed explanation of the OSI reference model and its layers can be found in Tanenbaum[7] and Martin.[8]

16.3.1   APPLICATION LAYER

The application layer provides network access to applications and protocols commonly used by end users. These applications and protocols include multimedia (audio/video, file system, and print services), file transfer protocol (FTP), electronic mail (SMTP), telnet, domain name service (DNS), and Web page retrieval (HTTP). Other higher-level issues such as security, privacy, user profiles, authentication, and data encryption also are handled by the application layer. In the case of *ad hoc* networks, the application layer also is responsible for providing location-based services.[9–12] The presentation layer is responsible for data representation as it appears to the end user, including character sets (ASCII/EBCIDIC), syntax, and formatting. The protocols associated with the presentation layer include Network Virtual Terminal (NVT), AppleTalk Filing Protocol (AFP), and Server Message Block (SMB). The session layer is responsible for data exchange between application processes, including session flow control and error checking.

16.3.2 Transport Layer

The purpose of the transport layer is to support integrity of data packets from the source node to the destination node (end-to-end). Transport protocols can be either connection-oriented or connectionless. Connection-oriented transport protocols are needed for ensuring sequenced data delivery. In order to ensure reliable sequenced delivery, the transport layer performs multiplexing, segmenting, blocking, concatenating, error detection and recovery, flow control, and expedited data transfer. Connectionless protocols are used if reliability and sequenced data can be traded in exchange for fast data delivery. The transport layer assumes that the network layer is inherently unreliable, as the network layer can drop or lose packets, duplicate packets, and deliver packets out of order.

The transport layer ensures reliable delivery by the use of acknowledgments and retransmissions. The destination node, after it receives a packet, sends the acknowledgment back to the sender. The destination node sometime uses cumulative acknowledgment wherein a single acknowledgment is used to acknowledge a group of packets received. The sender, rather than sending a single segment at a time, sometimes sends a group of segments. This group size is referred to as the window size. The sender increases its window size as the acknowledgments arrive. The Transmission Control Protocol (TCP),[7,8] which is the most commonly used reliable transport layer protocol, also takes care of congestion avoidance and control. The TCP protocol increases its throughput as acknowledgments arrive within a time period called the TCP time interval. If the acknowledgments arrive late, it assumes that the network is overloaded and reduces its throughput to avoid congesting the network. In contrast, User Datagram Protocol (UDP)[7,8] is a connectionless transport protocol used for applications such as voice-and-video transport and DNS lookup.

In the MANET environment, the mobility of the nodes will almost certainly cause packets to be delivered out of order and a significant delay in the acknowledgments is to be expected as a result. In a static MANET environment, the packet losses are mainly due to errors in the wireless channel. Retransmissions are very expensive in terms of the power requirements and also occur more often than in wired networks for the reasons explained previously. In the design of efficient transport layer protocol for MANETs, the following issues must be taken into consideration:

- Window size adjustments have to be made that not only take into account the channel errors and the end-to-end delays, but also should adjust based on the mobility dynamics of the network nodes. As pointed out earlier, in a stable *ad hoc* network packet losses are mainly due to errors in wireless channel and end-to-end delays.
- Cumulative acknowledgments can be both good and bad. Given the packet losses expected in MANETs, the loss of a single acknowledgment packet will result in retransmission of a large number of packets. In a static MANET environment, there is a significant advantage in using cumulative acknowledgments. Mechanisms to adjust the acknowledgment schemes based on the dynamics of the network should be taken into consideration.

- The time-out interval that dictates how long the protocol waits before beginning the retransmission should be adjusted based on the dynamics of the network. Clearly, a shorter time-out interval will increase the number of retransmissions, while longer time-out intervals decrease the throughput.
- The original TCP congestion control is purely based on acknowledgment delays. This does not necessarily work well in the case of MANETs, where the delays are attributed to channel errors, broken links caused by mobility, and the contention at the medium access control (MAC) layer that is not only dependent on the traffic in the network, but also on the degree (number of neighbors) of nodes in the network.

Research in the area of transport protocols for MANETs has focused on the development of feedback mechanisms that enable the transport layer to recognize the dynamics of the network, adjust its retransmission timer and window size, and perform congestion control with more information on the network. For example, when a session is initiated, the transport layer assumes that the route is available for a period of time. When the route changes, the transport layer is informed; then transmission freezes until a new route is established.[13] Several research efforts have examined the impact of various routing algorithms on TCP.[14–17] All of these studies have concluded that the route reestablishment time significantly and adversely impacts the throughput of the TCP.

For the robot team application, we presume that that the control packets sizes and hence the window sizes are small. The retransmission timer should be kept small and this will cause command and control packets to be sent constantly so as to ensure that the destination node receives it in the midst of constant route changes.

16.3.3 NETWORK LAYER AND ROUTING

The routing algorithms for MANETs have received the most attention in recent years, and many techniques have been proposed to find a feasible path between source and destination node pairs. In the wired environment, the routing protocol[7] can either be based on link state or distance vector. In link-state routing, each router periodically send a broadcast packet to all the other routers in the network that contains information about the adjacent routers. Upon receiving this broadcast message, each router has complete knowledge of the topology of the network and executes the shortest path algorithm (Dijkstra's algorithm) to determine the routing table for itself. In the case of distance-vector routing, which is a modification of the Bellman-Ford algorithm, each router maintains a vector that contains distances it knows at that point in time (initially infinity for all nodes other than its neighbors) between itself and every other node in the network. Periodically, each node sends this vector to all its neighbors and the nodes that receive the vector update their vectors based on the information contained in the neighbor's vector.

In MANETs, routing algorithms based on link state and distance vector face serious issues as outlined below:

- Executing a link-state protocol would require each node to send information about its neighborhood as it changes. The number of broadcast message sent by a node is related to the dynamics of the network. In a highly dynamic network, it is advisable that the nodes send updates based on the stability of a neighbor. It may be even useful to send information on only those neighbors that have been newly added or removed from its neighborhood since the last broadcast.
- Distributing the distance vector information in a highly dynamic environment is very ineffective. The distances to the nodes keep changing as the nodes move in the network. Constant updates are required for up-to-date distance information.
- Due to incorrect topological information, both algorithms produce routes containing loops.
- Both algorithms cause severe drain on batteries due to the excessive amounts of messages needed to construct routing tables at nodes.

Routing in MANETs involves two important problems: (1) finding a route from the source node to the destination node, and (2) maintaining routes when there is at least one session using the route. MANET routing protocols described in the literature[2] can be either reactive, proactive, hybrid (combination of reactive and proactive), or location based. In a proactive protocol, the nodes in the system continuously monitor the topology changes and update the routing tables, similar to the link state and distance vector algorithms. There is a significant route management overhead in the case of proactive schemes, but a new session can begin as soon as the request arrives. A reactive protocol, on the other hand, discovers a route as a request arrives (on-demand). The route discovery process is performed either on a per-packet basis or a per-session basis. When routes are discovered on a per-packet basis, the routing algorithm has a high probability of sending the packet to the destination in the presence of high mobility. A routing algorithm that uses the route discovery process for a session must perform local maintenance of severed or broken paths. Location-based routing protocols use the location information about each node to perform intelligent routing.

Dynamic Source Routing (DSR)[18] is a reactive algorithm similar in concept to source routing in IP. Before a packet is routed to the destination, the DSR algorithm initiates a route discovery process in which a broadcast packet is sent to all the nodes in the network that are reachable. A node receiving the broadcast packet appends its address and broadcasts it to its neighbors. When the destination node receives the broadcast packet, it uses source routing to send a route request reply packet back to the source. Each intermediate node receiving the route request reply packet simply forwards it to the next node in the route to the source node address contained in the packet. Upon discovering the route, the source sends the packets using route information gathered during the route discovery process. A route cache is maintained at nodes that use information gathered during reception of broadcast packets of the discovery process. To avoid the route discovery process for every packet that needs to be routed, a route maintenance process is initiated, which keeps track of route changes and makes local changes to the route cache.

The temporally ordered routing algorithm (TORA)[19] maintains a virtual network topology that is a directed acyclic graph (DAG). The source has an in-degree (number of arcs coming into the source) of zero. The height of node in the DAG is its distance from the source in the DAG. The entire algorithm is based on the maintenance of the DAG as the node moves and its height changes. The three phases of the algorithm are route creation, route maintenance, and erasing of invalid routes. Once the DAG is known, the packets are routed along the edges of the DAG.

In a MANET, certain routes are more stable than others because links on those routes are not severed for a period of time. It is important that the routing algorithms determine these routes. To select these routes, each node can advertise its presence to its neighbors from time to time. Nodes receiving this advertisement increment a counter associated with the node that sends the advertisement. The degree of stability is proportional to the value of the counter. Nodes prefer routing through nodes associated with a higher counter value. This concept has been used in associativity-based routing (ABR).[20] A similar concept based on the relative signal strength between nodes has been suggested in signal-stability-based adaptive routing (SSA).[21]

Destination-sequenced distance vector routing (DSDV)[22] is a proactive routing algorithm that maintains consistent routing information at all nodes in the network by propagating changes in links. Proactive protocols that build routing tables that contain next-hop information for each destination should be very concerned about the possibility of forming loops in certain routes. These loops are due to mobility of nodes that exchange distance vectors. DSDV is the standard distance vector protocol adapted for the MANET environment and is especially equipped to avoid loops in routes. The basic idea to avoid loops is the same as that used for effective flooding in wired networks. In each packet, a sequence number is placed and each node receiving the packet increases the sequence number by one and forwards it to its neighbor. If a node receives a packet from the same source with a sequence number smaller than the one it has seen so far, then the packet is not forwarded. However, care should be taken to purge the sequence number information at each node from time to time. *Ad hoc* on-demand distance vector routing (AODV)[23,24] is a routing algorithm that improves the performance of DSDV by minimizing the number of broadcast messages. This is done by on-demand route creation.

Proactive algorithms have been shown to perform effectively when the topology of the network is stable; reactive algorithms are highly effective in finding routes in the presence of high rates of topological changes. Reactive algorithms require a large number of broadcast packets to determine the destination. The dynamic spanning tree (DST) algorithm[5] is an efficient protocol that maintains a forest of trees containing the nodes and performs shortest path routing on the links of the trees. It has been shown that reactive routing on the forest of trees significantly reduces the number of messages required to find the path to the destinations. A novel concept called *connectivity through time* is introduced in the DST algorithm wherein if the path from the source node and the destination node may be absent currently but may be available in the future, then a path may be formed while the packet is in en route to the destination node. The DST algorithm also uses a concept termed *holding time,* where a packet is both forwarded and held for a period of time to allow the node to forward it later to new neighbors with which it may come in contact. The holding time allows the implementation of the concept of connectivity through time. The algorithm can be considered a hybrid protocol that is proactive in the sense that messages are to be exchanged to maintain the forest of trees and reactive when it comes to finding the path to the destination on the trees.

Zone Routing Protocol (ZRP)[25] is yet another hybrid protocol in which each node is associated with a zone of fixed radius r and contains all the nodes within a distance of r from it. ZRP uses proactive routing for routing within a zone and reactive routing to route between nodes belonging to two different zones. The size of the radius is adjusted according to the requirements of the application.

Change in the radius of influence (radius of the node's communication cell) directly affects the reachability of packets in MANETs. Ramanathan and Rosales-Hain[26] proposed a novel algorithm that determines the size of the radius of influence for each node to ensure connectedness, biconnectedness, and other levels of connectivity. The overall goals of the algorithm are to keep the radius of influence to a minimum because increase in the radius is directly proportional to the power expended. Power-aware routing protocols take into account available power at nodes and find paths containing nodes that have maximum power available in them.

Location-based routing algorithms use location information obtained from sources such as the Global Positioning System (GPS) to improve the efficiency of routing. Location-aided routing (LAR)[27] is an example of location-based routing. The LAR algorithm intelligently uses the location information of nodes to limit the search of routes in a MANET to a request zone. Two different methods of determining the request zone are supported in LAR.

Royer and Toh,[28] Broch et al.,[29] Das et al.,[30] and Racherla and Radhakrishnan[31] compare and contrast several proposed MANET routing algorithms using analytical modeling and simulation. Johnson[32] was one of the first to analyze issues involved in routing in MANETs. Obraczka and Tsudik[33] have made a similar analysis for multicast routing in MANETs. Mauve et al.[34] have surveyed several location-based routing algorithms. There have been a number of approaches proposed for the performance evaluation of *ad hoc* routing protocols, including simulation and analytical cost modeling.[35,36] The network simulator (ns-2)[37] and the GloMoSim/Parsec[38] are the two most popular MANET simulation tools. Simulation studies have examined the effectiveness of *ad hoc* routing protocols in terms of number of messages delivered, given different traffic loads and dynamics of the network. The other parameters that were evaluated include routing overhead, sensitivity of the protocols with increased network traffic, and choice of paths taken with respect to its end-to-end delay and other optimization characteristics. A detailed discussion on MANET routing protocol performance issues, quantitative metrics for comparing routing algorithm performance, and appropriate parameters to be considered can be found in the IETF MANET Charter[1] and Perkins.[2]

16.3.4 DATA LINK LAYER

The data link layer consists of the logical link control (LLC) and the medium access control (MAC) sublayers. The MAC sublayer is responsible for channel access, and the LLC is responsible for link maintenance, framing data unit, synchronization, error detection, and possibly recovery and flow control. The MAC sublayer tries to gain access to the shared channel so that the frames that it transmits do not collide (and hence get distorted) with frames sent by the MAC sublayers of other nodes sharing the medium. There have been many MAC sublayer protocols suggested in the literature for exclusive access to shared channels. Some of these protocols are centralized and others are distributed in nature. With the centralized protocols, there is a central controller and all other nodes request channel access from the controller. The controller then allocates time slots (or frequencies) to the requesting nodes. These are reservation-based protocols. The nonreservation-based protocols are purely based on contention and are very suitable for MANETs, where it is impossible to designate a leader that might be moving all the time. A detailed discussion on wireless MAC schemes can be found in Chandra et al.[39]

CSMA (Carrier Sense Multiple Access) is a distributed nonreservation-based MAC layer protocol that was used for packet radio networks wherein the MAC layer senses the carrier for any other traffic and sends the frame immediately if the medium is free. When the channel is busy, the MAC layer waits for a random time period (which grows exponentially) and then tries to resend the frame after sensing the medium. The CSMA scheme suffers from a few serious problems.[39] Assume that three nodes — A, B, and C — are in communication range, i.e., node A can hear node B's transmissions, B can hear A's as well as node C's transmissions, and C can hear B's transmissions. Assume that B is not transmitting any frames, while A and C have frames to be transmitted to B. A and C will find the channel free and send frames to B; this will result in a collision at B. This type of problem with CSMA is called the *hidden terminal* problem. Assume that there is a node D that is in communication with C. If B has frames to send to A, and C has frames to send to D, then both B and C will back off even though the frames sent by B and C can arrive at their respective destinations without distortions. This unnecessary backoff is termed the *exposed terminal* problem. These problems are depicted in Figure 16.4.

Multiple Access with Collision Avoidance (MACA)[40] is a distributed reservation-based protocol that tries to remove the two problems that exist in the CSMA protocol. Two control frames named RTS (request-to-send) and CTS (clear-to-send) are used for channel access and reservation. Each station (node) that has a frame to transmit sends RTS to the destination along with the length of the message it wants to send to the destination. It then waits for a CTS frame from the destination. All stations sending an RTS message on the medium defer transmission for collision-free delivery of CTS. If the other stations do not hear the CTS with a specified time interval, then they are allowed to send their requests. The stations hearing CTS defer transmission for a period of time required to transmit a frame of the size specified in the CTS. Because each RTS has a source and destination address, the exposed terminal problem can be eliminated. Because collisions occur at the destination, by the use of CTS the hidden terminal problem can be eliminated. The MACAW[41] protocol is an enhancement of MACA that allows stations to choose proper retransmission time and acknowledgments for LLC activity.

There has been a growing interest in the study of the IEEE 802.11 MAC standard which has gained popularity among vendors and users. The 802.11 MAC[42] provides access control functions such as addressing, access coordination, frame check

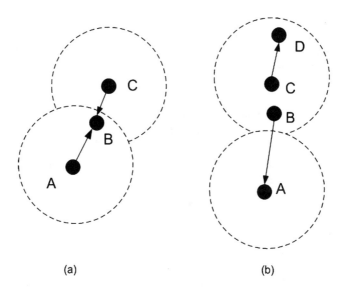

(a)   (b)

FIGURE 16.4 Terminal problems: (a) hidden; (b) exposed.

sequence generation, and frame checking. The MAC sublayer performs the addressing and recognition of frames in support of the LLC. This protocol expands on the
MACA protocol with link-level acknowledgments and performs collision avoidance,
thus falling into the category of CSMA/CA protocols. All nodes are assumed to be
time synchronized in this protocol. Time is divided into time slots which are divided
into two portions. In the first portion, nodes contend by the exchange of RTS/CTS
pairs and backing off if necessary for recontending. In the second portion of the time
slot, the node that has gained channel access sends the frame and all other nodes simply
wait for the beginning of the next time interval. It is important to note that the size of
the frame to be sent is fixed and its size is chosen in a way that it can be sent during
the allocated second portion of the time interval. If a node has a frame that is smaller
in length, then it is padded with additional blanks. Several research efforts[43,44] have
concentrated on extending the IEEE 802.11 MAC for MANETs. One of the main
concerns in trying to extend the IEEE 802.11 MAC is the problem of time synchronization in a multihop environment, as in the case of MANETs.

There has been growing interest in the design of power-efficient MAC layer
protocols, especially for MANETs. For example, the IEEE 802.11 has a built-in
power-saving feature that allows nodes to awaken themselves during the contention
period (the first portion of the time slot) and to go to sleep mode during the second
phase of the time slot if it is either not receiving or transmitting. The PAMAS (Power
Aware Multiaccess with Signalling) protocol[45] attempts to reduce contention and
hence power consumed by use of a special channel. Using this special channel,
nodes determine the status of the other channel before transmitting through the
regular channel. More information on efficient power usage in wireless systems can
be found in references 46 through 49.

In the case of the 802.15.3 draft,[50] the MAC is designed to support fast
connection times, *ad hoc* networking, data transport with QoS, security, dynamic

device membership, and efficient data transfer. The 802.15.3 standard is based on the concept of a piconet. Each piconet operates in a personal operating space (POS) that is defined as an approximate 10-m envelope around devices in all directions. Each piconet consists of a piconet coordinator (PNC) and devices (DEVs). The 802.15.3 is a combination of CSMA/CA and time-division multiple access (TDMA) MACs that use superframes for channel access. Each superframe consists of three parts: the beacon, the contention access period (CAP), and contention-free period (CFP). The beacon is used for setting the timing allocations and the channel to use, as well as communicating management. The CAP is used to communicate commands and uses CSMA/CA for access. The CFP is composed of guaranteed time slots (GTS) dedicated for data communication between a pair of DEVs.

16.3.5 PHYSICAL LAYER

The physical (PHY) layer[2,7–8] is a very complex layer that deals with the medium specification (physical, electrical, and mechanical) for data transmission between devices. The PHY layer specifies the operating frequency range, the operating temperature range, modulation scheme, channelization scheme, channel switch time, timing, synchronization, symbol coding, interference from other systems, carrier-sensing and transmit/receive operations of symbols, and power requirements for operations. The PHY layer interacts closely with the MAC sublayer to ensure smooth performance of the network. The PHY layer for wireless systems (such as MANETs) has special considerations that need to be taken into account:

- The wireless medium is inherently error-prone.
- The wireless medium is prone to interference from other wireless and RF systems in proximity.
- Multipath is important to consider when designing a wireless PHY layer, as the RF propagation environment changes dynamically with time. Multipath results in a composite received signal equal to the vector sum of the direct and reflected paths.
- Frequent disconnections may be caused because of the wireless link. The problem is compounded when the devices in the network are mobile because of handoffs and new route establishment.

PHY layer specifications and its parameters vary depending on the wireless system used. For example, the IEEE 802.11 Standard has a provision for three different PHYs.[42] Another example is the emerging draft standard of IEEE 802.15.3 for wireless personal area networks; Table 16.1 contains PHY parameters defined for IEEE 802.15.3.

16.4 MANET IMPLEMENTATION: RELATED TECHNOLOGIES AND STANDARDS

In this section, we explore related hardware, software, and networking technologies that may be used for the implementation of MANETs. These technologies provide some or all of the following *ad hoc* networks features:

TABLE 16.1
PHY Parameters for IEEE 802.15.3 Draft Standard

Parameter	Value/Range/Comments
Operating frequency	2.4 to 2.4835 GHz
Range	10 m
Modulation	Quadrature phase shift keying, eight-state trellis-coded modulation, 11 Mbps
Coding	Differential quadrature phase shift keying, none, 22 Mbps
Data rate	16/32/64 quadrature amplitude modulation, eight-state trellis-coded modulation, 33/44/55 Mbps
Operating temperature range	0 to 40°C
Base data rate	22 Mbps (uncoded differential quadrature phase shift keying)
PHY preamble	Multiple periods of 16 symbols constant-amplitude zero-autocorrection sequence
Symbol rate	11 Mbps ± 25 ppm
Clock accuracy	± 25 ppm
Power-on ramp	2 μs
Power-down ramp	2 μs
Maximum transmit power limit (United States)	50 mV/m at 3 m in at least 1-MHz resolution

- Distributed processing
- Collaborative computing
- Dynamic discovery of services and devices
- Mobility
- Detection of radio beacons and radio proximity
- Support for forming groups
- Support for wireless connectivity
- Self-administration

16.4.1 SOFTWARE TECHNOLOGIES

In this section, we look at software and software framework technologies that facilitate implementation of MANETs, and we discuss how these technologies contribute to building *ad hoc* networks:

- Java and Jini[51,52]
- Universal Plug and Play (UPnP)[53]
- Open Services Gateway Initiative (OSGI)[54]
- Home Audio Visual Interoperability (HAVi)[55]
- Peer-to-Peer (P2P) Computing[56]

16.4.1.1 Java and Jini

Java is more than a programming language; we explore some related Java technologies later in this section. For our purposes here, we note that the Java program paradigm consists of a set of tools and technologies that are amenable to the implementation of *ad hoc* networks, because Java provides:

- *Code mobility:* Software code implemented on one Java Virtual Machine (JVM) on one node can be moved across the network and directly run on another node without any change.
- *Platform/protocol independence:* Java is platform and protocol agnostic.
- *Remote method invocation (RMI):* Software modules can be executed remotely on a node.
- *Portability:* Java code is portable across all devices that support Java without any changes.
- *Security:* Java provides byte-code verification and other security features.
- *Dynamic load of code:* Java classes can be downloaded "on the fly."

Jini is a framework for distributed computing using a set of simple interfaces and protocols. Jini enables spontaneous networks of software services and devices to assemble into working groups of objects known as federations. Jini enables self-administration and self-healing when devices move dynamically from one federation to another. Jini uses RMI to pass entire Java objects and their code. In addition, RMI provides object serialization, transport, and deserialization of objects. RMI is robust and supports security protocols. Jini provides the following basic services:

- *Discovery service:* The discovery and join protocols can be used to join a group of services using a UDP multicast. Each service advertises capabilities and provides the required software drivers.
- *Lookup service:* This is a repository of available services. It stores each service as a Java object, and clients can download services on demand. The lookup service provides interfaces for registration, access, search, and removal of services.
- *Lease:* Leases are resource grants that are time-based between a grantor and a holder. Leases can be cancelled, renewed, and negotiated by third parties. Thus, resources are allocated only as long as needed. The network is self-healing as the resources are granted and released dynamically.
- *Event:* The Java network event model has been expanded in Jini to work in a distributed system. It supports several delivery models such as push, pull, and filter.
- *Transaction:* Jini's transaction model allows for distributed object coordination using two-phase commits.

Jini extends its architecture to allow a surrogate that is designed to deal with legacy and non-Jini devices and resource-limited Java-based devices. The Jini technology surrogate architecture specification defines interfaces and methodology by which these components, with the aid of a third party, can participate in a Jini network while still maintaining the plug-and-work model of Jini technology. Javaspaces is a Jini Service based on tuple-spaces, which uses a persistent object store for secure transactions in a simple fashion. It is a unified mechanism by which Java objects can be shared, dynamically communicated, and coordinated in the distributed object stores called spaces. This paradigm lends itself to parallel programming, distributed systems, and cooperating software entity groups.

16.4.1.2 UPnP

Universal Plug and Play is architecture for smart home networking and pervasive peer-to-peer connectivity of intelligent appliances, wireless devices, and PDAs. It is an extension of device plug and play supported by Microsoft®. It supports transparent networking also, and resource and service discovery. UPnP, like all the service discovery paradigms, aims to provide all these features to be exercised automatically without any user intervention. UPnP supports standard Internet and TCP/IP-based protocols with a view to providing interoperability with existing networks and infrastructure. UPnP is an open distributed network paradigm that does not define any APIs. The standard defines device and service descriptions based on common device architecture, thus keeping the device and service specificity away from the users.

16.4.1.3 OSGi

The Open Services Gateway Initiative is supported by more than 50 companies to develop services gateway architecture. The OSGi Forum is defining a set of APIs for this purpose and providing a reference implementation of services gateway architecture. A services gateway connects the external network with home-based internal networks and devices providing the user with transparency for service discovery. The services gateway adds to the usefulness of home networks by allowing service providers to deliver real-time, new and innovative value added services to the services gateway. The OSGi based gateway will provide distribution, integration and management of new and existing services. The OSGi forum is targeting the SOHO/ROBO (Small Office/Home Office and Remote Office/Branch Office) and residential users.

16.4.1.4 HAVi

Home audio/visual interoperability is a digital consumer electronics and home appliances communications standard. HAVi is specifically focused on digital audio/video (A/V) networking for home entertainment products. HAVi provides many of the semantics required for pervasive and *ad hoc* networking. An important tenet of the standard is interoperability among A/V devices from the major home entertainment consumer electronics companies. HAVi defines a middleware that manages A/V streams, and provides APIs for the development of home A/V software applications. The salient selling point of HAVi is that it provides highly optimized data communication between bandwidth-hungry A/V devices. The HAVi network standard has been architected to integrate seamlessly with other home networks. HAVi provides a distributed software architecture with support for network management, device abstraction, interdevice communication, and device user interface management.

16.4.1.5 P2P Computing

Peer-to-peer is a paradigm used for sharing of computing resources (information, CPU power, processing cycles, cache storage, and disk storage for files) and services by direct exchange between peer systems. In a P2P environment, computing devices have a dual nature and act as clients/servers, assuming the appropriate role required by the network. P2P lends itself to user collaboration, edge services (moving data closer to users across large geographic distances), distributed computing and resources, and intelligent agents. The Open P2P Initiative is a forum that provides more information on P2P technologies. Examples of P2P software implementations[56] include Napster, Gnutella, and Morpheus.

16.4.2 NETWORK TECHNOLOGIES

In this section, we study the following networking technologies that can facilitate implementation of *ad hoc* networks:

- Bluetooth[57,58]
- Ultra-Wideband (UWB)[59]
- HiperLAN/1 and HiperLAN/2[60,61]
- IEEE 802.11 Wireless LAN[42,62-63]
- IEEE 802.15.3 Wireless PAN[50]
- HomeRF[64]

16.4.2.1 Bluetooth

Bluetooth is a short-range radio technology originally intended as a wireless cable replacement to connect portable computers, wireless devices, handsets, and headsets. Today, Bluetooth is being used for deploying wireless personal area networks in homes and offices. Bluetooth business requirements make it necessary to produce a pair of units that are below $10. Other requirements include low power usage for running on batteries, a lightweight and small form factor.

Bluetooth devices operate in the 2.4-GHz ISM band. There are specifications for power and spectral emissions and interference to which Bluetooth devices must adhere.[58] It offers three different power classes for operation. The corresponding ranges for the power classes are 10 (lowest power range), 20, and 100 m (highest power range). Bluetooth uses the concept of a piconet, which is a MANET with a master device controlling one or several slave devices; it allows scatternets wherein a slave device can be part of multiple piconets. Bluetooth has been designed to handle both voice and data traffic. Bluetooth specifications provide different application profiles which are used for fine-tuning the implementation of the various applications. Bluetooth provides a service discovery protocol for discovering Bluetooth devices.

16.4.2.2 UWB

Ultra-Wideband, also known as baseband or impulse radio, is a carrier-free radio transmission technology that uses brief, low power pulses that spread the radio energy across a wide spectrum of frequencies. UWB radio can utilize a variety of modulation techniques. In one example, pulse position modulation (PPM) transmissions consist of precisely timed pulses. In PPM, the transmitter and receiver are tightly coordinated, and information is transferred using the position of the pulses. For FCC-compliant systems, the UWB signal level is comparable to background noise, and so interference with conventional carrier-based communication systems is unlikely. To qualify as a UWB communication, the transmitted signal must have a fractional bandwidth of more than 20 percent or occupy more than 500 MHz of spectrum.

UWB is an emerging technology that has strong advantages over conventional carrier-based communications. These advantages include low susceptibility to multipath fading, higher transmission security, low power consumption, simple architecture, and low implementation cost. In addition, UWB provides ranging information, and in a network can yield position data. UWB has been used for ground penetration radar and secure communications. The recent publication of the UWB Report and Order by the FCC[65] has given rise to a great deal of effort focused on commercial applications. Examples of UWB applications include collision avoidance radar, RF tagging, and geolocation and data communications in personal area network (PAN) and local area network (LAN) environments.

16.4.2.3 HiperLAN/1 and HiperLAN/2

HiperLAN/1 and HiperLAN/2 are wireless LAN (WLAN) standards developed by the European Telecommunications Standards Institute (ETSI). HiperLAN/1 is a wireless equivalent of Ethernet, while HiperLAN/2 has architecture based on wireless asynchronous transfer mode (ATM). Both standards use dedicated frequency spectrum at 5 GHz. HiperLAN/1 provides a gross data rate of 23.5 Mbps and a net data rate of more than 18 Mbps, while HiperLAN/2 provides gross data rates of 6, 16, 36, and 54 Mbps, and a maximum of 50 Mbps net data rate. Both standards use 10, 100, and 1000 mW of transmit power and have a maximum range of 50 m. Also, the standards provide isochronous and asynchronous services with support for QoS. However, they differ in their channel access and modulation schemes. HiperLAN/1 uses dynamic priority-driven channel access, while HiperLAN/2 uses reserved channel access. HiperLAN/1 uses Gaussian minimum shift keying (GMSK) and HiperLAN/2 uses orthogonal frequency division multiplexing (OFDM) plus binary phase shift keying (BPSK), quadrature phase shift keying (QPSK), and QAM (quadrature amplitude) modulation schemes. HiperLAN/1 and HiperLAN/2 support an *ad hoc* network mode of operation.

16.4.2.4 IEEE 802.11

This IEEE family of standards is primarily for indoor and in-building WLANs. There are several flavors of this standard. The current available versions are the 802.11a, 802.11b, and 802.11g (emerging draft standard) with other versions currently in the works. The 802.11 standards support *ad hoc* networking, as well as connections using an access point (AP). The standard provides specifications of the PHY and

TABLE 16.2
PHY Specifications for IEEE 802.11

PHY	Frequency Band	Data Rates	Modulation	Comments
Frequency hopping spread spectrum	2.4-GHz ISM band	1, 2 Mbps	2-level Gaussian frequency shift keying 4-level Gaussian frequency shift keying	50 hops per second 79 channels
Direct sequence spread spectrum	2.4-GHz ISM band	1, 2 Mbps	Differential binary frequency shift keying Differential quadrature phase shift keying	11-chip barker sequence spreading
Baseband IR	Diffuse infrared	1, 2 Mbps	16 pulse position modulation 4 pulse position modulation	Uses pulse position modulation

the MAC layers. The 802.11 standards have the same MAC sublayer specification, while their PHY specifications differ substantially. The MAC specified uses CSMA/CA for access and provides service discovery and scanning, link setup and tear down, data fragmentation, security, power management, and roaming facilities. The MAC provides for independent configuration (*ad hoc* network mode) and infrastructure configuration (using access points to increase the range). The 802.11 PHY specifications are shown in Table 16.2.

The 802.11a PHY is similar to the HiperLAN/2 PHY. The PHY uses OFDM (orthogonal frequency division multiplexing) and operates in the 5-GHz UNII band. 802.11a supports data rates ranging from 6 to 54 Mbps. 802.11a currently offers much less potential for RF interference than other PHYs (e.g., 802.11b and 802.11g) that utilize the crowded 2.4-GHz ISM band. 802.11a can support multimedia applications in densely populated user environments. The 802.11b standard, proposed jointly by Harris Corporation and Lucent Technologies, extends the 802.11 direct sequence spread spectrum PHY to provide 5.5 and 11 Mbps data rates. To provide the higher data rates, 802.11b uses 8-chip CCK (complementary code keying), a modulation technique that makes efficient use of the radio spectrum. The 802.11g specification uses the same OFDM scheme as 802.11a and will potentially deliver speeds on par with 802.11a. However, 802.11g operates in the 2.4-GHz frequency band that 802.11b occupies, and for this reason it should be compatible with existing WLAN infrastructures.

16.4.2.5   IEEE 802.15.3

The standard defines MAC and PHY (2.4 GHz) layer specifications for a wireless personal area network (WPAN). The standard is based on the concept of a piconet, which is a network confined to a 10-m personal operating space (POS) around a person or object. A WPAN consists of one or more collocated piconets. Each piconet is controlled by a piconet coordinator (PNC) and may consist of devices (DEVs).

The PNC's functions include the basic timing of the piconet using beacons, managing QoS, managing the power save modes, and security and authentication. The 802.15.3 PHY is defined for 2.4 to 2.4835 GHz band and has two defined channel plans. It supports five different data rates (11 to 55 Mbps). The base uncoded PHY rate is 22 Mbps. Table 16.1 provides other details of the 802.15.3 PHY specifications.

16.4.2.6 HomeRF

The HomeRF Working Group was formed to develop a standard for wireless data communications between personal computers and consumer electronics in a home environment. The HomeRF standard is technically solid, simple, secure, and easy to use. HomeRF networks provide a range of up to 150 feet, typically enough for home networking. HomeRF uses Shared Wireless Access Protocol (SWAP) to provide efficient delivery of voice and data traffic. SWAP uses digital enhanced cordless telecommunications (DECT), and the 802.11 FHSS technologies. SWAP uses a transmit power of up to 100 mW and a gross data rate of 2 Mbps. It can support a maximum of 127 devices per network. A SWAP-based system can work as an *ad hoc* network or as a managed network using a connection point.

16.4.3 HARDWARE TECHNOLOGIES

Following is a discussion of the hardware technologies that are helping implement *ad hoc* networks. These technologies offer low power/power aware hardware and miniaturization of memory, processor, and other peripherals.

16.4.3.1 Smart Wireless Sensors[66]

Smart wireless sensors have added substantially to the applications that MANETs execute. Sensor-based MANETs can be used in applications such as detecting chemicals, explosives, and toxins in hazardous areas; military reconnaissance; gathering geological data in difficult terrain, etc. Being able to make miniature sensors such as the ones used in Smart Dust[66] has shown that MANETs can be successfully scaled to deal with thousands of nodes. Sensor networks are exploring the limits of MANETs in terms of scalability, minimum resource requirements, network resilience, fault tolerance, and security. The emerging IEEE 802.15.4 (low rate WPAN) standard[70] has been proposed for several smart sensor applications.

16.4.3.2 Smart Batteries[67]

Mobile devices typically have strong battery and bandwidth constraints. Power conservation can be achieved on two different fronts: the device and the communication protocols. The power conservation of the device involves reducing the usage of the battery for all the hardware of the device, including the CPU, display, and peripherals. The communication protocols also can be power-aware designs. Smart batteries have low discharge rates, a long cycle life, a wide operating temperature range, and high energy density. Nickel cadmium (NiCad), nickel metal hydride (NiMH), and lithium ion (Li-ion) are the most commonly used for mobile devices. Li-ion batteries have the highest energy density among these technologies.

16.4.3.3   Software-Defined Radio[68]

Software-defined radio (SDR, or software radio) is a radio that can be controlled using software. In SDR systems, waveform generation, modulation techniques, wideband or narrowband operation, security functions, and frequency of operation can be adjusted in software based on the requirements. SDR systems, in essence, provide programmable hardware that increases the flexibility of use and development. SDR is the Holy Grail of radio design. A SDR system is designed to work with any existing or developing standard. SDR highlights the various trade-offs in the design of different radio architectures with a view to improving the radio performance, enhance the feature sets, and add new services resulting in a better user experience. SDR has a vital role in the implementation of MANETs with heterogeneous hosts that employ different radio technologies.

16.4.3.4   GPS[69]

Location awareness, as discussed earlier, can be valuable in establishing the routing topology. Location awareness of nodes distributed in a MANET requires the acquisition of information about each node with respect to an absolute or relative location reference. GPS has been used for obtaining location information of a node in a MANET. GPS is a global positioning system consisting of a group of satellites that continuously broadcast location and timing information while orbiting the earth. Using position triangulation, GPS receivers on Earth calculate the exact location of the receiver on an absolute global scale. The location information thus calculated is in reference to the latitude and longitude coordinate system.

16.5   CONCLUSION

In this chapter, we explored principles and practices related to MANETs. We presented MANET characteristics, their applications, and the issues related to the design and implementation of MANETs. There has been a significant amount of research done to address the various issues associated with MANETs. Many of the issues described in this chapter must be addressed to ensure successful implementation of a MANET. Clearly, many specialized applications such as the mobile robot application described in this chapter can benefit significantly by learning from the research in the area of MANETs. In this chapter, we studied related software, framework, hardware, and networking technologies that may contribute to the implementation of MANETs. These tools let users implement the various functional blocks of the MANETs.

ACKNOWLEDGMENTS

The authors gratefully acknowledge the support needed in writing this chapter, provided by the School of Computer Science, University of Oklahoma, and the Photonics Division of General Atomics, San Diego.

REFERENCES

1. IETF MANET Charter, available at www.ietf.org/html.charters/manet-charter.html, August 2002.
2. Perkins, C.E., Ed., *Ad hoc Networking,* Addison Wesley, Reading, MA, 2001.
3. MANET Paper Collection, available at www.ee.surrey.ac.uk/Personal/G.Aggelou/MANET_PUBLICATIONS.html, August 2002.
4. MANET Resources, available at students.cs.tamu.edu/youngbae/mcrelated.html, August 2002.
5. Radhakrishnan, S. et al., DST – A routing protocol for ad hoc networks using distributed spanning trees, Proc. IEEE International Conference on Wireless Communications and Networking (WCNC '98), 1998.
6. Racherla, G., Algorithms for routing and rerouting in mobile wireless and ad hoc networks, Ph.D. diss., University of Oklahoma, Norman, 1999.
7. Tanenbaum, A.S., *Computer Networks,* 3rd ed., Prentice-Hall, Englewood Cliffs, NJ, 1996.
8. Martin, M.J., *Understanding the Network: A Practical Guide to Internetworking,* New Riders, Indiana, 2000.
9. Camp, T., Boleng, J., and Wilcox, L., Location information services in mobile ad hoc networks, Proc. IEEE International Conference on Communications (ICC), 2002, pp. 3318–3324.
10. Capkun, S., Hamdi, M., and Hubaux, J.P., GPS-free positioning in mobile ad hoc networks, *Proc. Hawaii International Conference on System Sciences,* January 2001.
11. Niculescu, D. and Nath, B., Ad hoc postioning system, Internet draft, available at http://www.cs.rutgers.edu/dataman/papers/aps.ps, August 2002.
12. Li, J. et al., A scalable location service for geographic ad hoc routing, Proc. ACM Mobicom 2000, Boston.
13. Chandran, K. et al., A feedback based scheme for improving TCP performance in ad hoc networks, *IEEE Personal Communication Systems Magazine,* Special issue on Ad Hoc Networks, 8 (1), 34–39, 2001.
14. Ahuja, A. et al., Performance of TCP over different routing protocols in mobile ad hoc networks, Proc. IEEE Vehicular Technology Conference (VTC '2000), Vol. 3, Tokyo, May 2000, pp. 2315–2319.
15. Dyer, T.D. and Boppana, R.V., A comparison of TCP performance over three routing protocols for mobile ad hoc networks, Proc. ACM Symposium on Mobile ad hoc Networking and Computing, October 2001.
16. Gerla, M., Tang, K., and Bagrodia, R., TCP performance in wireless multihop networks, Proc. IEEE Workshop on Mobile Computing Systems and Applications (WMCSA), February 1999, pp. 41–50.
17. Holland, G. and Vaidya, N.H., Analysis of TCP performance over ad hoc networks, Proc. International Conference on Mobile Computing and Networking (MOBICOM), August 1999.
18. Johnson, D.B. and Maltz, D.A., Dynamic source routing in ad hoc wireless networks, in *Mobile Computing,* Imielinki, T. and Korth, H. Eds., Kluwer, Dordrecht, 1996, pp. 153–181.
19. Park, V.D. and Corson, M.S., A highly adaptive distributed routing algorithm for mobile wireless networks, Proc. IEEE INFOCOM, 1997, pp. 1405–1413.
20. Toh, C.-K. Associativity-based routing for ad hoc networks, *Wireless Personal Commun. J.,* Special issue on ad hoc Networks, 17 (8), 1466–1487, 1999.

21. Dube, R. et al., Signal stability based adaptive routing (SSA) for ad hoc mobile networks, *IEEE Personal Commun.*, 36–45, Feb. 1997.

22. Perkins, C.E. and Bhagwat, P., Highly dynamic destination-sequenced distance-vector routing (DSDV) for mobile computers, *Comput. Commun. Rev.*, 24 (4), 234–244, 1994.

23. Perkins, C.E. and Royer, E.M., Ad hoc on-demand distance vector routing, Proc. IEEE Workshop on Mobile Computing Systems and Applications (WMCSA), Feb. 1999.

24. Perkins, C., Ad hoc on demand distance vector (AODV) routing, Internet draft, draft-ietf-manet-aodv-00.txt, Aug. 2002.

25. Hass, Z. and Pearlman, M., The performance of a new routing protocol for the reconfigurable wireless networks, Proc. International Conference on Communications (ICC 98), June 1998, pp. 156–160.

26. Ramanathan, R. and Rosales-Hain, R., Topology control of multihop wireless networks using transmit power adjustment, Proc. IEEE InfoCom, Mar. 2000, pp. 404–413.

27. Ko, Y. and Vaidya, N., Location-aided routing (LAR) mobile ad hoc networks, Proc. MobiCom 98, Dallas, Oct. 1998.

28. Royer, E.M. and Toh, C.-K., A review of current routing protocols for ad hoc mobile wireless networks, *IEEE Personal Communications Magazine,* April 1999, pp. 46–55.

29. Broch, J. et al., A performance comparison of multihop wireless ad hoc network routing protocols, Proc. 4th Annual ACM/IEEE International Conference on Mobile Computing and Networking (MobiCom 98), ACM, Dallas, Oct. 1998.

30. Das, S.R., Perkins, C.E., and Royer, E.M., Performance comparison of two on-demand routing protocols for ad hoc networks, Proc. InfoCom, Mar. 2000, pp. 3–12.

31. Racherla, G. and Radhakrishnan, S., Survey of routing algorithms in ad hoc networks, Proc. 4th World Multiconference on Systemics, Cybernetics and Informatics (SCI 2000)/6th International Conference on Information Systems Analysis and Synthesis (ISAS 2000), Orlando, FL, July 23–26, 2000.

32. Johnson, D.B., Routing in ad hoc networks of mobile hosts, Proc. IEEE Workshop on Mobile Computing Systems and Applications, Dec. 1994.

33. Obraczka, K. and Tsudik, G., Multicast routing issues in ad hoc networks, Proc. IEEE International Conference on Universal Personal Communication (ICUPC 98), Oct. 1998.

34. Mauve, M., Widmer, J., and Hartenstein, H., A survey on position-based routing in mobile ad hoc networks, *IEEE Network Magazine,* 15 (6), 30–39, 2001.

35. Takai, M., Martin, J., and Bagrodia, R., Effects of wireless physical layer modeling in mobile ad hoc networks, Scalable Networks white paper, http://www.scalable-networks.com/pdf/mobihoc.pdf, Aug. 2002.

36. Golmie, N. and Mouveaux, F., Modeling and simulation of MAC protocols for wireless devices coexistence performance evaluation, Internet draft, Aug. 2002.

37. The network simulator — ns-2, http://www.isi.edu/nsnam/ns/, Aug. 2002.

38. GloMoSim home page, http://pcl.cs.ucla.edu/projects/glomosim/, Aug. 2002.

39. Chandra, A., Gummalla, V., and Limb, J.O., Wireless medium access control protocols, *IEEE Commun. Surv.,* www.comsoc.org/surveys, Second Quarter, 2000.

40. Karn, P., MACA: a new channel access method for packet radio, Proc. ARRL/CRRL Amateur Radio 9th Computer Networking Conference, Sept. 1990.

41. Bharghavan, V. et al., MACAW: a media access protocol for wireless LANs, Proc. ACM SIGCOMM, Aug. 1994, pp. 212–225.

42. IEEE 802.11 Specifications, grouper.ieee.org/groups/802/11/index.html, Aug. 2002.

43. Chowdary, N. and Radhakrishnan, S., A power efficient enhancement to IEEE 802.11 protocol for multihop wireless networks, Technical report, University of Oklahoma, Norman, 2002.

44. Lahiri, K., Raghunathan, A., and Dey, S., Battery efficient architecture for an 802.11 MAC processor, Proc. International Conference on Communications (ICC), New York, May 2002.

45. Singh, S. and Raghavendra, C., PAMAS – power aware multi-access with signaling for ad hoc networks, Proc. IEEE International Personal, Indoor and Mobile Radio Communications Conference, 1998, pp. 153–157.

46. Srisathapornphat, C. and Shen, C.C., Coordinated power conservation for ad hoc networks, Proc. IEEE International Conference on Communications (ICC), New York, April 28–May 2, 2002.

47. Agarwal, P., Energy Efficient Protocols for Wireless Systems, Proc. IEEE PIMRC 98, Sept. 1998.

48. Zorzi, M. and Rao, R,."Error control and energy consumption in communications for nomadic computing, *IEEE Trans. Comput.,* Mar. 1997.

49. Singh, S., Woo, M., and Raghavendra, C.S., Power-aware routing in mobile ad hoc networking, Proc. International Conference in Mobile Computing and Networking, 1998, pp. 181–190.

50. IEEE 802.15 Working Group for WPAN, http://grouper.ieee.org/groups/802/15/, Aug. 2002.

51. Jini Community Resource Page, http://www.jini.org, Aug. 2002.

52. Jini Networking Technology, http://wwws.sun.com/software/jini/, Aug. 2002.

53. Universal Plug and Play (UPnP) Forum, http://www.upnp.org/, Aug. 2002.

54. OSGI Official Website, http://www.osgi.org, Sept. 2001.

55. HAVi: Home Audio Video Interoperability home page, http://www.havi.org/, Aug. 2002.

56. Open P2P Project, http://www.openp2p.com, Aug. 2002.

57. Bray, J. and Sturman, C.F., *Bluetooth 1.1: Connect Without Cables,* 2nd ed., Prentice-Hall, Englewood Cliffs, NJ, 2002.

58. Bluetooth Official Website, http://www.bluetooth.com, Aug. 2002.

59. Ultra Wideband Working Group, http://www.uwb.org, Aug. 2002.

60. ETSI, www.etsi.org, Aug. 2002.

61. HiperLAN Global Forum 2, http://www.hiperlan2.com/, Aug. 2002.

62. Geier, J., *Wireless LANs,* MacMillan, New York, 2001.

63. Nader, S. et al., Ad hoc networks with smart antennas using IEEE 802.11-based protocols, Proc. IEEE International Conference on Communications (ICC), New York, April 28–May 2, 2002.

64. HomeRF Official Website, homerf.org, Aug. 2002.

65. FCC home page, www.fcc.gov, Aug. 2002.

66. Smart Dust, Autonomous sensing and communication in a cubic millimeter, http://robotics.eecs.berkeley.edu/~pister/SmartDust/, Aug. 2002.

67. Smart Battery Systems Implementers Forum, www.sbs-forum.org, Aug. 2002.

68. Software Defined Radio Forum, www.sdrforum.org/, Aug. 2002.

69. Trimble: All about GPS, http://www.trimble.com/gps/, Aug. 2002.

70. IEEE WPAN Task Group 4, IEEE 802.15.4 draft standard, http://www.ieee802.org/15/pub/TG4.html, Aug. 2002.

17 Managing Location in "Universal" Location-Aware Computing

Sajal K. Das, Amiya Bhattacharya, Abhishek Roy, and Archan Misra

CONTENTS

17.1 INTRODUCTION

Models of twenty-first century ubiquitous computing scenarios[1] depend not just on the development of capability-rich mobile devices (such as Web phones or wearable computers), but also on the development of automated machine-to-machine computing technologies, whereby devices interact with their peers and the networking infrastructure, often without explicit operator control. To emphasize the fact that devices must be imbued with an inherent consciousness about their current location and surrounding environment, this computing paradigm also is called sentient[2] (or context-aware) computing. Context awareness is one of the key characteristics of

0-8493-1502-6/03/$0.00+$1.50
© 2003 by CRC Press LLC

applications under this intelligent computing model. If devices can exploit emerging technologies to infer the current state of user activity (e.g., whether the user is walking or driving, at the office, at home, or in a public environment) and the characteristics of the user's environment (e.g., the nearest Spanish-speaking ATM), they can then intelligently manage both the information content and the means of information distribution.

Location awareness is the most important type of context, because the current (or future) location of users strongly influences their information needs. Applications in computing and communications utilize such location information in two distinct ways:

1. Location-aware computing: In this category, the information obtained by a mobile device or user varies with changes in the user's location. The most-common goal on the network side is to automatically retrieve the current or anticipated neighborhood of the mobile user (for appropriate resource provisioning); on the device side, the typical goal is to discover appropriate local resources. As an example of this category, we can consider the case where mobile users would be automatically provided with local navigation maps (e.g., floor plans in a museum that the user is currently visiting), which are automatically updated as the device changes its current position.
2. Location-independent computing: Here, the network endeavors to provide mobile users with a set of consistent applications and services that do not depend on the specific location of the users or on the access technology employed to connect to the backbone information infrastructure. Information about the user's location is required only to ensure the appropriate redirection of global resources to the device's current point of attachment; such applications are not usually interested in the user's absolute location but only in the point of attachment to the communications infrastructure. An example of this is cellular telephony, where mobility management protocols are used to provide a mobile user with ubiquitous and location-independent access.

While location-independent computing applications have a fairly mature history, location-aware computing is still at an early stage. Innovative prototypes of location-aware computing environments are still largely experimental and geared toward specific target environments. The location support systems of different prototypes, as a result, have been largely autonomous and have always remained at the disposal of the system designers. It is important, however, to realize that the full potential of location-aware computing can be harnessed only if we develop a globally consistent location management architecture that caters to the needs of both location-aware and location-independent applications, and allows the retrieval and manipulation of location information obtained by a wide variety of technologies. This is an interesting technological challenge, because location-aware and location-independent applications typically face significantly different scalability concerns. In general, location-aware applications do not generate significant scalability issues because they primarily involve local interactions; however, scalability is a critical concern for location-independent network services, which must support access to distributed content by a much larger user set.

In this chapter, we focus on identifying the various requirements that must be satisfied by such a universal location-management infrastructure. We also explain why we prefer that such location data be expressed in symbolic format, and then discuss the use of information-theoretic algorithms for effectively manipulating such location information. A symbolic representation of location data allows the management infrastructure to deal with an extremely heterogeneous set of networking technologies, with a wide variety of underlying physical layers and location sensor technologies. Indeed, the ability to accommodate device heterogeneity and technological diversity is vital to the success of a universal location-management scheme. As our survey of current trends will show, location information in various prototypes differs widely in their environment of applicability and the granularity of resolution. Moreover, we shall see how such symbolic information is more amenable to storage and manipulation across heterogeneous databases, and can be exploited to provide necessary functions such as location prediction, location fusion, and location privacy.

The rest of the chapter is organized as follows. Section 17.2 highlights some examples of location-aware applications and prototype systems. The various location-related functions that must be realized in a universal location-management framework designed for pervasive computing models are identified in Section 17.3. Based on this discussion, we explore also the relative merits of alternative schemes for global location representation. Section 17.4 discusses the novel concept of path update and outlines the LeZi-update algorithm that optimizes the signaling loads associated with location tracking. We present ongoing research that extends this algorithm to provide translation of location information across heterogeneous networks and multiple access technologies. Section 17.5 summarizes the chapter and discusses open problems.

17.2 LOCATION RESOLUTION AND MANAGEMENT TECHNIQUES IN PERVASIVE COMPUTING APPLICATIONS

To understand the functionality needed in any universal location-management architecture, it is helpful first to evaluate the various proposed application scenarios. It will become evident that these applications not only have differing performance requirements, but also exhibit significant diversity in the technologies that they use to obtain and track location information. Given the abundance of work in this area, we focus only on a selective subset that illustrates the main requirements and challenges.

17.2.1 IP MOBILITY SUPPORT OVER CELLULAR SYSTEMS

Wireless cellular networks offer the best example of location-management protocols for location-independent computing. The cellular architecture employs a two-level hierarchy for tracking the location of a mobile device as it roams across the cellular network, and redirects traffic to and from the mobile node's current point of attachment. The entire network is partitioned into a number of distinct registration areas (RA), with a home mobile switching center (HMSC) handling all the incoming and outgoing calls for the mobile terminals that are homed in its registration area. Each registration area has a database, called the visiting location register (VLR), which stores the precise location of all mobiles currently resident in that RA. A pointer to the current VLR of a mobile terminal is held at the home location register (HLR), located in the home registration area of the mobile terminal. The first level of mobility support is provided by having the HMSC redirect all arriving calls (after appropriate query resolution via the HLR and VLR) to the mobile switching center (MSC) serving the current point of attachment of the mobile node.

Because each RA comprises several cells, we need additional location-management techniques to identify the precise cell in which the mobile node is currently resident. The resolution of the mobile node's (MN) precise cell of attachment is performed using two complementary techniques, viz., (1) location update or registration, and (2) paging. Location update refers to the process by which the mobile node proactively informs the network element (the MSC) of its current position (or other information, such as future locations). Conversely, paging is the process by which the network element (MSC) initiates a search for the MN in all cells where the mobile has a nonzero residence probability. There has been a significant amount of research on improved paging and location update strategies, such as the use of distance-based location update strategies or selective paging mechanisms.[3]

The introduction of packet-based data services over cellular networks and the predicted move toward IP-based fourth-generation (4G) cellular networks have resulted in several efforts to introduce protocols for IP-based mobility support. Mobile IP[4] is the current standard for IP-based mobility management and provides ubiquitous Internet access to an MN without modifying its permanent IP address. Two entities analogous to HLR and VLR, namely, the home agent (HA) and the foreign agent (FA), are used to tunnel packets addressed to this permanent address to the MN's current point of attachment. The base Mobile IP protocol, however, suffers from several drawbacks, such as high signaling latency, in the absence of a hierarchical infrastructure. Accordingly, several protocols, such as cellular IP, HAWAII, or IDMP, have been recently proposed for introducing a location-management hierarchy in IP networks and thereby providing more scalable intradomain mobility support.[5]

All these schemes express the location of the MN in symbolic form: the MN location is essentially expressed in terms of the ID (e.g., IP address of the FA or the MSC identifier) of the network element to which it is currently attached. Location update and paging schemes operate on this symbolic representation of location; for example, popular paging strategies consist of a sequential search for an MN over a list of cell IDs. The cellular network illustrates the design of a global location resolution framework that combines hierarchical call and packet redirection with suitable paging and registration mechanisms. Of course, the use of such symbolic location information implies that the location of an MN is resolved only up to the granularity of the individual cell or subnet of attachment. In Section 17.4, we shall discuss the design of a provably optimal location-management algorithm, which uses such symbolic location information to establish optimal paging and location update strategies.

17.2.2 MOBILE INFORMATION SERVICES

Location-aware mobile information services are often touted as the "killer application" for the first generation of ubiquitous computing. These services are typically based on an information service infrastructure that retrieves the current location of a mobile device, and then provides the device with information and resources that are local to the MN's current location. The primary aim of such services is not to track individual users, but to ensure that a specific network resource is available to users who are currently "close" to the resource.

One example of this service is the Traveller Information Service being developed as part of the Advanced Traveler Information Systems (ATIS) initiative for smart highways in several countries (e.g., TravTek in Orlando, SMART Corridor in Santa Monica, and CACS in Japan). We outline the example of Genesis,[6] a fairly comprehensive system under development at the University of Minnesota for the Minneapolis–St. Paul area. The ATIS server in the Genesis system maintains the master database, with each road segment associated with a start and end node in the database. Nodes are essentially named objects with location attribute (x,y) coordinates expressed in geographical coordinates. Active databases are used with proper choice of triggers, such as traffic congestion, accidents, road hazards, and constructions and detours, to support a wide range of spatially correlated queries. Due to the geometric representation of location information, the local environment of a user is defined using simple spatial queries.

The Cyberguide Project at Georgia Tech[7] is an effort to develop an electronic tourist guide for both wide area and local environments (such as a building). The Cyberguide architecture uses explicit GPS-based positioning for outdoor environments and infrared-based positioning for indoor environments; the indoor location resolution technology does not scale well to large coverage areas. Lancaster University's GUIDE project[8] is another experimental prototype of local information services. It uses a cellular network arrangement, with IEEE 802.11 LANs providing short-range coverage within a single cell. By making the coverage areas deliberately discontinuous, GUIDE ensures that each cell caters only to mobile nodes within a specific zone. Because GUIDE simply broadcasts zone-specific information from a Linux-based cell server to all nodes within the corresponding zone, the system does not require any explicit location or positioning support and does not need to track the movement of individual nodes.

17.2.3 TRACKING SYSTEMS

Tracking applications differ from mobile information services in that they typically focus on the ability to continuously monitor the location of a mobile device. Present-generation tracking applications typically run as global services, where the location of the mobile nodes must be distributed over wide area networks. Fleet management applications are the most-obvious examples of present-day tracking systems. Most commercial fleet management systems (e.g., Qualcomm's OmniTRACS product) are based on the GPS technology, which provides the absolute location of a mobile device (relative to a geographical coordinate system) at varying levels of precision. Location update schemes in such systems employ dead reckoning, whereby the location information is extrapolated by the system based on velocity information; new updates are generated when the mobile object deviates from its predicted position by a distance threshold. A digital map database is maintained in a manner quite similar to that of ATIS, i.e., using path segments and nodes, along with their x,y coordinates. Depending on need, a portion of this database may be replicated in the memory of the on-board computers. Dynamic attributes and their indexing, spatio-temporal query languages, and uncertainty management are special features of such databases.

The Federal Communication Commission's (FCC) E911 initiative has made it mandatory for wireless cellular service providers to track the location of phones making emergency 911 calls. While GPS information provides the easiest way of determining location, most cell phones do not possess such technology. The location of a mobile user in such environments is often determined, typically in geographical coordinates, by triangulation technologies based on the relative signal strength of the cellular signal at multiple base stations. While GPS is indeed a popular technology for resolving location, it is applicable only to outdoor computing environments. Due to this limitation, as well as the fact that GPS technology cannot be embedded or is not available in all computing devices, GPS data cannot be used as the basis of a universal location representation scheme. Recently, several innovative research prototypes have focused on the problem of location tracking in indoor environments.

An example of such a research prototype is the Active Badge project,[9] originally conceived at the Xerox Palo Alto Research Center. Active badges are low-cost, low-power infrared beacon-emitting devices worn by employees in an office environment. Sensors are distributed in a pico-cellular fashion within the building, and the location of a badge is determined primarily by the identity of the sensor that reports the badge within its vicinity. Location management and paging algorithms are used to track the user's location, which is essentially expressed in symbolic form (based on the IDs of the neighboring sensors). While the infrared technology used in active badges can resolve device location up to the granularity of individual rooms, additional technologies are needed for finer location resolution. For example, Active Bats[10] have been developed to track both position and movement using ultrasonic technology; this approach can be considered the indoor analog of GPS because it expresses location in geometric coordinates. Follow-me applications in pervasive collaborate workspaces require such fine-grained location information; such applications also need efficient location prediction to ensure that computing and communication resources are available to a mobile device in an uninterrupted fashion.

Several other research prototypes have exploited alternative radio technologies for indoor location tracking. For example, MIT's Cricket Location Support System[11] requires the mobile devices to proactively report their locations. Such mobile devices use sophisticated triangulation mechanisms that monitor both RF and ultrasound signals emitted from wall- and ceiling-mounted beacons to resolve their geographical location information. Microsoft Research's RADAR system,[12] on the other hand, uses signal-to-noise ratio and signal strength measurements of IEEE 802.11 wireless LAN radios to resolve the location of indoor mobile nodes to a granularity of approximately 3-to-5-meter accuracy. While the accuracy of the resolution can suffer due to changes in the indoor layout (such as the moving of metal file cabinets), the approach offers the advantage of location resolution that piggybacks on the wireless networking infrastructure and does not require the extensive installation of new devices/radios. Pinpoint's 3D-ID performs indoor position tracking at 1-to-3-meter resolution using proprietary base station and tag hardware in the unregulated ISM band (also used by 802.11 LANs) to measure radio time of flight.

Research prototypes have also used alternative techniques for monitoring user location. Electromagnetic sensing techniques (e.g., Raab et al.[13]) generate axial magnetic-field pulses from a transmitting antenna in a fixed location and compute the position and orientation of the receiving antennas by measuring the response in three orthogonal axes to the transmitted field pulse, combined with the constant effect of the Earth's magnetic field. While they offer up to 1-mm spatial resolution, they suffer from limited tracking distances and steep implementation costs. Research projects also have used stereovision (e.g., Microsoft's Easy Living[14] project for indoor home environments) or ubiquitous pressure-sensing (e.g., Georgia Tech's Smart Floor proximity location system[15]) techniques to resolve the location of people in indoor environments. While such techniques may not be deployed universally, they do illustrate how the use of diverse location resolution and management techniques is a basic reality of pervasive computing architectures.

17.2.4 ADDITIONAL TECHNIQUES

We have recently witnessed research efforts in *ad hoc* location sensing, where user location in wireless environments is estimated without the use of static beacons or sensors that provide a fixed frame of reference. Mobile nodes in such *ad hoc* environments essentially act as peers, sharing sensory information with one another to progressively converge on a true representation of device location. Doherty et al.[16] have presented an algorithmic approach to this problem, as well as a framework for describing error bounds on the computed locations.

Another interesting area of location management research is sensor fusion, where the location information is obtained by simultaneously aggregating information from multiple hierarchical or overlapping sensing technologies. By integrating location tracking systems with different error distributions, we can often provide increased accuracy and precision beyond the capabilities of an individual system. An example of such fusion can be found in multisensor collaboration robot localization problems (e.g., Fox et al.[17]), where information from multiple sensors (such as ultrasonic and laser rangefinders, cameras, etc.) is integrated using Bayesian or Markovian learning techniques to develop a "map" of a building.

Table 17.1 shows a selective list of the location-management techniques employed in various pervasive computing contexts. We can see that location management prototypes use both geometric and symbolic representations to resolve, track, and predict the location of mobile devices.

TABLE 17.1
Examples of Location Management in Pervasive Computing Scenarios

Product/ Research Prototype	Primary Goal	Underlying Physical Technology	Techniques Employed	Location Represen tation
Cellular voice	Continuous global connectivity for mobile users	GSM, IS-95, IS-51, NA-TDMA, CDMA-2000, WCDMA (forthcoming for 3G)	Location updates, paging, HLR/VLR	Symbolic
Internet (IP) mobility	Roaming support for mobile nodes	Any technology supporting IP tunneling	HA/FA, packet tunneling	Symbolic
Genesis	Highway information services	GPS	Active databases, spatial queries	Geometric
GUIDE	Hot-spot information services	802.11 WLAN	Disconnected cellular topology	Symbolic
OmniTracks	Outdoor fleet management	GPS	Dead reckoning, paging	Geometric
Active Badge	Indoor tracking	Infrared	Vicinity-based reporting	Symbolic
ActiveBats	Follow-me indoor computing	Ultrasonic	Location updates, paging	Geometric
Cricket	Indoor location tracking	RF and ultrasound	Location updates	Geometric
RADAR	Indoor location tracking	802.11 WLAN	Triangulation, location updates	Symbolic
SmartFloor	Indoor user tracking	Foot pressure	Location updates	Geometric

17.3 PERVASIVE COMPUTING REQUIREMENTS AND APPROPRIATE LOCATION REPRESENTATION

The basic goal of pervasive computing is clear: develop technologies that allow smart devices to automatically adapt to changing environments and contexts, making the environment largely imperceptible to the user. However, the set of candidate applications and their underlying technologies is anything but uniform! Developing a uniform location-management infrastructure is thus a challenging task. We identify the following location-related features, which a universal architecture must support:

1. Interoperability across multiple technologies and resolutions: Current prototypes for pervasive applications typically choose a specific location-tracking technology that is suitable for their individual needs. A uniform location-management architecture must be capable of translating the location coordinates obtained by such systems into a universal format, which can be utilized by various application contexts. For example, cellular land-mobile systems will primarily need to resolve the location of a mobile device only up to the point of network attachment. Fleet management and tracking applications may, however, require explicit geometric information. The mobility management infrastructure should be capable of efficiently translating such location information between different representations, and also at different granularities (e.g., mobile commerce applications advertising E-coupons may not be interested in the precise room in which a user is located inside a hotel).

2. Prediction of future location: Predicting the user's future location is often the key to developing smart pervasive services. For example, the ATIS active database can be triggered more intelligently by predicting the most-likely routes, and by warning the client about adverse road conditions along those routes. Prediction of an individual's future position in an indoor office can be very helpful in aggressive teleporting (to support follow-me applications). In addition to this explicit service-oriented need for prediction, there is also an implicit need for predictive mobility tracking from the network infrastructure viewpoint. In several location-independent computing scenarios, the network must meet stringent performance and latency bounds as it ensures uninterrupted access to global information and services, even as the users change their location. For example, to provide quality-of-service (QoS) guarantees for multimedia traffic (such as video or audio conferencing) in cellular networks, appropriate bandwidth reservations must be made between the terminal and the serving base station (BS), as well as between the BS and the backbone network. To meet strict bounds on the handoff delay, the network also must proactively reserve resources at the cells where the mobile is likely to move. Because many of the tracking technologies do not themselves offer such predictive capabilities, the infrastructure must be capable of constructing such predictive patterns based on collective or individual movement histories.

3. Location fusion and translation: In several pervasive computing scenarios, location tracking is achieved through the combination of multiple technologies and access infrastructures. For example, an office application can resolve the location of a user at different levels of granularity using different technologies. Thus, the specific building could be identified through the current wireless LAN cell where the mobile currently resides, whereas an additional ultrasonic system (such as Cricket[11]) may be used to identify the precise orientation and room location of the mobile user. Because the user's complete location reference is obtained only by combining these distinct location management systems, our global location-management framework must efficiently fuse and merge location information from two or more distinct network technologies.

    The intelligent management of vertical (or intersystem) handoff, on the other hand, often requires the ability to translate the mobility and location-related information from one frame of reference to another. For example, when a user switches from a wireless LAN to an overlaid PCS network, the network must be able to translate the mobility patterns and location-prediction attributes from one system to the other, independent of the representation format imposed by each individual system.

4. Scalable and near-optimal signaling traffic: The desire for efficient and provably optimal location update and paging strategies is not new; there has indeed been a great deal of work on efficient location-management strategies, especially for cellular systems. The pervasive world will however see a quantum jump in the number of mobile nodes (from millions of cell phones to billions of autonomous pervasive devices) and an even greater variation in the capability (such as power or memory constraints) of individual devices. We must therefore develop efficient and near-optimal signaling mechanisms that minimize any unnecessary signaling load on both the devices and the networking infrastructure.

5. Security and privacy of location information: Security and privacy management are key challenges in pervasive networking environments; notwithstanding the availability of advanced devices and location-resolution technologies, users will not embrace a pervasive computing model until a scalable infrastructure for appropriately protecting such location information is in place. The problem is not one of simply making such location information either visible or invisible to specific networks; we must allow the user to dynamically configure the scope of location visibility, possibly in multiple representation formats, to individual pervasive services and applications. For example, a user may wish to expose his precise GPS coordinates to emergency response applications (such as 911), but only a much coarser view (perhaps at a granularity of 20 miles^2) to insurance companies trying to monitor his driving profile. Alternatively, the user may specify his network point of attachment (symbolic information), but not his precise in-building location (geometric coordinates) to a pervasive enterprise application.

17.3.1 Geometric or Symbolic Representation?

While different pervasive location tracking and management systems resolve the location of a mobile node at different granularities, they can all be classified into two classes* (as per the taxonomy of Leonhardt and Magee[4]) based on the way in which they represent the location information of a mobile device:

1. Geometric: The location of the mobile object is specified as an absolute n-dimensional coordinate, with respect to a geographical coordinate system that is independent of the network topology. The most-common form of geometric data representation in location-aware computing systems is the use of GPS data, which resolves the latitude and longitude of a mobile on the earth's surface using a satellite-based triangulation system.
2. Symbolic: The user location data is specified not in absolute terms, but relative to the topology of the corresponding access infrastructure. This form of representation is in widespread use in current telecommunications networks. For example, the PCS/cellular systems identify the mobile phone using the identity of its serving MSC; in the Internet, the IP address associated with a mobile device (implicitly) identifies the subnet/domain/service provider with which it is currently attached.

The choice between a geometric and symbolic representation is one of the fundamental decisions in the development of a universal location-management architecture. We believe that the symbolic representation is the preferred form, primarily due to its structured nature. The main advantage of geometric representation is that it is invariant: because the location information is an intrinsic property of the mobile device, it can be uniformly interpreted across heterogeneous environments, and does not depend on the topology of the associated networks. In spite of this seeming attractiveness, geometric representation is not appropriate for a universal location-management infrastructure. For one thing, the same reference coordinate system is not universally applicable. As an example, GPS may be appropriate outdoors but does not apply indoors, where ultrasonic or infrared-based indoor positioning systems may use different location coordinates. Moreover, we have demonstrated how different pervasive applications and environments require the location of a mobile device at different levels of granularity. Thus, while GPS information may be accurate up to 5-m resolution, certain in-room pervasive applications may require tracking at submeter resolutions. Because we cannot practically mandate the universal deployment of a technology that provides location at the finest granularity (the tracking costs would become prohibitive), we must allow for the coexistence of different networks and access technologies, providing location information at varying resolutions. Finally, geometric location-resolution technologies are inapplicable to a large

---

* There is also the semisymbolic (or hybrid) model,[4] which essentially consists of both geometric and abstract (symbolic) representations. While such a model is more expressive, it suffers from the same drawbacks as the geometric one (the main problem being the need for location-specific hardware on the pervasive device itself).

category of pervasive devices, which may not possess location-resolution hardware (such as GPS devices) due to restrictions on cost and form factors. In contrast, symbolic location information (such as the point of attachment to the network) can be obtained solely from the capabilities of the infrastructure.

Our preference for the symbolic form of location representation is based on the observation that most location-independent applications, and a significant number of location-aware ones, are interested primarily, not in the absolute location of the mobile device, but only its position relative to the networking infrastructure. More importantly, the location-independent applications are typically global in scope and cut across multiple network and access technologies. Accordingly, scalability concerns for the location-management infrastructure apply primarily to the location-independent component of the pervasive application space. The interaction between mobile devices and applications that require the explicit geometric location of such devices (such as map-based interactions), is often local and restricted to the access network. It thus makes sense to base the universal infrastructure on the symbolic representation, allowing each access network to make the appropriate translation to geometric coordinates whenever necessary (rather than the reverse). For example, consider applications such as wireless Internet access that need to resolve the location of a mobile device only up to the granularity of the point of attachment. Even apparently location-aware services, such as the Electronic Tourist Guide, are really interested in knowing the user's location relative to the access infrastructure; a museum information system needs to know only the current access point serving the mobile visitor to provide appropriately tailored local content. Similarly, follow-me applications are primarily interested in predicting the device's future point of attachment (rather than its absolute position), because the final objective is to make advance reservations on the network path to the future point of attachment. Furthermore, location data is much more amenable to database storage and retrieval if it is a named object — such an object hierarchy is possible only when location data is expressed in symbolic form relative to other objects. Because an object hierarchy also simplifies the computational burden associated with multiresolution processing, the translation of location data across different systems and location databases is more efficient when stored in symbolic format.

Of course, we must not lose sight of applications, such as dynamic floor maps, which do need geometric location information. Geometric coordinates are clearly better suited for answering spatial queries related to physical proximity and containment (e.g., is my device physically located within a designated building?). As stated earlier, we believe that such specialized geometric queries (e.g., directions to the nearest ATM or restroom facilities) typically involve "local" resources and interact with server applications lying within the access domain, especially in pervasive environments where access networks will have considerably greater intelligence. In the future, a user currently located on a street in New York City is likely to obtain the location of the nearest ATM from a local tourist-guide server, rather than relay his request back to a mapping software located on a server in San Francisco. While such queries may need to express location in geometric form, it is better to obtain such information either from "local" access-specific technologies,

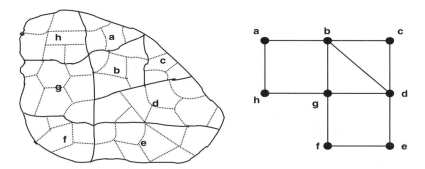

FIGURE 17.1 A hierarchical map and its top-level graph representation.

or by appropriate mapping from the universal symbolic location format. It is important also to not lose sight of the fact that many pervasive applications generate queries related to topological proximity and containment, where the query issuer is interested in resources relative to the network topology (i.e., which is the closest [fewest hops or least congested] video server? or, does this printer belong to the research division?). Thus, both geometric and symbolic representations appear to be equally balanced from a query suitability standpoint, with the geometric format better suited to spatial queries and the symbolic one more appropriate for topological queries.

The hierarchical nature of communication networks implies the imposition of a logical hierarchy on the symbolic location representation (which expresses location relative to the network layout) as well. As an example, we will consider an IEEE 802.11 wireless LAN infrastructure at a university campus, which is overlaid by the wide area cellular PCS infrastructure. For the sake of simplicity, we assume that each PCS cell consists of multiple 802.11 LANs.* We can then construct a symbolic positional hierarchy based on the coverage area of each technology, which yields the neighborhood graph shown in Figure 17.1. The top level (corresponding to the cellular network) has eight zones, a, b, c, d, e, f, g, h, connected by neighborhood relationship as in the graph shown next to it. The second level zones which correspond to the 802.11 LANs may be named a1, a2, …; b1, b2, …; c1, c2, …, where a1, a2, … are subzones in the zone a and so on.

We now focus on evaluating the suitability of using symbolic information to satisfy the five requirements enumerated at the beginning of this section. We have already seen how symbolic information is better suited to requirement 1, because it does not need any special support from the wireless access technology. In the rest of this chapter, we focus on features 2 and 4, showing how we can develop a path-based location-prediction algorithm (based on symbolic representation) that is provably optimal for stationary mobility patterns. While we are currently working on requirement 3, the issue of configurable universal location security and privacy,

---

* Of course, in general, algorithms for storing and manipulating such symbolic information must allow for hierarchies with partial or incomplete overlap (e.g., a WLAN may span multiple PCS cells).

although a very interesting problem area in itself, is essentially beyond the scope of this chapter. However, symbolic location should clearly be more amenable to location privacy; because the user location is specified only relative to the topology of the network infrastructure, precise location of a mobile node is not possible without a knowledge of the physical network topology.

17.4 "OPTIMAL" LOCATION TRACKING AND PREDICTION IN SYMBOLIC SPACE

In this section, we first show how LeZi-Update,[18,19] a path-based location update mechanism, can be used to provide efficient location predicting and tracking of mobile devices in a symbolic location space. This path-based approach is asymptotically optimal and outperforms earlier location-management algorithms based on location areas (LA), because it exploits the mobility profile associated with individual users. Moreover, we can store the relevant details of the user's location history in a compact data structure, and also derive accurate predictions of the relative likelihood of future locations of the mobile device. Finally, we shall describe our ongoing work on enhancing this algorithm with an efficient location-translation capability for transferring mobility profiles across heterogeneous systems.

Most mobility management solutions employ a position update technique for location management, where the mobile device simply updates its current location (e.g., cell ID or GPS coordinates) whenever it crosses a cell boundary (or other thresholds such as time or distance). Position update schemes can be viewed as a lossy sampling of the true trajectory of the mobile object. Such position-based schemes do not, however, use these location update samples to construct or predict the user's path; in effect, the schemes do not correlate across multiple sample points to learn the "pattern of device movement." The performance of conventional paging and location update schemes thus depend heavily on the precise parameters of the user mobility model; different algorithms perform better for different mobility models. A generic location management scheme must, however, perform well, independent of the individual mobility patterns followed by different users. Such a generic model must be based on the weakest set of assumptions on the mobility behavior of individual users or devices, and must incorporate some form of learning that uses the past history of the mobile node to optimize the signaling associated with location tracking.

We assume that the user mobility is "well-behaved," in that users/devices typically move on some definitive paths that are based on the lifestyle of the mobile user. According to the activity-based model, trips are considered the basic element of a user's long-term mobility profile. Trips in both outdoor and indoor environments are categorized by the purpose behind them, such as going to and coming back from work, shopping, a walk to a colleague's cubicle, a lunch-hour visit to the cafeteria, etc. Each trip in the symbolic space then appears as a phrase of symbols. For example, if the location of a user is sampled successively, the mobility profile of a user over the graph of Figure 17.1 may be expressed by the stream of symbols "aaababbbb-baabccddcbaaaa...." From a computational standpoint, the mobility profile of any user can then be represented by a user-specific stationary distribution over the

ENCODER

```
initialize dictionary : = null
initialize phrase w : = null
loop
  wait for next symbol v
  if (w.v in dictionary)
    w : = w.v
  else
    encode <index(w),v>
    add w.v to dictionary
    w : = null
  endif
forever
```

DECODER

```
initialize dictionary : = null
loop
  wait for next codeword <i,s>
  decode phrase : = dictionary [i].s
  add phrase to dictionary
  increment frequency for every prefix of phrase
forever
```

FIGURE 17.2 Encoder at the mobile, decoder at the network element.

generation of the symbol stream. Because neither the lifestyle nor the stationary pattern of mobility remains the same for one person, it is more realistic to conjecture that the symbolic capture of the movement pattern of a well-behaved pervasive device is stationary or piecewise stationary. The assumption of stationarity is weak enough to accommodate a wide variety of mobility models (from random to piecewise deterministic), yet strong enough to prove the performance of our LeZi-update location-tracking algorithm.

17.4.1 THE LEZI-UPDATE ALGORITHM

The LeZi-update algorithm[18,19] is based on the observation that location update is essentially equivalent to the transmission of the generated symbol sequence. Optimizing the signaling associated with location updates is then functionally equivalent to maximally compressing the symbol stream through the use of appropriate encoding schemes. The LeZi-update algorithm is thus based on an incremental parsing and compression technique, the LZ78 algorithm,[20] which parses the outgoing symbol stream in a causal manner to adaptively construct the optimal transmission code.

We now briefly describe the fundamental functioning of the LeZi-update algorithm using the encoder-decoder duo shown in Figure 17.2. The encoder part, residing in the mobile terminal, intercepts any combination of primitive dynamic update (distance/movement/time-based) treating the zone-ids as input symbols. The coded update message is sent to the system (MSC/VLR for cellular systems) where the decoder resides, which on receipt of the coded update message, decodes it into the original symbol sequence and updates their relative frequencies. For example, the symbol sequence aaababbbbbaabccddcbaaaa... considered earlier, gets parsed as "a, aa, b, ab, bb, bba, abc, c, d, dc, ba, aaa, ..." by the encoder, where commas indicate the points of updates separating the updated path segments.

The symbol sequences (actually user path segments) can easily be maintained in a structure known as a trie (shown in Figure 17.3), which captures all the relevant history of the user in a compact form. In addition to representing the dictionary, the

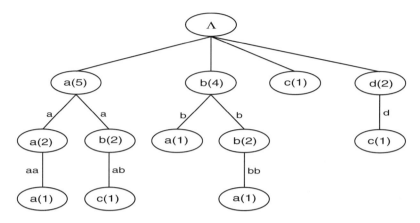

FIGURE 17.3 Trie for the classic LZ symbolwise model.

trie can store statistics for contexts explored, resulting in a symbol-wise model for LZ78. Using the stored frequencies, the trie can be used to predict the probability of future occupancy in the cell geometry (symbolic space). The paging operation then consists of a progressive search for the mobile node in a decreasing sequence of the occupancy probabilities. By storing individual mobility profiles, the LeZi-update algorithm adaptively learns the optimal update and paging scheme for each individual mobile node.

The algorithm is asymptotically optimal if the user mobility profile remains stationary. Information theory, in fact, shows the existence of a lower bound (known as the entropy rate) on the transmission rate of a stationary sequence of symbols. No lossless compression scheme can reduce the symbol stream to a lower rate; moreover, this lower bound can only be reached asymptotically (using infinitely long sequences of generated symbols). The LeZi-update algorithm, in essence the LZ78 compression scheme, can be proved to asymptotically converge to the entropy rate, as long as the mobile moves randomly according to a stationary distribution.

17.4.2 Translation of Mobility Profiles during Vertical Handoffs

The LeZi-update algorithm discussed in the previous section leads to efficient and intelligent tracking as long as the user moves within a specific symbolic space (equivalently, access technology). To predict the location attributes of a mobile node when it moves to a different access technology (e.g., when a vertical handoff occurs across access technologies), we need to translate the mobility profile from the current to the new symbolic space representation. Figure 17.1 provides an illustration of such a scenario. As long as the user moves across the LAN infrastructure, its path update and trie information are stored in the form of strings such as a1, a2, b1, c1, etc. When a vertical handoff occurs (from the LAN to the PCS network), the network needs to transfer the mobility profile it has learned after translating it to the new symbol space a, b, c, ... The new space, in general, need not directly overlap with the old space; for example, we could have a1, b1, and d2 all map to a, and g2, f3, and h4 map to b.

We are currently working on such a translation mechanism, called hierarchical LeZi-update. The hierarchical algorithm is based on the observation that the entire relevant movement history of the mobile node in its original symbolic space is captured in the stored user-specific trie, Given a mapping from the old space to the new symbolic space, we should thus be able to manipulate the trie to obtain an equivalent movement history in the new symbolic space. We can then express the mobile's movement history in a new trie, which refers to the symbolic space associated with the new access infrastructure. The problem of location translation across heterogeneous symbolic location coordinates thus reduces to the construction and communication of a modified trie structure (using a mapping between the new and old coordinate systems) between heterogeneous networks.

17.5   CONCLUSION

In this chapter, we survey the various ways in which context-aware pervasive computing applications are likely to exploit and manage location information; we use this understanding to debate whether a universal location-management infrastructure should store location information in a topology-dependent (symbolic) or topology-independent (geometric) format. Our analysis of both location-aware and location-independent applications reveals three important points: (1) different systems and prototypes use a wide variety of location-resolution technologies, (2) a significant number of location-based applications are primarily interested in resolving the location of a mobile node only relative to the connectivity infrastructure, and (3) obtaining geographical location coordinates requires varying levels of hardware that are absent in many pervasive devices. We thus conclude that the universal location-management infrastructure should manipulate location information primarily in a structured, symbolic form. In cases where the geographical coordinates are needed, they may be obtained through the use of access-specific technologies or via appropriate mapping.

We then consider the objectives of pervasive computing and enumerate the desirable features of a universal location-management infrastructure. In particular, we believe that location prediction, location translation, signaling optimality, and location privacy are four "must-haves" in a practical pervasive infrastructure. While the problem of location privacy is beyond our current scope, we consider the problem of location prediction and signaling optimality in greater detail. We explain how the LeZi-update algorithm uses adaptive learning to optimize the signaling associated with location update and paging in a symbolic domain. By treating the movement of a mobile device as a sequence of strings generated according to a stationary distribution, the algorithm is able to efficiently store a mobile's entire movement history, and also predict future location with asymptotically optimal cost. We finally turn to the problem of location translation, and give an overview of our ongoing development of a hierarchical LeZi-update that permits efficient translation of location profiles between heterogeneous systems.

Our immediate plans for future work include the development and performance testing of the hierarchical LeZi-update algorithm. We are interested also in the problem of efficiently translating between symbolic and geometric coordinates in practical systems. A great deal of work also is needed to standardize protocols for location fusion and translation in real-life environments. We hope that the findings of this chapter serve as a useful starting point for the design and specification of formats for specifying user location, the architecture of location databases, and the development of intelligent location-reporting protocols.

ACKNOWLEDGMENT

The work of S.K. Das, A. Bhattacharya, and A. Roy is supported by NSF grants EIA-0086260, EIA-0115885, and IIS-0121297.

References

1. Weiser, M., The computer for the 21st century, *Sci. Am.,* 265 (3), 94–104, 1991.
2. Hopper, A., Sentient computing, The Royal Society Clifford Patterson Lecture, 1999, http://www-lce.eng.cam.ac.uk.
3. Yong, V.W.S. and Leung, V.C., Location management for next-generation personal communications networks, *IEEE Network,* 14(5), 18–24, Sept.–Oct. 2000.
4. Leonhardt, U. and Magee, J., Toward a general location service for mobile environments, Proc. Int. Workshop on Services in Distributed and Networked Environments, Macau, June 1996, pp. 43–50.
5. Das, S. et al., IDMP: an intradomain mobility management protocol for next generation wireless networks, *IEEE Wireless Commun.* (formerly *IEEE Personal Commun.*), 9 (3), 38–45, 2002.
6. Shekhar, S. and Liu, D., Genesis and Advanced Traveler Information Systems (ATIS): killer applications for mobile computing?, Proc. NSF MOBIDATA Workshop on Mobile and Wireless Information Systems, Rutgers University, NJ, Nov. 1994.
7. Abowd, G.D. et al., Cyberguide: a mobile context-aware tour guide, *ACM/Baltzer Wireless Networks,* 3 (5), 421–433, 1997.
8. Cheverst, K. et al., Experiences of developing and deploying a context-aware tourist guide: the GUIDE project, Proc. 6th Ann. Int. Conference on Mobile Computing and Networking, pp. 1–12, Aug. 1999.
9. Want, R. et al., The Active Badge location system, *ACM Trans. Inf. Syst.,* 10 (1), 91–102, 1992.
10. Harter, A. et al., The anatomy of a context-aware application, Proc. 5th Ann. Int. Conference on Mobile Computing and Networking, pp. 59–68, August 1999.
11. Priyantha, N., Chakraborty, A., and Balakrishnan, H., The Cricket location support system, Proc. 6th Ann. Int. Conference on Mobile Computing and Networking, Boston, pp. 32–43, Aug. 2000.
12. Bahl, P. and Padmanabhan, V., RADAR: an in-building RF-based user location and tracking system, Proc. IEEE Infocom, IEEE CS Press, Los Alamitos, California, pp. 775–784, 2000.
13. Raab, F. et al., Magnetic position and orientation tracking system, *IEEE Trans. Aerospace Electron. Syst.,* 15(5), 709–718, 1979.

14. Krumm, J. et al., Multi-Camera Multi-Person Tracking for Easy Living, Proc. 3rd IEEE Int. Workshop on Visual Surveillance, IEEE Press, Piscataway, NJ, pp. 3–10, 2000.

15. Orr, R.J. and Abowd, G.D., The Smart Floor: a mechanism for natural user identification and tracking, Proc. Conference on Human Factors in Computing Systems, ACM Press, New York, 2000.

16. Doherty, L. et al., Convex position estimation in wireless sensor networks, Proc. Infocom 2001, IEEE CS Press, Los Alamitos, CA, 2001.

17. Fox, D. et al., A probabilistic approach to collaborative multi-robot localization, *Autonomous Robots,* 325–344, June 2000.

18. Bhattacharya, A. and Das, S.K., LeZi-update: an information-theoretic approach to track mobile users in PCS networks, Proc. 6th Ann. ACM Int. Conference on Mobile Computing and Networking (MobiCom), pp. 1–12, Aug. 1999.

19. Bhattacharya, A. and Das, S.K., LeZi-update: An information-theoretic framework for personal mobility tracking in PCS networks, *ACM/Kluwer Wireless Networks J.,* 8 (2-3), 121–135, 2002.

20. Ziv, J. and Lempel, A., Compression of individual sequences via variable-rate coding, *IEEE Trans. Infor. Theory,* 24 (5), 530–536, 1978.

21. Perkins, C., IP mobility support, RFC 2002, Internet Engineering Task Force, Oct. 1996.

22. Satyanarayan, M., Pervasive computing: vision and challenges, *IEEE Personal Commun.,* 8 (4), 10–17, 2001.

Part IV

Applications

18 Mobile and Wireless Internet Services: From Luxury to Commodity

Valerie A. Rosenblatt

CONTENTS

18.1   INTRODUCTION

It is now common knowledge that the Internet has radically changed the way people do business and live day-to-day life. The Internet has opened new channels of communication within society in very much the same fashion as stone plates, papyrus, and books did. Today, people communicate with their business partners and clients through e-mail, check inventory and fulfill orders, make travel reservations, take care of their Christmas shopping, conduct research, store information, and accomplish many other vital tasks faster and more efficiently using the Internet.

Similar to the Internet, over the past few years mobile phones have turned into a mass-market sensation, opening new ways of communicating. People are no longer tied to their office desks, home phones, and telephone booths when they need to make a call. The widespread cellular coverage, at least in Europe, Asia, and North America, allows one to make a phone call from a mobile phone practically from anywhere, and services are only getting better and more accessible. There are already over 600 million mobile phone users in the world, and by 2004 this number is expected to grow to as many as 1 billion.

But the technological advancement never stops. Humans are constantly trying to raise the bar and continue to create new ideas and tools to improve their lives. A new technology uniting wireless capabilities and the Internet concept has emerged — the mobile Internet. Mobile Internet gives users access to data and applications "anytime, anywhere" using mobile devices and wireless networks. Although mobile Internet has yet to live up to the hype and show true benefits and returns that will prove its viability and secure its permanent space in everyday life, mobile Internet already boasts several success stories that got the attention of the big players in the technology industry. The enthusiasm has been accompanied by the introduction of new products and services designed to leverage the best attributes of mobile devices' form factor and wireless connectivity. It is undeniable that at a certain time, there will be more users using wireless technologies than users using PC desktops to connect to the Internet. The main question is when the proliferation of the wireless Internet services will hit mass-market capacities. According to many sources, this revolution is not that far away. Based on the evaluation of data published by several major research houses, by 2005 the number of devices accessing Internet services wirelessly will exceed the number of devices accessing the Internet using wired technologies, opening a whole new world in business as well as personal growth opportunities.

This chapter focuses on the subject of proliferation of the wireless Internet in the enterprise and consumer markets. Starting first by going over the main historical events in the world of the wireless Internet services, the chapter traces the evolution and underlying ground for wireless Internet services technology. Similar to any technological advancement, wireless Internet boasts many success stories along with fallouts from which lessons were learned; you will find examples of both in this chapter. Yet, technology alone is relevant only in science books, and cannot be successful without proper applications. This chapter explores how various applications make wireless Internet viable, and not just another over-hyped sensation.

18.2 EVOLUTION OF MOBILE INTERNET SERVICES

The idea of wireless Internet has been around for a while. Many companies have invested heavily in developing wireless Internet applications and services, but have failed to find the silver bullet due to the lack of the most important aspect that could turn it into an everyday aspect of life: fast, reliable broadband connection. At the dawn of the wireless Internet, many companies emerged with a grand idea of extending existing Internet content and services to mobile devices, making personal and corporate data and applications services available "anytime, anywhere." While the idea promised grand success, as was proved by the amount of venture capital invested in companies focused on development of mobile and wireless platforms, applications, and services, the most important component was missing: the connection medium with speeds capable of satisfying the end users. Among the companies that will go down in history as wireless Internet pioneers will be Metricom, Palm, GoAmerica, OmniSky, and Phone.com.

The evolution of the mobile Internet very much replicates the evolution of the tethered Internet services. The pioneering wired Internet services were very slow and expensive. The first computer modems capped at 9.6 kbps data transfer rates. The services have been offered on per-minute bases at very expensive rates, basically making early Internet services unavailable to the wide audience. As time passed, PC modem technology improved, and today individuals enjoy fast T1 and T3 connections averaging 10 Mbps data transfer rates in their offices and at least 100-kbps broadband Internet services on home PCs. Mobile Internet technology is following a similar pattern. It started with slow speeds and poor coverage and has been slowly growing to higher bandwidth and more dense coverage. From the plain old cellular AMPS (Advanced Mobile Phone System) networks that became the base for the first wireless data communications, all the way to fast, newer standards like NTT DoCoMo and WiFi, wireless Internet evolved providing faster, more-secure, and more-reliable data connections with each new technology offering.

18.3 SLOW MOTION OVER PLAIN OLD CELLULAR

In 1969, engineers at Bell Labs developed the cellular telephone technology known as Advanced Mobile Phone System (AMPS). This system uses the 800-MHz frequency band and has been widely deployed in North and South America for mobile voice communications. Although the AMPS cellular network was designed primarily for voice transmission, techniques have been developed to send data over the network. In order for the mobile worker to do a database query or check e-mail over a dial-up circuit-switched connection, it was necessary to dial in; establish the communication channel through the cellular network, server, mainframe, and database; and stay connected while the application is launched and information is retrieved. The process was slow and cumbersome, and oftentimes sending a fax was faster and easier. In that sense, the AMPS, in much the same way as the "plain old telephone system" used to access the Internet with a trusty 28.8-kbps analog modem, was prone to data loss and high and variable propagation delay, impeding reliability and reducing effective throughput.

AMPS wireless data service was similar to a standard cellular phone call, using the same channels and the same frequency as the cellular voice call, but with specialized protocols used by the modems on each end for circuit switched cellular data. The mobile device required a modem, such as SpeedPaq 336 offered by Compaq, which connected to a cellular phone and supported the necessary cellular protocols. To send a data signal using AMPS over-the-air protocols, a temporary dedicated path was established for the duration of the communication session. All signals flow continuously over the same path, and billing for AMPS data service was generally a function of airtime used, typically in 1 minute increments, with charges based on the user's selected rate plan. And, just as with a normal phone call, all applicable long distance charges, roaming charges, and taxes also were billed.

In 1991, U.S. cellular operators initiated an activity to see if they could offer a digital data service for uses like email and telemetry. The analog cellular worked, but the operators did not see the expected return from the subscribers and the technology cost was too high. The data throughput offered over the AMPS networks was very slow, ranging from 2.4 to 14.4 kbps, affected by interference, noise, fading, and overall channel degradation, common RF-related affects, and varied from one location to another. Dropped calls also were common. Security was another issue. There was a limited amount of the content available for the wireless Internet users. Mainly the AMPS wireless networks were used to connect to the custom build corporate gateways that were designed to serve data often in proprietary format. Browsers also were primarily proprietary, and limited standards existed for the Mobile Data communications.

18.4 WEB CLIPPING OVER PAGER NETWORKS

In late 1998, Palm came out with the Palm VII device containing a small radio unit that accessed content servers operated by Palm via the BellSouth Data Network. Because of the high cost wireless transmission, Palm decided to provide only clipped content and launched Palm.net service that supported small software apps known as PQAs (Palm Query Applications) that acted as an interface to the Internet. PQAs were small programs usually 3K in size that a user could load on the Palm VII devices. The Palm VII shipped with a slew of PQAs installed, which included software from ABC News, MapQuest, USA Today, Ticketmaster, The Weather Channel, and *The Wall Street Journal.* In addition, users could download any of the other 400 applications provided by the Palm.Net services, allowing them to track stocks, schedule or reschedule flights, track UPS packages, find restaurants and hotels, find phone numbers, get directions, find ATMs, or look up words in the dictionary. Furthermore, Palm.Net encouraged developers to develop new applications, supporting developer programs.

Palm was very smart to use the pager network provided by the BellSouth Data Network, because it was widely available in major cities. However, the page networks offered very small bandwidth, at very expensive prices. The original Palm.Net services were offered at $9.99 for the 50 K of data per month or $24.99 for a roomier 150 K per month and, depending on the amount of data queries made, for many users the bills went up into the hundreds of dollars in just one week of normal use.

In addition, the early device design was also clumsy and inefficient. The unit required AAA batteries, which normally lasted a week at most and added additional maintenance cost to the device. Preloaded PQAs provided only generic services, and to download additional PQAs, Palm VII users had to establish very expensive data connection to the Palm.Net servers. Further, the 2 MB available memory on the device did not provide much room for the PQAs, and users often found themselves short of memory. Mainly, the price of the device was set at a very high $599. Yet, Palm still got a lot of applause for launching the Palm.Net services and paving the road of the early mobile Internet services.

18.5 PRIMITIVE DIGITAL DATA OVER PACKET-
SWITCHING NETWORKS

In the early 1990s, the cellular carriers concerned with the decreasing revenue per subscriber saw a considerable opportunity in the provision of the mobile data services. The operators saw the increasing demand for Internet services and growing trends for mobile devices supported by virtual and mobile corporate working environments. In 1992, all of the leading cellular carriers formed a group to develop a digital service that was in line with the Internet protocols to provide data. It was to become CDPD (Cellular Digital Packet Data) and it was designed to address critical mobile data issues such as roaming, billing, security, and authentication.

CDPD was intelligently designed to use spare radio channels in the AMPS spectrum to carry data in packet form (IP packets). When the end user created a request to send or receive data, the data was segmented into small sequence-numbers packets by the modem, and sent separately on different paths toward the nearest modem, where the receiver assembled the packets according to the sequenced order. User charges were typically based on the number of packets transmitted and received, but some carriers offered flat rates with unlimited data. The maximum data rate of the CDPD data transmission capped at 19.2 kbps.

The goal of the CDPD service providers was to offer nationwide, seamless, wireless data service, combining the services provided by multiple carriers through appropriate intercarrier and partnership agreements. Among the carriers participating are Ameritech Cellular, AT&T Wireless Services, Bell Atlantic, NYNEX, Mobilem, and GTE Mobilnet (PCSI). In addition, some major equipment manufacturers have participated in the CDMA initiative, including Hughes Network Systems, Motorola, Inc., and Sierra Wireless, Inc. Ten years after its conception, CDPD was found in over 209 markets, including 123 metropolitan areas, 43 rural areas, and 43 international markets, with coverage extending to nearly 39 million people in the United States, almost 55 percent of the population. GoAmerica, OmniSky and Tellus were all wireless service resellers using the same CDPD hardware and network configurations from the leading carriers. The differences were in the included software, the customized Web subsets that each offered and, of course, the price of the plans.

The primary advantage of the CDPD wireless Internet was the full Web-browsing capability, and not just Web-clipping services. Not only did CDPD offer raw data rates of 19,200 bps, but also it provided full-duplex communications, allowing a radio modem to talk and listen at the same time. This allowed CDPD to handle real-time interactive applications that competing packet networks like RAM and ARDIS could not support due to their half-duplex nature. An ARDIS or RAM radio modem must switch between transmit and receive, taking up valuable time.

Another packet switching network, ARDIS, started out in the 1980s when Motorola built a custom solution for IBM's nationally distributed technical field-service crew. In the early 1990s, when packet-switching technology caught the eye of fast-growing cellular companies, IBM tried to reposition ARDIS as a public wireless data network, but never attained the mainstream appeal it was looking for. In 1998, it sold its entire majority position to American Mobile Satellite Corporation (AMSC), which soon was renamed Motient. With a 19.2-kbps access architecture that has a presence in 430 of the top 500 U.S. wireless markets, Motient had inherited substantial network assets. The slow acceptance of wireless data overall was a mixed blessing for the company, which had the most success in the corporate world, especially in the financial verticals.

Motient played its cards right when it teamed up with a Canadian company, Research in Motion (RIM). Together they created something of a wireless phenomenon with Blackberry devices, which put the power of the "always-on" e-mail into a form factor as small as a wireless pager. RIM worked well because it was a small device with long battery life and great usable design for its purpose as a mobile e-mail device. But end-user needs evolved, the expectations were changed by the introduction of high-level color Pocket PC devices with larger, easier to read and browse Internet screens, multimedia capabilities, easy synchronization with desktops, and even faster wireless capabilities. Motient was not able to realize revenues to cover the costs and was forced to file Chapter 11. Many blame it on the introduction of the new 2.5G and 3G networks.

Although the early packet-switching CDPD and Motient networks were a definite upgrade from the AMPS technology, wireless Internet services that it offered were still very primitive. The radio environment that CDPD and Motient relied on was just as delicate as the AMPS, and if the user was out of range of the base station, the radio connection could suddenly be lost. Applications and wireless Internet developers were forced to design an application that could handle intermittent connections, which increased the system development and maintenance cost. Performance was another important issue. With the channel rate of 19.2 kbps, the actual throughput to the end wireless Internet user was averaging 10 kbps. Moreover, CDPD and Motient networks were still too expensive to be widely accepted by the end users in the consumer and enterprise market.

Yet, these early packet-switching networks whet the appetite for the wireless Internet in the consumer and enterprise markets. Carriers definitely caught on the interest that the wireless data services instigated; however, they realized that in order to maximize the return on investment, wireless Internet had to become faster and cheaper.

18.6 MODERATE SPEEDS OVER WIRELESS WANS

In 2000, Metricom surfaced with a nationwide advertising campaign convincing individuals to use their Ricochet wide area networks. The goal of the Ricochet system was to provide Internet access through wireless mode at moderate speed close to 100 kbps at competitive rates. The Ricochet packet radio network enabled coverage through the deployment of a large number of inexpensive packet radios on pole tops, which routed modem packets to a wired access point (WAP), which then connected to the wired Internet. Ricochet employed spread spectrum, frequency hopping technology across 160 channels in the license-free Part 15 902 to 928 MHz ISM band.

The company was founded in 1985 by Paul Baran, who in 1962 helped create the Internet. Metricom was able to attract significant capital investment, including over $500 million from MCI WorldCom and Microsoft cofounder Paul Allen, who once talked Bill Gates into dropping out of Harvard to begin a software company. Originally, Metricom provided utility companies with a way to automatically read gas and electric meters; however, the company soon changed course, choosing to offer wireless Internet access to mobile users. In 1999, the company began to upgrade its 28.8 kbps wireless network to 128 kbps speeds, more than twice as speedy as the fastest dial-up Internet connection. Metricom managed to wire 17 cities/markets, including Manhattan, San Francisco, and other major metropolitan areas, totalling about 30 million people under the coverage umbrella.

The first version of Ricochet, sold by Metricom through its Web site, operated at 28.8 kbps and cost subscribers $29.95 per month. The faster system was later sold through a number of providers, such as MCI WorldCom, Juno, SkyTel, and UUNet, and cost subscribers between $60 and $100 per month. Modems were priced between $220 and $250. While most users enjoyed the system's excellent coverage within the 17 metropolitan areas in which Metricom operated, and found the performance to be adequate for most Web-based applications, Metricom was never able to overcome negative market perceptions. By the end of the first quarter of 2001, Metricom counted only about 40,900 subscribers. Among the long laundry list of problems were limited coverage, very high costs to expand infrastructure, submegabit performance, and consumer price sticker shock at $80 per month.

It was clear that Metricom's wireless Internet access product was viable, and its Ricochet service offering provided the fastest mobile wireless communications solution on the market at the time, but unfortunately for Metricom, the company turned into another lesson in mobile Internet history, filing for Chapter 11 in late 2002. Consumer markets, which Metricom heavily targeted, were not willing to pay the $80 per month, and Metricom could not find anyone to finance the already-high $1 billion debt accumulated only 2 years after the new mobile Internet services launch.

Other experts in the field suggest that the slow launch and slow spread of Metricom wireless Internet access services is not a problem of the technology and its high cost, but unsuccessful execution by Metricom's executives. In other words, the technology was ready for launch and the right customers existed, but the company failed to find the right market and distribution strategy. Many blame Metricom's dismal performance on the company's wholesale model, which left Metricom at the mercy of a handful of resellers, such as WorldCom and Juno. Others said Metricom was going after the wrong customers the whole time. Metricom was trying to convince consumers to use mobile Internet services in place of their broadband Internet connection services in their homes and offices. The problem was that those wired Internet services were cheaper and offered higher speed at that time. Metricom would have been more successful in positioning their service as a mobile extension and field service for mobile workers. So often the success or failure of the mobile Internet services depended not on the quality and cost of the technology, but company execution.

18.7 2.5G: HALF-STEP FORWARD TO WIRELESS BROADBAND

The switch from the circuit-switched networks to packet-switched networks provoked the carriers to heavily invest in another new generation technology: 2.5G. Based on the digital transmission protocols, the 2.5G is not a single wireless standard, but a collection of several. Bolting on to existing 2G infrastructure built on the operational GSM, CDMA, TDMA, and PHS standards among others, 2.5G CPRS and CDMA2000 standards are expected to provide faster data speeds up to 171 kbps.

However, among the most-important attributes that 2.5G wireless Internet technologies can offer is wide area coverage. GSM/GPRS is already available in most of the United States and is widely available throughout Asia and Europe. AT&T expected the national rollout of GSM/GPRS to be completed by the end of 2002. If all goes as planned, in the near future mobile phone users will be able to dial from their phones anywhere they travel without worrying about coverage areas. Wireless carriers are putting major roaming agreements in place that will break down the regional use barriers. GSM and GPRS are fairly easy and affordable for the wireless carriers to deploy, and almost become necessary for their survival. After all, maximum throughput will not matter if users cannot access the network. Carriers are expected to spend some $2 billion over the next several years, with about 95 percent of this spending earmarked for GPRS.

Although GPRS and CDMA2000 definitely offer higher speeds, they still do not provide the true broadband that mobile end users and mobile content providers expect. While specifications suggest data rates of 144 kbps for CDMA2000 and 171 kbps for GPRS, these speeds are theoretical maximums. The important thing to keep in mind is that the maximum 171 kbps throughput is for an entire channel, and each channel has to support multiple callers, and within each channel there are multiple frames, and within each frame there are eight time slots. So, in reality users usually hit the maximum throughput of 33 kbps. Yet, 2.5G has already created success stories in the mobile Internet services area. In August 2002, Audiovox launched Thera in concert with Verizon Wireless — the first American Pocket PC with a built-in phone. Thera was the first PDA with the built-in connectivity to one of the fastest next-generation wireless networks in the United States, offered by Verizon Wireless. The Verizon Wireless Express Network, using the first phase of the CDMA2000 technology was designed to provide effective data rates of 40 to 60 kbps. With Thera, users can make phone calls, access e-mail and the Internet or network data, as well as use mobile services applications on demand.

For the most part, at the time of this writing, the 2.5G standards are looked upon as an intermediary step on the way to the true fast-speed wireless Internet access promised by 3G technologies. Yet, there are many positives that carriers and wireless Internet service providers can capitalize on. Some believe that in its glory, 2.5G may prove to be enough for the consumer, given that the right application will be offered. For consumers, technological advancement is not the main driver for adoption. Instead, viable applications and ease-of-use are two main prerequisites for mass-market acceptance. 2.5G does offer many features to the end user that make it more usable. With Instant IP access or "always-on" service offered by GPRS, users no longer need to dial up every time they connect to the wireless data network. They can be instantly notified of new messages or information according to their own preset preferences. In addition, the "always-on" feature can be used to add location/proximity and personalization services to customers. Moreover, based on the packet-switched versus circuit-switched data technologies, with 2.5G users will pay for data volume instead of air time, offering better value. Finally, unlike other earlier technologies, Metricom's Ricochet being a good example, 2.5G rollout is timed well with other supporting technologies, such as location services through GPS and network-based location; biometrics offering personalization; miniaturization allowing integration of more memory, energy, and processing power in portable devices; voice recognition offering easy access and interface; Bluetooth; Wi-Fi; and others. Many see 2.5G as a great market experiment powerful enough to open new business models, new entrants, and a whole slew of new business and consumer products and services.

18.8 I-MODE: WIRELESS INTERNET PHENOMENON

In 1999, Japanese NTT DoCoMo launched the i-mode, which is today considered the first wireless Internet service on the way to the 3G mobile systems. i-mode began as a "WAP-like," text-based mobile information service provided by NTT DoCoMo. By 2001, there were over 1600 DoCoMo endorsed i-mode sites; some were free, others charged up to $2.50 per month. DoCoMo handles the billing for the official sites and keeps 9 percent of their revenues. In addition, as of 2001 there were over 40,000 "unofficial" i-mode sites. i-mode is a brand, not a technology. The technology is packet-switched overlay, as opposed to circuit-switched digital voice, and offers "always-on" and "on-demand" access to the Internet without users having to dial up. The technology offers high-speed transmission of data at a reasonable cost. The transmission rate is currently only 9.6 kbps, but i-mode already offers multimedia applications, well-suited 3G devices with color displays, sound, and other multimedia-supporting features, and a common billing system for all service subscribers.

The i-mode mobile Internet access service has enjoyed phenomenal success in Japan, winning more than 12 million subscribers in a year and a half after its launch and reaching an unbelievable 23 million subscribers by mid-2001, surpassing the number of fixed line subscribers.

i-mode wireless Internet service offers a broad variety of consumer services, including entertainment (games, download of music and ring tones, horoscopes, karaoke), multimedia messaging, information (news, weather, market quotes, transportation schedules), financial services (bank statements, money transfers, bill paying), database queries (phonebooks, dictionaries, restaurant guides, city information), and M-commerce (movie tickets, shopping, video rentals). i-mode has attracted 700 partner and 30,000 nonaffiliated content/applications providers to its platform, equally to as many as WAP content providers throughout the world. Contrary to popular misconceptions, i-mode does not attract the youth market only; a mere 7 percent of subscribers are teenagers, although their revenue per subscriber is higher. E-mail, messaging, and voice are still the driving applications for i-mode.

Several factors, including i-mode's design, content strategy, business model, and technology, have contributed to its success. Simple and functional handsets with easy-to-read screens, easy navigation through content, ability to prioritize and personalize most-popular content gave users easy access to wireless data and services. Flexible billing systems did not stop end users from using the services, and i-mode in return capitalized on the transaction service fees. Content providers were encouraged also to provide more services. There were no slotting fees, and anyone could become a partner, but only the most-attractive content providers were bound to receive premium placement.

i-mode is evolving and getting ready to jump into the next 3G stage. There are speculations that it will make its way into the U.S. market, but as things stand now, there are many cultural and technological differences between the U.S. and Japanese markets that will have negative effects on i-mode implementation in the United States.

18.9   3G: REDEFINING WIRELESS INTERNET SERVICES

Most wireless carriers treat 2.5G as a short-term solution toward the ultimate high-speed 3G mobile data networks. The visions of 3G networks are still evolving, growing in both scope and complexity, considering that it is being defined by all members of the wireless value chain, including network operators, service providers, equipment manufacturers, government agencies, and others. The broad definition of 3G is focused on the global telecommunications infrastructure that is capable of supporting voice, data, and multimedia services over a variety of mobile and fixed networks. Multimedia support is perhaps the largest, most-important differentiator of the 3G networks from its wireless data networks predecessors. Among the key objectives of the 3G networks are high data-transmission rates from 144 kbps in high mobility context to 2 Mbps for stationary wireless connections, interoperability with fixed-line networks, worldwide roaming capability, common billing/user profiles, location services, and ability to support high-quality multimedia services.

While the switch from the 1G cellular to the 2G digital networks was far more noticeable from the technology point of view, with the industry focusing on adjusting to a major technological paradigm shift, the move from the 2G to the 3G networks is still a little blurred, with the industry focusing more on the qualitative service provision characteristics, and thus making it harder to agree on specific quantitative standards. In reality, the promise of 3G does not lie in the technical sophistication of the system, but rather in the benefits that consumers and providers are hoping to derive from it. The benefits of 3G to consumers focus primarily on two dimensions: convenience and cost. With 3G services in place, consumers will obtain access to wider quantity and variety of information and applications from their mobile devices. The 3G devices that will enable access to the 3G services will be enabled with much-richer multimedia features, location-based services, and other instrumental functions that will allow the end users to have the best possible experience with the broadband connection and plethora of content that 3G networks will be able to offer.

Economically, 3G services will be more reachable for mobile end users. One of the main complaints from the end users of the previous mobile Internet services based on the CDPD or Metricom technology was very high cost associated with data transactions. 3G systems are being designed to get the most efficient use of the spectrum, and the tight competition created in the 3G services providers' field will most likely result in lower costs and prices.

Three classes of 3G networks are expected to emerge: EDGE, W-CDMA, and CDMA2000. What technology-specific carriers will choose will mostly depend on the type of current networks that the carriers have. AT&T and most likely the SBC/BellSouth joint venture will follow the EDGE network, which is built on GPRS. Sprint and Verizon, currently using CDMA, are planning to move to CDMA2000.

W-CDMA is the standard that will most likely be implemented in Europe and Asia, and will come to the United States only if the Asian or European telecoms will move to the U.S. market through mergers and acquisitions. Unlike CDMA2000, W-CDMA is not backward compatible with the 2G CDMA networks. The incompatibility ensures that Japan and Europe will move to 3G more quickly than the United States. In addition, 3G rollout in the United States will be slower, considering the fragmented market lacking the nationwide infrastructure that Europe and Japan have developed, the widespread and lower-density concentration of mobile users (due to most Americans living in rural areas), and the patched network with at least three competing standards.

Japan will lead the 3G revolution, with the 3G services rollouts by NTT DoCoMo and its competitor J-Phone. In Europe, the 3G rollouts are planned for the 2003–2004 timeframe. Vodaphone plans to offer 3G services in the United Kingdom and H3G is planning to start providing 3G services in Italy by the end of 2002. In the United States, both Sprint PCS and Verizon Wireless have announced plans to roll out 3G services using CDMA2000 technology in late 2002 or early 2003. However, 3G technology carries certain technological implications associated with the network's upgrade that could slow down the 3G conversion by a year or two.

18.10 HIGH-SPEED WI-FI: A DIFFERENT TYPE OF WIRELESS

Despite the excitement created by the 2.5G and 3G connectivity standards for wireless data services, it is impractical to expect that the 3G revolution will happen tomorrow, mainly because the 2.5G and 3G technologies are costly, scarce, not well tested, and still being defined. However, the market is already filled with mobile devices, such as laptops and PDAs. According to IDC, close to 17 million handheld companion units were shipped in 2001. Palm and PocketPC devices experienced great success in the enterprise and consumer markets. At the same time, people realized the efficiency gains brought by the Internet and access to the vast amounts of organized data that it provides. As the workforce is getting more mobile and people are realizing the benefits of receiving instant information, the demand for "anytime, anywhere" access to corporate and personal data is bound to increase.

A new wireless standard came into play. 802.11, also called Wi-Fi, has become the most popular standard for wireless Internet access technology. Using radio frequency connections between a base station and devices with add-on or built-in 802.11 wireless cards at roughly 1000 feet radius, Wi-Fi gives access to the Internet and remote corporate and personal data without using the wires and cables of a conventional local area network in public places, homes, and offices. The global push to adopt 802.11 is based largely on its high bandwidth of 11 Mbps and rich user experience that is comparable to being on a wired company LAN. This standard is open, unlicensed, internationally adopted, interoperable, and supported by every major player in the wireless LAN industry. Wireless Ethernet options are available today for most consumer devices, and the next generation of laptops, handheld PCs and PDAs will be wireless Ethernet enabled.

Enterprises have taken the most-prominent role among the early adopters of Wi-Fi wireless LAN technologies. Vertical markets and enterprises accounted for the majority of shipments and will continue to do so. Wi-Fi technology serves as a practical extension to existing broadband and high-bandwidth wired LAN technologies. As enterprises become more convinced that wireless LAN technology adds hard and soft dollars to the return on investment, and as Wi-FI devices reach the IT market at lower costs and in larger shipments, the industry will see a huge increase in 802.11 adoption in corporate offices, plants, campuses, and other premises. Public access is tagging right behind the corporations, and in the beginning could even outrun the corporations while they are ramping up. The main venues for public 802.11 access points include coffee houses, with Starbucks leading the pack; hotels with, Four Seasons and Hilton as the earliest adopters; airports; train terminals; restaurants; and universities.

As in the telecom business, distinct camps of players have formed quickly to take advantage of the unlicensed frequency that Wi-Fi services are using. On the smaller scale side of business, a number of wireless network companies, also called "microcarriers"by the industry tycoons, are actively building 802.11 networks in public spaces installing equipment and leasing space from the landlords. Three-year-old wireless LAN service provider Wi-Fi Metro Inc. expanded on the "hot-spot" concept, providing a large area of wireless Internet connectivity unrestricted by physical boundaries. The first hot-spot covers roughly an eight-block area of downtown Palo Alto, allowing Wi-Fi Metro subscribers to log on whether they are in their favorite cafe or out on the sidewalk.

Then there are service aggregators, who purchase from 802.11 microcarriers on a wholesale basis, integrate these networks together, and sell a single service to customers. Boingo, who at launch had the largest wireless broadband footprint in the world, focuses on the complex integration of hundreds of Wi-Fi wireless Internet providers around the world into a single service, providing marketing services, customer support, and billing. On the larger scale, this market of course will not be missed by the carriers. VoiceStream, who recently acquired MobileStar and took over its large network known for offering the Wi-Fi services at Starbucks, is the first carrier to move into the Wi-Fi space. Under the name of T-mobile, VoiceStream started to offer wireless Internet services in California and Nevada, and plans to be in 45 of the top 50 U.S. markets, following with similar branding campaigns of T-Mobile International's subsidiaries in Germany, the United Kingdom, Austria, and the Czech Republic.

The popularity of laptop computers and handheld devices is fueling demand for wireless LANs. Many manufacturers, such as IBM, Toshiba, and Sony, are shipping laptops with built-in Wi-Fi hardware, allowing these machines to connect to a WLAN straight out of the box. IBM has become a leader in constructing wireless LANs, using its unrivalled size to capture market share through its global services division. Already, Microsoft's Windows® XP operating system supports Wi-Fi, and Microsoft announced plans to make a wireless portable monitor that uses Wi-Fi technology to link to the terminal and keyboard.

All this is increasing consumer awareness of WLAN products, accelerating chip sales, and creating demand for WLAN infrastructure. Poor market conditions and the lack of next-generation handsets, which has forced mobile operators to delay the launch of their 3G networks, also gives a boost to Wi-Fi.

18.11 APPLICATIONS ARE KEY TO WIRELESS INTERNET GROWTH

Just like the Internet strategy became indispensable for companies in the early 1990s, wireless strategy is becoming more important for businesses of all types, from small home-office operations to large Fortune 100 companies. Declining prices for wireless access and services, changing socioeconomics supporting transformation to an information-based society, Internet penetration offering users real-time information, handheld devices becoming mainstream, increasing use of mobile phones, higher transmission rates and bandwidths, introduction of new bandwidth-intensive mobile data applications, and convergence of fixed and wireless communications platforms — all contribute to amplified wireless Internet adoption.

Following the build-out of the mobile Internet infrastructure, new mobile applications will drive unprecedented growth. However, content providers have already discovered that the mobile Web is not the same as the desktop Web, and unfortunately the wheel will have to be reinvented in the wireless Internet services implementation. To give an example of how different the conventional desktop Internet is from the wireless Internet, it is worth analyzing the most-important premise that both services are built on. One of the great things about the conventional desktop Internet is that it disregards location, making the same data accessible no matter where the customer is logging on. Wireless Internet, on the other hand, will become heavily reliant on location, offering services and data based on the customer's location.

As in the case of the conventional Internet, before the users will be able to fully understand the value of the wireless Internet, applications will have to be built that offer improved or new ways of accomplishing day-to-day tasks, offer entertainment, and make work and business processes more efficient. End-user surveys show that among the most-useful wireless Internet applications are e-mail; location-based directions and mapping; location-based Yellow Page services; content delivery, including stocks, news, sports, and weather; instant messaging; and receiving discounts and promotions based on location. Location, of course, plays a very important role.

A new concept of the wireless Internet Services has started to evolve in the last few years. Qualcomm saw enormous opportunity in wireless Internet services, and debuted the new wireless development platform, Qualcomm's Binary Runtime Environment for Wireless (BREW). BREW is an open, end-to-end solution that provides tools services for applications developers, device manufacturers, and network operators to lower time-to-market barriers and efficiently develop, deploy, buy, sell, manage, and maintain wireless data applications.

Developers use BREW to build wireless applications quickly, spending minimal resources. Operators use the BREW solution to deploy, manage, maintain, and support applications; to provide applications discovery services; and to bill users. BREW reduces costs and risk to network operators and enhances their operational efficiency by lowering infrastructure and integration costs, reducing time-to-market with an end-to-end solution, and increasing operational efficiencies for operators. In early 2002, Verizon Wireless launched the BREW application services, and immediately saw a 9-percent increase in average data revenue per user.

For mobile phone consumers, the built-in "Mobile Shop" offered with the new BREW-enabled phones allows users to easily find, add, and remove applications with just a few clicks. Applications written for BREW offer excellent graphics, speed and action, and real-time interactivity. Already, mobile users can play a game of golf, file an expense report, access Zagat's Restaurant Guide, find a destination map, and get directions, while sitting on the train, in the cab, walking to the office, strolling in the park, or sitting at the beach. BREW offers a categorized search, making it easier for end users to find any application they need.

As was proved by i-mode in Japan, platforms such as BREW are key to bringing wireless data to the masses in the same fashion that Yahoo!, AOL (the first Internet Service Provider to offer consumer services in the United States), CompuServe, and other Internet service providers were instrumental in delivering the Internet to the masses. Carriers see that as well: Verizon Wireless is already offering the 1G black-and-white and color Internet-enabled phones to consumers, and is the first carrier in North America to offer downloadable applications to consumers nationwide. Applications consumers can download over the air on a phone are available nation-ally, and Verizon believes that these services will help it reach its goal to morph the wireless phone into a valuable resource for consumers who want up-to-the minute information to help them manage their life.

19 Wireless Technology Impacts the Enterprise Network

Andres Llana, Jr.

CONTENTS

19.1 INTRODUCTION

Wireless technology has been with us for many years; however, the application of this technology did not begin a very real advance until the mid-1990s. Much of the success of this technology can be traced to the rapid deployment of wireless technology in European countries. In these areas, the deployment of wireless local loop (WLL) systems made it possible to provide an alternative to the lack of a dependable copper infrastructure. In some countries where subscribers waited years for a telephone, the availability of wireless technology reduced the wait time to weeks. Later, as GSM networks began to proliferate, the concept of greater mobility (i.e., mobile handsets) enabled many more subscribers to move onto the public network without

0-8493-1502-6/03/$0.00+$1.50
© 2003 by CRC Press LLC

the requirement for even a terminal in their homes, as was the case with the WLL systems.

Due to the growing penetration of cellular services, the International Telecommunications Union projects that by 2008 there will be more mobile than wireline subscribers, perhaps as many as a billion cellular subscribers. The fast-paced growth in global wireless services has greatly impacted the expansion of wireless data communications. This is not to say that wireless systems in support of data communication requirements have not been around for some time; it just was not embraced as an enterprise network solution.

However, with the success of wireless technology in European countries and around the world, more viable wireless solutions have made their way into the marketplace. Broadly speaking, the driving forces for change can be seen in the growth of the Internet, increased user mobility, and pervasive computing, where computer chips now play a greater role in the monitoring and control of various service devices. Mobile telephones and pagers have accomplished a great deal in supporting the remote worker's requirement for maintaining a meaningful information exchange with corporate headquarters. Applications such as voice messaging, online fax, and online information access have driven wireless data transmission to the next tier. These applications have served to give the new-age "road warrior" a definite advantage as a remote worker.

19.2 WIRELESS COMMUNICATIONS

Wireless communications in the United States extend back to the early 1950s when the Rural Electrification Administration (REA) sought ways to provide telephone service to remote farms and ranches. Early efforts bore little fruit and, as late as 1985, the REA was still trying to get a system into operation. However, by the mid-1990s, a rush of new products resulted following the successful deployment of global analog cellular mobile telephone service. The most common form of wireless telephones came with the application, the CT-10 cordless telephone and later the CT-2 digital phone. Wireless internal telecommunications became fairly commonplace when AT&T, Ericsson, Nortel, NEC, and Rolm introduced wireless adjunct systems for installed PBX systems. These adjunct systems linked to a PBX via separate station line cards based on the standard 2500 nonelectronic desk telephone (see Figure 19.1). These add-on systems supported an RF controller that used the ISM (Industrial-Scientific-Medical) 900-MHz frequencies. Remote RF controllers were positioned around the user's premises to receive transmissions from roving users with mobile handsets.

Today, AT&T (Lucent), Ericsson, NEC, Nortel, and Rolm (Siemens) have introduced an entirely new generation of wireless PBX products that allows the end user to establish a totally integrated wireless voice and data network. For example, Lucent has introduced its Definity Wireless Business Systems as well as the TransTalk 9000 system. This latter system can be either a dual-zone or single-zone system and can support up to 500,000 square feet. A similar two-zone system can be used to support

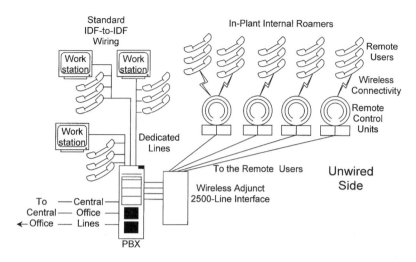

FIGURE 19.1 Wireless internal communications.

a multilevel building or a combination of several closely coupled buildings (i.e., warehouse, manufacturing, etc.). The Definity Wireless DECT (digital enhanced cordless telecommunications) system, which operates in the 1880- to 1900-MHz range, has similar capabilities and is marketed outside the United States. The Nortel Companion system is a similar wireless system that works off the Meridan I (Option 11 C) system. The Companion system supports all of the same station features as found on a standard electronic desk telephone.

These new wireless PBX systems can be integrated directly to the corporate LAN or WAN and function as centralized communications servers. For example, the Ericsson MD 110 system, when configured with an IP gateway unit, serves to interface the MD 110 PBX to an IP network, allowing voice traffic to share bandwidth with data over the IP network (see Figure 19.2).

19.3 WIRELESS OFFICE SERVICES (WOS)

Office complexes, manufacturing warehouses, and other facilities that are spread out and supported with disbursed populations of employees offer ideal opportunities for the wireless service provider. Motorola, which introduced its M-Cell GSM (Global System for Mobile Telecommunications) access product at the 1998 GSM World Congress, was able to provide attendees with support for over 16,000 calls during the three-day conference. This system is essentially an internal telephone system that functions like any other PBX system except that it is supported by a localized GSM wireless network operator. In this environment, building distributed RF (radio frequency) units linked to cluster controllers support internal interoffice calling. When a user leaves the office, his or her calls are then seamlessly linked via the local GSM wireless network. Once General Packet Radio Service (GPRS) support is added to the network, nonvoice services also can be supported.

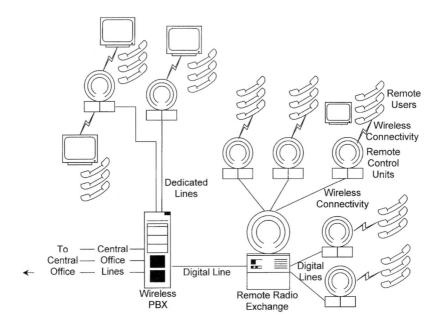

FIGURE 19.2 Wireless PBX system.

In the United States, service providers are now pursuing wireless office services (WOS) as a new market niche. In this environment, the service provider establishes a distributed radio system (DRS) throughout the office or multitenant facility in much the same way that a PBX system or wireless LAN is configured. In this scenario, mini base stations (MBS) are interfaced to distributed antennas (DAS), forming the basic infrastructure. The MBS units are linked in much the same way as in building data networks, which in turn are linked to a central radio. The advantage of these carrier-provided solutions is the transparent mobility of the end users in the system. While in the building or corporate facilities, the end users do not incur any per-minute billing; however, once they leave the premises, they are treated like regular mobile users and billed accordingly.

In this arrangement, the end user is never out of touch and always within reach, as one assigned telephone number follows the end user both on and off premises. Cellular One on the West Coast is currently offering this service in the San Francisco area.

Sprint has begun to offer wireless data service over its PCS network, which comprises over 11,000 base stations. This network exceeds the BellSouth Wireless Data service and ARIDS combined data networks. The Sprint data network will work through Sprint PCS smart phones, such as the Nokia, Motorola, or Qualcomm, that support smart set displays. Further, these smart sets, when configured with microbrowsers, can be used to access the Internet for e-mail and other abridged services. This new data service also provides access to stock quotes and other time-critical information. Kits are available to provide Internet access for laptops or PDAs at 14.4 kbps.

19.4 MORE INTEGRATION

Wireless data communications using a packetized data standard called Cellular Digital Packet Data (CDPD) has been getting more use as more wireless applications are being deployed. However, this service is limited to low speeds of 19.2 kbps or less, and has been implemented on D-AMPS IS-136 networks. CDPD technology serves to enhance the existing AMPS cellular infrastructure by detecting unused cellular channel space in which to transmit data. This allows the operator to maximize the use of the available physical cell site infrastructure.

While 19.2 kbps may seem slow, it does answer a broad requirement for low-speed transactions aimed at one-way data collection for meter or device reading. This application of CDPD has made it possible to offer many new data-collection-type applications for electric, gas, and water meter reading.

To meet the growing demand for wireless data applications, newer CDPD modems have made their appearance. For example, Novatel has introduced the Minstrel modem for applications with Palm computing devices. These modems have their own IP address and can be used to access the Internet. The modems also support a built-in TCP/IP stack that can be used for custom software development using the Palm OS®. The Minstrel modem is configured with SmartCode software, Hand-Mail™ and HandWeb™ software, and a modem management package. This new technology has resulted in a number of sales terminal applications, field technician applications, as well as mobile applications in transportation (e.g., fleet and vehicle management, public safety, and disaster recovery). Handheld terminal applications also have been aided by the introduction of Windows CE software configured with utilities such as Pocket Excel, Pocket Word, Internet Explorer, Scheduler, E-Mail, Calendar, and Task Manager. All of these packages allow mobile workers to become more efficient with their time while in transit.

19.4.1 WIRELESS LOCAL AREA NETWORKS

Some of the earliest wireless LAN products were slow by comparison with today's products. For example, in the early 1990s Motorola introduced a product that was developed around a microcellular design using the 18- to 19-GHz frequency. The system used an intelligent six-sector antenna, which was used for both data reception and transmission. The antenna supported a scanning system that was used to select the best transmission path from its associated terminal to the next terminal in the network. A high-performance RF digital signal processor was used to handle the modulation and demodulation of the 18-GHz carrier using four-level frequency shift keying (FSK). This would ultimately support 10 Mbps Ethernet, which was considered fast for the early 1990s.

Wireless LAN technology in the early 1990s was slow to catch on as many networks were hardwired; it was not until changes were made in office and facility arrangements that wireless technology gained acceptance. Because the early products were unlicensed, they could be used to cover short distances (several hundred feet) within buildings and under a mile between buildings. A good example of such a wireless network can be seen in the Jacob Javits Convention Center in New York.

In this application, a wireless LAN was tailored to cover 1.5 million square feet of convention center floor space. Distributed smart antennas, which act like mini base stations, are spread around the facility and allow transmission of voice and data throughout the facility.

19.5   A NEW STANDARD

In 1997, the IEEE standard for wireless networking was finally ratified, establishing an interoperability standard for all vendor products. Essentially, the 802.11 standard made it possible for companies to introduce a higher performing wireless LAN product that offers a degree of interoperability. These new products provide wireless connectivity starting at the mobile PC level and include products to interface a wired LAN with wireless desktop PCs and peripherals. Also new to wireless LANs are firewalls that protect against unauthorized access into the corporate LAN. These wireless LAN security devices are based on an IP network layer encryption using the IPsec (IP security) standards. Also incorporated as part of these systems are a range of authorization keys, authentication policies, and automatic security procedures.

As a group, 802.11 products operate in the 2.4-GHz ISM band with a bit rate of up to 2 Mbps and a fall-back rate of 1 Mbps. Many vendor products can go higher; for example, Ericsson introduced an 802.11 product line in 1998 that provides a data rate of 3 Mbps.

The Wireless Ethernet Compatibility Alliance (WECA) is developing a series of interoperability tests that will allow vendors to test their products to determine if they are interoperable. This is seen as a vital step toward ensuring that when the new 802 High Rate Direct Sequence (HRDS) standard (2.4 GHz at 11 Mbps) is agreed upon, the WECA will be able certify products for enterprise deployment. HRDS products have been announced by several vendors; for example, Cabletron has announced an 11-Mbps product for its RoamABOUT wireless LAN product line.

In some sectors, work is in progress on a HyperLAN/2 standard product that will support data rates of up to 54 Mbps. These devices will operate in the 5-GHz ISM band. Ericsson plans to offer a HyperLAN/2 product that will support an end-user data rate of 20 to 25 Mbps.

Many vendors now offer wireless bridges that provide the capability to link wireless LAN islands into a contiguous wireless/wired LAN network. Many of these devices operate in the ISM band and offer the network administrator a cost-effective means of linking remote "line-of-sight" locations for up to 20 km. A good example of such a class of terminals can be seen in Wireless, Inc.'s MicroLink microwave radio terminal. This device operates in the 2.4-GHz ISM band and supports two models: a low-end model at 64 to 256 kbps and a high-end model at 512 and 1024 kbps. The terminal can operate at distances of up to 20 km and integrates both voice and data traffic between locations (see Figure 19.3).

Other vendors with similar terminal products include ADTRAN, which recently introduced its Tracer terminal that will go up to 30 miles and support dual T1s. IOWAVE also provides a similar terminal that supports links of up to 20 miles for about $12,000 per link.

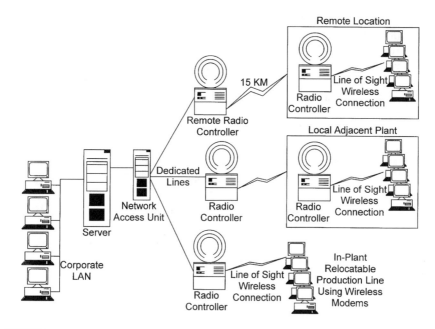

FIGURE 19.3 MicroLink microwave radio terminal.

19.6 WIRELESS INTERNET ACCESS

In some areas, broadband access to the Internet is gradually getting away from the ISDN or dial-up access model. This can be attributed in part to the Federal Communications Commission (FCC), which released 300 MHz of spectrum for the Unlicensed National Information Infrastructure (U-NII). The U-NII band is broken down into three bands: (1) 5.15 to 5.25 GHz for indoor application, (2) 5.25 to 5.35 GHz for campus application, and (3) 5.75 to 5.85 GHz for local access of up to 10 miles. This new spectrum has resulted in the introduction of a new generation of wireless Internet routers, also referred to as Internet radios. Internet radios can be set up on rooftops by an ISP to provide direct Internet access via the ISP Internet hub. These terminals can be configured in a point-to-point or a point-to-multipoint configuration. A good example of such terminals can be seen in Wireless Inc.'s WaveNet IP series, which can be used by an ISP to set up a point-to-multipoint Internet access arrangement completely outside of the public utility. By controlling the cost of local loop access, the ISP can offer better rates and higher-speed access. The WaveNet IP arrangement is sometimes referred to as W-DSL because a network can support DSL-like access with speeds of up to 512 kbps of symmetrical bandwidth.

19.7 BROADBAND INTERNET ACCESS

Broadband Internet access is now being offered via licensed 38-GHz Local Multipoint Distribution Services (LMDS) and Local Multipoint Communications Systems

(LMCS) license holders. These fixed wireless service providers are able to support fiber-optic network bandwidth without the physical fiber being in place. A good example of a broadband wireless LMDS system can be seen in the TRITON Invisible Fiber product line used to deploy a network of rooftop terminals in a consecutive point network.

These networks are capable of supporting a 20- to 40-square mile geographic area, providing local broadband service for an entire metropolitan area. MaxLink Communications of Ontario, Canada, has launched an LMDS service in Canada using a Newbridge LMDS system to offer IP over ATM. Home Telephone, a successful bidder in the 1998 FCC LMDS spectrum auction, is offering LMDS service in the Charleston, South Carolina, basic trading area (BTA), using the Newbridge LMDS system. A similar service is being trialed in San Jose using the TRITON Invisible Fiber product. Initially, this service will be limited to a select user group within an office park and expanded from there.

LMDS broadband services provide the enterprise network designer with a potentially more cost-effective option where broadband services are required to support multimedia, video, and IP data transport requirements.

19.8 WHO USES WIRELESS TECHNOLOGY?

Some of the largest users of wireless technology can be seen in the transportation and shipping industry; Federal Express and United Parcel are good examples. Another area is that of automated vehicle location systems that are supported through a combination of satellite and landline systems coupled with the Internet.

19.8.1 CONSUMER APPLICATIONS

A good example of a consumer-level system can be seen in the OnStar product being offered as an option with some high-end General Motors products, such as their Cadillac automobile product line. The OnStar system is combined with a cellular service and the GPS tracking system. The system provides a series of end-user services that includes travel directions, emergency road services, automobile enabling services, personal notification, and theft notification.

The OnStar system uses a GPS tracking device that is installed on the vehicle and allows the OnStar control center to locate a subscriber's vehicle. Through a cellular link with an on-board computer, the control center can detect if the car's airbags have been deployed. If so, the control center detects a change, and a call to the subscriber is made to determine if there is a need for assistance. The control center also can remotely open the car doors if the subscriber has locked himself out of the car.

19.8.2 TRANSPORTATION

Qualcomm offers a multilevel vehicle location and monitoring service for large trucking and transport companies. This service is supported through a combination of satellite, cellular, and landline services. Trucks with special roof-mounted units can be tracked and monitored anywhere within the United States and Canada. Monitoring includes truck system performance, loading and unloading events, as well as redirection of vehicles for new load pickups. Drivers are able to communicate with the control center via messaging or cellular wireless contact. Through landline contact with the Qualcomm control center, dispatchers are able to dispatch and manage all company assets deployed on the nation's highway network.

19.8.3 HEALTH CARE

A surprisingly large number of health care service providers have taken advantage of wireless technology. Good examples of the application of wireless technology can be seen at Austin Regional Clinic, Indiana Methodist Hospital, St. Joseph Hospital, Wausau Hospital, and Winthrop-University Hospital, to name a few. All of these facilities have essentially the same problem: getting to patient information, where and when needed. Many found that they had to take handwritten notes to the nearest nurse station and enter the information manually into a computer terminal. As a result, administrators had to come up with a more-efficient way to operate.

Austin Regional Clinic elected to supply its medical professionals with mobile handheld computers to record and retrieve patient information in real-time. These terminals were linked to the clinic's Novell Netware LAN using PCMCIA modem cards. A series of wireless distributed access points located throughout the clinic provided a direct link to the LAN via a corresponding link in the clinic's communications server. The portable computers used were grid pad, pen-based portables configured with application screens, and allowed medical professionals simplified data entry and retrieval. This system eliminated large amounts of paperwork, thus allowing the professionals to function in a paperless environment.

19.8.4 MANUFACTURING

In some manufacturing plants, sensors and programmable logic controllers (PLCs) are used to control many of the processes related to product manufacturing. In many places, these devices are hardwired into high-maintenance networks that need frequent attention. In many plants, these networks have been fitted with Ethernet interfaces as part of a plantwide LAN. However, many plant managers have found that they can refit with wireless adapter cards that provide an RF link to wireless access points located around the plant. These arrangements link the PLCs directly into the wired LAN and the server, ensuring timely monitoring of all devices.

Avon Products, Inc. faced an expensive problem in extending the LAN in a Chicago-area plant's factory floor. In this facility, production lines were not static and subject to regular reconfiguration. Furthermore, operator mobility required to support 50 production lines along 500 linear feet confounded the problem of rewiring print stations to support the operators with barcode labels. Instead of rewiring, a series of printers configured with wireless modems were set up to receive barcode label files from print servers. The plant has a series of distributed base stations (terminal servers) that are linked to the LAN and a host system that supports the wireless link between the wireless printers and the LAN. The print servers, which are linked to the LAN Ethernet, receive barcode files from a VAX computer. As product is being manufactured, barcode information can be sent to the appropriate print server, where it can then be routed to the proper remote wireless printer.

19.8.5 FINANCIAL

The Pacific Exchange (on the West Coast) and Hull Trading (headquartered in Chicago) both opted to deploy wireless terminals on the trading floor to simplify the trading process. Instead of walking to a static terminal to enter trade information, traders can now do that from their handheld terminals. This innovation permits much faster trades, while eliminating many manual steps and the reliance on handwritten notes.

19.9 SEARCHING FOR A WIRELESS SOLUTION

In planning for the migration to a wireless network arrangement, the planner must be certain of his or her plan. Wireless applications require antennas and base stations to receive and transmit wireless signals between a mobile terminal and a mini base station. The planner must be certain that antenna coverage can be established throughout the area(s) to be served by the wireless terminals.

While most wireless modems and RF base stations work, many may not be interoperable between vendor equipment. Because there are so many vendors offering products, the planner needs to be certain of the vendor's commitment to the market. Now that the 802.11 standard has been accepted, the planner should not consider proprietary systems to avoid early obsolescence; many products and vendors of the early 1990s that had great products are no longer with us.

Wireless network arrangements provide a great deal of flexibility, but the planner should limit the migration to a wireless network arrangement to those applications that will produce a reasonable savings in terms of reduced manpower.

Application software requires careful review because much of the software designed to function over a LAN with standard PCs may not work the same way with a laptop PC. Furthermore, because many mobile terminals are configured with Windows CE software, one needs to be aware of the differences and their interface to the LAN operating system. Where the opportunity for the application of wireless technology is limited, the planner may find opportunities for direct linking of facilities to avoid central office dedicated circuit costs for voice and data transport. With many of the newer systems on the market, the planner can gain greater reach than before to link company facilities. By working with an ISP provider, many times the planner can arrange for a rooftop Internet radio to link the ISP hub directly with the corporate network hub, thus providing much-higher-speed access to the Internet for corporate network users.

19.10 SUMMARY

Wireless technology has opened up a new range of possibilities for linking the enterprise network than previously available to the network planner. The keys to success are proper preplanning and selection of equipment, adherence to established standards with an eye toward the future, and the availability of future systems with higher throughput options. Careful alignment of applications software is another important issue as some tailor-made software may be necessary to link legacy applications.

Enhancing the corporate network to bring it into line with the state of the art should not be the end-all, but rather an opportunity to reduce operating costs and improve overall corporate productivity.

20 An Efficient WAP-Enabled Transaction Processing Model for Mobile Database Systems

G. Radhamani and Mohammad Umar Siddiqi

CONTENTS

ABSTRACT

Advances in computer and telecommunication technologies have made mobile computing a reality. In the mobile computing environment, users can perform online transaction processing independent of their physical location. As mobile computing devices become more and more popular, mobile databases have started gaining popularity. Hence, software applications have to be redesigned to take advantage of this environment while accommodating the new challenges posed by mobility. A new class of multidatabase (a collection of autonomous and heterogeneous databases) that provides access to a large collection of data via a wireless networking connection is called mobile data access system. The proposed WAP-enabled transaction model allows timely and reliable access to heterogeneous databases while coping with the mobility issue. This model is implemented as a software module based on the multilayered approach capable of supporting preexisting global users on the wired network in addition to the mobile users. Various mobility applications are discussed briefly and a simple illustrative application is described in detail. Performance of the proposed WAP model is evaluated through simulation using the Nokia WAP simulator.

20.1 INTRODUCTION

Advances in wireless communications technology make it possible to realize a data processing paradigm that eliminates geographical constraints from data processing activities. Multidatabase can be viewed as a database system formed by independent databases joined together with a goal of providing uniform access to the local database management systems (DBMS). The mobile data access system (MDAS) is a multi-database system (MDBS) that is capable of accessing a large amount of data over a wireless medium. It is necessary that the MDAS provide timely and reliable access to shared data. The transaction technique presented in Pitoura and Bhargava[1] is based on an agent-based distributed computing model. Agents may be submitted from various sites including mobile stations. There are several multidatabase concurrency control schemes such as site-graph locking[2] and V-Lock.[3] None of the reviewed techniques handle mobile transactions and long-lived transactions in a multilayered approach, and hence global transactions are not executed as consistent units of computing.

In this chapter, we propose a WAP-enabled transaction model for the MDAS that addresses the deficiencies that exist in the current literature. The model is built on the concept of global transactions in multidatabases. In Bright and coworkers,[4] the model uses a hierarchical structure that provides an incrementally concise view of the data in the form of schemas. The hierarchical data structure of the Summary Schemas Model (SSM) consists of leaf nodes and summary schema nodes. Accessing a lower-most node takes time in the hierarchical structure. However, these systems have not been designed to cope with the effects of mobility. The goal of this chapter is to present a WAP-enabled transaction model for the MDAS that handles mobile transactions and long-lived transactions in a multilayered approach.

Section 20.2 gives the background material on Wireless Application Protocol (WAP), mobile computing environment, and mobile database systems. Various mobility applications are discussed in Section 20.3. The WAP-enabled transaction model is introduced in Section 20.4. An implemented sample application of the proposed model is given in Section 20.5. Simulation results of the proposed model are presented in Section 20.6. Conclusions are given in Section 20.7.

20.2 BACKGROUND

The WAP architecture provides a scalable and extensible environment for application development for mobile communications devices. This is achieved through a layered design of the entire protocol stack. Security in the WAP architecture should enable services to be extended over potentially mobile networks while also preserving the integrity of user data. Figure 20.1 summarizes the WAP as a series of layers.[5]

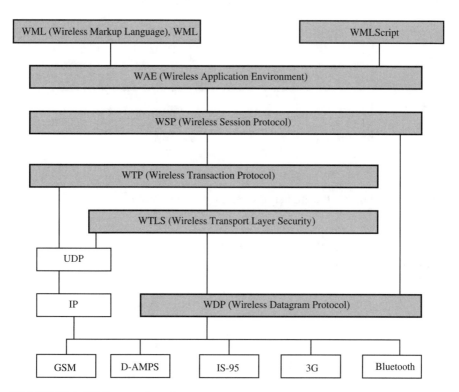

FIGURE 20.1 WAP protocol stack.

The Wireless Application Layer (WAE) is most likely concerned with the deployment of WAP applications. The Wireless Session Protocol (WSP) provides a consistent interface to WAE for two types of session services: a connection mode and a connectionless service. When providing connection mode, the WSP utilizes the Wireless Transaction Protocol layer. In the case of connectionless mode, the WSP takes the advantage of the Wireless Datagram Protocol layer.

Wireless Transaction Protocol (WTP) provides transaction services to WAP. It manages different classes of transactions for WAP devices: unreliable one-way requests, reliable one-way requests, and reliable two-way requests. Wireless Datagram Protocol (WDP) provides a consistent interface to the higher layers of the WAP architecture so that they need not concern themselves with the exact type of wireless network the application is running on. Among other capabilities, WDP provides data error correction. Bearers or wireless communication networks are at the WAP's lowest level. Additionally, each layer is allowed to interact with the layer above and below it.

According to WAP specifications, WTLS is composed by the record protocol, which is a layered protocol. The WTLS record protocol accepts the raw data from the upper layers to be transmitted and applies the selected compression and encryption algorithms to the data. Moreover, the record protocol takes care of the data integrity and authentication. Received data is decrypted, verified, and decompressed,

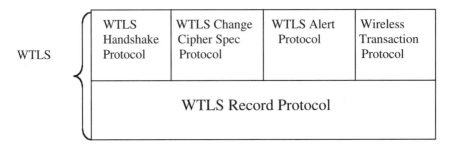

FIGURE 20.2 WTLS architecture.

and then handed to the higher layers. The record protocol stack is shown in Figure 20.2. The record protocol is divided into four protocols:

1. Change Cipher Spec Protocol (CCP)
2. Handshake Protocol (HP)
3. Alert Protocol (AP)
4. Wireless Transaction Protocol (WTP)

The change cipher spec is sent to a peer either by the client or the server. When the change cipher spec message arrives, the sender of the message sets the current write state to the pending state, and the receiver also sets the current read state to the pending state.[6] The change cipher spec message is sent during the handshake phase after the security parameters have been agreed on.

Wireless Application Protocol provides a universal open standard for bringing Internet content and advanced services to mobile phones and other wireless devices. Figure 20.3 shows the WAP architecture. Whenever a mobile phone uses WAP, a connection is created via Wireless Session Protocol (WSP) between the mobile phone and the gateway. When the user enters the address of the WAP site, the gateway is sent a request for the device's microbrowser using WAP.[7] The gateway translates the WSP request into a Hypertext Transfer Protocol (HTTP) request and sends it to the appropriate origin server (or Web server). The Web server then sends back the requested information to the gateway via HTTP. Finally, the gateway translates and compresses the information, which can then be sent back to the microbrowser in the mobile phone.

The mobile computing environment is a collection of mobile units (MU) and a fixed networking system.[3,8] A mobile unit is a mobile computer that varies in size, processing power, and memory, and is capable of connecting to the fixed network via a wireless link. A fixed host is a computer in the fixed network, which is not capable of connecting to a mobile unit. A base station is capable of connecting with a mobile unit and is equipped with a wireless interface. They are known also as mobile support stations (MSS). Base stations, therefore, act as the interface between mobile computers and stationed computers. Each MSS can communicate with MUs that are within its coverage area (a cell). Mobile units can move within a cell or between cells, effectively disconnecting from one MSS and connecting to another. At any point in time, an MU can be connected to only one MSS.

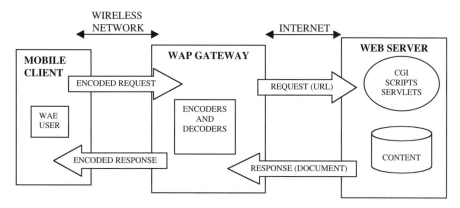

WAE - Wireless Application Environment
URL - Uniform Resource Locator
CGI - Common Gateway Interface

FIGURE 20.3 WAP architecture.

Mobile databases are gaining popularity and are likely to do quite well in coming years as portable devices become increasingly popular. The databases are connected to MSS via the fixed network and wireless digital communications networks. The architecture of the mobile database system[9] is shown in Figure 20.4. Many new database applications require data from a variety of preexisting databases located in various geographical locations. Users of mobile computers may frequently query databases by invoking a series of operations, generally referred to as a transaction.

20.3 MOBILITY APPLICATIONS

Mobility applications are applications that handle transmission of data and user interaction between wireless devices and a central repository. Broadly, we can divide the mobile applications as synchronization, real-time, and hybrid applications.

- Synchronization applications (SA) handle data transfers in situations where the user is not connected to the central database on a real-time basis. Instead of an instant data transfer, the user occasionally synchronizes the data residing on a mobile device with either a PC application or (potentially) a server. In this scenario, the user must schedule when the synchronization occurs and must be connected to a network of some type at that time. SAs work with a thick client that contains most or all of the data. During the synchronization process, the mobile device receives new or updated data, or the device sends its data to the PC or server. SAs are clearly very effective because they allow the user to access and carry critical information while on the go.[10] Synchronization in many cases should be the preferred means of data transmission, given the current wireless Internet networks' limited bandwidth and connection reliability, especially for business applications that are not time critical or that do require large amounts of data to be transferred back and forth from the device.

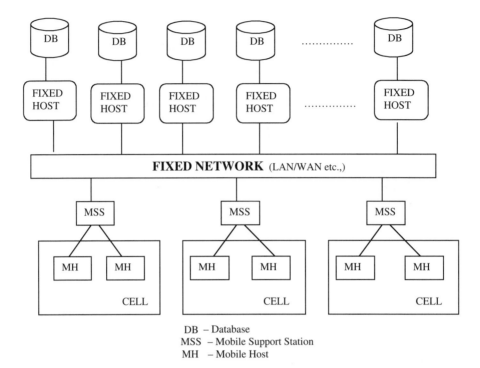

DB – Database
MSS – Mobile Support Station
MH – Mobile Host

FIGURE 20.4 Mobile database system architecture.

- Real-time applications connect in real-time only. Examples include WAP connections over cell phones, PDAs with wireless Internet connectivity, or applications that are available if and only if there is a connection at the exact moment that the information is needed. The distinction here is that limited data is kept on the mobile device and the data that the user wants to see is accessible only at the time that a reliable network connection can be established. This type of application is the one we commonly think of when using the term mobile commerce, as the term implies that the activities are happening in real-time exclusively. Many applications that contain time-critical elements and low data volumes, such as financial trading applications, bidding in auctions, and status tracking, require such connectivity to function reliably and meaningfully.
- Hybrid applications are only sporadically connected. These applications typically have a thick client, one that will process interactions in real-time if the real-time connection can be established, but queue the transaction and do something else (in the context of the business process) when the connection is not available. When the connection becomes available again, the queried interactions are processed at that time.
- Many of the most-valued hybrid applications automatically detect the availability of a connection and then choose whichever means of connectivity is most applicable. The important part here is that the user of these applications does not have to know whether the applications are connected

or not. The software performing the applications will take care of the connectivity and render it mostly invisible to the user. Obviously, if we are in need of real-time access to satisfy a query but are not connected, then we will receive an error message. However, most information can be fetched ahead; it can be synchronized, and it can be available with some degree of currency although the wireless connection may not be available at the exact time of processing.

Each application being designed today falls into one of the categories listed in this section. It depends on the situation and the real business needs that drive the application itself. If a user can satisfy his or her needs with an occasionally connected application, developing such an application is usually much less expensive and complex than building a real-time solution. Having the wireless Internet, real-time connectivity certainly has some benefits in terms of the data being current, but one most weigh the disadvantages of such an approach, mostly stemming from the fact that connectivity is not always available in key locations. Handheld devices come in all shapes and sizes, with many more currently being developed in manufacturers' research labs around the globe.

20.4 THE WAP-ENABLED TRANSACTION MODEL

The WAP-enabled transaction model is based on the multilayered approach capable of supporting preexisting global users on the wired network, in addition to mobile users. The proposed model is implemented as a software module on top of the preexisting multidatabase management system. In reality, a mobile transaction is no different from a global transaction as far as the MDBS layer is concerned.[11] However, a number of factors make it sufficiently different enough to consider it as a separate type in the MDAS:

- Mobile accessing requires the support of fixed hosts for computations and communications.
- Mobile accessing might have to split the processing, with one part executing on a mobile unit and the other part executing on a fixed host.
- Mobile transactions might have to share state and data. This is a violation of the revered ACID (atomicity, consistency, isolation and durability) transactions-processing assumptions.
- Mobile accessing tends to be long lived. This is a consequence of the frequent disconnection experienced by the mobile client and the mobility of the mobile client.

The MDAS, as we envision it, consists of a software module, called a mobile accessing manger (MAM), implemented above the MDBS layer. The two layers combine to form the MDAS. The MAM is responsible for managing the submission of mobile transaction to the MDBS layer and its execution. Thus, the MAM acts as a proxy for the mobile unit, thereby establishing a static presence for the mobile unit on the fixed network.

Our approach is based on the principle that the computation and communication demands of an algorithm should be satisfied within the static segment of the system to the extent possible.[12] In this chapter, we attempt to:

- Localize the communication between the fixed host and a mobile host within the same cell
- Reduce the number of wireless messages by downloading most of the communication and computation requirements to the fixed segment of the network so that the database access is faster

In the proposed WAP model, communication occurs through the exchange of messages between static and mobile hosts. In order to send a message from a mobile host to another host, either fixed or mobile, the message is first sent to the local MSS over the wireless network. The MSS forwards the message to the local MSS of the other mobile host, which forwards it over the wireless network to the other mobile host if it is meant for a mobile host; otherwise the message is directly forwarded to the fixed host. The location of a mobile host within the network is neither fixed nor universally known in the network. Thus, when sending a message to a mobile host the MSS that serves the mobile host must first be determined. Each MSS maintains a list of mobile hosts' IDs that are local to its cell. When a mobile host enters a new cell, it sends a join message to the new MSS. The join message includes the ID (usually the IP address) of the mobile host. To change location, the mobile host also must send a left message to the local MSS. The mobile host neither sends nor receives any further messages within the present cell once the left message has been sent. When the MSS receives the left message from the mobile host, it removes the mobile host ID from its list of local mobile hosts.

Disconnection is often predictable by a mobile host before it occurs. Therefore, in order to disconnect, the mobile host sends a disconnect message to the local MSS. The disconnect message is similar to the leave message; the only difference is that when a mobile host issues a leave message, it is bound to reconnect at some other MSS at a later time. A mobile host that has issued a disconnect message may or may not reconnect at any MSS later.

To initiate a mobile access to the database, the mobile host sends a start message to the MAM. The MAM acknowledges the request by returning a transaction number. Each MSS has a MAM associated with it, and the transaction numbers are assigned in a distributed manner among the MAMs in the system using any distributed ordering algorithm.[12] The mobile unit tags each accessing request with an ID, which is composed of the mobile host ID and the transaction number. The access request message is composed of the mobile host ID, the transaction number, and the transaction operations. To signify the completion of an accessing request, a stop message is triggered to the MAM, in order to guarantee that the entire transaction as a whole is submitted to the MDBS.

The WAP-enabled transaction model workflow is as follows:

1. The mobile unit initiates a transaction request. The message is received by the MSS, and is forwarded to the associated MAM.

2. The MAM receives the request from the MSS. This request is logged and the transaction ID (transaction number along with the mobile host ID) is placed in the ready list. A transaction proxy is created to execute the transaction.
3. Now the proxy removes a transaction ID from the ready list and inserts it into the active list. The proxy translates the transaction request and then submits the transaction to the MDBS for execution.
4. The request is executed at the MDBS layer, and the results and data are returned to the proxy.
5. The proxy places the transaction ID in the output list along with the results and data to be returned to the mobile host.
6. The MAM initiates a search for the location of the mobile host and the results are transferred to the mobile unit if it is still connected, and then the transaction ID is removed from the ready list.

In applying the proposed WAP model to the MDAS, we may derive the following benefits:

- The proposed model decouples the effects of mobility from the MDAS. Hence, any developed concurrency control-and-recovery mechanism can be readily adopted into our model.
- The MDAS layer does not need to be aware of the mobile nature of some nodes. The accessing speed increases because the mobile transactions are submitted to the MDBS interface by the transaction proxies. The MDBS interacts with the transaction proxy as though it were the mobile unit. In the case of a mobile transaction, most of the communication is within the fixed network and, as far as the MDBS is concerned, a static host has initiated the transaction.
- The operations of nonmobile users are unaffected by the transactions of mobile users. The effects of long-lived transactions can be effectively and efficiently handled. Delegating the authority to commit or abort a transaction on behalf of the mobile host to the transaction proxy can minimize the effects of long-lived transactions. Thus, transactions initiated by nonmobile users will experience less conflict and, as a consequence, system throughput and response times are not severely affected.

In Section 20.5, we present this approach through a sample application to show how it can be put to work in the real world.

20.5 A SAMPLE APPLICATION

Our work embodies most of the suggestions in the previous section. In this section, we describe a sample application. Using this application, we show that in order to create WAP services or applications in a PC, we have to install a System Developers Kit (SDK) and simulator, such as those from Nokia, Ericsson, or Phone.com. For WAP applications that require database, a WAP-enabled transaction-processing model is more efficient than other models described in the literature because it is implemented as an additional layer on top of the MDBS that handles mobile accessing and long-lived transactions in a multilayered approach. In integration, a local server may be needed, such as the Microsoft Personal Web Server (PWS), IIS, Apache, Xitami, or any other type of server for testing the access speed in LDBS. A range of database integration tools can be used for WAP development, such as Microsoft Active Server Pages (ASP), Perl, etc. We have installed Nokia Mobile Internet Toolkit, Version 3.0, which supports WAP standards, authored by the WAP Forum, as well as other specifications authored by other organizations, and used ASP for programming.

Normally, the mobile phone emulator is capable of viewing WAP content in different ways. It can either load files that are stored in the PC directly or access files stored on Web servers and pretending to be a gateway or access files via a real WAP gateway. We can create a profile that specifies a connection to a particular WAP gateway in connectionless mode, another to the same gateway in secure connectionless mode, a third to the same gateway over a proxy server, and a fourth to a different-origin server using a direct HTTP connection.

In an attempt to explore the system, application, and user issues associated with the development of such mobile applications, we have considered the hotel booking system in Malaysia. As this research is relatively new, the default values of various parameters are educated guesses. Message transmission time has been calculated assuming that the static network is a 100-Mbps Ethernet. Current cellular technology offers a limited bandwidth on the order of 10 kbps, whereas current wireless LAN technology offers a bandwidth on the order of 10 Mbps; these numbers are most likely to change in the future. The hotel booking system consists of interaction with other hotel databases as our system needs to access other participating hotel databases to get room availability and also to send confirmation to members who book through the system.

The hotel booking process is done through WAP-enabled mobile phones by getting input from the member, such as check-in date, duration of stay in the hotel, number of rooms to book, etc. A member makes a booking to the hotel of his or her choice. The system will automatically check for room availability. If the system check finds that there are no rooms available, then the system will cancel the booking, inform the mobile user of the unsuccessful booking, and request him to book again. We have tested our WAP-enabled transaction model with a few existing hotel databases. The size of local database at each site, which has a direct effect on the overall performance of the system, can be varied. The global workload may consist of randomly generated global queries, spanning over a random number of sites. Each operation of a subtransaction (read, write, commit, or abort) may require data or acknowledgments sent from the local DBMS. The frequency of messages depends on the quality of the network link. In order to determine the effectiveness of our WAP-enabled transaction-processing model, several parameters are varied for different simulation runs.

20.6 SIMULATION RESULTS

Comparative evaluation of the proposed model has been carried out through simulation using the Nokia Mobile Internet Toolkit 3.0.[13] The Toolkit window, along with a snippet of ASP coding, is shown in Figure 20.5. Toolkit keeps a time record of all

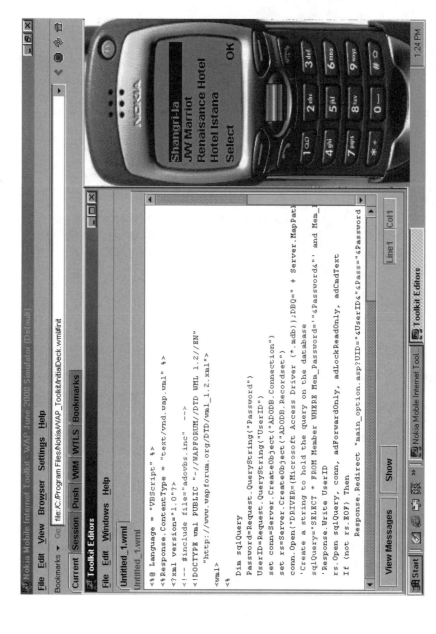

FIGURE 20.5 Nokia mobile Internet toolkit window with a snippet of ASP coding.

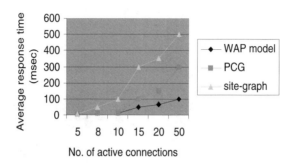

FIGURE 20.6 Comparison of average response time with number of active connections.

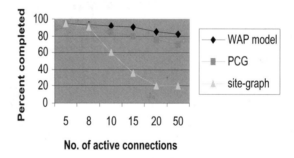

FIGURE 20.7 Comparison of the percentage of completion with number of active connections.

outgoing requests, measured in milliseconds, starting from the time of the initial request to the time a response is received. A time-out failure occurs if Toolkit does not receive a response within a specified number of milliseconds (the time-out value). By default, the time-out value is set to 2000 milliseconds. This value can be increased for slower network environments.

We can restrict the number of users in the WAP gateway. On separate simulation runs, the simulator measured and compared the response time taken with the number of active connections. The results are shown in Figure 20.6. The WAP model is seen to have a much-better response time than the Potential Conflict Graph (PCG) method and the site-graph method,[2,3] particularly for a large number of active connections.

The simulator also measured the percentage of active users during a certain period of time for PCG, site-graph, and our WAP model. The result of these measurements, shown in Figure 20.7, again suggest that the performance of our proposed WAP model is better, especially when the number of active connections increases, as compared to the other two analytical models reported in Breitbart et al.[2] and Lim and coworkers.[3]

20.7   CONCLUSION

The proposed multilayered WAP model uses the concept of proxy server to manage the execution of mobile transactions. To provide support for mobile transactions, a software layer, the Mobile Accessing Manager (MAM), is implemented above the preexisting multidatabase system. Performance of the system has been evaluated by varying the number of simultaneous databases using a hotel booking system. This work can be extended to global transactions in a larger context. There is a need for an extensive study on the effects of changing the distribution of data, processing I/O, caching, and communication in the near future.

ACKNOWLEDGMENTS

We would like to thank Dr. Borko Furht for initiating and encouraging this contribution.

References

1. Pitoura, E. and Bhargava, E., A framework for providing consistent and recoverable agent-based access to heterogeneous mobile databases, *SIGMOD Record,* pp. 44–49, 1995.
2. Breitbart Y. et al., On rigorous transaction scheduling, *IEEE Trans. Software Engineering,* 17 (9), 954–959, 1991.
3. Lim, J.B., Hurson, A.R., and Ravi, K.M., Concurrent data access in mobile heterogeneous systems, Hawaii Conference on System Sciences, pp. 1–10, 1999.
4. Bright, M.W., Hurson, A.R., and Pakzad, S., Automated resolution of semantic heterogeneity in multidatabases, *ACM TODS,* 19 (2), 212–253, 1994.
5. Liang, J. et al., Research on WAP clients supports SET payment protocol, *IEEE Wireless Commun.,* 1 (1), 90–95, February 2002.
6. Radhamani, G. et al., Security issues in WAP WTLS protocol, *IEEE Int. Conf. Commun. Circuits Syst.,* 483–487, July 2002.
7. Komnini, N. and Honary, B., Modified WAP for secure voice and video communication, *IEEE 2nd Int. Conf. 3G Mobile Commun. Technol.,* pp. 33–37, 2001.
8. Dirckze, R.A. and Gruenwald, L., Nomadic transaction management, *IEEE Potentials,* 17 (2), 31–33, 1998.
9. Chung, I. et al., *Efficient Cache Management Protocol Based on Data Locality in Mobile DBMSs,* Springer-Verlag, LNCS 1884, pp. 51–64, 2000.
10. Cap Gemini Ernst & Young, *Guide to Wireless Enterprise Application Architecture,* John Wiley & Sons, New York, 2002.
11. Segun, K. et al., A Transaction Processing Model for the Mobile Data Access System, LNCS 2127, pp. 112–127, 2001.
12. Badrinath, B.R. et al., Structured distibuted algorithms for mobile hosts, *Conf. Distributed Computing Syst.,* pp. 21–38, 1994.
13. Nokia Mobile Internet Toolkit, Version 3.0: User's guide, Nokia Ltd., www.forum. nokia.com.
14. Dirckze, R.A. and Gruenwald, L., A pre-serialization transaction management technique for mobile multidatabases, *ACM Mobile Networks Appl.,* 5 (4), 311–321, 2000.

21 Mobile Video Telephony

Igor D.D. Curcio*

CONTENTS

21.1 INTRODUCTION

Video telephony is not a new technology. It was proposed two decades ago for home usage; however, it has not been as successful or accepted as anticipated for technical reasons, or because of incorrect marketing strategies (including pricing) or unfamiliarity of users with the technology.

Today, wide usage of Internet technology for searching, browsing, etc., has educated users toward an increased image and video data fruition and, in general, toward a multimedia scenario where audio, still images, video, text, and other data are presented together.

---

* The opinions expressed in this chapter are those of the author and not necessarily those of his employer.

Mobile communications and devices are becoming more and more multimedia oriented. In addition, video and mobile network technologies are mature enough to be considered a single technology: mobile video technology.

During recent years different standardization organizations, such as the ITU-T (International Telecommunications Union, Telecommunications sector), IETF (Internet Engineering Task Force), and 3GPP (Third-Generation Partnership Project), have made enormous efforts to specify mobile multimedia network architectures, protocols, and codecs. Two main applications are enabled by those technology and research efforts: (1) mobile multimedia streaming, which has been described in Chapter 4 of this book; and (2) mobile video telephony. This chapter is about the state of the art in mobile video telephony.

A mobile video telephony application or a conversation multimedia application, as defined in 3GPP terminology, brings a new set of challenges:

1. The end-to-end delay requirements are very tight (compared to multimedia streaming).
2. Low-delay requirements restrict the range of techniques that can be used to provide good error resilience.
3. Mobile devices must have a consistent processing power to run speech and video encoders and decoders simultaneously, in order to process outgoing and incoming media flows.

In our framework, a mobile video telephony application includes (in addition to multiple bidirectional media, i.e., speech and video) the use cases where only one medium is used, i.e., the case where speech only is transmitted and received (Voice over IP), and the case where video only is transmitted and received (Video over IP).

This chapter is organized as follows: Section 21.2 describes the end-to-end system architecture for mobile video telephony systems. Section 21.3 briefly introduces the mobile networks for mobile video telephony. Section 21.4 introduces the current standards for mobile video telephony, based on ITU-T H.324 (for circuit-switched video telephony) and on IETF SIP (Session Initiation Protocol) (for packet-switched video telephony). Section 21.5 contains some performance and quality-of-service (QoS) considerations for implementations. Section 21.6 concludes this chapter.

21.2 END-TO-END SYSTEM ARCHITECTURE

A mobile video telephony system is a real-time system of the conversational type. It is real-time because the playback of continuous media, such as audio and video, must occur in an isochronous fashion. A video telephony application is different from a streaming application because the former has the following properties:

1. Bidirectional data transfer: The media flow is always carried from a source mobile videophone to a destination videophone, and vice versa. In this perspective, the flow of data is symmetric between the two end-points.

2. Real-time media encoding: Each videophone must have encoding and decoding capabilities. Speech and video signals must be encoded and transmitted in real-time to the other peer end. This requirement implies that mobile devices need a higher processing power because of the additional encoding capability (devices for mobile streaming require only decoding capability). Real-time encoding must be performed efficiently and with the shortest delays.

3. Delay sensitivity: Mobile video telephony systems are real-time with conversational features. This implies that a high level of interactivity between the two endpoints is a must to guarantee that the system is usable for speech and video conversations. A conversation can be held only if the end-to-end delays are very tight and preferably constant. For instance, the characteristic of conversationality and dialog interactivity between two parties would be lost in the case of end-to-end delays larger than few hundred milliseconds. This is the most-critical success factor for a mobile video telephony service. In order to guarantee low end-to-end delays, both network and mobile stations must be optimized for processing of conversational traffic. A very important factor in mobile videophone systems is error resilience: any mechanism for error detection and correction/concealment must be run within the maximum delay budget allowed. For this reason, retransmission algorithms at the network or application level cannot normally be used, and forward error correction (FEC) or error concealment algorithms are the only possible choice for providing error resilience against bit errors (or packet losses) produced by the air interface.

A mobile video telephony system consists mainly of two mobile videophones, used by the end users, and the mobile network. Figure 21.1 describes the high-level architecture of a typical mobile video telephony system over an IP-based mobile network. We will follow an end-to-end approach, analyzing the system in its different parts.

Mobile videophone A is connected to the mobile network through a logical connection established between the network and the mobile station addresses called Packet Data Protocol (PDP) context. PDP uses physical transport channels in the downlink and uplink directions to enable data transfer in the two directions. The mobile device has the capability to roam (i.e., upon mobility, change the network operator without affecting the received service), provided there is always radio coverage to guarantee the service. The mobile videophone is equipped with ordinary telephony hardware (microphone and speaker) and video hardware (camera and display).

The speech and video content is created in a live fashion from the microphone and camera input. This is encoded in real-time by the mobile device and transmitted in the uplink direction toward the network and the other end user. Speech and video data in the opposite direction (downlink) is conveyed from the network to mobile videophone A, which performs data decoding and display/playback of video and speech data. In addition, the videophone sends and receives information for session

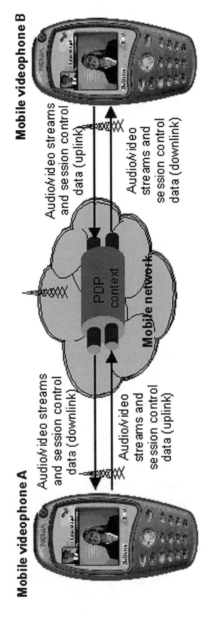

FIGURE 21.1 A typical mobile video telephony system.

TABLE 21.1
Mobile Network Channels for Video Telephony

Mobile Network	CS/PS	Theoretical Maximum Bit Rates (kbps)	Layer 2 Configuration
HSCSD	CS	57.6	Transparent mode
ECSD	CS	64.0	Transparent mode
UMTS (UTRAN) Release 99 and Release 4	CS	64.0	Transparent mode
UMTS (UTRAN) Release 5	PS	2048.0	Unacknowledged mode
UMTS (GERAN) Release 5 Gb mode	PS	473.6	Unacknowledged mode

establishment, QoS control, and media synchronization. The videophone may react promptly upon reception of QoS reports, taking appropriate actions for guaranteeing the best possible media quality at any instant.

The mobile network carries conversational multimedia and control traffic in the uplink and downlink directions, allowing real-time communication between the two mobile videophone users.

Mobile videophone B is placed at the other end of the architecture shown in Figure 21.1. Its functionality is symmetrically identical to that provided by mobile videophone A.

21.3 MOBILE NETWORKS FOR VIDEO TELEPHONY

In this section we review briefly the mobile network architectures and options that enable mobile video telephony. The considerations made in Chapter 4, Section 3 remain valid also for the case of mobile video telephony. However, because this type of application is more challenging in terms of end-to-end delays, not all the network configurations presented in Chapter 4 are suited for mobile video telephony. Therefore the main purpose of this section is to select the mobile network channels that enable video telephony.

Mobile channels can be divided into two categories:

1. Circuit-switched (CS) channels
2. Packet-switched (PS) channels

Table 21.1 shows a summary of network channels that can enable mobile video telephony (the bit rates indicated are maximum, and practical mobile videophone terminal implementations can have even lower maximum bit rates). In this table, we find neither GPRS Release '97 nor EGPRS networks. The reason is that they are not capable of sustaining conversational real-time traffic because of the high delay bounds compared to those required to support video telephony services.

For implementing mobile video telephony both CS and PS bearers can be used. In either case the transmission channel must be transparent. This implies that no retransmissions or mechanisms that produce additional delays must be employed at layer 2 of the mobile network (data link layer). In fact, layer 2 protocol data units (PDUs) are required to have the smallest header overhead, in order to reduce (or totally avoid) processing delays induced by complex PDU encapsulation and decapsulation. Unacknowledged mode is generally used at the data link layer in packet-switched connections. The PDUs used are slightly more complex than those used for the transparent mode, but light enough to allow fast data delivery between the two layer-2 peer entities.

UMTS networks allow theoretical maximum bit rates of 2048 kbps. However, the tested CS connections for video telephony for Release '99 and Release 4 networks are up to 64 kbps, as defined in 3GPP.[1,46] In these specifications, the recommended bit rates for video telephony services are 32 and 64 kbps, whereas the offered residual BERs are in the order of 10^{-4} or 10^{-6}. For PS connections, the maximum bit rates indicated in Table 21.1 are just theoretical. In practice, the tested and implemented maximum bit rates will be much smaller (in the order of 384 kbps).

The QoS profile for conversational traffic is defined in the 3GPP specification.[2] It is very similar to the profile defined for streaming traffic in Chapter 4, Section 3.1.2. However, due to the more-stringent delay requirements for the conversational traffic, two key parameters need to be defined differently:

1. Service data unit (SDU) error ratio. The maximum value for this parameter is defined as 10^{-2}. In other words, whenever erroneous packets are not delivered to the higher protocol layers and are considered lost the maximum packet loss rate is equal to 1 percent. The corresponding value defined in the QoS profile for streaming traffic is ten times larger, i.e., 10 percent. The rationale behind this parameter selection is that a higher packet loss rate can be allowed for streaming traffic. However, making use of higher-layer retransmissions, which can be implemented because streaming traffic can tolerate larger end-to-end delays, can reduce this error rate. Whenever no retransmissions are allowed, such as in the case of conversational traffic (i.e., video telephony traffic), a smaller maximum SDU error rate is more appropriate, and conservatively it helps in yielding a better application QoS.

2. Transfer delay. Because the end-to-end delay requirements for mobile video telephony are more stringent than for streaming service, the QoS profile defined for conversational traffic includes delay values that are more challenging than those allowed for streaming (where lower bounds are equal to 280 ms). For conversational traffic, the lower bound for UMTS bearers (i.e., between the mobile terminal and Core Network gateway) is 100 ms, while for Radio Access Bearers (i.e., between the mobile terminal and the Core Network edge node) it is 80 ms.[2]

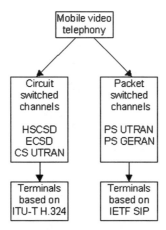

FIGURE 21.2 Standards for mobile video telephony.

After a short review of the mobile network channels for video telephony, in the next section we will describe the protocols and codecs standardized for circuit-switched and packed-switched video telephony.

21.4 STANDARDS FOR MOBILE VIDEO TELEPHONY

3GPP has specified standards for mobile video telephony, taking into account the nature of the mobile network channel. In fact, as introduced in the previous section, two different types of channels can enable mobile video telephony applications: circuit-switched and packet-switched channels. Following this dual approach, 3GPP has defined two different sets of standard specifications:

1. Specifications for CS mobile video telephony are based on the ITU-T H.324 standards for video telephony terminals over circuit-switched channels. H.324-based terminals also can be implemented over GSM-based CS channels (HSCSD, ECSD).
2. Specifications for PS mobile video telephony are based on the IETF SIP standard for video telephony over packet-switched channels.

Figure 21.2 summarizes the mapping between the mobile network channels and the standards for mobile video telephony defined in 3GPP.

A more-detailed description of the standards for mobile video telephony for CS and PS networks is given in the following sections.

21.4.1 CIRCUIT-SWITCHED MOBILE VIDEO TELEPHONY

H.324 terminals for 3GPP circuit-switched mobile video telephony are essentially ITU-T. H.324 terminals with Annex C[3] and with modifications specified by 3GPP[4] since Release '99. In 3GPP, these are called 3G-324M terminals.

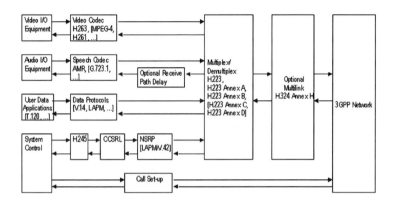

FIGURE 21.3 System architecture of 3G-324M terminals.

The system architecture of a 3G-324M terminal is depicted in Figure 21.3.[5] The mandatory elements of this architecture are a wireless interface, the H.223 multiplexer with Annex A and B,[6] and the H.245 system control protocol (version 3 or successive).[7] 3G-324M terminals are specified to work at bit rates of at least 32 kbps.

We will give an overview of the basic building blocks of a 3G-324M terminal, considering also some implementation guidelines, as described in 3GPP TSGS-SA.[8] The reader interested in the differences between H.324 and 3G-324M terminals can find more information in References 3 and 4. Here we will not emphasize these differences.

21.4.1.1 Media Elements

3G-324M terminals can support a wide set of media. They can be either continuous media (speech and video) or discrete media (real-time text). Among the former set, the following codecs can be supported in a mobile terminal:

- AMR (Adaptive MultiRate) narrowband is the mandatory speech codec for 3G-324M terminals,[9] if speech is supported. Speech is encoded at 8 kHz sampling frequency and at eight different bit rates ranging from 4.75 to 12.20 kbps.
- G.723.1 is the recommended speech codec supported.[10] It encodes speech at two bit rates, 5.3 and 6.3 kbps. The G.723.1 codec is needed if inter-operation against GSTN (General Switched Telephone Networks) is a requirement.[8]
- H.263 video Profile 0 Level 10 is the mandatory codec, if video is supported.[11]
- MPEG-4 Visual is an optional codec that can be supported at Simple Profile Level 0.[12]
- H.261 is another optional video codec[13] that can be supported by 3G-324M terminals.

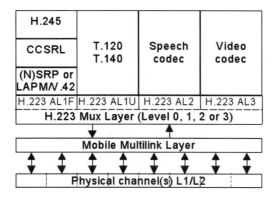

FIGURE 21.4 3G-324M protocol stack.

The discrete media defined in 3GPP specifications of circuit-switched video telephony terminals are in the framework of the optional user data application:

- T.120[14] is a protocol that allows multipoint data conferencing for transfer of data, images, and sharing of whiteboard and applications.
- T.140[15] is a protocol that allows real-time text conversation between two 3G-324M terminals. Text sessions can be opened in a stand-alone fashion or simultaneously with speech, video, and other data applications. Further information about this capability is available in Reference 16.

21.4.1.2 System Control and Multiplexing

In this section a general description of the system control and the multiplexing is given. Figure 21.4 shows a more detailed view of the 3G-324M protocol stack.

The control protocol H.245[7] provides end-to-end signaling for proper operation of a 3G-324M terminal, capability exchange, and messages to open and fully describe the content of logical channels. Most of the control signaling occurs at the beginning and at the end of the terminal call. The needed bandwidth for H.245 signaling is always allocated on-demand by the H.223 multiplexer.[5] This ensures that most of channel bandwidth is effectively used by the media.

H.324 Annex C[3] introduces also the Control Channel Segmentation and Reassembly Layer (CCSRL), which is used to split large control channel packets. The segmentation is required because successful transmission of large packets at high error rates may be difficult, and the connection set up may even fail without CCSRL.

Control messages can make use of retransmission for providing guaranteed delivery. H.324 uses the (Numbered) Simple Retransmission Protocol, or (N)SRP,[3] for this functionality.

The multiplex protocol H.223[6] multiplexes audio, video, data, and control streams into a single bit stream, and demultiplexes the received bit stream into separate bit streams. H.223 should support at least 32-kbps speed toward the wireless interface. However, also lower bit rates are possible, especially over GSH-based channels (HSCSD, ECSD). The multiplexer consists of an adaptation layer (AL) that exchanges information between the higher layers (i.e., audio/video codecs and system control), and a lower layer called the multiplex layer (MUX) that is responsible for transferring information received from the AL to the eventual mobile multilink layer and the physical layer(s). The AL handles the appropriate error detection and correction, sequence numbering, and retransmission procedures for each information stream. Three different ALs are specified in the H.223 Recommendation, each targeted to a different type of data:

1. The AL1 adaptation layer is designed primarily for transfer of data or control information, which is relatively delay insensitive but requires full error correction. However, AL1 does not provide any error control or retransmission procedure, but it relies on higher layers (i.e., (N)SRP) for this functionality. AL1 works in framed (AL1F) and unframed (AL1U) mode. The former is used for transfer of control data, while the latter is used for user data transfer, such as chat-data or other T.120- or T.140-enabled applications.
2. The AL2 adaptation layer is intended primarily for digital audio, which is delay sensitive, but may be able to accept occasional errors with only minor degradation of performance. AL2 receives data from its higher layer (i.e., an audio codec) and transfers it to the MUX layer after adding an 8-bit CRC (Cyclic Redundancy Check) and optional 8-bit sequence numbers which can be used to detect missing or misdelivered data.
3. The AL3 adaptation layer is designed for the transfer of digital video. It appends a 16-bit CRC to the data received from its higher layer (i.e., a video encoder), and it passes information to the MUX layer. AL3 includes optional provision for retransmission and sequence numbering by means of an 8- or 16-bit control field. 3GPP recommends encapsulating one MPEG-4 video packet into an AL3-SDU (Service Data Unit). To avoid additional delays caused by possible retransmissions, video data can be transferred using the AL2 that uses a smaller packet overhead and does not allow retransmission procedures.[8]

The MUX layer is responsible for mixing the various logical channels from the sending ALs (e.g., data, audio, video, and control) into a single bit stream to be forwarded to the physical layer for transmission. All MUX layer packets are delimited using HDLC flags, and include an 8-bit header, which contains, among other data, a 3-bit CRC for error detection. The variable-length information field of each MUX packet can contain 0 or more octets from multiple (segmentable) logical channels. To guarantee error resilience and a low delay, MUX packets are recommended to be between 100 and 200 bytes (for speech data, this means to encapsulate 1 to 3 speech frames into a MUX packet).[4]

To provide higher error resilience for data transmission over mobile networks, four different H.223 multiplexer levels are defined,[6] offering progressively increasing error robustness at the cost of progressively increasing overhead and complexity. The different levels are based on a different multiplexer packet structure:

- H.223 Level 0 describes the basic functionality as defined in Recommendation H.223. All 3G-324M terminals should be able to interwork using this level.
- H.223 Level 1 is described in Annex A of Recommendation H.223. The HDLC flag used to delimit multiplex packets in the MUX layer of H.223 is replaced with a longer flag, and HDLC zero-bit insertion (bit stuffing) is not used.
- H.223 Level 2 is described in Annex B of Recommendation H.223. In addition to the features of H.223 Level 1, a 24-bit (optionally also 32-bit) header describing the multiplexer packet is used. The header includes error protection (using Extended Golay Codes) and packet length fields.
- H.223 Level 3 is described in Annexes C and D of Recommendation H.223. The level includes the features of H.223 Level 2. Furthermore, additional error protection and other features are provided to increase the protection of the payload. For instance, H.223 Level 3 define changes not only to the MUX layer, but also to the AL layer, so that the various ALs in Figure 21.4 are replaced with more robust ones that make use of Reed-Solomon codes.

Two 3G-324M terminals establish a connection at the highest level supported by both terminals. This ensures the interoperability also with GSTN H.324 terminals. A dynamic level change procedure can be used to adjust error resilience when channel conditions vary during a connection. The levels can be used independently in receiving and transmission directions.

The optional Mobile Multilink Layer (MML)[3] usage has been introduced in Release 4 of 3GPP 3G-324M specifications. It allows the data transfer along up to eight independent physical connections, which provide the same transmission rate, in order to yield a higher aggregate bit rate. The MML provides the split functionality toward the lower protocol stack layers (HSCSD, ECSD, or CS UTRAN mobile networks) and the aggregation functionality toward the upper protocol stack layers.

Call setup issues in circuit-switched networks and capability for HTTP content downloading of 3G-324M terminals are not addressed here. The interested reader can find additional details respectively in Curcio and coworkers[17] and Annex I of ITU-T Recommendation H.324.[3]

21.4.2 PACKET-SWITCHED MOBILE VIDEO TELEPHONY

Mobile video telephony applications have been included in the framework of packet-switched conversational multimedia applications of 3GPP Release 5 specifications. A conversational multimedia application is any application that requires very low delays and error rates. For instance, a Voice over IP (VoIP) application or a one- or two-way multimedia application with the mentioned quality requirements belongs to this category.

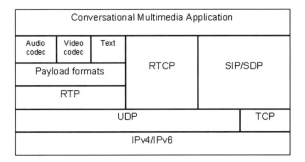

FIGURE 21.5 Protocol stack for PS conversational multimedia applications.

Release 5 3GPP specifications for video telephony are tightly connected to the 3GPP network specification. In fact, the call control mechanism in the IP Multimedia Subsystem (IMS) of 3GPP Network Release 5 is based on the SIP protocol defined by IETF. This is the same protocol used for the control plane of mobile videophones, defined in the framework of packet-switched conversational multimedia applications in 3GPP. Figure 21.5 shows the protocol stack for PS mobile videophones. In the next sections, a brief description of the codecs and protocols depicted in Figure 21.5 will be given.

21.4.2.1 Media Elements

The codecs and payload formats used for mobile video telephony are described in the specification.[18] Media either can be continuous (speech and video) or discrete (real-time text). For interoperability issues, 3GPP has ensured that the mandatory codecs for PS video telephony are the same codecs defined for CS video telephony (3G-324M). However, different codecs than the mandatory or recommended ones can be used, and these must be signaled and negotiated through SIP/SDP.

The codecs for continuous media are:

- AMR narrowband is the mandatory speech codec,[9] if speech is supported in PS videophones. AMR speech is packetized using the payload format described in Sjoberg et al.[19]
- AMR wideband is the mandatory speech codec[20] whenever wideband speech is supported in the terminal. AMR wideband speech is packetized according to the payload format in Sjoberg et al.[19]
- H.263 baseline is the mandatory codec when video is supported.[11] H.263 video is encapsulated following the payload format defined in Bormann et al.[21]
- H.263 Version 2 Interactive and Streaming Wireless Profile (Profile 3) Level 10 is an optional codec to be supported by the terminals.[22] It provides a better coding efficiency and error resilience in a mobile environment, compared to the baseline H.263, because of the use of the video codec Annexes I, J, K, and T. The packetization algorithm is the same defined for the H.263 baseline.[21]

- MPEG-4 Visual is an optional codec that can be supported at Simple Profile Level 0.[12] Encapsulation of MPEG-4 video is done according to the payload format defined in Kikuchi et al.[23]

Whenever static media are available in mobile videophone terminals, T.140 is the real-time text conversation standard to be optionally supported[15] for chat applications. Packetization of text data follows the formats defined in Hellstrom.[24]

The protocol used for the transport of packetized media data is the Real-Time Transport Protocol (RTP).[25] RTP provides real-time delivery of media data, including functionalities such as packet sequence numbers and time stamping. The latter allows intermedia synchronization in the receiving terminal. RTP runs on the top of UDP and IPv4/v6.

RTP comes with its control protocol (RTCP) that allows QoS monitoring. Each endpoint receives and sends quality reports to and from the other endpoint. The quality reports carry information such as number of packets sent, number of bytes sent, fraction of packets lost, number of packets lost, and packet interarrival jitter. Further details about RTCP will be given in Section 21.5.

21.4.2.2 System Control

The Session Initiation Protocol (SIP) defined in IETF[26] is an application layer control protocol for creating, modifying, and terminating sessions with one or more participants. SIP performs the logical bound between the media streams of two video telephony terminals. As shown in Figure 21.5, SIP can run on the top of TCP and UDP (other transport protocols also are allowed). However, UDP is assumed to be the preferred transport protocol in 3GPP IPv4- or IPv6-based networks.[27]

SIP makes use of the Session Description Protocol (SDP)[28] to describe the session properties. Among the parameters used to describe the session are IP addresses, ports, payload formats, types of media (audio, video, etc.), media codecs (H.263, AMR, etc.), and session bandwidth.

A simple IETF SIP signaling example between two video telephony terminals is presented in Figure 21.6.

A SIP call setup is essentially a three-way handshake between caller and callee. For instance, the main legs are INVITE (to initiate a call), 200/OK (to communicate a definitive successful response) and ACK (to acknowledge the response). However, implementations can make use of provisional responses, such as 100/TRYING and 180/RINGING when it is expected that a final response will take more than 200 ms. 100/TRYING indicates that the next-hop server has received the request and that some unspecified action is being taken on behalf of this call (for example, a database query). 180/RINGING indicates that the callee is trying to alert the user.

After the call has been established, the actual media transfer (speech and video) can take place. The release of the call is made by means of the BYE method, and the successful call release is communicated to the caller through a 200/OK message.

Quality of service of signaling is an important issue when measuring the performance of terminals for mobile video telephony. In Section 21.5 of this chapter we will clarify the concepts of Post Dialing Delay (T1), Answer-Signal Delay (T2),

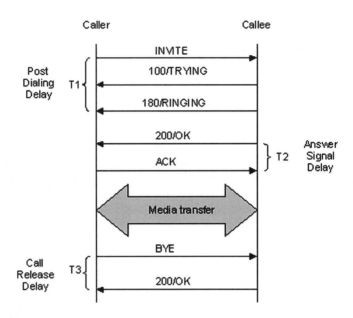

FIGURE 21.6 Call setup and release using SIP.

and Call Release Delay (T3) shown in Figure 21.6. The next section addresses SIP signaling in 3GPP networks.

21.4.2.3 Call Control Issues

SIP-based mobile applications based on IETF signaling can be implemented in 3GPP Release '99 and 4 networks. In this case, only the mobile applications resident in the mobile terminals run the SIP protocol, while the network is not aware of it.

A further step has been made in 3GPP Release 5 specifications, where SIP has been selected to govern the core call-control mechanism of the whole IP multimedia subsystem. Here, both the network and the mobile terminal implement the SIP protocol and exchange SIP messages for establishing and releasing calls. This choice has been made to enable the transition toward all-IP mobile networks. The SIP protocol in 3GPP Release 5 networks is more complex than the IETF SIP, because of factors such as resource reservation or the increased number of involved network elements. For a deeper understanding of the call control in 3GPP networks, you are refer to 3GPP.[27,29,30] Here we will give an example of SIP signaling for call setup and release between a mobile terminal and a 3GPP network (see Figures 21.7 and 21.8).[27]

The mobile terminal (or UE, user equipment) initiates a call toward the mobile originated (MO) network. The UE sends the first INVITE (1) message to the P-CSCF (proxy-call session control function) that works as a call router toward other network elements and the destination mobile terminal. Before the 180/RINGING (19) message is received by the UE, the messages (11–17) are exchanged mainly to allow resource reservation in the network and PDP context activation between the

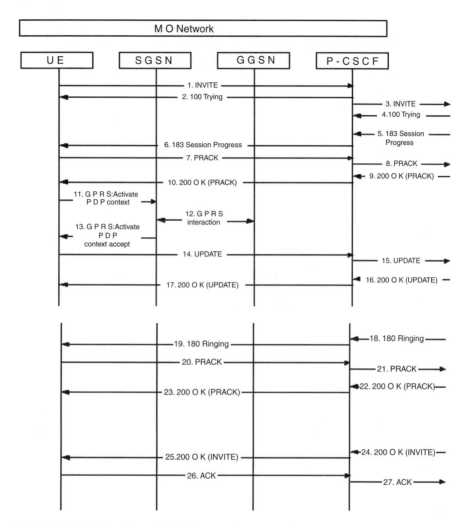

FIGURE 21.7 SIP call setup in 3GPP networks.

UE and the network. PRACK messages[31] play the same role as ACK, but they apply to provisional responses (such as 183/SESSION PROGRESS or 180/RINGING) that cease to be retransmitted when PRACK is received (more details about reliability of SIP messages are available in Section 21.5).

In a 3GPP network, the total number of SIP messages exchanged by the UE for establishing a call is 12 (plus resource reservation), while a simple IETF call setup requires 5 SIP messages.

Call release signaling is shown in Figure 21.8. The number of SIP messages exchanged by the UE is 2 (the same number as in IETF SIP call release), plus the required signaling to release the PDP contexts resources. In this scenario, messages (2–3) can occur even before BYE (1) and in parallel with procedure 4 (remove resource reservation).

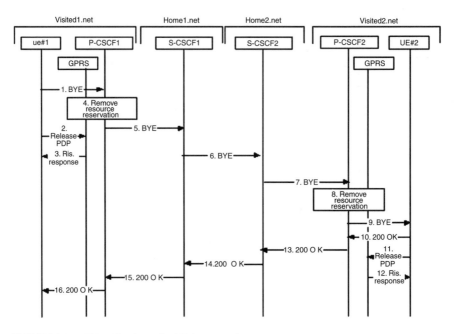

FIGURE 21.8 SIP call release in 3GPP networks.

21.5 PERFORMANCE ISSUES IN MOBILE VIDEO TELEPHONY

After having given an overview of the standards for mobile video telephony, including CS/PS terminals and call control issues in 3GPP networks, in this section we will make some considerations for and remarks on performance. In particular error resilience, QoS profiles for conversational service, QoS metrics for video, video quality results for 3G-324M terminals, SIP signaling delay, and RTCP reporting capability aspects will be analyzed.

21.5.1 ERROR RESILIENCE AND QoS

In mobile video telephony special attention must be paid to error resilience issues. Because an efficient system must operate with minimal end-to-end delays, often there is not enough time for media reparation, whenever media is hit by errors due to the lossy characteristics of the air interface. In most of the cases, forward error correction (or redundancy coding) is the only means to provide error resilience within the imposed delays. In addition, to reduce the impact of data corruption and packet losses on the received media, some special shrewdness also can be taken. Here we will focus on PS video telephony systems.

When encoding a video signal using the H.263 Profile 3, the achieved error resilience is higher than the baseline H.263. MPEG-4 visual offers also advanced tools for error resilience, such as data partitioning, RVLC (Reversible Variable Length Codes) and resynchronization markers. To guarantee low delay, the specification[32] recommends that the video packets must be no larger than 512 bytes. In general, the smaller the packets, the smaller the amount of video data lost (and the visual quality loss) in case of packet losses. On the other hand, too-small packets produce excessive RTP/UDP/IP header overhead. The choice of the right packet size is a trade-off between error resilience, delay, and bandwidth occupancy. The packet size also can be changed dynamically on the fly, based on the condition of the network link.

When encoding and packetizing speech data with AMR or AMR-WB, the specification[32] mandates (or forbids) the use of certain codec options:

- Speech data must be packetized using bandwidth-efficient operations. The encapsulation algorithm[19] offers both bit and byte alignment of data. The former is more efficient in terms of bandwidth usage.
- Encapsulation of no more than one speech frame into an RTP packet to keep the delay at the minimum. One AMR speech frame is of 20-ms duration. This implies that the packet rate at the videophone terminal is 50 packets per second for both incoming and outgoing RTP flows.
- The multichannel session shall not be used.
- Interleaving shall not be used. This causes an increase in delay.
- Internal CRC shall not be used. Data correction is performed in the lower layers of the protocol stack. This saves bandwidth.

For the transmission of real-time text using T.140, the use of redundancy coding is recommended to provide a better error resilience.[18]

At the network level, 3GPP specifications offer the possibility to configure the QoS profile for a conversational multimedia application running over a conversational PDP context. The specification[32] defines the recommended target figures for error rates and delays:

- SDU error ratio (or packet loss rate): 0.7 percent or less for speech and 0.01 percent for video.
- Transfer delay: 100 ms for speech and 150 ms for video.

In CS networks, errors in the air interface produce single bit errors in the video packet payload. A video decoder is generally resilient to bit error rates up to 10^{-3}. In PS networks, errors in the air interface produce erroneous packets that generally are not forwarded to the higher protocol layers than IP. So, they are regarded as lost packets. In this case, SDU error ratios as indicated previously can be used to provide enough media resilience from packet losses.

21.5.2 VIDEO QoS METRICS

Adequate techniques for objective and subjective speech and video quality assessment must be adopted to guarantee that a given mobile videophone implementation fulfills a minimum set of QoS requirements. This section focuses on video quality metrics.

When developing mobile video telephony applications the need is to decide which fundamental quality parameters should be selected as key parameters in QoS assessment of video. For this purpose both subjective and objective metrics must be used, because they can be considered complementary. You are referred to Curcio[33] for details about subjective metrics. Regarding objective quality metrics, standardization bodies have defined some methods. For example, the ANSI[34] and ITU[35] standards describe some metrics. However, for some of the metrics described in these documents, the implementation is not straightforward. Despite the effort of standardization bodies to define common video quality metrics, often the most-used objective method for video quality assessment is the PSNR (Peak Signal-to-Noise Ratio), because it is the easiest to apply to the metrics available. However, other useful quality metrics can be put to use when developing mobile videophone terminals. For further details on the metrics computation methods, please refer to Curcio.[33,36]

The quality metrics are categorized into six classes, depending on the type of information they can provide:

1. Frame-based: This set of metrics gives information about the number of frames that have been processed end-to-end. The metrics are
   - Number of encoded frames
   - Number of decoded frames
   - Number of dropped frames
   - Drop frame rate
   - Encoding frame rate
   - Decoding frame rate
   - Display frame rate
   - Size of the first INTRA-coded frame
2. Bit rate-based: The objective of these metrics is to provide information about the repartition of the channel bandwidth. This information is precious for optimizing system performance. The metrics are
   - Audio bit rate (obtained by the audio codec)
   - Video bit rate
   - Packetization overhead
   - Application total bit rate (computed as a sum of the above values)
3. Packet-based: These metrics give information about the packets that are generated by the RTP packetizer or the H.223 multiplexer:
   - Number of packets per frame
   - Size of the packets
4. Loss- or corruption-based: These metrics provide information about the amount of packets lost, or the amount of correctly/incorrectly delivered data:
   - Packet loss rate
   - Correctly delivered data rate
   - Misdelivered data rate
   - Bit error rate computation
5. PSNR-based. PSNR is a measure of the difference between the original frame and the corresponding encoded (or decoded) frame. PSNR-based metrics are
   - PSNR of the video sequence

- Standard deviation of PSNR
- PDF (Probability Density Function) and CDF (Cumulative Density Function) of PSNR
- Representative run for subjective evaluation (when multiple simulation runs are considered)

6. Delay-based. Delay is a very critical issue in mobile video telephony. Because end-to-end delay is made up of different components, one approach would be to measure the different delays and try to optimize them separately. A set of measurable delays are

- Capturing delay
- Initial video encoding delay (time required to encode the first INTRA frame)
- Encoding delay for video frames (minimum, average, and maximum)
- Packetization delay
- Transmission delay (related to the network)
- Depacketization delay
- Decoding delay for video frames (minimum, average, and maximum)
- Display delay
- End-to-end delay
- PDF and CDF for any of the delay components above
- Out of delay constraints rate (to measure the percentage of delay violation over a fixed threshold T of time)
- Delay jitter computed for different delays above (a particularly interesting value is the frame rate jitter)

21.5.3 VIDEO QUALITY RESULTS FOR 3G-324M

To provide an idea of the performance of a videophone in mobile environment; we have implemented a PC version of 3G-324M terminal and made mobile-to-mobile calls between two 3G-324M PC terminals through a simulated circuit-switched WCDMA network at 64 kbps. Table 21.2 summarizes the main simulation parameters used in our tests.

The error patterns were injected two times into the bit stream to simulate the case of mobile-to-mobile connection, where two radio links are involved (this is the reason the bit error rates are doubled in Table 21.2).

The results obtained were measured in terms of average PSNR over 10 runs, standard deviation of PSNR, frame rate, delay, bandwidth usage, and visual quality. The reader interested in details about performance of 3G-324M terminals at different bit rates with service flexibility for WCDMA and HSCSD networks may refer to Curcio and coworkers,[17] Hourunranta and Curcio,[37] and Curcio and Hourunranta.[38] Average PSNR results are shown in Table 21.3.

The maximum quality loss achieved at the higher BER is below 0.5 dB, with a maximum standard deviation below 0.2 dB. The average encoding frame rate was 10.2 frames per second. The end-to-end delay from encoding to display (excluding capturing and network delay) was 140 ms, of which 98 ms was for shaping delay[37] and 42 ms was the processing delay. The bandwidth usage is reported in Table 21.4.

TABLE 21.2
3G-324M Simulation Parameters

Speech	Preencoded AMR speech stream with average bit rate of 4.9 kbps (silence suppression is used)
Video codec	H.263+ with Annex F, I, J, T
Input frame rate	30 fps
Frame size	QCIF (176 × 144 pixels)
Original video sequence	Carphone concatenated three times
	Original (382 frames, 12.7 seconds)
	Concatenated (1146 frames, 38.1 seconds)
WCDMA channel bit rate	64 kbps
Mobile speeds	3 and 50 kmph
Bit error rates (BERs)	64 kbps, 3 kmph: 2*7E-05 and 2*2E-04
	64 kbps, 50 kmph: 2*6E-05 and 2*2E-04
Frequency	1920 MHz
Chip rate	4.096 Mbps
Transmission direction	Uplink
Interleaving depth	40 ms
Coding	1/3-rate turbo code, 4 states
Duration of each error pattern	180 seconds
Multiplexing	H.223 Level 2
Number of simulations	10 for each error pattern file (each time starting from a different random position of the file)

TABLE 21.3
PSNR for Video over 3G-324M

Speed/BER	Average PSNR (dB)	Standard Deviation of PSNR (dB)
Error free	32.12	0.02
3 kmph 2*7E-05	32.01	0.07
3 kmph 2*2E-04	31.65	0.19
50 kmph 2*6E-05	31.89	0.14
50 kmph 2*2E-04	31.64	0.19

Finally, Figure 21.9 shows the average visual quality under the worst of the conditions tested (BER = 2*2E-04 at 3 kmph). The picture has been selected in a way that the PSNR of the sample picture is as close to the average PSNR as possible (31.65 dB). As it can be seen, the picture does not show critical degradations, and its quality is fairly good.

TABLE 21.4
Bandwidth Repartition for 3G-324M

Type of Data	Percentage of Occupancy on the Total Bandwidth
Video data	84
Audio data	8
H.223 multiplexer overhead	8

FIGURE 21.9 Carphone 64 kbps BER = 2*2E-04 3 kmph.

21.5.4 SIP SIGNALING DELAY

One factor that influences the overall user QoS, in addition to media quality, is the call setup delay. This is important when globally evaluating user satisfaction for a certain service. We take this issue into consideration in this section, evaluating the performance of a SIP user agent (UA) signaling with video telephony capabilities that we have implemented.

When SIP is used over UDP on a mobile network, the call setup time between two terminals can vary because of the following factors:

1. Lossy nature of the channel: If SIP packets are lost during call establishment, these are retransmitted.
2. Size of the channel: Smaller network bandwidths yield higher call setup delays than larger bandwidths.
3. Processing delays in the network: Each network element takes some time to process the requests made by the endpoints.
4. Congestion in the network path along the two end-points.

The message reliability system defined in SIP[26] is made in such way that it can cope with packet losses and unexpected delays within the network. The basic idea is that if a SIP message is not received within a certain specified time, it is retransmitted by the protocol itself. In the following, the retransmission rules for the different SIP messages exchanged in a session between two SIP UAs such as one in Figure 21.6 are explained:

- INVITE method. A SIP UA should retransmit an INVITE request with an interval that starts at T1 seconds, and doubles after each packet transmission. T1 is an estimate of the round-trip time (RTT). The client stops retransmissions if it receives a provisional (1xx) or definitive (2xx) response, or once it has sent a total of seven request packets. A UA client may send a BYE or CANCEL request after the seventh retransmission (i.e., after 64*T1 seconds). In our implementation the value of T1 is set at 0.5 seconds.
- BYE method. In this case, a SIP client should retransmit requests with an exponential backoff for congestion control reasons. For example, if the first packet sent is lost, the second packet is sent T1 seconds later, and eventually the next one after 2*T1 seconds (4*T1 seconds, and so on), until the interval reaches a value T2. Subsequent retransmissions are spaced by T2 seconds. T2 represents the amount of time a BYE server transaction will take to respond to a request, if it does not respond immediately. If the client receives a provisional response, it continues to retransmit the request, but with an interval of T2 seconds (this is done to ensure reliable delivery of the final response). Retransmissions cease when the client has sent a total of 11 packets (i.e., after T1*64 seconds), or it has received a definitive response. Responses to BYE are not acknowledged via ACK. In our implementation the values of T1 and T2 are set to 0.5 and 4 seconds, respectively.
- ACK method. ACK is not retransmitted, but in case of loss the UA server retransmits the 200/OK.
- Informational (provisional) responses (1xx). UA servers do not transmit informational responses reliably. For instance, our implementation does not retransmit informational responses (100/TRYING, 180/RINGING). However, the UA server, which transmits a provisional response, will retransmit it upon reception of a duplicate request.
- Successful responses (2xx). A UA server does not retransmit responses to BYE. In all the other cases a UA server, which transmits a final response, should retransmit it with the same spacing as the BYE. Response retransmissions cease when an ACK request is received or the response has been transmitted 11 times (i.e., after 64*T1 seconds). The value of a final response is not changed by the arrival of a BYE or CANCEL request.

In 3GPP Release 5 networks, the timers T1 and T2 are set to different and more-conservative values.

The tests we have run have been performed over a 3GPP Release '99 network emulator. The results will be expressed in terms of the following metrics:

- Postdialing delay (PDD). It also is called postselection delay or dial-to-ring delay. This is the time elapsed between the caller clicking the button of his terminal to make the call and hearing the terminal ringing. In our case the PDD corresponds to the time T1 (see Figure 21.6).
- Answer-signal delay (ASD). This is the time elapsed between the phone being picked up and the caller receiving indication of this. In our case the ASD corresponds to the time T2 (see Figure 21.6). It must be noted that

TABLE 21.5
Call Setup Times for SIP Signaling

Call Setup Metric	Delay (ms)
Postdialing delay	62
Answer-signal delay	45
Call-release delay	50

TABLE 21.6
Postdialing Delay for SIP Signaling with Limited Bandwidth

Network Bandwidth (kbps)	Delay (ms)
2	981
5	427
9.2	287
16	164
32	119
64	78

the caller receives notification that the callee has picked up the phone when the first receives the 200/OK. However, the call-signaling handshake is completed when the callee receives the ACK from the caller. This is the reason we have considered the ASD in this way.

• Call-release delay (CRD). This is the time elapsed between the phone being hung up by the releasing party (the caller in our example in Figure 21.6) and a new call can be initiated/received (by the same party). In our tests the CRD corresponds to the time T3.

Results of simulations are shown in Table 21.5. No signaling compression algorithms were used. For comparisons between calls over 3GPP Release '99 networks and calls in Intranet or WLAN environment, and for further details about SIP signaling delays, the reader can refer to Curcio and Lundan.[39,40]

Table 21.6 contains results for SIP call set-up times in the case of restricted bandwidths. Also in this case, no signaling compression was used. Table 21.5 results assumed a bandwidth of 384 kbps; however, in many cases it is better to assume, as we did, that the bearer reserved for SIP signaling is a dedicated one, and of smaller size. Also, when running the tests with restricted bandwidth we have injected 2 percent packet losses using the NISTNET[41] simulator. The figures show that there is an increase in postdialing delays up to almost one second for network bandwidths as narrow as 2 kbps.

The results presented in this section are not related to SIP signaling within Release 5 of 3GPP specifications, where SIP is part of the call control in the IMS. In this case, the SIP signaling delays are estimated to be larger than those shown, due to the increased complexity of the whole network system.

21.5.5 RTCP PERFORMANCE

The RTCP protocol basics have been introduced in Chapter 4, Section 4.5. We recall the fact the RTCP is used by a receiver to provide QoS information to the transmitting party in order to repair the media transmitted or, in general, to take some (possibly prompt) action to adjust or improve the QoS toward the receiver.

RTCP packets are normally sent with a minimum interval of 5 seconds. However, some applications may benefit from sending a more-frequent feedback. Video telephony can certainly benefit from a faster feedback, because this allows a faster reaction of the sender terminal to provide a better QoS to the receiving terminal. One possible action is to change on the fly the encoding parameters when the packet loss rate increases. This action should be taken as early as possible in the transmitting terminal, and a 5-second interval could be too long a time window to allow a fast reaction, especially if the media to repair is a speech stream.

The new RTP specifications[42,43] define a more-flexible use of the RTCP data flow, allowing more-frequent feedback by reducing the transmission interval to a value lower than 5 seconds or by fixing the percentage of the RTP session bandwidth reserved for RTCP traffic.

We have run some tests for 1-minute speech and video streams. The former was encoded using the AMR codec at 12.2 kbps with silence suppression. The latter was encoded using the H.263+ video codec at 64 kbps. For the speech session the maximum RTCP packet length was 168 bytes (including UDP/IP headers, a sender report and full SDES), while for the video session the RTCP packet length was 88 bytes (including UDP/IP headers, a receiver report and SDES). Results for different RTCP bandwidth percentages are shown in Tables 21.7 and 21.8.

The leftmost column in Tables 21.7 and 21.8 contains the RTCP bandwidth as percentage of the RTP session bandwidth, which includes media and headers overhead (including RTP/UDP/IP headers). The second column contains the RTCP bandwidth in kilobits per second. The third column of data is the computed average RTCP interval between two QoS reports. The last column contains the number of RTCP packets sent by a receiver during 1 minute of data reception.

The reader can see that the minimum bandwidth occupied by RTCP is below 0.2 kbps when a 5-second transmission interval is used. In this case the receiver sends only 12 QoS reports. When increasing the bandwidth reserved for RTCP, more-frequent feedback can be sent. For example, for speech traffic, a feedback message every 2.3 seconds would allow 26 QoS reports in one minute. In the same way for video traffic, a feedback message every 375 ms would allow 160 QoS reports in one minute. This would let the transmitting terminal have more possibilities to adjust the error resilience properties of the video stream. Theoretically, it could be possible to have 160 QoS reports for the speech stream as well. However, this would imply an RTCP bandwidth of about 2.4 kbps, i.e., over 13 percent of the RTP session

TABLE 21.7
Results for Different RTCP Bandwidth Percentages (AMR Speech)

RTCP Bandwidth (%)	RTCP Bandwidth (kbps)	Average RTCP Interval (ms)	Number of RTCP Packets
1.0	0.19	5000	12
1.6	0.28	3158	19
1.9	0.35	2609	23
2.2	0.39	2308	26

TABLE 21.8
Results for Different RTCP Bandwidth Percentages (H.263+ Video)

RTCP Bandwidth (%)	RTCP Bandwidth (kbps)	Average RTCP Interval (ms)	Number of RTCP Packets
0.2	0.14	5000	12
0.3	0.18	4000	15
0.5	0.34	2069	29
1.0	0.67	1053	57
1.2	0.82	857	70
1.9	1.31	536	112
2.4	1.63	432	139
2.8	1.88	375	160

bandwidth for speech. The reader interested in details about RTCP traffic can refer to Curcio and Lundan.[44,45]

21.6 CONCLUSIONS

Mobile video telephony is enabled by different mobile networks (HSCSD, ECSD, and UMTS). Standards also have been developed by ITU-T, IETF, and 3GPP to allow circuit-switched and packet-switched video telephony (3G-324M and SIP-based).

Video is a new media dimension that affects the current usability paradigms of mobile devices. In this perspective, new usage models are envisaged for the users of mobile video telephony. Finally, the success and widespread use of this new type of application will be certainly influenced not only by technical challenges, but also by end-user requirements.

ACKNOWLEDGMENTS

The author would like to express sincere gratitude to his colleagues Ari Hourunranta, Ville Lappalainen, and Miikka Lundan for the cooperation offered.

References

1. 3GPP TSG-T, Common test environments for user equipment (UE), Conformance testing (Release '99), TS 34.108, v.3.10.0 (2002–12).
2. 3GPP TSGS-SA, QoS concept and architecture (Release 5), TS 23.107, v.5.7.0 (2002–12).
3. ITU-T, Terminal for low bit-rate multimedia communication, Recommendation H.324, March 2002.
4. 3GPP TSGS-SA, Codec for circuit switched multimedia telephony service, Modifications to H.324 (Release '99), TS 26.111, v.3.4.0 (2000–12).
5. 3GPP TSGS-SA, Codec for circuit switched multimedia telephony service, General description (Release 4), TS 26.110, v.4.1.0 (2001–03).
6. ITU-T, Multiplexing protocol for low bit rate multimedia communication, Recommendation H.223, July 2001.
7. ITU-T, Control protocol for multimedia communication, Recommendation H.245 (Version 8), July 2001.
8. 3GPP TSGS-SA, Codec(s) for circuit switched multimedia telephony service, Terminal implementor's guide (Release 4), TS 26.911, v.4.1.0 (2001–03).
9. 3GPP TSGS-SA, Mandatory speech codec speech processing functions, AMR speech codec, General description (Release 5), TS 26.071, v.5.0.0 (2002–06).
10. ITU-T, Dual rate speech coder for multimedia communications transmitting at 5.3 and 6.3 kbit/s, Recommendation G.723.1, March 1996.
11. ITU-T, Video coding for low bit rate communication, Recommendation H.263, February 1998.
12. ISO/IEC, Information technology – Coding of audio-visual objects – Part 2: Visual, 14496-2, 2001.
13. ITU-T, Video codec for audiovisual services at p x 64 kbits, Recommendation H.261, March 1993.
14. ITU-T, Data protocols for multimedia conferencing, Recommendation T.120, July 1996.
15. ITU-T, Protocol for multimedia application text conversation, Recommendation T.140, February 1998.
16. 3GPP TSGS-SA, Global text telephony, Stage 1 (Release 5), TS 22.226, v.5.2.0 (2002–03).
17. Curcio, I.D.D., Lappalainen, V., and Mostafa, M.-E., QoS evaluation of 3G-324M mobile videophones over WCDMA networks, *Comput. Networks,* 37 (3-4), 425–445, 2001.
18. 3GPP TSGS-SA, Packet switched conversational multimedia applications, Default codecs (Release 5), TS 26.235, v.5.1.0 (2002–03).
19. Sjoberg, J. et al., RTP payload format and file storage format for the Adaptive Multi-Rate (AMR) and Adaptive Multi-Rate Wideband (AMR-WB) audio codecs, IETF RFC 3267, March 2002.
20. ITU-T, Wideband coding of speech at around 16 kbits/s using Adaptive Multi-Rate Wideband (AMR-WB), Recommendation G.722.2, January 2002.
21. Bormann, C. et al., RTP Payload format for the 1998 version of ITU-T Recommendation H.263 (H.263+), IETF RFC 2429, October 1998.
22. ITU-T, Video coding for low bit rate communication, Profiles and levels definition, Recommendation H.263 Annex X, April 2001.
23. Kikuchi, Y. et al., RTP payload format for MPEG-4 Audio/Visual streams, IETF RFC 3016, November 2000.

24. Hellstrom, G., RTP Payload for Text Conversation, IETF RFC 2793, May 2000.

25. Schulzrinne, H. et al., RTP: A Transport Protocol for Real-Time Applications, IETF RFC 1889, January 1996.

26. Rosenberg, J. et al., SIP: Session Initiation Protocol, IETF RFC 3261, March 2002.

27. 3GPP TSG CN, Signaling flows for the IP multimedia call control based on SIP and SDP, Stage 3 (Release 5), TS 24.228 v.5.3.0 (2002–12).

28. Handley, M. and Jacobson, V., SDP: Session description protocol, IETF RFC 2327, April 1998.

29. 3GPP TSG-SSA, IP Multimedia Subsystem (IMS), Stage 2 (Release 5), TS 23.228 v.5.7.0 (2002–12).

30. 3GPP TSG CN, IP multimedia call control protocol based on SIP and SDP, Stage 3 (Release 5), TS 24.229 v.5.3.0 (2002–12).

31. Rosenberg, J. and Schulzrinne, H., Reliability of Provisional Responses in the Session Initiation Protocol (SIP), IETF RFC 3262, June 2002.

32. 3GPP TSGS-SA, Packet switched conversational multimedia applications, Transport protocols (Release 5), TS 26.236, v.5.7.0 (2002–12).

33. Curcio, I.D.D., Mobile video QoS metrics, *Int. J. Comput. Appl.,* 24 (2), 41–51, 2002.

34. ANSI, Digital Transport of One-Way Video Signals – Parameters for Objective Performance Assessment, T1.801.03, 1996.

35. ITU-T, Multimedia communications delay, synchronization and frame rate measurement, Recommendation P.931, December 1998.

36. Curcio, I.D.D., Practical Metrics for QoS Evaluation of Mobile Video, Internet and Multimedia Systems and Applications Conference (IMSA 2000), Las Vegas, 9–23 November 2000, pp. 199–208.

37. Hourunranta, A. and Curcio, I.D.D., Delay in Mobile Videophones, IEEE 7th Mobile Multimedia Communications Workshop (MoMuC 2000), Tokyo, 23–26 October 2000, pp. 1-B-3–1/1-B-3–7.

38. Curcio, I.D.D. and Hourunranta, A., QoS of Mobile Videophones in HSCSD Networks, IEEE 8th International Conference on Computer Communications and Networks (ICCCN '99), Boston, 11–13 October 1999, pp. 447–451.

39. Curcio, I.D.D. and Lundan, M., SIP Call Setup Delay in 3G Networks, IEEE 7th Symposium on Computers and Communication (ISCC '02), Taormina, Italy, 1–4 July 2002, pp. 835–840.

40. Curcio, I.D.D. and Lundan, M., Study of Call Setup in SIP-Based Videotelephony, 5th World Multi-Conference on Systemics, Cybernetics and Informatics (SCI 2001), Orlando, 22–25 July 2001, Vol. IV, pp. 1–6.

41. NIST, NISTNet, http://www.antd.nist.gov/nistnet/.

42. Schulzrinne, H. et al., RTP: A Transport Protocol for Real-Time Applications, IETF draft, Work in progress, November 2001.

43. Casner, S., SDP Bandwidth Modifiers for RTCP Bandwidth, IETF draft, Work in progress, November 2001.

44. Curcio, I.D.D. and Lundan M., Event-Driven RTCP Feedback for Mobile Multimedia Applications, IEEE 3rd Finnish Wireless Communications Workshop (FWCW '02), Helsinki, 29 May 2002.

45. Curcio, I.D.D. and Lundan, M., On RTCP Feedback for Mobile Multimedia Applications, IEEE International Conference on Networking (ICN '02), Atlanta, 26–29 August 2002.

46. 3GPP TSG-T, Common Test Environments for User Equipment (UE), Conformance Testing (Release 4), TS 34.108, V.4.5.0 (2002–12).

22 WAP: Transitional Technology for M-Commerce

Mahesh S. Raisinghani

CONTENTS

ABSTRACT

Wireless Application Protocol (WAP) is the most-popular Internet-enabling technology being adopted en masse by handset manufacturers and service providers alike. The International Data Corporation promises 1 billion cellular telephones worldwide by 2004, with half of them Internet-enabled. This chapter discusses the fast-growing trend for WAP tools to access the Internet versus the current predominant use of personal computers. The chapter describes also the importance of WAP in the field of E-commerce, with its popularity leading E-commerce into mobile commerce (M-commerce). Because it works in already existing networks, WAP needs little modification in Web content and can be available with ease. Already, there are numerous companies providing E-commerce services through WAP around the world, and with the huge mobile telephone subscriber base, the potential for M-commerce is tremendous.

22.1 INTRODUCTION

Trade developed through many stages, from barter in the old days to E-commerce today. What will be the tool for transactions tomorrow? In the past half decade, the Internet has revolutionized the practice and procedure of trade, giving birth to the new world of E-commerce. Now people can buy or sell goods and services practically 24 hours a day, 7 days a week, if they have access to the Web. Vendors have been able to tap into markets that were impossible to reach due to remote geographic location or other reasons. It is this "anytime, anywhere" technology that has fueled the new economy.

Although much has been accomplished toward this goal of being able to trade "anytime, anywhere," personal computer laptops are too bulky for M-commerce. The obvious choice is to empower the mobile telephone to be the preferred tool for M-commerce. The M-commerce phenomenon is centered in Asia and Europe (not the United States), where mobile telephony is further advanced and PC usage is much lower. Nokia, Ericsson, Motorola, and NTT DoCoMo, to name a few — as well as giants in banking, retail, and travel, including Amazon and Schwab — are developing their mobile E-sites; and all are settling on WAP.

WAP works with all major wireless networks — code division multiple access (CDMA), Global System for Mobile Communications (GSM), time division multiple access (TDMA), and Cellular Digital Packet Data (CDPD) — via circuit switched, packet, or short messaging service. It can be built into any operating system, including Windows CE, Palm OS®, Epoc, or JavaOS. The Japanese mobile operator DoCoMo is the leader, with the first-mover advantage in bringing mobile Internet services to market by attracting 10 million subscribers to its i-mode service in less than one year. The Palm VII personal digital assistant (PDA) from 3COM can deliver wireless e-mail and information access service in the United States and the United Kingdom. Most current mobile Internet services are based on the WAP standard. Microsoft, which came a bit late to the game, gave its grudging approval recently by redoing its cellular telephone browser for WAP.[1,2]

Analysts say such personalized services will be the meat of M-commerce. According to Gartner's research vice president, Phillip Redman, "the personalization of content and services that help consumers make their purchasing decisions" will be pivotal. Information is key to the overall success of M-commerce, and Cellmania and BroadVision are two wireless applications based on that premise. Cellmania's mEnterprise is intended to help companies bolster customer relations in part by increasing the productivity of traveling employees. mEnterprise integrates with a company's infrastructure and powers field-service and sales-force automation applications and mobile portals. BroadVision is offering BroadVision Mobile Solution to help businesses get a better line on the content customers want pushed to handheld devices by capturing customer data. It also can create home pages that site visitors can customize to their needs.

22.2 WILL WAP-ENABLED PHONES DOMINATE THE PERSONAL COMPUTER MARKETPLACE?

Everyday there is some news about WAP-enabled phones and their growing use toward the Internet. Is this growth going to sustain or even surpass people's expectations to become the most-important medium for communication and commerce? Will it lead the static wired E-commerce to wireless M-commerce? Although the only Internet-enabling technology being adopted en masse by handset manufacturers and service providers is WAP, there are other options such as J2ME (Java 2 Micro Edition), a mobile ASP (application service provider), a Citrix terminal solution, and an OracleMobile solution, all of which totally ignore cellular telephones and promise to satisfy all of your mobile Internet business needs over a pager. In addition, there are issues with WAP's Wireless Markup Language (WML), which cannot be read on an HTML browser and vice versa. Is Sun's J2ME, which allows a small application to run on the telephone so it can be used even when disconnected, a good solution; or is it too small and does it lack too many of the Java standard-edition components needed to create usable applications, as reported by Internet service vendors (ISVs)? In Sun's defense, Motorola displayed applications such as expense reports, e-mail, and calendaring on a Motorola iDEN cellular telephone running J2ME.[3]

In the business-to-business (B2B) environment, real-time mobile access to online exchanges, virtual communities, and auctions can be facilitated by M-commerce. Mobile workers such as sales reps, truck drivers, and service personnel will be able to use the mobile Internet. Medical doctors will be able to use their handheld PDAs to access patient information, information on available drugs, and online ordering and scheduling of prescriptions, clinical tests, and other procedures. Unified messaging services will allow mobile workers to use a single device for all their communications and interactions; and ubiquitous computing will use online connections to communicate exception reports, performance problems, and errors to service personnel.[2] Most IT executives are still on the fence, whereas a few early adopters have settled on proprietary technologies. One example is a women's accessory company, NineWest, which has a non-WAP client/server solution for its field reps and buyers deployed into older Nokia 9000 cellular telephones. Developed by the Finnish company Celesta, it creates smart forms using Short Message Service (SMS) rather than going through an ISP. This solution has reportedly been profitable for NineWest because it alerts headquarters in real-time, rather than through weekly batch files, when a store carrying its line needs to be restocked.

Similarly, NeoPoint of La Jolla, California, a developer of Web telephones, has created a wireless service called myAladdin.com that, among other abilities, can monitor information such as airline flights or stock performance, and alert a user when a flight is delayed or a stock price drops. InfoMove of Kirkland, Washington, integrates the Global Positioning System (GPS) and text-to-speech technologies to create a private-label information service that has been sold to DaimlerChrysler and Paccar, a heavy truck manufacturer. Tekelec makes equipment for wired and wireless telecommunications suppliers to enable them to offer value-added services to their customers. Because the Federal Communications Commission requires that if you switch or move, your telephone company must let you keep your old telephone market, Tekelec's local number portability (LDP) software is the best on the market and with its reseller networks such as Lucent and Tellabs, Tekelec is a strong takeover candidate.

22.3 WAP: A GLOBAL STANDARD

WAP is a format for displaying Web and other data on the small screens of handheld devices, specifically cellular telephones. WAP is a set of specifications, developed by the WAP Forum, that lets developers using WML build networked applications designed for handheld wireless devices. WAP is a standard, similar to the Internet language HTML, which translates the Web site into a format that can be read on the mobile's screen. The data is broadcast by the telephone's network supplier. WAP v1.1 constitutes the first global transparent *de facto* standard to be embraced by well over 75 percent of all relevant industry segments. WAP's key elements include

1. The WAP programming model
2. Wireless Markup Language and WML Script
3. A microbrowser specification
4. Wireless Telephony Application
5. The WAP stack.[4]

WAP is designed to work with most wireless protocols such as CDPD, CDMA, DataTAC, DECT, FLEX, GSM, iDEN, Mobitex, PDC, PHS, ReFLEX, TDMA, and TETRA.

22.4 OPERATING SYSTEMS FOR WAP

WAP is a communications protocol and an application environment. It can be built on any operating system including PalmOS®, EPOC, Windows CE, FLEXOS, OS/9, and JavaOS. WAP provides service interoperability even between different device families. WAP uses existing Internet standards, and the WAP architecture (illustrated in Figure 22.1) was designed to enable standard off-the-shelf Internet servers to provide services to wireless devices.

In addition to wireless devices, WAP uses many Internet additions when communicating standards such as XML, UDP, and IP. WAP wireless protocols are based on Internet standards such as HTTP and Transport Layer Security (TLS), but have been optimized for the unique constraints of the wireless environment. Internet standards such as HTML, HTTP, TLS, and TCP are inefficient over mobile networks,

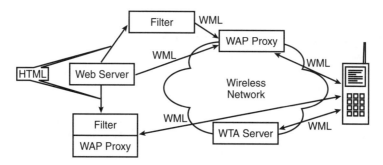

FIGURE 22.1 WAP architecture.

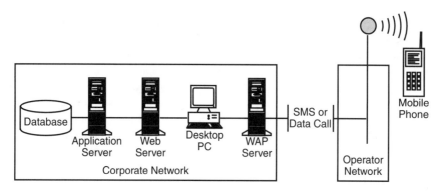

FIGURE 22.2 WAP and the Web.

requiring large amounts of mainly text-based data to be sent. Standard HTML Web content generally cannot be displayed in an effective way on the small screens of pocket-sized mobile telephones and pagers, and navigation around and between screens is not easy in one-handed mode. HTTP and TCP are not optimized for the intermittent coverage, long latencies, and limited bandwidth associated with wireless networks. HTTP sends its headers and commands in an inefficient text format instead of compressed binary. Wireless services using these protocols are often slow, costly, and difficult to use. The TLS security standard requires many messages to be exchanged between client and server which, with wireless transmission latencies, results in a very slow response for the user. WAP has been optimized to solve all these problems, utilizing binary transmission for greater compression of data, and is optimized for long latency and low-to-medium bandwidth. WAP sessions cope with intermittent coverage and can operate over a wide variety of wireless transports using IP where it is possible and other optimized protocols where IP is impossible. The WML used for WAP content makes optimum use of small screens; allows easy, one-handed navigation without a full keyboard; and has built-in scalability from two-line text displays through to the full graphic screens on smart telephones and communicators.[5] Figure 22.2 illustrates the relationship between WAP and the Web.

22.5 WAP FORUM

The WAP Forum is the industry association comprised of more than 200 members that has developed the *de facto* worldwide standard for wireless information and telephony services on digital mobile telephones and other wireless terminals. The primary goal of the WAP Forum is to bring together companies from all segments of the wireless industry value-chain to ensure product interoperability and growth of the wireless market. WAP Forum members represent more than 95 percent of the global handset market carriers, with more than 100 million subscribers, leading infrastructure providers, software developers, and other organizations providing solutions to the wireless industry (http://www.wapforum.org/).

22.6 ARGUMENTS FOR WAP

WAP is efficient at coping with the limited bandwidth and connection-oriented nature of today's wireless networks due to its stripped-down protocol stack.[2] WAP works with all major wireless networks and can be built into any operating system, including Windows CE, PalmOS®, Epoc, or JavaOS. WAP applications are available over second-generation Global System for Mobile (GSM) networks albeit only at 14.4 kbps. WAP services, however, also work on other platforms, including 2.5G (data-enhanced second generation) networks offering up to 128 kbps beginning in 2001. WAP low data rate services, already available in many European national markets, include SMS wireless e-mail, which can interconnect with the Internet. New products and services that use the WAP format provide instant access to personal financial data, flight schedules, news and weather reports, and countless shopping opportunities. Finally, WAP gateway's flexibility enables operators to introduce and bill for new services easily without having to make changes to existing billing systems.

22.7 ARGUMENTS AGAINST WAP

Although WAP has drawn a tremendous amount of attention in the business and technology sector, its huge popularity also has drawn criticism that leads one to think that WAP will not develop into a major force impacting business and life. According to David Rensin, CTO at Aether Systems, a handheld infrastructure developer in Owings Mills, Maryland, "WAP is dead." Chief among his complaints was the necessity for rewriting Web sites in WML for every device a WAP-enabled Web site is sent to. WML is used as a technique to get content from an HTML Web site using WAP to small-screen devices. "You have to rewrite the same Web site for a four-line cell phone display and again for an eight-line display," and "the problem [with WAP] is content. Redoing a Web page for multiple sites on different devices is a nightmare," according to Rensin.[1]

Handheld devices are more limited than desktop computers in several important ways. Their screens are small, able to display only a few lines of text, and they are often monochrome instead of color. Their input capabilities are limited to a few buttons or numbers, and entering data takes extra time. They have less processing power and memory to work with, their wireless network connections have less bandwidth, and they are slower than those of computers hard-wired to fast LANs.[6] Web applications are traditionally designed based on the assumption that visitors will have a desktop computer with a large screen and a mouse. A smart telephone cannot display a large color graphic and does not have point-and-click navigation capabilities. As some analysts say, these limitations will hinder WAP as the choice for tomorrow's technology.

22.8 ARE MOBILE TELEPHONES HAZARDOUS TO HEALTH?

All mobile telephones and wireless LAN devices emit microwave radiation at the same frequencies used to cook food. Now scientists are trying to determine whether end users are at risk. "We have evidence of possible genetic damage," says Dr. George Carlo, chairman of Wireless Technology Research LLC (Washington, D.C.), which has been conducting research into cellular telephones for 6 years. His study found that "using mobile phones triples the risk of brain cancer."[7] Dr. Kjell Hansson Mild in Sweden studied radiation risk in 11,000 mobile telephone users. Symptoms such as fatigue, headaches, and burning sensations on the skin were more common among those who made longer mobile telephone calls. At the same time, there are a growing number of unconfirmed reports of individuals whose health has been affected after chronic, frequent use of mobile telephones, presumably from radiation effects on cells.

There is no evidence so far of mobile phone radiation causing tumor formation or memory impairment in humans. Much more research is needed before any firm conclusions can be drawn. Whatever the effects of using mobile telephones may be in humans, the health risk to an individual user from electromagnetic radiation is likely to be very small indeed, but some individuals may be more prone to radiation side effects than others (http://www.globalchange.com/radiation.htm).

22.9 POOR SECURITY?

Furnishing full protection in a wireless world involves three types of code: (1) encryption algorithms to scramble data, (2) digital certificates to restrict access, and (3) antivirus software. Encryption, the most demanding of the three, follows a fairly simple equation: the larger the algorithm, the stronger the security, and the more CPU cycles needed. WAP-enabled telephones do not have the horsepower to handle the bulky security software designed for PCs. At this point all handheld devices, including PDAs, are vulnerable to any virus that comes along. It is worth noting that there are currently no known viruses that attack wireless gear, but as mobile IP gains popularity, it will become an increasingly attractive target. "It's conceivable one could have a worm virus similar to explore.zip that could spread to every person's device in a matter of a few seconds," says Nachenberg.[7]

22.10 WAP AND M-COMMERCE

The average mobile telephone is essentially a dumb device: good for allowing people to chat, but hopeless when it comes to managing the information that makes people's lives go round. For the past few years, the wireless industry has been engaged in a gargantuan effort to change this. The idea is to create a single smart gadget that will allow people to check their e-mail, consult the Internet, plan their schedule, and, of course, make telephone calls; in other words, a combination of an electronic organizer, a personal computer, and a mobile telephone.

Toward M-commerce applications, Sonera of Finland, which has implemented an Apion WAP gateway, is the world's first telecom operator to launch WAP services (Spring 1999). In addition to providing its own services, the telco/cellco is actively and rapidly creating partnerships with companies such as Finnair, CNN Interactive, Yellow Pages, Tieto Corporation, and Pohjola.[4]

In April 2000, a company in California called Everypath started to deliver a new era of freedom in mobility and convenience which enabled a user to shop, purchase gift certificates, bid on auctions, trade stocks, play games, pay bills, purchase fine wines, get driving directions, check the calendar, reserve a hotel room, track home prices, plan a vacation, stay in touch, or order tickets from the palm of the hand or with the sound of the voice, regardless of the user's location.

In Japan, NTT DoCoMo has sold more than 1 million of its Internet-based i-mode telephones in the six months since they were launched, and received remarkably few complaints. The rest of the world's producers are getting ready for a surge in demand as they release their products over the next few months.

Internet content providers are already tailoring their products for telephone users: getting rid of power-hungry pictures, for example, and distilling long-winded news stories into the bald facts. Nokia has an alliance with CNN to provide news that has been specifically designed for telephones. NTT DoCoMo reports that there are already more than 1000 companies providing Web pages for its telephones.[8]

22.11 CRITICAL SUCCESS FACTORS FOR M-COMMERCE

22.11.1 SPEED

Today, most digital cellular users are limited to circuit-switched data at about 9.6 kbps, sufficient for text-based messaging and limited file transfer. This is where desktop Internet users were in 1994, when there were just 4 million host computers on-net compared with more than 60 million Internet hosts worldwide in October 1999. The next move in the circuit-switched world is high-speed circuit-switched data (HSCSD), running at 57.6 kbps. This is sufficient for fully functional Web browsing. However, as underlined by analysts such as Gartner Group's Dataquest, HSCSD is an early adopter scenario that gives operators a competitive edge with corporations. Essentially, it is profiled for bulky data transfers.

Conversely, General Packet Radio Service (GPRS) is quick and agile. As a packet-switched bearer, it promises "always-on" service at up to 115 kbps (for practical purposes). At the same time, it sits comfortably on the migration path to

Enhanced Data for GSM Evolution (EDGE), running at up to 384 kbps. So, although speed may be a concern for WAP surfers now, technology will enhance that in the very near future.[4]

22.11.2 Billing

The WAP gateway has been profiled to gather extensive billing detail for each transaction, e.g., the download of content (both volume and time), universal resource locators (URLs) visited, and other typical events during a WAP session. This information is stored in a generic, flexible format in a billing log. This, in turn, interfaces to a mediation platform, which translates it into valid call detail records (CDRs) and passes them to the billing agency or credit card company's billing system. The billing could be

1. Transaction-based, where the services are paid according to service usage, with different prices possible for different services
2. Subscription based, with a monthly fee
3. Flat rate, with one price for all
4. Free, where the content provider may pay the operator for the airtime
5. A combination of the four billing options

The billing log receives "billable events" from the event manager. The gateway's billing data interface requires only minor tuning to adjust its data formatting for different billing systems. In short, the WAP gateway's flexibility enables operators to introduce and bill for new services easily without having to make changes to their existing billing systems. However, service roaming is difficult if transaction-based billing is used. The Holy Grail is turning the handheld device into a payment device or the equivalent of an electronic wallet. As we move toward the third-generation (3G) mobile standard, also known as Universal Mobile Telecommunications System (UMTS), an International Telecommunication Union (ITU) standard for voice, video, and Internet services licensed in Europe in 2000 and deployed in 2002, airtime is packet-based with an emphasis on content. The billing possibilities are

1. Monthly fee (similar to the Internet model)
2. Amount of data, or time based
3. Commercials
4. Service transactions
5. A combination of these options

Billing is a very market-sensitive problem and one solution is not possible. Without a doubt, the biggest change will be more choices, and in the end, markets will decide between free versus price for M-commerce.

22.11.3 Security

Security is optional in the WAP standard, but is clearly mandatory for E-commerce providers and users. It may be implemented initially at the Wireless Transport Layer

Security (WTLS) level of the WAP stack. This is the wireless version of industry-standard Transport Layer Security (TLS), equivalent to the widely deployed Secure Sockets Layer (SSL) 3.1. As a recent Baltimore Technologies white paper notes, it provides a secure network connection session between a client and a server, and it most-commonly appears between a Web browser (in WAP's case, the handset microbrowser) and a Web server, which can be an existing Internet server that is also WAP-enabled.

Full participation in E-commerce requires that the additional security elements of verified authentication, authorization, and nonrepudiation aree addressed. In real terms, this implies integration with public key infrastructure (PKI) systems that are already deployed and with new systems in the future. In the wireless arena, these systems will be defined in WAP.[4] Citing the growth in usage of wireless devices, Richard Yanowitch, VeriSign's Vice President of Worldwide Marketing, said that his company plans to provide "a complete trust infrastructure to the wireless world." Key to the plan is an arrangement whereby Motorola will include VeriSign technology in the browsers that run on Motorola mobile telephones. Other companies endorsing VeriSign's plan include RSA Security, BellSouth, Sonera SmartTrust, and Research In Motion. These companies will leverage the technologies in their own products and services. For instance, technologies and services available from VeriSign include:

- Microclient Wireless Personal Trust Agent code for embedding in handheld devices to enable seamless use of private keys, digital certificates, and digital signatures available to device manufacturers now.
- Short-lived wireless server certificates, "mini-digital certificates," according to officials, that are optimized for authentication of wireless devices and services.
- A gateway-assisted Secure Sockets Layer (SSL) trust model to enable network service providers to substitute wireless certificates for SSL certificates.
- A gateway-assisted public key infrastructure roaming model to enable small-footprint devices to digitally sign transactions.
- Subscriber trust services for secure messaging and transactions using wireless handheld devices.
- Server/gateway trust services designed to allow electronic businesses operating wireless servers and gateways to deliver secure applications.
- Developer trust services for digitally protecting downloadable content.
- Enterprise trust services for wireless, B2B, and B2C applications such as banking, brokerage, healthcare, and messaging.
- Service provider platforms for network operators and applications service providers to offer VeriSign wireless trust services.

Transaction services offered include Wireless Validation Services for real-time certificate validation, and Wireless Payment Services, which enable wireless payment applications.[9]

22.12 FUTURE IMPACT: GENERATION "W"
IN A WIRELESS WORLD

A new study by International Data Corporation predicts that the number of wireless Internet subscribers will jump from 5 million in 2000 to nearly 300 million in just 3 years. That would account for more than half of all Internet users worldwide. WAP's impact on mobile data would be similar to what Netscape's impact was for the Internet: to provide an attractive and notionally transparent portal to the cyber world, which had more than 200 million users in September 1999, in addition to thousands of corporate intranets. For E-commerce providers, that portal provides a potential user base of more than 400 million mobile subscribers worldwide because the Internet is ultimately about E-commerce. Although it includes a vast range of so-called "free" services — e-mail, social networks, consumer networks, a range of educational tools, computer games, and more — it is all about global economic activity and productivity. For the vast majority of fixed-network Internet users, E-commerce is still essentially only 1 to 3 years old; Amazon.com was not a household word in 1996. Internet banking, brokering, and financial services were not yet deployed into the mass market.

Yet this E-commerce world of B2B, retail banking, brokering, insurance, financial services, and purchase of almost any good or service is commonplace. There is no reason why the Internet space should not be embraced by mobile users in the same manner, subject to some differences in their marketing profile.[4] Salespeople, for example, are provided, through a wireless database access, the information needed to close a deal on the spot. Prices and delivery dates can be checked, orders can be entered, and even payments can be made without stepping outside the customer's office. That boosts the hit ratio, eliminates paperwork (and low-level administrative positions), improves customer service, and speeds cash flow.

Similar to the Internet revolution, this mobile makeover will change forever the way companies do business. Out of the office will no longer mean out of touch. In fact, remote employees may make wireless a way of life, so they do not have to dial in for e-mail and other information. Companies will be able to reinvent business processes, extending them directly to the persons in the field who deal directly with customers. Ultimately, companies and carriers could deploy wireless LANs to hotels and other public places, creating hot spots of high-speed connectivity for M-commerce. In the future, the ideal mobile device will be a single product suited for standard network access and services to handle tasks that extend the use of the device beyond its hardware-based limitations.

A U.K.-based consultancy's analyst predicts that 70 percent of current cellular users in developed countries may be using advanced data services by 2005, with the value of the cellular data market overall set to reach $80 billion, from a very low base in 1999. The takeoff of cellular data is attracting a host of new players to the mobile communications market, including Internet-based companies such as Netscape, Amazon, Excite, Microsoft, IBM, and Cisco. Media companies such as CNN, Reuters, and ITN are examples of earlybird providers.[4]

As for the United States, the number of people using cellular telephones for wireless data skyrocketed from 3 percent of the online population to 78 percent over the 12 months from January through December 2001. The main reason for the increase is that employers are starting to pay for these services, according to a survey released by New York-based Cap Gemini America and Corechange, Inc., a wireless portal provider based in Boston. Currently, 33 percent of the U.S. online population uses cellular phones for business purposes. Of that 33 percent, 11 percent (or 3 percent of the total online population) uses them for data applications such as e-mail and news, the companies say. By the end of 2001, 78 percent of the U.S. online population will be using cellular telephones for data. According to this survey of 1000 U.S. Internet users, which was conducted by Greenfield Online, Inc. on behalf of Cap Gemini, 47 percent of those who will begin using cellular telephones to access data in the year 2001 said they would do so because someone else, mainly their employer, would begin paying for it. "This was the most important reason for adoption of the new technology," said David Ridemar, head of Cap Gemini America's E-Business Unit. Of those who will start accessing data with their cell phones in 2001, 52 percent said they will use the functionality for a mix of e-mail, personal data, and business information, 24 percent will use it for e-mail and personal data, and 13 percent will use it for e-mail only.[10]

Jupiter Communications forecasts a jump in consumer-to-consumer auctions from $3 billion in 1999 sales to $15 billion in 2004. These numbers are significant because auctions are a natural match for wireless providers for the following reasons:

- Wireless auctions require much less bandwidth and data than a typical E-commerce Web site.
- The time-sensitivity of auctions makes it much easier to access over WAP-enabled phones or PDAs such as the Palm VII (compatible with eBay) or Research in Motion's 957 wireless handheld compatible with Bid.com.

Indeed, it is suggested by some analysts that cellular subscriber numbers will top 1 billion by 2004, a substantial number of them WAP-enabled. Clearly, giving mobile users the same mobile data connectivity that fixed network Internet users enjoy could more than double the potential global Internet market at a stroke.

The Gartner Group's Nigel Deighton maintains that, given current penetrations of mobile and Internet markets, the stage is set for a global boom in M-commerce that could largely ignore the PC in favor of mobile devices. He predicts further that some 30 to 50 percent of B2B E-commerce will be carried out via a mobile device by 2004.[4] Motorola, for example, estimates that by 2005 the number of wireless devices with Internet access will exceed the number of wired ones. These smart new telephones will not only give another boost to the sale of mobiles, but they will change the nature of the Internet economy, making personal computers far-less important, yet at the same time tempting many more people onto the information superhighway.[8]

I strongly believe that trade cannot be tied to wires. As so much research indicates, a major part of the workforce is heading toward location independence. The PC-based Internet has already redefined the nature of doing business, giving birth to popular E-commerce. However, to be truly location independent and to be "anytime, anywhere," the PC is not the choice for B2B and B2C M-commerce.

Necessity is the mother of all invention. M-commerce is already becoming a necessity in this age of the digital economy. In conclusion, the world is betting on M-commerce, in a manner reminiscent of the 1999 United States bet on Internet commerce. We can safely predict many losers, and a few winners, from the worldwide run to mobile Internet services.

References

1. Schwartz, E., Wireless Application Protocol draws criticism, available at http://www. cnn.com/2000/tech/computing/03/14/wap.critics.idg/index.html, March 14, 2000.

2. Herman, J., The coming revolution in M-commerce, *Business Communications Review,* October 2000, pp. 24–24.

3. Schwartz, E., WAP: The technology everyone loves to hate, available at www.infoworld. com, June 23, 2000.

4. Murphy, D., The mobile economy becomes a reality, *Telecommunications,* 33 (11), 31–34, 1999.

5. Johnson, A.H., WAP, *Computerworld,* November 1, 1999, pp. 33–44.

6. Furchgott, R., Web to go — sort of: today's net phones are O.K. for e-mail, but surfing is a chore, *Business Week,* February 14, 2000, p. 144.

7. Saunders, S. et al., Wireless IP: Ready or Not, Here It Comes, *Data Commun.,* 28 (12), 42–68, 1999.

8. Woolridge, A., Survey: telecommunication — in search of smart phones, *Economist,* October 1999, 353(8140), pp. 12–16.

9. Krill, P., Verisign aims to secure wireless transactions, available at http://www. cnn.com/2000/TECH/computing/01/19/verisign.secure.idg/index.html, January 19, 2000.

10. Trombly, M., Web access via cell phone to skyrocket this year available at http://www.cnn.com/ 2000/TECH/computing/04/18/data.cell.idg/index.html and http:// www.cnn.com/2000/TECH/computing/ 01/19/verisign.secure.idg/index.html, 2000.

23 Wireless Internet in Telemedicine

Kevin Hung and Yuan-Ting Zhang

CONTENTS

0-8493-1502-6/03/$0.00+$1.50
© 2003 by CRC Press LLC

23.1 INTRODUCTION

23.1.1 DEFINITION OF TELEMEDICINE

The term *telemedicine* consists of two parts. The first part, "tele," means "at a distance," so basically telemedicine is the practice of medicine at a distance. The evolution and emergence of various communications technologies, such as the telephone, television, computer network, and wireless communication, have been enhancing the feasibility and diversity of telemedicine applications over the past few decades. Hence, the meaning of telemedicine also is being refined as time goes by.

One of the current definitions of telemedicine is as follows:

> Telemedicine involves the use of modern information technology, especially two-way interactive audio/video telecommunications, computers, and telemetry, to deliver health services to remote patients and to facilitate information exchange between primary care physicians and specialists at some distance from each other.[1]

Some might have come across the term "telehealth" and wonder what this has to do with telemedicine. Telemedicine can be considered a subset of telehealth. According to the World Health Organization (WHO), telemedicine is oriented more toward the clinical aspect, while telehealth is generally the delivery of health care services at a distance.[2]

23.1.2 AREAS OF TELEMEDICINE APPLICATIONS

There are two modes of operation for telemedicine: real-time and store-and-forward.[3,4] In real-time mode, the information, which can be any combination of audio, images, video, and data, is transmitted to the remote terminal immediately after acquisition, thus allowing real-time interaction between patients and health care personnel. Some examples are consultation with a remote doctor, ambulatory electrocardiogram (ECG) monitoring, and instant transmission of ultrasound images. Because real-time applications involve continuous, time-critical information exchange, high transmission bandwidths are often required, resulting in relatively high operation cost.

In store-and-forward mode, the acquired information is viewed or analyzed at a remote terminal at a later time, so it is less demanding in bandwidth. Diagnosis and clinical management are the main applications of store-and-forward telemedicine.[4] Some examples of these are transfers of previously recorded ECG and computer tomography (CT) scans. Table 23.1 shows some areas of telemedicine.

23.1.3 THE NEED FOR TELEMEDICINE

Why do we need telemedicine? Obviously, it is the only way to deliver medical services in cases of emergency in remote, isolated areas where no medical professional is present. Immediate medical advice received at the patients' side of a teleconsultation system would be valuable and would save lives. Many countries do

---

TABLE 23.1
Areas of Telemedicine

Telemedicine Applications

Teleradiology	Telepsychiatry
Telepathology	Telecardiology
Teledermatology	Tele-homecare
Teleoncology	Tele-surgery
Remote patient monitoring	Teleconsultation

---

not lack medical facilities; instead they face the problem of uneven distribution of these resources. This problem has triggered the development of telemedicine systems for sharing resources between areas. Such systems not only can reduce cost, but also can eliminate the need for patients or health care staff to travel. For example, many hospitals today are already using a picture archiving and communication system (PACS) and a hospital information system (HIS) for information exchange between hospitals and clinics.[5]

Another motivation for telemedicine development is the world's aging population. The number of persons aged 60 years or older is projected to be almost two billion by 2050, and it will be the first time in human history that the population of older persons will be larger than that of children (0 to 14 years).[6] These statistics have triggered the United Nations to discuss many social development issues related to meeting the specific needs of the elderly, including requirements for health care.[7] The prevalence of chronic conditions among the elderly is generally higher than among younger persons.[8,9] Because symptoms for chronic illnesses tend to be more subtle and vague, recognition and diagnosis of disease in the elderly requires a high degree of alertness. From these it can be seen that telemedicine applications such as home-based teleconsultation and patient monitoring would definitely be useful in health care support for the large elderly population.

23.1.4 CHAPTER OVERVIEW

The following sections of this chapter include a brief history of telemedicine and a review of how Internet-based telemedicine emerged. Some past and current telemedicine applications based on wireless Internet are then described, along with a recent example in detail. Finally, issues in practicing telemedicine with wireless Internet are discussed.

23.2 TELEMEDICINE APPLICATIONS

23.2.1 BRIEF HISTORY OF TELEMEDICINE

Primitive forms of telemedicine had already been practiced hundreds of years ago. An example is the lepers' use of bells to warn others to stay way. In the Middle

Ages, information about bubonic plague was transmitted across Europe using bon-fires.[4] Some wealthy families back then also sent urine samples to their doctors for diagnosis.[10] The telegraph was used to transmit casualty lists and medical supplies orders in the American Civil War,[11] and also in the early 1900s for medical consul-tations, diagnosis, and transmission of dental x-rays.[12,13] A description of using the telephone to transmit amplified sounds from a stethoscope for remote auscultation appeared in 1910.[4,10] In 1920, the Seaman's Church Institute of New York probably became the first organization to provide medical care using radio.

Telemedicine using modern communication technologies only appeared in the past few decades. The first practices of true telemedicine occurred around the 1950s. An article in 1950 described the transmission of radiologic images between West Chester and Philadelphia using the telephone. Based on this, radiologists at Jean-Talon Hospital, Montreal, set up a teleradiology system in the 1950s. Telemedicine using interactive video started in 1959, when a two-way closed circuit television (CCTV) was used to transmit neurological examinations and other information at the University of Nebraska.[11] In 1964, they established a link with the Norfolk State Hospital to provide services such as speech therapy, neurological examinations, diagnosis of difficult psychiatric cases, case consultations, and education and train-ing. By the 1960s, the National Aeronautics and Space Administration (NASA) and the U.S. Indian Health Service deployed a satellite-based telemedicine system, which included mobile examination rooms, x-ray imaging, and ECG facilities, in the Papago Indian reservation. Another early example of mobile telemedicine is the Alaska ATS-6 Satellite Biomedical Demonstration, which assessed the viability of improving village health care in Alaska using satellite for video consultation.

23.2.2 INTERNET-BASED TELEMEDICINE APPLICATIONS

Telemedicine has advanced tremendously in the past few years, due to the growth of the Internet and availability of low-cost personal computers (PC). By the mid-1990s, many had started to explore medical applications on the Internet. To begin, PC applications for Web browsing, e-mail access, and file transfer were proposed for medical information exchange.[14] The first telemedicine applications using the Internet operated in store-and-forward mode. As an early example, in 1993 a medical team in London transmitted ultrasound images of a fetus to the Fetal Treatment Program in San Francisco for surgical opinion.[15] In 1996, Yamamoto et al. demon-strated digitization and transmission of radiographic and medical images using a scanner, a digital camera, and PCs connected to the Internet.[16] Scanned radiographic images and digital medical images were stored as JPEG files in a PC, and then transmitted over the Internet between Hawaii, Tennessee, and Texas using a file transfer protocol (FTP) program, the World Wide Web, and Prodigy. Connections established via high-speed local area network (LAN) and via 14.4k baud modem were tested. Many health care professionals today are still using e-mail and other forms of store-and-forward technologies for low-cost telemedicine.[17,18]

The Internet has been used also in remote patient monitoring. Magrabi et al. developed a Web-based longitudinal ECG monitoring system that stored recorded ECGs at a Web server for remote offline analysis and viewing.[19] Park et al. set up a real-time patient-monitoring system based on 100 Base-T Ethernet LAN.[20] Besides providing patients' general information, it allowed doctors to monitor patients' ECG, respiration, temperature, blood oxygen saturation, and blood pressure in real-time. Teleconferencing facilities also were available. There are other similar Internet-based systems that acquire vital signs through a short-range wireless link.[21-23]

23.2.3 IMPORTANCE OF MOBILITY IN TELEMEDICINE

Advances in telecommunications and mobile computing have stimulated numerous researches and developments in mobile telemedicine applications over the past few years. These applications are becoming more and more feasible as technologies evolve. Telemedicine is no longer limited to usage only within hospitals and clinics. Its coverage is expanding into homes, workplaces, outdoors, airplanes, and even into outer space. Mobility is obviously a trend in telemedicine.

Telemedicine is probably the only option in remote, isolated areas where no medical facility is available. Equipment used would not be stationary, but should be designed for long-distance transport and easy setup. Efforts in providing an integrated health care service to patients have eventually lead to emphasis in home health care, patient monitoring, and other patient-centered telehealth programs. In these cases, the patients or medical staffs often are not in the hospital, and telemedicine based on mobile communications acts as a bridge between the two parties. Mobile telemedicine has been used also in emergencies, where immediate attention and consultation are needed.

23.3 WIRELESS INTERNET IN TELEMEDICINE

23.3.1 TELEMEDICINE USING CELLULAR TECHNOLOGIES

As mentioned in the last section, mobility has become a factor determining the feasibility of telemedicine in various cases. Because many of these applications are based on the Internet, and mobility is required, wireless Internet appears as an attractive option. Modems for cellular data transmission were soon available in the 1990s after cellular phones hit the market. A study in 1995 presented the possibility of wireless teleradiology with some wireless modems commercially available at that time.[24] Files containing computer tomography (CT) and x-ray images that were scanned and stored in a PC were sent to a remote portable notebook via cellular modem. A Motorola Digital Personal Communicator cellular phone connected the portable wireless modem of the receiving side to the cellular communication system. They tested with Motorola's pocket modem, based on the CELLect protocol; AT&T's Keep-In-Touch cellular modem, utilizing the Extra Throughput Cellular (ETC) Protocol; and Megahertz's card-type modem, which used the Microcom Networking Protocol (MNP-10) for circuit-switched connection.

Cellular Digital Packet Data (CDPD), the first digital data application to use packet data for cellular phones, came out in 1992 to provide wireless data service. Based on TCP/IP protocols, it provides packet data communication at 19.2 kbps. Starting in 1999, Yamamoto's group attempted viewing CT images on a remote pocket computer equipped with a wireless digital modem that used the CDPD data network.[25,26] In one demonstration they downloaded five sets of CT images, saved in JPEG format, from a Web server to a Hewlett Packard 620LX and to a Sharp Mobilon 4500, using the Sierra Wireless Air Card 300. Device turn-on time plus download time ranged from 4 to 6 minutes, and the image quality was satisfactory. Then in 2001, the group performed similar tests with downloading 12-lead ECG recordings, which were saved as either JPEG or Internet fax, from the Web server to a Hewlett-Packard Jornada 680 pocket computer.[27]

Many GSM phones now have built-in traffic-channel modems. Via a local cable, infrared or Bluetooth™ link to the phone, a notebook computer or a personal digital assistant (PDA) can wirelessly connect to the Internet. A phone equipped with a browser application also can access the Internet itself. Numerous telemedicine applications have used GSM-based cellular data modems for data transmission. The following are just some examples.

In 1997, Giovas et al. in Greece investigated the feasibility of store-and-forward ECG transmission from a moving ambulance to a hospital-based station for prehospital diagnosis.[28] An ambulance was equipped with an ECG recorder connected to a notebook computer, which coupled to a GSM telephone via a PCMCIA data card. Data rate was 9.6 kbps. Curry and Harrop in the United Kingdom also had a similar idea of mobile telemedicine in the ambulance.[29] They tested a telemedicine ambulance installed with three cameras and a transmitting module, which also was based on GSM phone data connection at 9.6 kbps. A frame of the digitized video was sent to the hospital every 4 seconds. These pictures were received and displayed by a PC with modem at the A&E department.

The AMBULANCE project in 1998 went a step further.[30] Pavlopoulos' group developed a portable emergency telemedicine device that supported real-time transmission of critical biosignals as well as still images of the patient. The mobile station consisted of a notebook computer with CCD camera, a GSM modem from Siemens, and a biosignal monitor. Through TCP/IP over GSM and data rate of 9.6 kbps, three-lead ECG, blood pressure, oxygen saturation, heart rate, temperature, and still images were transmitted from the mobile station to the hospital consultation unit.

In the same year, Reponen et al. demonstrated CT examinations on a remote notebook computer that wirelessly connected to a computer network via a GSM cellular phone.[31] The notebook was equipped with a PCMCIA digital cellular data card that interfaced the computer to the phone. CT images, each 256 kb in JPEG format, were stored in a network directory in a Linux-based PPP server, which provided TCP/IP connections between the notebook and the LAN of the Department of Radiology of a hospital in Finland. After dialing into the PPP server, images were downloaded with an FTP program. At a nominal data rate of 9.6 kbps, average transfer time for a single CT slice was 55 seconds. Neuroradiologists' diagnoses from the images at the notebook were the same as that from original images in 66 cases and slightly different in two. Two years later, the group carried out similar tests with a GSM-based wireless PDA.[32] They downloaded the CT images using a Nokia 9000 Communicator equipped with FTP software. This time the PPP server was set up using Windows® NT remote access service (RAS). The PDA was found to be suitable for the reading of most common emergency CT findings for consultation purposes.

Besides common cellular networks such as the GSM, other proprietary wireless networks also have been used in telemedicine. In 2000, Karlsten and Sjöqvist described an information management system that utilized a network called Mobitex™, which was developed by Swedish Telecom.[33] The system was integrated into the emergency ambulance service in Uppsala County, Sweden, for in-ambulance and prehospital use. It consisted of stationary and mobile workstations that communicated via Mobitex™ on the 80 MHz channel at 1200 bps or via GSM. One function of the system was transmission of ECG and other data from mobile ambulance workstations to the stationary hospital workstations at predefined intervals.

23.3.2 TELEMEDICINE USING LOCAL WIRELESS NETWORKS

Thus far we have highlighted telemedicine applications that used cellular devices and networks. However, another technology often used in forming a wireless Internet link is a local wireless network. Zahedi et al. described a mobile teleconsultation system for video communication between a ward within a hospital and a remote physician situated outside the hospital.[34] Video stream captured by a camera was converted into IP packets by a software and Web server running in a notebook computer at the patient module. The wireless spread spectrum link between this patient module and the ISDN modem in the relay module connecting at 128 kbps allowed connection from the outside. At the physician side was a multimedia desktop PC equipped with an ISDN modem and a Web browser.

The Georgia Tech Wearable Motherboard™ (GTWM), developed at the Georgia Institute of Technology, was a vest that could be used to monitor vital signs, such as ECG, body temperature, and respiration. In 2000 Firoozbakhsh et al. set up a prototype wireless link between the GTWM and a LAN.[35] Acquired ECG waveform was digitized at a notebook terminal, and transmitted across an IEEE 802.11 Wave-LAN wireless network. Besides being accessed by other terminals connected to the LAN, the system could also be expanded to enable remote access over the Internet. Wireless LANs also have been utilized in other medical informatics systems for storing and retrieving medical images.[36]

23.3.3 TELEMEDICINE USING SATELLITE COMMUNICATION

In cases where telemedicine is practiced at places that are beyond the reach of wireless networks or even wired telecommunications services, satellite communication becomes the only option for Internet access. Satellite-based telemedicine is commonly practiced worldwide. For example, Dr. Bernald Lown started SatelLife in 1989, and initiated a medical information-sharing network called HealthNet.[37] Utilizing relatively cheap, low earth orbit satellites, the service provides store-and-forward Internet access to health professionals around the world. Whenever the satellite comes in range of a ground station, it exchanges messages with it. Messages received by the satellite are stored and later delivered to SatelLife's headquarters in Boston, where they are forwarded to other HealthNet users or via Internet to other

Internet users. If users want to surf the Internet, they use a special Web-browsing software to issue requests, which are later processed at headquarters. The desired information is then sent to the users in subsequent message exchanges between satellite and users' stations.

Another recent example is a Web-based PACS developed by Hwang et al. in 2000.[38] An asymmetric satellite data communication system (ASDCS) allowed a remote hospital to download medical information from the telemedicine center's server. The client side had a PC installed with equipment capable of Ku-band and C-band satellite links. The radiological images and patient information could be viewed on a typical Web browser with the help of a Web application written in Java.

23.4 CASE STUDY: WAP IN TELEMEDICINE

23.4.1 OBJECTIVE

As seen in the last section, numerous telemedicine applications are based on the Internet and the mobile phone. Some of the recent examples even reflect the convergence of wireless communications and computer network technologies.[39] This trend also is realized with the emergence of new mobile phones, PDAs, cellular modems, wireless infrastructure and networks, and mobile application programming languages and protocols. There are various application protocols, such as Wireless Application Protocol (WAP) and i-mode; application technologies, including Java Version 2 Micro Edition (J2ME™); and operating systems, such as Symbian, Pocket PC 2002, and Palm OS®, aimed at supporting various wireless devices.

WAP-enabled devices are now commonly available. As WAP also will be a feature found in various future handheld devices, it is worthwhile to investigate its possible use in telemedicine. This section describes the implementation and experiences with a WAP-based telemedicine system recently developed.[40–41] It was tested with an emulator and with a WAP phone using wireless connections provided by a mobile phone service provider in Hong Kong. Store-and-forward monitoring and analysis of wireless ambulatory ECG and other parameters also were demonstrated.

23.4.2 METHOD

23.4.2.1 System Specification

The WAP programming model is shown in Figure 23.1. A handheld WAP device communicates with a content server, which stores information and responds to the users' requests. A WAP gateway translates and passes information between the device and the server.[42] To access an application stored at the content server, the device's WAP browser first initiates a connection with the gateway before requesting for content. The gateway converts these requests into HTTP format for the server. After processing, the server sends the content to the gateway, which then translates it into WAP format for the WAP device. The layers of WAP protocols that govern the communication are shown in Figure 23.2.

To determine which telemedicine applications are feasible with WAP, it is important to examine the capabilities of a typical WAP device. Such a device has limited

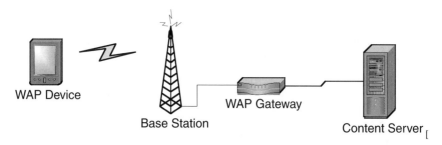

FIGURE 23.1 WAP programming model.

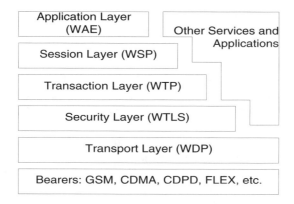

FIGURE 23.2 WAP architecture.

processing power, memory, battery life, display size and resolution, and entry capability. Compared to wired networks, most currently used wireless networks for WAP have low bandwidth, resulting in perceptible delay between request and response at the mobile device. Due to the nature of such a network, requests and responses are required to be concise for minimal latency. This latency depends on the type of bearer used. With a GSM network, some possible bearers are SMS (short message system), CSD (circuit-switched data), and GPRS (General Packet Radio Switching).[43] WAP over SMS is the most time consuming of the three. A CSD connection requires several seconds for initial setup before data transfer, and the typical data rate is 9.6 kbps. GPRS, which provides data rates up to 171.2 kbps, is already a commercially available service. It does not need the long connection time as with CSD. When a GPRS phone is switched on, it is always online and is ready to start receiving and sending data in less than one second.

WML (Wireless Markup Language) is designed for creating WAP applications, and is user-interface independent. It supports text, images, user input, variables, navigation mechanisms, multiple languages, and state and management-server requests. WML has been designed for the high latency and narrow band of the wireless network, so connections with the server should be avoided unless necessary. As a result, WAP is mainly intended for displaying text content. Wireless Bitmap

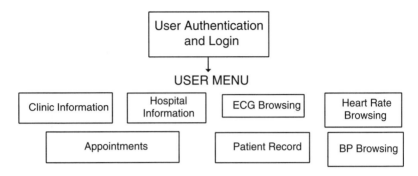

FIGURE 23.3 General features of the WAP-based telemedicine system.

Format (WBMP) is a graphics format optimized for efficient transmission over low-bandwidth networks and minimal processing time in WAP devices.

These technical capabilities suggest that use of current WAP devices in telemedicine is feasible in applications operating in a store-and-forward, client/server, and low-bandwidth fashion. The displayed information is limited to text and low-resolution WBMP static images. When displaying graphical information, it is better to first construct the image at the server, thus reducing the usage of memory and processing time at the device.

Based on these requirements, a WAP-based telemedicine system has been developed. Applications include viewing of general patient information, previously captured BP and heart rate readings, and recorded ECG waveforms. Other functions are remote request for doctors' appointments and general inquiries on clinic and hospital information. Targeted users are both doctors and patients. Figure 23.3 shows general features of the system.

An ECG browsing function is included because of the increasing need for ambulatory ECG monitoring. In 1999, heart disease accounted for 30 percent of all deaths worldwide.[44] Monitoring services would allow early detection and diagnosis of pathological symptoms, and thus lead to earlier treatment. A major concern for displaying the ECG is the small screen size and low display resolution of a WAP phone. A group has tried displaying ECG on a 160 × 144 pixels LCD gray-scale display of a handheld video game platform, and it was shown that some basic features such as R-R intervals were still recognizable with the user's selection of the lead displayed and time scale used.[45] Noting that the display resources on a WAP device are similar to those of the game platform, ECG browsing with the WAP-based system should be possible.

23.4.2.2 Overall Architecture

The developed system was set up for testing the feasibility of telemedicine with WAP. Figure 23.4 shows the architecture for the connection between a WAP device and the content server. Applications were stored in the content server. Part of the user interface was written in WML and WMLScript, which executed at the WAP device after being downloaded from the server. The other part of the application,

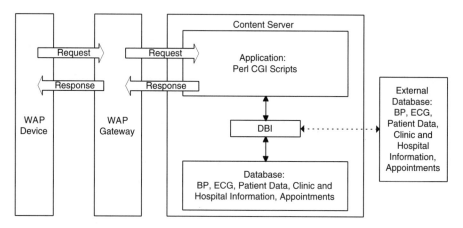

FIGURE 23.4 Structure of the system.

written in Perl, executed within the Linux-based content server.[46] It provided the common gateway interface (CGI) for more-complex tasks. Using the GD and CGI modules, the Perl program could dynamically create WBMP graphics and WML decks upon requests from the WAP device. Patient data that the applications accessed and manipulated were stored in a relational database system.

23.4.2.3 Relational Database

A relational database is made up of tables and columns that relate to one another. MySQL is a relational database management system (DBMS) that can handle multithreaded operations.[47] It also has many application programming interfaces (API), including Perl, TCL, Python, C/C++, JDBC, and ODBC, thus enabling access to the database by applications written in various languages. MySQL uses the Structured Query Language (SQL) to manipulate, create, and display data within a database.

For the telemedicine system, a MySQL database system consisting of two databases at two different sites was set up to store data, including BP and heart-rate readings, patient records, clinic and hospital information, doctors' appointments with patients, and recorded ECG. One database resided in the content server, and another in a remote PC. During WAP access, data was retrieved by the applications through Perl's Database Interface (DBI), as shown in Figure 23.4. By specifying permissions given by each database, the application could access data not only in the content server, but also data in remote databases. Figure 23.5 shows the entity-relationship (ER) model, which is a high-level conceptual representation of data contained within the database.[48]

23.4.2.4 Program for WAP Access

The flow of the program starts when the user accesses the first WML deck at a predefined site. The following WML decks that the user interacts with are then generated by Perl. The user first logs into the WAP site and loads the first card (login.wml), which prompts for username and password. Using CGI with method

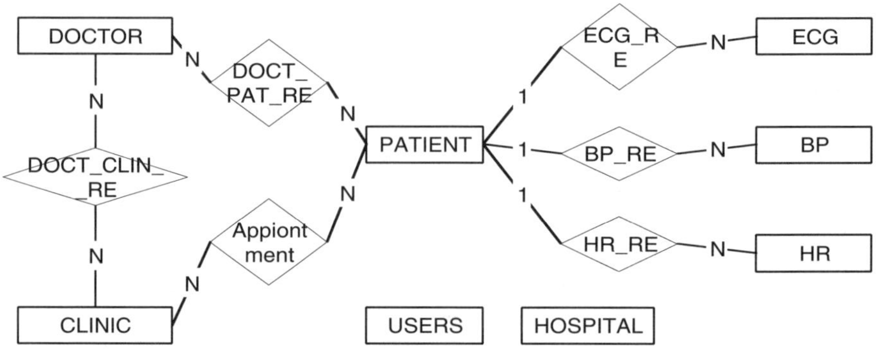

FIGURE 23.5 Entity-relationship model of the database.

'post,' Perl first takes the user inputs. It then accesses the TABLE USER of the database through the DBI. If the user chooses to view patient information, the patient ID must be entered. A menu is then generated if the patient ID is valid and accessible. The user has the choice of viewing general information, a single blood pressure reading, a day log of blood pressures, ECG browsing, and heart-rate reading. The process flow is shown in Figure 23.6.

The menu for single blood pressure reading works in a similar way. After a list of BP recording sessions is displayed, the user chooses the session. Perl then searches through TABLE BP, and charts out the date, time, pulse rate, and the systolic, diastolic, and mean pressure values in WML format. The day log for blood pressure allows the user to view all the pressure values of a day in chart or graphical form. To display BP graphs, the program first searches through TABLE BP and stores the necessary values in a temporary array, which is then used along with the GD module to create a WBMP file called by the WML card. Scrollable graphs are presented by updating the array with a new set of data from the database whenever the user chooses to scroll forward or backward.

When a user browses through the ECG waveform for a specified recording session, the program first searches through TABLE ECG for details, such as sampling rate, recording time, and duration. Recorded ECG data for each session is stored in a separate ECG data table named according to the names in TABLE ECG. To display the ECG, the program traverses through the ECG data table with a pointer, and loads the necessary data points into a temporary array. According to the information in TABLE ECG, it creates a WBMP for each new frame of ECG waveform. The user also can view the waveform with the choice of time-scale, scroll direction, and scroll distance. This flow is summarized in Figure 23.7. Inquiry service for hospital and clinic information also is available by accessing TALBES CLINIC and HOSPITAL.

23.4.2.5 ECG Browsing and Feature Extraction

The heart serves as the pump for the circulatory system. The well-coordinated pumping action of its four chambers are a result of a series of electrical depolarizations and repolarizations over different regions of the heart, and these activation sequences establish conduction fields which also extend to the body surface. The

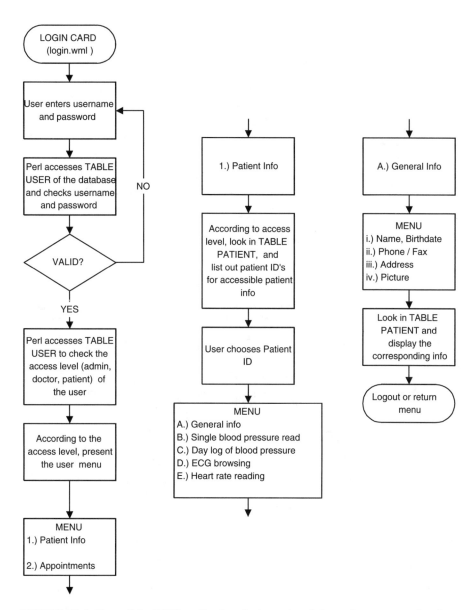

FIGURE 23.6 Flow of the WAP application; login, patient information menu, and patient general information.

heart can be viewed as an electrical equivalent generator, and at each instant of time, the electrical activity of the heart can be represented by a net equivalent current dipole located at a point of the heart.[49] The thoracic medium can be considered a resistive load, resulting in attenuation of the field with increasing distance from the source, as well as potential drops measured between two points measured on the body surface. Measurement of the resulting electrical potentials between different

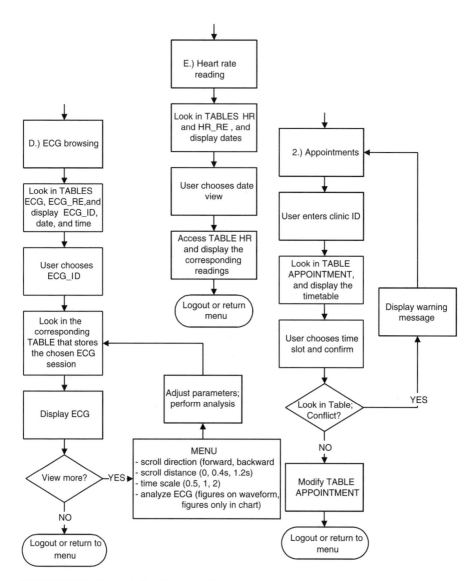

FIGURE 23.7 Flow of the WAP application; ECG browsing, heart rate reading, and appointments.

sites on the body surface is the ECG, which provides information on the condition of the heart. Its dynamic range is from 10 μV to 5 mV, and its bandwidth is from 0.05 to 1000 Hz; however, 0.05 to 80 Hz is adequate for most monitoring purposes.[50]

For initial testing of ECG browsing, a Lead I waveform from a subject was sampled at 250 samples per second by a data acquisition unit, stored as delimited numbers in a file, and loaded into the content server. A Perl program then extracted the sample points from the file and stored them inside an indexed table in the database. Recording sessions were stored in ECG data tables. Because the display size and resolution are limited on a WAP device, performing feature extraction would further enhance the feasibility of using WAP in viewing the acquired data. To demonstrate this, a QRS detection program was written in Perl, providing estimation of QRS occurrence times and R-R intervals in the recorded ECG. The algorithm used was based on amplitude and first derivative.[51]

Upon receiving a request for estimating the QRS occurrence times, the program retrieves the ECG data of the specified part of the recording session from the database, and puts it into a one-dimensional array of sample points of the ECG. For example, for 9000 sample points, the array is in the form

$$X[n] = X[0], X[1], X[2], \ldots X[8999]$$

The first derivative, $Y[n]$, is then calculated at each point of $X[n]$:

$$Y[n] = X[n + 1] - X[n - 1], 1 < n < 8998$$

A QRS candidate occurs when three consecutive points in the $Y[n]$ array exceed a predefined positive slope threshold, TH_POS, and are followed within the next 100 ms by two consecutive points which exceed a predefined negative slope threshold, TH_NEG.

$$Y[i], Y[i + 1], Y[i + 2] > TH\_POS$$

and

$$Y[j], Y[j+1] < TH\_NEG$$

where $(i + 2) < j < (i + 25)$. The value of 25 is based on the sampling rate of 250 samples per second. Each sample interval is

$$1/250 = 0.004 \text{ s}$$

Therefore, the number of samples that corresponds to 100 ms is

$$0.1/0.004 = 25$$

Once such a QRS candidate is detected, all $X[n]$ data points that are between the onset of the rising slope and before the end of the descending slope must exceed the amplitude threshold, TH_AMP, in order to be considered a valid QRS complex.

$$X[i], X[i + 1], \ldots, X[j + 1] > TH\_AMP$$

Finally, the occurrence times of the highest points in the QRS complexes are put into an array, which is then used in the dynamic construction of chart or graphical display in WML and WBMP format (see Figure 23.16).

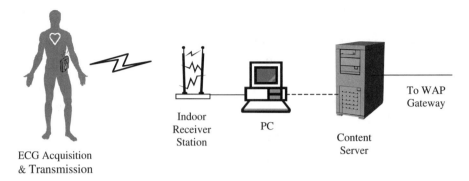

FIGURE 23.8 Block diagram for the wireless ECG connection.

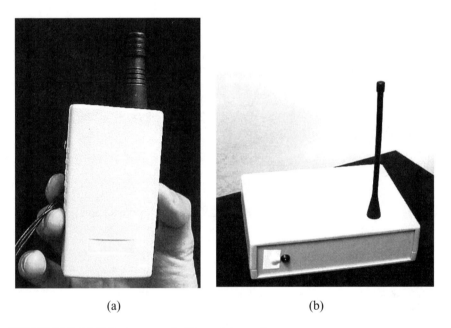

(a)   (b)

FIGURE 23.9 (a) Patient-worn unit; (b) receiving unit.

23.4.2.6 Wireless Subsystem

An indoor, wireless subsystem has been built for recording ECG from a mobile subject to demonstrate using WAP in patient monitoring. The data was immediately stored in the PC-based database after each recording session, and was available for viewing and analysis on a remote WAP device. The subsystem, as shown in Figure 23.8, consisted of a patient-worn unit, a receiving unit, and a PC. Photographs of the transmitting and receiving units are shown in Figure 23.9.

The patient-worn unit, depicted in Figure 23.10, was a portable device consisting of circuits for one-lead ECG acquisition and RF transmission. The biopotential

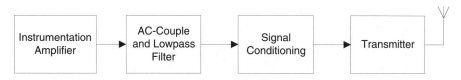

FIGURE 23.10 Patient-worn unit.

FIGURE 23.11 Receiving unit.

sensed by Ag-AgCl prejelled electrodes was fed into an instrumentation amplifier with gain of 1000, followed by AC coupling and a Butterworth lowpass filter having cutoff frequency at 150 Hz. After conditioning, the analog signal was input to a two-stage, SAW-controlled, 433 MHz FM transmitter operating on 3V supply voltage with transmitting power of 10 mW. A 24-turn helical antenna was used. Figure 23.11 shows the receiving unit, which consisted of a double conversion FM Superhet receiver operating on a 3V supply voltage, an 8-bit analog-to-digital converter (ADC), an 8-bit microcontroller unit (MCU), and interfacing circuits for connection to a PC via serial port.

The receiver was connected to a 1/4 wavelength whip antenna, and drew 14 mA when receiving. Output of the receiver was digitized by the ADC at 240 samples per second, fed to the MCU, and pushed into the PC's serial port through an RS232 transceiver at a baud rate of 19,200. The program that resided in the PC was written in Visual Basic 6.0. During each recording session, assigned a unique identifier, the program would read data from the serial port and save it into the PC-based database through ODBC DBI. The program also provided an interface for direct access to the database. The program interfaces are presented in Figure 23.12.

23.4.3 RESULTS

23.4.3.1 Emulation

All the applications were first tested with an emulation software before using actual WAP phones. The Nokia™ WAP Toolkit was used on a Windows platform to emulate how the applications would appear on a WAP phone. Applications were loaded directly from the server through the Internet. The setup is shown in Figure 23.13. Figures 23.14 to 23.17 are some screenshots of the interface.

23.4.3.2 Experience with WAP Phone

An actual WAP 1.1-compliant phone was used at GSM 1800 MHz through CSD to connect to the same WAP site. The gateway used in the link was provided by a

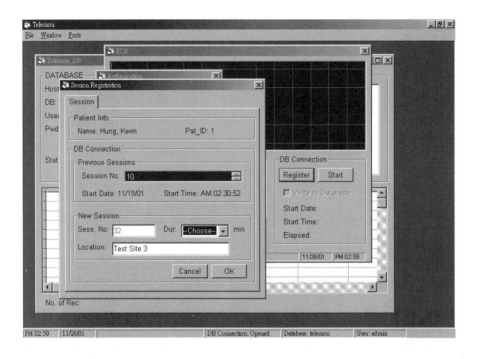

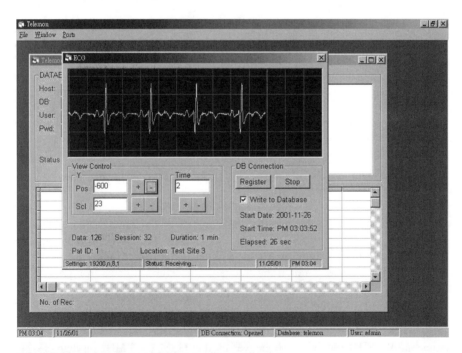

FIGURE 23.12 Interfaces of ECG recording application.

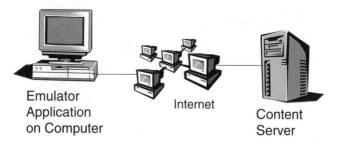

FIGURE 23.13 Setup for accessing WAP applications with emulation software.

(a)	(b)

FIGURE 23.14 (a) Login menu; (b) patient data menu.

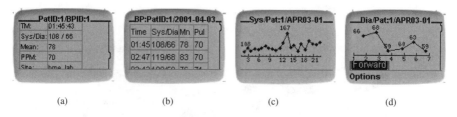

(a)	(b)	(c)	(d)

FIGURE 23.15 Display of blood pressure readings. (a) readings from a single blood pressure measurement; (b) readings within a day; (c) graphical display of systolic pressures within a day; (d) graphical display of diastolic pressures within six hours.

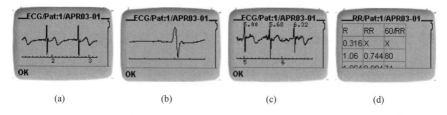

(a)	(b)	(c)	(d)

FIGURE 23.16 ECG browsing. (a) ECG browsing with a 2-sec window; (b) ECG browsing with a 0.5-sec window; (c) ECG browsing with QRS occurrence times estimation function activated; (d) Chart for estimated QRS occurrence times and R-R intervals.

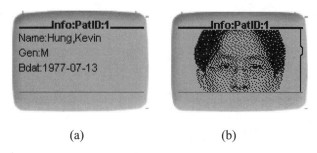

(a)   (b)

FIGURE 23.17 Display of patient general data.

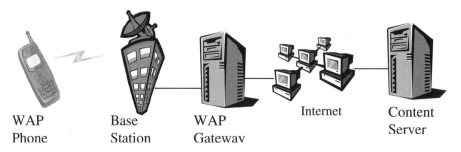

WAP   Base   WAP   Internet   Content
Phone   Station   Gateway   Server

FIGURE 23.18 Setup for accessing WAP applications with WAP phone.

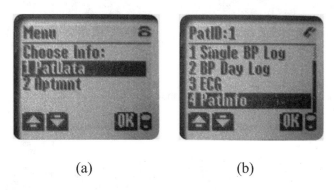

(a)   (b)

FIGURE 23.19 User menus.

mobile phone service provider in Hong Kong. Figure 23.18 describes the setup, and Figures 23.19 to 23.22 show some screenshots of the interfaces.

The time required for establishing a connection to the site was about 10 to 12 seconds. Starting from the time of request for information at the phone, the time required for database query, dynamic generation of WML and WBMP, and display of new information at the device ranged from 3 to 5 seconds on average.

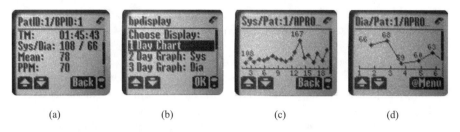

(a) (b) (c) (d)

FIGURE 23.20 Display of blood pressure readings. (a) readings from a single blood pressure measurement; (b) BP menu; (c) graphical display of systolic pressures within a day; (d) graphical display of diastolic pressures within six hours.

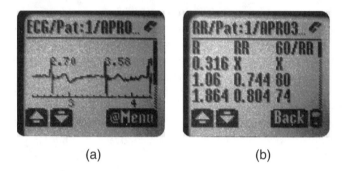

(a) (b)

FIGURE 23.21 ECG browsing. (a) ECG browsing with estimated QRS occurrence times; (b) Chart for estimated QRS occurrence times and R-R intervals.

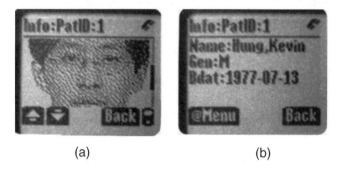

(a) (b)

FIGURE 23.22 Display of patient general data.

23.4.4 DISCUSSION

Viewing and analyzing patient data have been demonstrated on a WAP phone. This included interactive feature extraction of medical data. Although response time was long, the feasibility of such a system is expected to improve in the future, as newer versions of the WAP specification will be integrated into 3G mobile phones, which operate at a much higher data rate and have more on-board resources.

An issue of concern, as in all telemedicine applications, is security. The security features of a WAP-based system are implemented at several levels. WAP implements most of its security in WTLS (Wireless Transport Layer Security) protocol, which is the wireless equivalent of TLS (Transport Layer Security) protocol. Because data are encrypted between the phone and the gateway (at which point they are decrypted by the gateway before being reencrypted and sent on to the content server over a TLS connection over the Internet), the gateway has access to all of the data in decrypted form. Therefore, using a WAP gateway hosted by a third party is not recommended for telemedicine applications. The solution is to set up a private WAP gateway for the application.

There is still much room for improvement in this system. One area is to utilize the push technology provided by WAP 1.2 and higher. WAP-based push can use SMS or cell-broadcast as the bearer to transmit packets over the wireless network. This messaging service will further enhance the feasibility of using WAP for patient monitoring. When a real-time analysis program detects pathological abnormality in the recorded data, the content server will be able to send a short message to the WAP devices carried by the doctor or the patient.

The wireless subsystem is being upgraded to use Bluetooth™ for transmission. Wireless medical sensors based on conventional infrared or radio frequency transmission have been developed by others.[52–54] However, the small size and low power consumption of the radio module, use of 2.4 GHz frequency hopping spread spectrum, and the design for short-range transmission makes Bluetooth™ an attractive option in dynamic monitoring of physiological signals for telemedicine. Some parties have already integrated the technology into medical applications.[55,56] Further tests on the telemedicine system will be accessing the site with PDAs and PDA-phones, and connecting with GPRS. The system can be expanded also to a network of databases for resource sharing. Currently, another content server that uses Java Servlet and Extensible Markup Language (XML) has been set up.

23.5 ISSUES TO BE RESOLVED

Because telemedicine involves crucial information exchange over communication networks, security has always been a prominent concern. Security measures should be provided at the network and access levels, and a separate, private network often is required. As seen in the examples presented, data transmitted across a telemedicine system can be huge files such as high-resolution images, or time-critical information, including video streams and real-time physiological signals. A major bottleneck in using wireless Internet for applications like these is the low data transfer rate, because bandwidth provided by wireless communication is still generally lower than that provided by wired communication. This problem can even defeat the purpose of using telemedicine if the situation is not evaluated carefully before system implementation. One solution is to compress the data before transmission. Various lossy and lossless compression algorithms specifically for telemedicine already have been developed. In cases with lossy compression, the decompressed data should contain enough information in order to be qualified for proper clinical diagnosis. All these challenges become even more complicated as new wireless and medical technologies continue to emerge.

Inevitably, we are lead to the point where standards for medical information exchange are needed to ensure reliability, interoperability, and widespread use of telemedicine. Medical equipment industries and health care organizations have attempted to develop different standards, such as Digital Imaging and Communications (DICOM) and Health Level Seven (HL7), over the past few decades.[57] Despite all these efforts, there is still a need for a set of standards that are globally accepted. The IEEE 1073, currently still under construction, is a set of standards for medical equipment communication. Among these, transport standards for wireless communication also are investigated. The third generation of cellular communication is on its way. Data rate will be significantly higher, and 3G devices will have multimedia capability. This will undoubtedly bring telemedicine into a new era. The major challenge here is merging it with the existing telemedicine systems and at the same time conforming to the standards.

REFERENCES

1. Bashshur, R.L., Telemedicine and the health care system, in *Telemedicine: Theory and Practice,* Bashshur, R.L., Sanders, J.H., and Shannon, G.W., Eds., Charles C Thomas, Springfield, IL, 1977.
2. Darkins, A.W. and Cary, M.A., *Telemedicine and Telehealth: Principles, Policies, Performance, and Pitfalls,* Springer, New York, 2000.
3. Lin, J.C., Applying telecommunication technology to health-care delivery, *IEEE Eng. Med. Biol. Mag.,* 18 (4), 28, 1999.
4. Craig, J., Introduction, in *Introduction to Telemedicine,* Wootton, R. and Craig, J., Eds., Royal Society of Medicine Press, Glasgow, 1999.
5. Huang, H.K, *PACS: Basic Principles and Applications,* John Wiley & Sons, New York, 1999.
6. United Nations, *World Population Ageing: 1950–2050,* United Nations Publications, New York, 2001.
7. United Nations, *The World Aging Situation: Strategies and Policies,* United Nations Publication, New York, 1985.
8. Kart, C.S., Metress, E.K., and Metress, S.P., *Human Aging and Chronic Disease,* Jones and Bartlett, London, 1992.
9. Timiras, P.S., Ed., *Physiological Basis of Aging and Geriatrics,* CRC Press, Boca Raton, FL, 1994.
10. Sosa-Iudicissa, M., Wootton, R., and Ferrer-Roca, O., History of telemedicine, in *Handbook of Telemedicine,* Ferrer-Roca, O. and Sosa-Iudicissa, M., Eds., IOS Press, Amsterdam, 1988.
11. Field, M.J., *Telemedicine: A Guide to Assessing Telecommunications in Health Care,* National Academy Press, Washington, D.C., 1996.
12. Holland, J.J., Classic episodes in telemedicine: treatment by telegraph, *J. Telemed. Telecare,* 3 (4), 223, 1997.
13. Kenny, E., Classic episodes in telemedicine: diagnosis by telegraph, *J. Telemed. Telecare,* 4 (4), 223, 1998.

14. McKinney, W.P. and Bunton, G., Exploring the medical applications of the Internet: a guide for beginning users, *Am. J. Med. Sci.,* 306 (3), 141, 1993.

15. Fisk, N.M. et al., Intercontinental fetal surgical consultation with image transmission via Internet, *Lancet,* 341, 1601, 1993.

16. Yamamoto, L.G. et al., Telemedicine using the Internet, *Am. J. Emerg. Med.,* 14 (4), 416, 1996.

17. Yamamoto, L.G. and Suh, P.J., Accessing and using the Internet's World Wide Web for emergency physicians, *Am. J. Emerg. Med.,* 14 (3), 302, 1996.

18. Della Mea, V., Internet electronic mail: a tool for low-cost telemedicine, *J. Telemed. Telecare,* 5 (2), 84, 1999.

19. Magrabi, F. et al., Web based longitudinal ECG monitoring, in Proc. 20th Ann. Int. Conf. IEEE EMBS, Chang, H.K. and Zhang, Y.T., Eds., Hong Kong, 20 (3), 1155, 1998.

20. Park, S. et al., Real-time monitoring of patients on remote sites, in Proc. 20th Ann. Int. Conf. IEEE EMBS, Chang, H.K. and Zhang, Y.T., Eds., Hong Kong, 20 (3), 1321, 1998.

21. Kong, K.Y., Ng, C.Y., and Ong, K., Web-based monitoring of real-time ECG data, *Comput. Cardiol.,* 27, 189, 2000.

22. Pollard, J.K., Rohman, S. and Fry, M.E., A Web-based mobile medical monitoring system, in International Workshop on Intelligent Data Acquisition and Advanced Computing Systems: Technology and Applications, Foros, 32, 2001.

23. Bai, J. et al., A portable ECG and blood pressure telemonitoring system, *IEEE Eng. Med. Biol. Mag.,* 18 (4), 63, 1999.

24. Yamamoto, L.G., Wireless teleradiology and fax using cellular phones and notebook PCs for instant access to consultants, *Am. J. Emerg. Med.,* 13 (2), 184, 1995.

25. Yamamoto, L.G., Instant pocket wireless telemedicine consultations, *Pediatrics,* 104 (3), 670, 1999.

26. Yamamoto, L.G. and Williams, D.R., A demonstration of instant pocket wireless CT teleradiology to facilitate stat neurosurgical consultation and future telemedicine implications, *Am. J. Emerg. Med.,* 18 (4), 423, 2000.

27. Yamamoto, L.G. and Shirai, L.K., Instant telemedicine ECG consultation with cardiologists using pocket wireless computers, *Am. J. Emerg. Med.,* 19 (3), 248, 2001.

28. Giovas, P. et al., Transmission of electrocardiograms from a moving ambulance, *J. Telemed. Telecare,* 4 (1), 5, 1998.

29. Curry, G.R. and Harrop, N., The Lancashire telemedicine ambulance, *J. Telemed. Telecare,* 4 (4), 231, 1998.

30. Pavlopoulos, S. et al., A novel emergency telemedicine system based on wireless communication technology — AMBULANCE, *IEEE Trans. Inf. Technol. Biomed.,* 2 (4), 261, 1998.

31. Reponen, J. et al., Digital wireless radiology consultations with a portable computer, *J. Telemed. Telecare,* 4 (4), 201, 1998.

32. Reponen, J. et al., Initial experience with a wireless personal digital assistant as a teleradiology terminal for reporting emergency computerized tomography scans, *J. Telemed. Telecare,* 6 (1), 45, 2000.

33. Karlsten, R. and Sjöqvist, B.A., Telemedicine and decision support in emergency ambulances in Uppsala, *J. Telemed. Telecare,* 6 (1), 2000.

34. Zahedi, E. et al., Design of a Web-based wireless mobile teleconsultation system with a remote control camera, in Proc. 22nd Ann. IEEE EMBS Int. Conf., Enderle, J.D., Ed., Chicago, 2000.

35. Firoozbakhsh, B. et al., Wireless communication of vital signs using the Georgia Tech Wearable Motherboard, in 2000 IEEE Int. Conf. on Multimedia and Expo, 3, 1253, 2000.

36. Pedersen, P.C. et al., Low cost wireless LAN based medical informatics system, in Proc. 1st Joint BMES/EMBS Conf., Blanchard, S.M. et al., Eds., Atlanta, 1225, 1999.

37. Groves, T., SatelLife: getting relevant information to the developing world, *Br. Med. J.,* 313 (7072), 1606, 1996.

38. Hwang, S. et al., Development of a Web-based picture archiving and communication system using satellite data communication, *J. Telemed. Telecare,* 6 (2), 91, 2000.

39. Coiera, E., *Guide to Medical Informatics, the Internet and Telemedicine,* Chapman & Hall, London, 1997.

40. Hung, K. and Zhang, Y.T., On the feasibility of the usage of WAP devices in telemedicine, in Proc. 2000 IEEE EMBS Int. Conf. on Information Technology Applications in Biomedicine, Laxminarayan, S., Ed., Arlington, VA, 2000.

41. Hung, K., Zhang, Y.T., Web-based telemedicine applications, in Ann. Conf. of Eng. and the Physical Sci. in Med. and Asia Pacific Conf. on Biomed. Eng., Fremantle, 2001.

42. Mann, S., *Programming applications with the Wireless Application Protocol: the complete developer's guide,* John Wiley & Sons, New York, 1999.

43. Arehart, C., et al., *Professional WAP,* Wrox Press, Birmingham, 2000, 34.

44. World Health Organization, Death by cause, sex and mortality stratum in WHO Regions, estimates for 1999.

45. Rohde, M.M., Bement, S.L. and Lupa, R.S., ECG boy: low-cost medical instrumentation using mass-produced, handheld entertainment computers, *Biomed. Inst. Tech.,* 32 (5), 1998, 497.

46. Cozens, S. and Wainwright, P., *Beginning Perl,* Wrox Press, Birmingham, 2000.

47. DuBois, P., *MySQL,* New Riders, Indianapolis, 2000.

48. Elmasri, R. and Navathe, S.B., *Fundamentals of Database Systems,* Benjamin/Cummings, Redwood City, CA, 1994.

49. Webster, J.G., *Medical Instrumentation: Application and Design,* 3rd ed., John Wiley & Sons, New York, 1998.

50. Kenedi, R.M., *A Textbook of Biomedical Engineering*, Blackie & Son, East Kilbride, 1980.

51. Friesen, G.M. et al., A comparison of the noise sensitivity of nine QRS detection algorithms, *IEEE Trans. Biomed. Eng.,* 37 (1), 85, 1990.

52. Leung, S.W., Wireless electrode for electrocardiogram (ECG) signal, M.Phil. thesis, The Chinese University of Hong Kong, 1999.

53. Santic, A., Theory and application of diffuse infrared biotelemetry, *CRC Crit. Rev. Biomed. Eng.,* 18 (4), 289, 1991.

54. Yang, B., Rhee, S., and Asada, H.H, A Twenty-four hour tele-nursing system using a ring sensor, in Proc. 1998 IEEE Int. Conf. Robotics and Automation, Leuven, Belgium, 387, 1988.

55. Berggren, M., Wireless communication in telemedicine using Bluetooth and IEEE 802.11b, Master's thesis, Dept. of Computer Systems, Uppsala University, 2001.

56. Ortivus, http://www.ortivus.com

57. Blair, J.S., Overview of Standards Related to the Emerging Health Care Information Infrastructure, in *The Biomedical Engineering Handbook,* 2nd ed., Bronzino, J.D., Ed., CRC Press, Boca Raton, FL, 2000.

24 Delivering Music over the Wireless Internet: From Song Distribution to Interactive Karaoke on UMTS Devices

Marco Roccetti, Paola Salomoni, Vittorio Ghini, Stefano Ferretti, and Stefano Cacciaguerra

CONTENTS

ABSTRACT

The pace of developments in wireless technology is enabling a wide range of exciting applications, including location-aware systems, wearable computers, wireless sensor networks, and novel use of cellular telephony systems. Many of those applications may play the role of key drivers of future wireless technology provided that they may guarantee affordable access, ubiquitous reach, and an effective service-delivery model. In this challenging scenario, a growing demand is emerging for delivering modern multimedia entertainment services to wireless handheld devices on a large scale. In this chapter, we report on our experience in developing a wireless Internet application designed to deliver advanced musical services to mobile consumers over Universal Mobile Telecommunications System (UMTS) links. In essence, our application allows mobile users to listen to songs and karaoke clips on UMTS-enabled devices by exploiting the Internet as a large musical storehouse. Field trials have been conducted that confirm that an adequate structuring of the wireless Internet application, along with the use of 3G (third generation) mobile network technologies, may be effective for the delivery of modern musical-entertainment services to mobile consumers.

24.1 INTRODUCTION

With the advent of the twenty-first century, market forces are accelerating the pace of wireless technology evolution. It is widely anticipated that future mobile users will enjoy a near-ubiquitous access to high-bandwidth wireless networks.[1] Probably, mobile phone services represent the most-prominent example of widespread access to wireless communication networks. Recently, notable efforts such as i-mode[2] and WAP[3] have been carried out to exploit phone-based wireless technology along with the Internet, to provide Web services to mobile users. As of today, unfortunately, those efforts have resulted in little more than e-mail access and limited Web browsing. Although this situation is slowly improving, it is easy to envisage that old pricing schemes, along with content not suitably formatted for mobile phones, may limit the interest of mobile consumers in wireless services.

However, in parallel with the recent improvements of high-bandwidth wireless communications, the wired Internet's progress has determined a significant innovation among music distribution paradigms. The maturing distributed file-sharing technology implemented by systems like Napster has enabled the dissemination of musical content in digital form, allowing consumers to link to stored music files from around the world.

Hence, with users adopting Internet-enabled cellular phones and similar handheld devices we may expect that a growing demand may emerge for wireless services that enable mobile consumers to access music content from the large musical storehouse represented by the Web.[4]

In this context, the 3G UMTS technology promises to be one of the milestones in the process of building an adequate large-scale wireless infrastructure for delivering musical services to mobile consumers. Important advantages of the UMTS technology amount to the fact that packets originating from UMTS devices can be directly transmitted to IP networks (and vice versa), while specific quality of service guarantees may be provided over the wireless links. Additionally, UMTS offers higher data rates (up to a few Mbps) and an increased capacity. These data rates plus compression techniques will allow users to access HTML pages and video/audio streaming, as well as enhanced multimedia services for laptops and smaller devices.[5]

Nevertheless, it is well known that even with the adoption of the UMTS technology various technical communication problems may arise due to:

- The time-varying characteristics of the UMTS links
- Possible temporary link outages
- Protocol interference between the UMTS radio link level protocols and the Internet transport protocols
- Typical high bit error rates of the UMTS radio links

As a result, all the above-mentioned limitations may keep some older applications designed for music distribution over the wired Internet from working efficiently and effectively when deployed over UMTS-based mobile communications scenarios.

In this chapter, we describe a modern Internet wireless application we have designed to support the widespread delivery of musical services to Internet-enabled mobile phones and similar handheld devices over UMTS links. In particular, the prototype implementation of our wireless application allows mobile consumers equipped with UMTS-enabled devices to exploit the Internet as a vast storehouse of music resources, where two different musical services are provided, namely:

1. A mobile song-on-demand distribution service
2. A mobile karaoke distribution service

The mobile song-on-demand distribution service permits to mobile users to download and to listen to MP3 files[6] on UMTS devices. Specifically, our wireless application exploits the background traffic class of UMTS to provide its users with (1) a simple and rapid Internet-based mobile access to a music-on-demand download service, (2) a robust and widely available music-on-demand distribution system based on the technique of replicated Web servers, and (3) the possibility of interactively customizing the use of the system.

From a user's perspective, it is worth noticing that different types of clients may exploit the developed service. In particular, our wireless application may be exploited by the following categories of users:

- Music listeners may search for their favorite songs over the Internet, download them onto their UMTS devices, and play them at their convenience.
- Music producers may wish to exploit our system to distribute their own music. (At the current state of the art, this type of user needs a regular wireline Internet connection in order to upload musical resources to the system.)

- Musical service providers may exploit our system to organize, build, and maintain structured repositories of musical resources over the Internet for use by UMTS-enabled consumers.

Karaoke is an MTV-type multimedia entertainment service that has gained much popularity in Asia, the United States and Europe. With the advent of the wireless Internet and of Internet-enabled cellular phones, a growing demand is emerging for delivering such multimedia entertainment services to wireless handheld devices on a large scale. In this context, the mobile karaoke distribution service we have developed allows mobile users to download and to play multimedia karaoke clips constructed out of synchronized multimedia resources, such as music, text, and video. A karaoke clip is represented by means of a SMIL (Synchronized Multimedia Integration Language) file containing the formal specification of the media scheduling and synchronization activities concerning all the audio, video, and textual resources that compose a karaoke clip.[7]

The important experiences of P2P systems, such as Napster,[8] Freenet,[9] and Gnutella,[10] are at the basis of our music delivery services, but our Internet application is essentially new in the sense that it allows a reliable and distributed music sharing service over wireless UMTS links. As previously mentioned, the choice of adopting UMTS as the key technology for wireless access to the Internet has posed a set of technical challenges which will be discussed later in the chapter.

The reminder of this chapter is organized as follows. In Section 24.2, we discuss some technical obstacles we have surmounted for integrating wireless UMTS access technology with the Internet. In Section 24.3, alongside the operational description of our wireless Internet application we illustrate the design principles we have followed to design our system. Section 24.4 reports on a large set of performance results we have gathered from experimental trials conducted on the field. Section 24.6 provides an insight into some important research areas that have influenced our work. Finally, Section 24.7 concludes the chapter.

24.2 SYSTEM ISSUES

The aim of this section is to discuss some technical issues which are at the basis of the system we have developed to support the distribution of mobile musical services to UMTS-enabled devices.

Some of the most prominent technical issues for the development of our system are those related to the problems of integrating UMTS wireless access technology with the Internet. It is well known that choosing UMTS wireless technology as a means to provide mobile access to the Internet poses a number of obstacles. In this context, the first problem is to decide if advanced TCP/IP (Transmission Control Protocol/Internet Protocol)-based applications will behave well over UMTS-type radio communications protocols.[11,12] With regard to this fact, it is important to notice that the Internet TCP/IP protocol stack has not been especially designed for wireless communications. The standard (TCP) provides a sliding window-based ARQ (automatic repeat request) mechanism that incorporates an adaptive time out strategy for guaranteeing end-to-end reliable data transmissions between communicating peer nodes over wired connections. Because the ARQ mechanism of TCP essentially uses a stop-and-resend control mechanism for ensuring connection reliability, the question here is whether this mechanism may trigger a TCP retransmission at the same time the radio link level control mechanism is retransmitting the same data.

An even more significant problem of mobile wireless is that of temporary link outages. If a user enters an area of no signal coverage, there is no way that the standard TCP protocol may be informed of this link-level outage.[13]

After having considered all these challenges, an additional problem is strictly related to the internal architecture of those advanced Internet-based applications that should be accessed through radio interfaces. Those applications, in fact, must exhibit a high rate of robustness and availability, because mobile access to those applications should not be influenced by possible problems occurring on the Internet side.

To overcome these obstacles we adopted a number of important strategies:

1. In order to ensure both the availability and the responsiveness of our mobile musical service, we have structured our application according to the special technology of replicated Web servers.[14] Following this technology, a software redundancy has been introduced on the Internet side by replicating the multimedia resources across a certain number of Web servers distributed over the Internet. A typical approach to guarantee service responsiveness consists of dynamically binding the service client to the available server replica with the least-congested connection. An approach recently proposed to implement such an adaptive downloading strategy on the Internet side amounts to the use of a software mechanism, called the client-centered load distribution (C^2LD) mechanism.[15] With this particular mechanism, each client's request of a given multimedia resource is fragmented into a number of subrequests for separate parts of the resource. Each of these subrequests is issued concurrently to a different available replica server, which possesses that resource. The mechanism periodically monitors the downloading performance of available replica servers and dynamically selects at run-time those replicas to which the client subrequests can be sent, based on both the network congestion status and the replica servers' workload.

2. Our wireless application has been structured following an all-IP approach, where a wireless session level has been developed additionally (on the top of the standard TCP protocol) to guarantee connection stability in case of possible temporary link outages.

3. As the download of musical resources over wireless links may experience long duration (and high costs), our wireless application has been designed to exploit the UMTS background service class. This is the UMTS service class with lower costs, because it has been designed for supporting non-interactive, best-effort traffic.

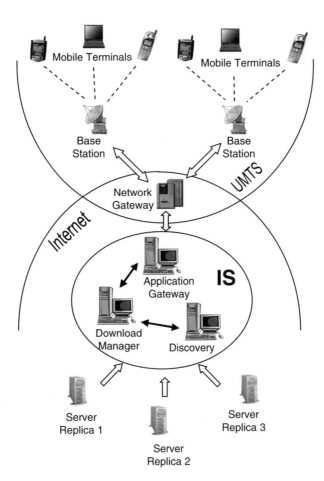

FIGURE 24.1 Wireless application architecture.

24.3 A WIRELESS INTERNET APPLICATION FOR MUSIC DISTRIBUTION

A comprehensive visual representation of the architecture of the wireless Internet application we have developed is reported in Figure 24.1. The client part of our software application, running on the UMTS device, provides users with the possibility of searching, downloading, and playing out the desired musical resources.

Musical resources may be of different types, based on the musical service that is selected by the user. If a consumer wants to enjoy the song-on-demand delivery service, then musical resources are represented by simple musical files (typically encoded with the MP3 format). Those musical files are stored in different Web servers, geographically distributed over the Internet, which act as music repositories. In general, different Web servers can be managed and administrated by music service providers, and also may offer different sets of replicated songs. Simply stated, this replication scenario can be thought of as a loosely coupled replication system where, potentially, different servers support different sets of musical resources. Needless to say, each single song may be replicated within a number of different Web servers.

If the selected service amounts to the mobile karaoke, the corresponding musical resource is represented by a more-complex set of data constituting a karaoke clip. A karaoke clip, in practice, is represented by a textual SMIL-based file with pointers to all the multimedia resources that compose that given clip. All the multimedia resources (specified in the SMIL description file) are stored on different replicated Web servers, geographically distributed over the Internet. As in the case for the song-on-demand delivery service, those Web servers perform as redundant repositories for the multimedia resources that compose the karaoke clip of interest. As seen from Figure 24.1, an intermediate software system (IS) has been interposed between the mobile clients and the Internet-based multimedia repositories. The responsibilities of the IS are

- Providing each UMTS device with a wireless access point to the Internet-based music distribution service
- Discovering and downloading all those multimedia resources (songs or karaoke clips) that are requested by a user

In essence, the IS is constructed of three main subsystems: (1) an application gateway subsystem, (2) a discovery subsystem, and (3) a download manager. Those subsystems cooperate to support the download of musical resources to the mobile consumer, as detailed in the following discussion.

24.3.1 SEARCH AND DOWNLOAD OF MUSICAL RESOURCES

The client part of our application represents the interface between consumers and services, and provides a set of search-and-download functions. Taking into account that a typical mobile device (such as a UMTS telephone or a PDA) possesses a very limited memory capacity and disk size, we have made the decision to move all the search-and-discovery functions to the IS side. This means that the client part of our application needs the collaboration of all the IS software subsystems to determine which musical resources are available, and where they are located. This implies also that users must perform the search-and-download activity by stepping through several different phases.

As an example, a complete search-and-download session for karaoke clips steps through three different phases. In the first phase, a user issues a request for a given karaoke clip from his UMTS device to the application gateway subsystem. The request may refer either to a specific song title or author. The gateway subsystem passes this request down to the download manager. The download manager asks the discovery subsystem for the complete list of all the available karaoke clips matching the request issued by the user. The discovery subsystem performs the search of the clips requested by the client, and proceeds as follows.

First, the discovery subsystem tries to establish a relationship between the titles of the songs requested by the user and the SMIL description files that represent the requested songs. (Note that different clips that match the request issued by the user may be stored in different Web servers.)

Once this activity is completed, the discovery subsystem passes to the user (via the application gateway) the list of all the clips (and correspondent SMIL files) that match the initial request. Upon receiving this list, the user chooses one of the clips. This choice activates an automatic process to download the corresponding SMIL file. It is the download manager, now equipped with all the relevant information needed to locate it, that downloads the SMIL file, and finally delivers it to the software application running on the UMTS device. Upon receiving the SMIL file, the software application running on the UMTS device examines that file and, following the specified schedule, calls again for the help of the discovery subsystem and the download manager in order to locate and download all the multimedia resources specified in the SMIL file.

This represents the beginning of the third phase of the discovery/download activity, at the end of which all the multimedia resources are delivered to the UMTS device that can perform playback according to the time schedule defined in the SMIL file. It goes without saying that during this final phase it is the responsibility of the discovery subsystem to individuate the Web locations of all the required multimedia resources, while the download manager carries out the download activity by engaging all the different replica servers that maintain a copy of the requested multimedia resources.

Needless to say, in order for the system to work correctly, a preliminary phase must be carried out, where each karaoke server announces the list of the clips it can make available for sharing. Each karaoke server that wants to add its own repository to our IS system may do that by running a software application called the data collector, which is in charge of communicating to the discovery subsystem the list of the clips along with the associated multimedia resources.

When the song-on-demand service is used, a similar activity is carried out by the system, with the only difference being that the intermediate phase when a SMIL file is searched, downloaded, and interpreted is not needed. Simply put, the mobile user may activate the automatic download of a given song after he has chosen from the list submitted to him at the end of the first phase. Figure 24.2 illustrates the progression of the above-mentioned process (the interface is in Italian), which has been developed in our prototype implementation by resorting to the Visual C++ programming language on a Microsoft® Windows® Pocket PC platform.

24.3.2 DESIGN PRINCIPLES

In this section, we highlight the technical attributes that were significant for the development of all the software subsystems we have previously introduced.

We have already mentioned that our wireless Internet application exploits a standard TCP-IP stack for carrying out the communications between the application gateway subsystem and the client part of our application. According to the adopted approach, the client part of our application works as any other Internet-connected

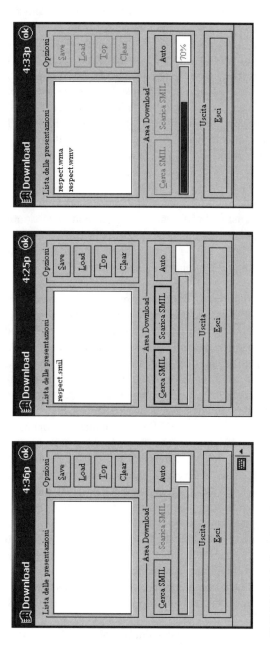

FIGURE 24.2 Search and download over the wireless Internet.

device, and the end-to-end connection is guaranteed by using the standard TCP protocol. However, to circumvent all possible problems due to the time-varying characteristics of the wireless link, our wireless application incorporates a session layer developed on the top of the TCP stack. This additional protocol layer provides stability to the download session, which may suffer from possible link outages.

With this in view, Figure 24.3 shows the protocol stack we have designed and developed to support all gateway-related communications. In particular (as shown in the left-most side of the figure) the gateway subsystem communicates with the client part of the application over a UMTS link. As seen in the figure, on the UMTS protocol stack an IP layer, based on the Mobile IP (Version 4) protocol, is implemented. On the top of this Mobile IP level, a standard TCP layer has been built. Finally, to circumvent all the network problems due to the radio link layer, the application layer built on the top of TCP has been designed as constructed out of two different sublayers:

- A session layer devoted to organize and manage a download session which may possibly consist of different subsequent communication patterns, in the face of possible link outages
- An application layer in charge of supporting the different connections needed to search and download songs

It is worth noting that our designed session layer provides users with the possibility of resuming a session that was previously interrupted due to temporary link outages. The session management mechanism we have designed and implemented has a greater importance for the full success of the download activity of musical resources onto a UMTS device. It is easy to understand that very large files (e.g., songs of about 5 Mbps) must be delivered to the UMTS terminal, and this must be carried out in the presence of a wireless cellular access, which typically exhibits scarce connection stability and unpredictable availability. Stated simply, our session layer works as follows: when the UMTS client application opens a connection to the application gateway, the gateway assigns a unique identifier to this new session. If the gateway eventually detects a network failure (i.e., the TCP connection is closed), the download status is saved on the gateway side. In particular, a pointer (to the last received byte of the musical resource) is saved, along with the session identifier. At the same time, the identifier of the suspended session is saved at the client side of the application. As soon as the mobile client application is able to open a new TCP connection to the gateway, the client application tries to resume the interrupted session by exploiting the session identifier that was previously saved.

As a final note, it is important to remember that the session management mechanism we have developed is suitable for recovering sessions that are interrupted due to temporary link outages, but it is not adequate to recover from system failures occurring at the UMTS terminal or at the application gateway subsystem.

The download manager is the real agent responsible for the download process. It has been incorporated in our application built on top of the HTTP protocol. With the aim of maximizing service availability and responsiveness, it makes use of the Web server replica technology[14] along with the client-centered load distribution

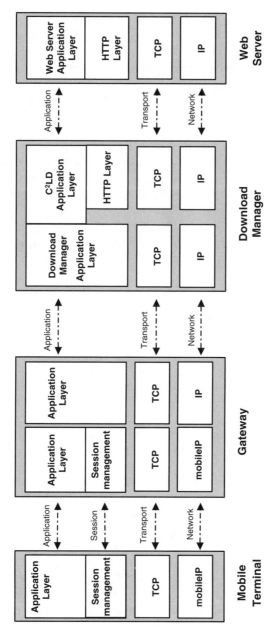

FIGURE 24.3 Wireless Internet application protocol stacks.

(C²LD) mechanism.[15] The C²LD mechanism implements an effective and reliable download strategy that splits the client's requests into several subrequests for fragments of the needed resource. Each of these subrequests is issued concurrently to a different available replica server. The C²LD mechanism is designed so as to adapt dynamically to state changes both in the network and Web servers; in essence, it is able to monitor and select at run-time those replicas with best downloading performances and response times. Figure 24.3 shows the download manager protocol stack. As seen from the figure, the download manager has to communicate with each different Web server replica, and then forks into different processes for each requested resource. Each process uses a C²LD application protocol to carry out download activities from different Web server replicas. It is worth noting that the use of the C²LD mechanism does not force music providers to organize musical repositories, which are all perfect replicas of the same list of musical resources. A musical resource, in fact, may be replicated within only some of the available server replicas of our system.

The software component of our developed system that stores and indexes relevant information about musical resources is the discovery subsystem. The main responsibility of the discovery subsystem is that of performing a sort of naming resolution for musical resources that are requested by clients. In particular, it carries out the folowing activities:

- Accepts users' requests to establish a formal relationship between them and the corresponding musical resources stored in the system
- Locates the exact Internet location where a given musical resource is stored throughout the entire system composed of replicated Web servers

To carry out this activity, the discovery subsystem calculates for each of the multimedia resources embedded in the system a 32-bit-long identifier (called the checksum). This value is computed on the basis of the file content and other information (such as file name, file creation time, file length, song title, and author). Two different indexes are maintained by the discovery subsystem: one is needed to resolve users' requests and the other is used to locate the corresponding musical resources.[16] We have devised a decentralized method for performing the calculation of the checksum. According to this scheme, each host server computes the checksum of its musical files and communicates the results to the discovery system. To minimize both the computational and traffic overheads, each server has to run the data collector locally to provide the possibility to add or delete the referenced musical resources by the discovery system. In essence, the data collector locally performs the checksum computation, and after having computed the checksum of all the files that a given music provider wants to distribute, opens a TCP connection toward the discovery subsystem. As a final task, the data collector application uploads the computed checksums to the discovery subsystem. It is worth noting that the data collector is implemented as a Java application to enhance software portability, and also it meets standard security constraints, as it can only read from the local file system, but it cannot execute local write operations.

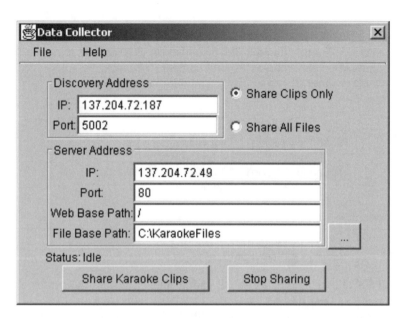

FIGURE 24.4 Screenshot of the data collector application.

Figure 24.4 shows a screenshot of the data collector application. As seen from the figure, two different kinds of information must be specified: the address of the discovery subsystem (under the form of IP and port numbers) and the complete address of the server where the musical resources are stored (including the data path within the local file system).

24.3.3 STRUCTURING KARAOKE CLIPS

The SMIL mark-up language is an XML-derived technology designed to integrate continuous media into synchronized multimedia presentations.[7] SMIL allows one to (1) manage the timing behavior of the presentation, (2) manage the layout of the presentation on the device screen, and (3) associate hyperlinks with media objects. The design of a SMIL-based presentation is performed according to two different phases: first, the author creates spatial regions that will contain the associated multimedia objects, then those multimedia objects are specified along with the timing schedule of their presentation. A SMIL file contains two main elements: a header (between <head> and </head>) and a body (between <body> and </body>). An SMIL header may specify spatial areas by using the <region> tag. (In a SMIL header, it is possible also to define meta tags that allow one to insert meta-information.) In a SMIL body, it is possible to define which multimedia objects are to be loaded in specific regions. To this aim, tags such as <video> for video files, <audio> for audio files, and <text> for text strings are exploited. The SMIL body is used also to schedule the synchronization of different multimedia objects. Two basic synchronization methods are provided:

```
<smil>
   <meta id="meta1" name="Titolo" content="A little respect" />
   <meta id="meta2" name="Autore" content="Wheatus" />
   <meta id="meta3" name="Key1" content="Wheatus" />
   <head>
      <layout>
         <root-layout/>
         <region id="region1_1" top="76%" left="2ì
                 height="24%" width="100%"/>
         <region id="region1_2" top="1" left="15"
                 height="75%" width="100%"/>
      </layout>
   </head>
```

FIGURE 24.5 Header of a karaoke SMIL file.

- Parallel (`<par>`,`</par>`): All multimedia objects are executed concurrently in their own regions
- Sequential (`<seq>`,`</seq>`): With the sequential method, each multimedia object is executed in its own region according to a predefined sequential time schedule

By using the SMIL technology, it is easy to specify a karaoke clip that includes audio, video, and the text that periodically flows following the song melody. An example of a karaoke clip, specified by using SMIL, is presented in Figures 24.5 and 24.6 (the title of the song is "A little respect," by Wheatus). In particular, Figure 24.5 reports a code fragment with the SMIL header. As shown in the figure, two different regions are defined `region1_1` and `region1_2`, respectively. Three meta tags are used to specify title, author of the song, and title of the album that contains the song. Figure 24.6 shows a code fragment representing the body of the SMIL file in which the following three different multimedia objects are executed in parallel:

1. An audio file (respect.wma)
2. A video file (respect.wmv), loaded in `region1_1`
3. A sequence of textual information flowing on the screen in `region1_2` for a limited duration of time, which is specified in seconds by using the `dur` attribute

Summing up, the wireless karaoke service we have provided exploits the SMIL technology, thus allowing users to enjoy:

- A search session where karaoke clips may be searched by simply indicating a part of the song title or a part of the author name
- A download session during which the SMIL file and, subsequently, the associated multimedia resources are downloaded

```
<body>
    <par>
        <video src="respect.wmv"
                region="region1_2" fit = "slice" repeatCount="6">
        </video>
        <audio src="respect.wma"></audio>
        <seq>
            <text begin="6s" dur = "7s" region="region1_1">
                I tried to discover a little something to make me
            </text>
            <text dur = "10s" region="region1_1">
                sweeter Oh baby refrain from breaking my heart
            </text>
            ...
            <text dur = "4s" region="region1_1">I hear you calling</text>
            <text dur = "17s" region="region1_1">
                Oh baby please give a little respect to me.
            </text>
        </seq>
    </par>
</body>
</smil>
```

FIGURE 24.6 Body of a karaoke SMIL file.

- A playout session when the SMIL player plays back the multimedia objects according to the time schedule specified in the SMIL file

24.4 AN EXPERIMENTAL STUDY

This section introduces an experimental study we have conducted in order to assess the effectiveness of our proposed musical services. We carried out several experiments (about 4000) consisting in the download of either a set of MP3-type songs or a set of different karaoke clips, according to the selected service. During the experimentation of the song-on-demand distribution service, four different replicated Web servers were exploited at the Internet side as song repositories, which were dispersed throughout the world; for the mobile karaoke service, we used three different replica servers that were located within a metropolitan scenario.

The communication between the IS and the mobile client was simulated by means of an UMTS simulator which was able to produce the transmission delay time of each frame at the UMTS radio link layer. Detailed information concerning the experimental models we adopted for our experiments are discussed in the following sections.

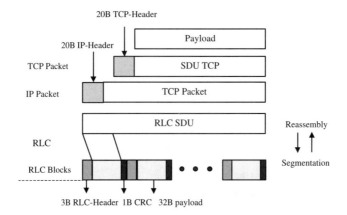

FIGURE 24.7 Segmentation of an IP packet into RLC data blocks.

24.4.1 UMTS SIMULATION MODEL

Currently, no real measurements of UMTS wireless data are available; hence, in our experiments the communication between the IS system (on the Internet side) and the mobile client application was carried out through a simulated UMTS network (running the background traffic class). To this aim, a UMTS simulator provided by the Fondazione Marconi (an Italian public foundation for wireless computing) was exploited. The UMTS network simulator we used was able to return Wireless Network Transmission Time (WNTT) values after simulations, i.e., the time spent to download musical resources, as computed at the UMTS radio link control (RLC) layer.

The simulated transmission of an IP packet over an air interface is illustrated in Figure 24.7. The RLC layer received a PDCP frame composed of an IP data packet to which various headers were added for each different level of the protocol stack (indeed, the PDCP layer was not simulated, for the sake of simplicity).

We conducted experiments with IP packets of different sizes (160, 480, and 960 bytes) coming from the Internet. The resulting PDCP frames were then segmented into RLC data blocks, each of which was 36 bytes. The result of this segmentation activity at the RLC level was that 5, 15, and 30 RLC data blocks were needed to transmit IP packets with the dimensions of 160, 480, and 960 bytes, respectively. In summary, if the transmission slot was available, 10, 30, and 60 milliseconds were needed to transmit over the air interface IP packets with the dimensions of 160, 480, and 960 bytes, respectively. It goes without saying that the obtained WNTT values depended on some operational parameters, such as the amount of traffic present in the simulated cell, the number of active clients and their speeds. WNTT measurements included the time spent for possible retransmissions at the RLC level.

The main problem that derives from adopting an approach where simulated results for RLC frames (traveling over the wireless link) are combined with the real measurements obtained for the TCP segments (traveling over the wired Internet) is that segment errors and resulting retransmissions at the TCP level are not taken into account.

To circumvent this problem, our experiments included the possible retransmission time delays incurred at the TCP level obtained by exploiting an external delay management mechanism. This external delay management mechanism was designed to take into account the typical TCP error recovery method based on received ACKs. Simply put, the delay mechanism compared the WNTT values obtained through the UMTS simulation against the time out values computed by TCP. If the simulated WNTT value was larger than the correspondent TCP time out value, then a retransmission must have occurred at the TCP level. In such a case, the WNTT value of that particular TCP segment was augmented by an additional value chosen as equal to the next WNTT value extracted from the set of the UMTS-based simulated values. Consequently, the TCP time out value was updated as follows: If a retransmission at the TCP level was detected according to the method mentioned previously, then the subsequent TCP time out value was calculated as the double of the previously computed value. If no retransmission at the TCP level was detected, then the traditional adaptive formula for the calculation of the TCP timeout value, as proposed by Jacobson, was followed.[17]

The next section presents the two different measurement architectures (along with obtained results) we adopted on the Internet side to evaluate the song-on-demand distribution service and the mobile karaoke service we have implemented.

24.4.2 SONG-ON-DEMAND: MEASUREMENT ARCHITECTURE AND RESULTS

To evaluate the efficacy of the song-on-demand distribution service we have developed, we used four different Web servers, geographically distributed over the Internet, providing the same set of 40 different songs. The four replica servers were located in Finland, Japan, the United States, and New Zealand (see Figure 24.8). Our designed intermediate system, located in Bologna, Italy, was running over a Pentium 3 machine (667 MHz, 254 MB RAM) equipped with the Windows 2000 Server operating system. The UMTS device, on which the client of our application was running, was emulated by means of a Pentium 2 computer (266 MHz, 128 MB RAM) equipped with the Windows CE operating system.

In order to provide the reader with an approximate idea of the transmission times experienced over the considered Internet links, it is worth mentioning that the round trip times (RTTs), obtained with the ping routine, between the client and the four servers measured 70 (Finland), 393 (Japan), 145 (United States) and 491 (New Zealand) milliseconds. As far as the downloading process is concerned, we took two basic assumptions:

1. MP3 file dimension: In our experiments we used 40 different MP3-based songs, whose corresponding file dimension ranged from 3 to 5 MB, which corresponds to the average file dimension of the songs maintained in the Napster system.
2. Number of downloads activities: Our software application provides support to two different types of song download services, the first consisting of downloading a single song, the other consisting of downloading a

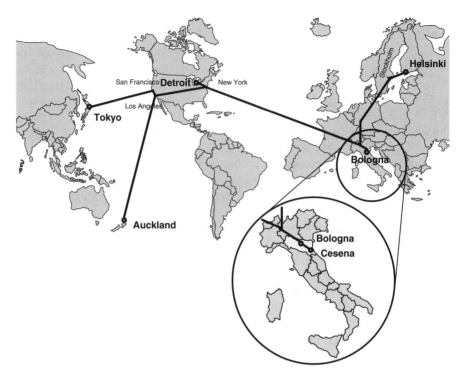

FIGURE 24.8 Web server replicas and clients (big picture: song-on-demand; small picture: mobile karaoke).

complete set of songs (song compilation). To evaluate the performance of our system under both circumstances, we conducted the following experiments:

- A set of independently replicated experiments consisting in the download of single songs.
- A set of independently replicated experiments, with each one consisting of the download of a set of songs. The number of songs for each compilation ranged in the set of 3, 5, and 10. These three values were chosen based on the consideration that the average disk capacity of typical MP3 players never exceeds 50 MB.

For the sake of brevity, in this chapter we only report the results obtained for the download of single songs. If interested, you may refer to the work of Roccetti and cowokers[18,19] for further details on the results we obtained when song compilations were downloaded.

In the following, we report on a large set of results obtained during many experimental trials based on the previously mentioned models.

In particular, we begin by presenting the measurements we obtained for our wireless application on the Internet side. In essence, Wireline Network Transmission

TABLE 24.1
Song-On-Demand WLNTT Results

	C2LD (4 Servers)		HTTP	
	Finland	USA	Japan	New Zealand
Download time (seconds)	32.547	47.889	122.191	248.740	624.195
C^2LD improvement (percent)	—	32	73.4	86.9	94.7

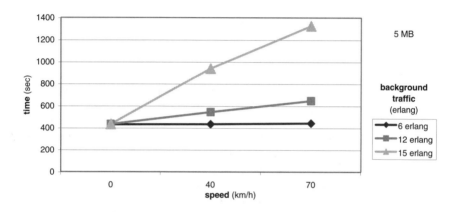

FIGURE 24.9 Song-on-demand WNTT results for 5-MB-sized songs.

Time (WLNTT) values refer to the time spent over the wired Internet links to download a requested song from the replicated Web servers toward the IS on the Internet side. These measurements have been compared with those that may be obtained by downloading the same MP3-based song with a standard HTTP GET method. The first row of Table 24.1 reports those results for MP3 files whose size is 5 MB. The second row shows the average WLNTT percentage improvement obtained by our system that exploits the previously mentioned C^2LD mechanism, with respect to the standard HTTP protocol. As shown in the table, our system obtains an average percentage improvement over the fastest HTTP replica, which is equal to 32 percent.

As already mentioned, Wireless Network Transmission Time (WNTT) values refer to the time spent for downloading songs to the mobile devices through the wireless links. (It is worth remembering that these values have been obtained through UMTS simulations.) Figures 24.9 and 24.10 show the WNTT values for 5 MB- and 3 MB-sized songs, depending on the following traffic parameters:

1. The speed at which users move throughout the cell (expressed in km/h)
2. The additional traffic in the cell (expressed via Erlang values)

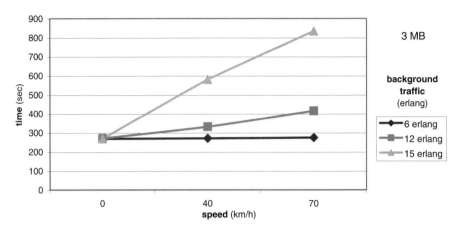

FIGURE 24.10 Song-on-demand WNTT results for 3-MB-sized songs.

In addition, it is worth noting that the WNTT values we have plotted have been obtained by averaging all the experiments conducted with IP packets of different dimensions (i.e., 160, 480, and 960 bytes). The main considerations that derive from an analysis of the results presented in Figures 24.9 and 24.10 are (1) the larger the traffic in the cell (and the users' speed), the larger the corresponding WNTT values, and (2) the best WNTT result may be obtained when the mobile device is completely still. (In such a case, a data rate of about 12 KBps may be obtained.) Additionally, it is worth noting the impact that the download time values, obtained on the Internet side, have on the total time requested to download songs on UMTS terminals. The obtained average download delays on the Internet side (about 33 seconds) seem to be quite irrelevant if compared with the WNTT values that have been experienced on the wireless links (ranging from 250 to 1325 seconds, i.e., from about 4 to 22 minutes). This optimal result on the Internet side is probably due to the use of the adopted Web replication technology along with the use of our distribution mechanism (C^2LD). Note that if we try to download songs from a single Web server (such as the New Zealand Web server) with the standard HTTP, this can lead to an increase of the WLNTT value by about 600 seconds (10 minutes).

24.4.3 MOBILE KARAOKE: MEASUREMENT ARCHITECTURE AND RESULTS

To evaluate the efficacy of the mobile karaoke distribution service we have implemented, three different Web replica servers were used. They all maintained the same set of karaoke clips, along with the associated multimedia resources. As shown in the small picture of Figure 24.8, out of these three Web servers, two were located on the same LAN at the Department of Computer Science of the University of Bologna (a 100-Mbps Ethernet). The third server and the IS system were deployed on a different 100-Mbps Ethernet LAN, located at a remote site of the University of Bologna (the Computer Science Laboratory of Cesena). The two LANs were approximately 10 hops each from other, interconnected through a 34-Mbps link.

The IS application was running over a Pentium 3 machine (800 MHz, 512 MB RAM) equipped with the Windows 2000 Professional operating system. Finally, the client side of our mobile application was emulated on a Pentium 3 machine (667 MHz, 512 MB RAM) equipped with the Microsoft Pocket PC operating system emulator. (It is worth mentioning that the round trip times, obtained with the ping routine, between the emulated client and the three Web servers measured about 10 milliseconds on average.)

As far as the downloading process is concerned, we took the following basic assumptions:

1. SMIL files: We used SMIL files with dimensions ranging from 3 to 4 kb. SMIL files of such dimensions are typically large enough to specify complete karaoke clips.

2. Multimedia resources: SMIL files were used which typically pointed to two different multimedia objects: (1) a WMA (Windows Media Audio) file, and (2) a WMV (Windows Media Video) file. For audio files, we used a set of WMA files with dimensions ranging from approximately 1.5 to 2.0 MB, corresponding to songs sampled at 64 kbps (and lasting approximately from 3.5 to 4 minutes). For video files, we used WMV video clips lasting approximately 30 seconds, with a quality needing a data rate of 190 kbps, thus yielding file dimensions ranging from 750 to 850 kb. As previously explained, the execution of audio and video resources were synchronized (along with textual information) by using SMIL commands. In the case when an audio file had a duration longer than the video file, the execution of the video file was scheduled to be repeated through the conclusion of the music file.

As far as the obtained results are concerned, the first consideration is the time spent over the wired links to download karaoke clips from the replicated Web servers toward the IS (i.e., WLNTT results). Those WLNTT results amounted to quite small values. Indeed, 0.2 seconds on average were needed to download the SMIL files, while 5/6 seconds on average were measured to download the corresponding multimedia objects.

Much-larger values were measured for the WNTT results experienced over the wireless link. As in the case of simple song distribution, those measurements were taken depending on the two traffic parameters: (1) the speed at which users moved through the cell, and (2) the additional traffic in the cell.

The WNTT values needed for delivering SMIL files to the mobile device were as much as 1 second on average. Instead, as the example in Figure 24.11 shows, the WNTT values are reported that were measured to deliver over the wireless link the multimedia resources of two different karaoke clips, respectively: "Losing My Religion" by REM (hereinafter referred to as song 1) and "A Little Respect" by Wheatus (hereinafter referred to as song 2). In particular, song 1 was comprised of a WMA audio file of 2.15 MB and a WMV video file of 765 kb; song 2 was comprised of a WMA audio file of 1.6 MB and a WMV video file of 850 kb. Figure 24.11 presents three different graphs for each song, with each graph plotted for a different Erlang

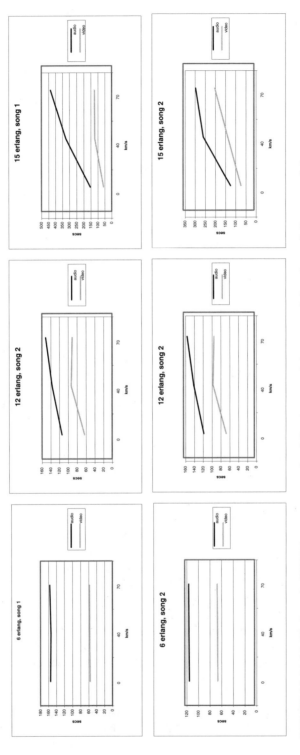

FIGURE 24.11 Mobile karaoke WNTT measurements for delivering song 1 (upper graphs) and song 2 (lower graphs).

value. In each graph, the WNTT values needed to deliver the audio and the video files are presented separately, through two different curves. Each curve varies depending on the user speed. (Again, it is worth noting that the WNTT values we plotted were obtained by averaging all the experiments conducted with IP packets of different dimensions.)

As in the case of simple song distribution, it is easy to notice that (1) as the traffic in the cell increases, the corresponding WNTT values increase, and (2) the lowest WNTT results may be obtained when the mobile device is still. In addition, the karaoke clip for song 1 was available at the handheld device after an average time interval ranging from about 3.5 to 9.5 minutes (220 to 570 seconds), while the karaoke clip for song 2 was delivered to the UMTS device after an average time interval ranging from about 3 to 8 minutes (180 to 490 seconds).

To conclude this section, it is worth noting that the WNTT results obtained for the mobile karaoke service appear to be better than those obtained with the song distribution service due to the fact that in the case of mobile karaoke multimedia resources of smaller size were exploited in the field trials. (Further details on the field trials conducted for the mobile karaoke service may be found in Roccetti et al.[20])

24.5 RELATED WORK AND COMPARISON

This section discusses some issues of interest at the basis of the most-relevant design choices we have made. The issues which are of paramount importance for the development of our proposed music services are (1) distribution of multimedia resources across the Internet and P2P networking, (2) wireless network access to the Internet, and (3) multimedia stream synchronization for delivering karaoke clips.

24.5.1 DISTRIBUTION OF MULTIMEDIA RESOURCES
OVER THE INTERNET

Recently, there has been much emphasis about the possibility of an effective, secure, and reliable access to multimedia information on the Internet from mobile terminals. This has determined the evolution of architectural solutions and technologies based on content. In essence, so-called content networks deal with the routing and forwarding of requests and responses for content using upper-level application protocols. Typically, data transported in content networks amount to images, movies, and songs which are often very large in dimension.[21–25] Simply put, a content distribution network (CDN) can be seen as a virtual network overlay of the Internet that distributes content by exploiting multiple replicas. A request from a client for a single content item is directed to a good replica, where "good" means that the item is served to the client quickly compared to the time it would take if that item were fetched from the original server. A typical CDN has some combinations of a content-delivery infrastructure, a request-routing infrastructure, and a distribution infrastructure. The content-delivery infrastructure consists of a set of replicated servers that delivers copies of content to users who issue requests for a certain content. The request-routing infrastructure consists of mechanisms that enable the connection of a given client with a selected replica. The distribution infrastructure consists of mechanisms that copy content from the origin server to the replicas. Finally, a set of software architectural elements constitute the core of the content distribution internetworking (CDI) infrastructure that uses commonly defined protocols to share resources so as to reach to the most-distant participants.

It is easy to recognize that the architecture of the wireless Internet application we have developed resembles the above-mentioned CDI technology because it interconnects a CDN, located in the Internet, with the UMTS network. At the basis of our CDI infrastructure we have set the application gateway, which manages all the interactions between the UMTS terminals and the wired Internet. Our developed intermediate system (IS), along with the set of all the replica servers, constitutes a real CDN. The content-delivery infrastructure is implemented by means of the replica servers that store multiple copies of musical resources. The search functionalities of the discovery subsystem, integrated with the C²LD downloading mechanism that operates by engaging all the available replicas in supplying fragments of the requested song, provide the request-routing infrastructure.

Another important issue related to the design of our wireless Internet application is concerned with the use of P2P technologies. Modern P2P technology embraces a class of applications that take advantage of resources, computing cycles, content, and human presence available at the edges of the Internet. Traditionally, a P2P architecture comprises a decentralized system where all peers communicate symmetrically and exhibit equal roles.[26]

It is possible to observe that our designed system resembles traditional P2P systems because it aims at sharing multimedia resources anywhere, anytime. However, because accessing decentralized musical resources from UMTS devices entails operating in an environment of unstable connectivity, our system rests on a centralized entity (the application gateway and the download manager), which operates like a standard wired client with respect to the decentralized replicas. Additionally, our architecture embeds a centralized discovery subsystem that collects the references for a requested song. This centralized architectural solution provides the advantage to permit songs to be shared even in the presence of musical content of large dimensions, as well as with devices with scarce computational capacity.

24.5.2 WIRELESS ACCESS TO THE INTERNET

It is well known that a typical approach to providing wireless access to the Internet amounts to selecting a specific protocol especially designed for the wireless environment. A protocol gateway uses this specific wireless protocol to enable the interaction of the wireless device with the Internet. An example of this type of solution is the Wireless Application Protocol (WAP), incorporating a protocol gateway able to translate requests from the wireless protocol stack to the Web protocols. Moreover, instead of using HTML, WAP uses the Wireless Markup Language (WML), a subset of XML, to recode the Internet content for the wireless device.[3]

It is important to observe that the application gateway embodied in our proposed architecture performs different functions with respect to the protocol gateway of the WAP solution. The WAP-based gateway performs translations from HTML-based content to the proprietary format that is understandable at the mobile terminal.

Contrariwise, our application gateway does not recode content, but simply provides interconnection between two different CDNs.

Beyond WAP, microbrowser technology continues to move forward with innovative solutions such as, for example, i-mode and the Pixo Internet Microbrowser.[2,27] Those protocols are specifically aimed at the wireless Internet, because they recode Internet content for wireless devices and utilize Compact HTML (CHTML) or Extensible Hypertext Markup Language (XHTML) as their markup languages.

Unlike WAP and similar approaches, middleware often offers an alternative to manually replicating content. Its basic purpose is to transparently transcode content on the fly without maintaining Web content in multiple formats.[27] The Parlay Project,[28] the micro version of Java (J2ME),[29] the Mobile Execution Environment (MexE),[30] the micro edition of JINI (JMatos),[31] Online Anywhere,[32] and Proxinet,[33] along with the use of the Relational Markup Language (RML), are all examples that fall in the category of middleware-based approaches.

We conclude this overview by mentioning the JXTA technology.[34, 35] This is a set of open peer-to-peer protocols that allows any connected device on the network to communicate according to a peer-to-peer pattern. The focus of JXTA protocols is on creating a virtual network overlay on top of the Internet, allowing peers to directly interact independently of their network location, programming language, and different implementations. At the heart of JXTA technology we can find advertisements (XML documents) that are exploited to advertise all network resources (from peers to content). Advertisements are exploited to provide a uniform way to publish and discover network resources.

In this context, a final comment is due regarding our choice to design all our protocol architecture following an all-IP approach. This approach has the advantage of allowing mobile terminals to function as any other Internet-connected device. However, this choice requires that the end-to-end protocol function continuity be preserved in the wireless segments, and we must admit that many are the problems of providing such seamless internetworking between wired and wireless worlds with Internet protocols. Nevertheless, our decision of resorting to an additional session layer, along with the intense experimental monitoring we have conducted, have shown that the all-IP choice does not cause too many interferences (in terms of packet retransmissions) between TCP and the radio link layer. In addition, an all-IP approach overcomes the interoperability problems which may arise in the case of proprietary protocol solutions.

24.5.3 MULTIMEDIA SYNCHRONIZATION FOR DELIVERING KARAOKE

One of the main concerns in the design of karaoke systems is the adopted synchronization strategy. In fact, it is clear that because a karaoke playout consists of a presentation of synchronized multimedia files, an underlying model is needed for specifying the synchronization rules to be adopted by different media streams. To accomplish this goal, we exploited the SMIL technology (and a SMIL player), but other solutions exist, e.g., the FLIPS model.[36] FLIPS is a model developed for specifying coarse synchronization for flexible presentations supporting a wide range of temporal synchronization specifications. It provides algorithms for attaining a consistent and coherent presentation state in response to user interaction and other state-changing events. Another traditional technology for playing back synchronized digital data is the MIDI technology that can also be used to play back karaoke clips. In essence, a computer program can play a MIDI-based karaoke file containing musical data, as well as the lyrics that are displayed on a computer monitor. Hence, MIDI karaoke files are standard MIDI files that may be executed on desktop computers. However, the most modern technology for the synchronized playback of multimedia data is the MP3 technology. Many vendors today produce MP3 players that are designed to display lyrics and other graphics while songs play out. For example, the Irock 680 player from Motorola plays out songs in both MP3 and MP3i formats.[37] (MP3i is the new interactive format that integrates graphical data with digital music files.) This allows content such as lyrics, artwork, text notes, photographs, and videos to be displayed as music plays back on a device.

Another important networked technology that has been extensively exploited to synchronize multimedia streams over the Internet and implement Internet-based karaoke systems is the RealMedia technology.[38] This is a client/server technology for streaming synchronized media on the Internet. For example, Karaoke/SureStream is a RealSystem feature that allows the RealServer to dynamically adjust the stream for each listener, depending on the dynamic network conditions of the user's connection.[39] SureStream manipulates media streams by providing an encoding framework allowing multiple streams at different bit rates to be simultaneously encoded and combined into a single file. Additionally, it provides a client/server mechanism for detecting changes in bandwidth and translating those changes into combinations of different streams. Karaoke Online uses the audio streaming technology provided by RealNetworks to deliver music and lyrics to a Web browser.[40]

Other interesting research experiences are those discussed in Lee and coworkers [41] (the SESAME project was presented where scalability issues for karaoke systems were investigated), and in Liu et al.[42] and Tseng and Huang[43] (client-server karaoke systems were proposed for video and audio streams that allowed a wired access through the public switched network). Many are the karaoke societies that use SMIL, SureStream, plus other synchronization technologies to implement karaoke systems on a client/server basis for the Internet. Relevant examples are Cyber-Karaoke-On-Demand,[44] Karaoke Jukebox,[45] StreamKaraoke,[46] and finally Streaming21.[47] We conclude by mentioning that all the cited experiences refer either to the wired Internet or to small-sized wireless LAN environments.

24.6 CONCLUDING REMARKS

In this chapter, we have reported on our experience in implementing a wireless Internet application designed to support the large-scale distribution of musical resources (from simple songs to synchronized karaoke clips, as shown in Figure 24.12). Our application allows mobile consumers to listen to songs/karaoke clips on handheld UMTS-enabled devices by exploiting the Internet as a vast storehouse of music resources. Experimental results were obtained that show that fast, large-scale wireless musical services may be provided by exploiting the UMTS

FIGURE 24.12 A screenshot of the mobile karaoke service.

technology. Measurements have been taken that confirm that both songs and karaoke clips (composed by audio, video, and scrolling text) may be downloaded from the Internet to UMTS devices in a few minutes on average. Whether the role of wireless networks is limited to extending the Internet reach or whether new applications may be enabled by wireless access are subjects of much discussion.[1,48,49] We claim that our wireless application demonstrates that exciting musical services may be implemented profitably using the wireless technology available today.

ACKNOWLEDGMENTS

We wish to thank the Italian MIUR and CNR, the Fondazione Marconi of Bologna, the Department of Computer Science of the University of Bologna, and Microsoft Research Europe for the partial financial support of this work.

References

1. Bhagwat, P. and Sreenan, C.J., Eds., Future Wireless Applications, *IEEE Wireless Commun.,* 9 (1), 6–59, 2002.
2. All about i-mode, http://www.nttdocomo.com/index.html, 2001.
3. WAP Architecture Specification, http://www1.wapforum.org/tech/terms.asp?doc= WAP-100-WAPArch-19980430-a.pdf.
4. Macedonia, M., Distributed file sharing: barbarians at the gates?, *IEEE Comput.,* 33(8), 99–101, August 2000.
5. UMTS Forum, What is UMTS?, http://www.umts-forum.org/what_is_umts.html.
6. MP3 resources by MPEG.ORG, http://www.mpeg.org/MPEG/mp3.html.
7. W3 Recommendation, Synchronized Multimedia Integration Language (SMIL) 2.0 Specification, http://www.w3.org/TR/smil20/, 2001.
8. Napster official site, http://www.napster.com/
9. The Freenet Project, http://freenet.sourceforge.net.
10. Gnutella official site, http://gnutella.wego.com/.
11. Staehle, D., Leibnitz, K., and Tsipotis, K., QoS of Internet Access with GPRS, in Proc. 4th ACM International Workshop on Modeling, Analysis and Simulation of Wireless and Mobile Systems, Rome, 2001, 57–64.
12. Kalden, R., Meirick, I., and Meyer, M., Wireless Internet access based on GPRS, *IEEE Personal Commun.,* 7 (2), 8–18, 2000.
13. Huston, G., TCP in a wireless world, *IEEE Internet Computing,* 5(2), 82–84, March–April, 2001.
14. Ingham, D., Shrivastava, S.K., and Panzieri, F., Constructing dependable Web services, *IEEE Internet Computing,* 4 (1), 25–33, 2000.
15. Ghini, V., Panzieri, F., and Roccetti, M., Client-centered load distribution: a mechanism for constructing responsive Web services, in Proc. 34th IEEE Hawaii International Conference on System Sciences, Maui, 2001.
16. Roccetti, M. et al., The structuring of a wireless internet application for a music-on-demand service on UMTS device, in Proc. ACM Symposium on Applied Computing, ACM Press, Madrid, 2002, 1066–1073.
17. Jacobson, V., Berkeley TCP evolution from 4.3-Tahoe to 4.3-Reno2, in Proc. 18th Internet Engineering Task Force Meeting, University of British Columbia, Vancouver, 1990, 365.
18. Roccetti, M., Ghini, V., and Salomoni, P., Distributing music from IP networks to UMTS terminals: an experimental study, in Proc. 2002 SCS EUROMEDIA Conference, Roccetti, M., Ed., The Society for Modeling and Simulation International, Modena, 2002, 147–154.
19. Roccetti, M. et al., Bringing the wireless Internet to UMTS devices: a case study with music distribution, to appear in the *International Journal of Multimedia Tools and Applications,* Kluwer, 2003.

20. Roccetti, M. et al., MoKa: a wireless Internet application for delivering mobile karaoke on UMTS devices, in Proc. IASTED International Conference on Communications, Internet and Information Technology, M.H. Hamza, Ed., St. Thomas, VI, November 2002, pp. 346–351.

21. Barbir, A. et al., Known CDN request-routing mechanisms, draft-cain-cdnp-known-request-routing-02.txt, in progress, June 2001.

22. Cain, B. et al., Request-routing requirements for content internetworking, draft-cain-request-routing-req-02.txt, in progress, July 2001.

23. Day., M., A model for content internetworking, draft-day-cdnp-model-05.txt, in progress, March 2001.

24. Day, M., Gilletti, D., and Rzewskip, P., CDN peering scenarios, draft-day-cdnp-scenarios-03.txt, in progress, March 2001.

25. Green, M. et al., Content internetworking architectural overview, draft-green-cdnp-gen-arch-03.txt, in progress, March 2001.

26. Flammia, G., Peer to peer is not for everyone, *IEEE Intelligent Systems,* 16(3), 78–79, May-June, 2001.

27. Saha, S., Jamtgaard, M., and Villasenor, J., Bringing the wireless Internet to mobile devices*, IEEE Comput.,* 34 (6), 54–58, 2001.

28. The Parlay Group, www.parlay.org.

29. Java J2ME, http://www.java.sun.com/j2me.

30. MexE Forum, http://www.mobilmexe.com.

31. JINI Network Technology, http://www.sun.com/jini/index.html.

32. Online Anywhere, http://www.onlineanywhere.com/.

33. Proxinet, http://www.pumatech.com/proxinet.

34. Gong, L., JXTA: a network programming environment, *IEEE Internet Computing,* 5(3), 88–95, May–June, 2001.

35. Project JXTA, http://www.jxta.org.

36. FLIPS, http://www.cs.umn.edu/Research/dmc/html/dlearn.html.

37. Irock 680 player from Motorola, http://www.motorola.com.

38. Real Media, http://www.real.com.

39. Real Networks Inc., http://www.realnetworks.com/.

40. Karaoke On Line, http://www.thinks.com/karaoke/.

41. Lee, Y., Du, D.H.C., and Ma, W., SESAME: A Scalable and ExtenSible Architecture for Multimedia Entertainment, in Proc. IEEE 20th International Computer Software and Applications Conference, Seoul, 1996.

42. Liu, C. et al., The construction of a multimedia application on public network, in Proc. SPIE High-Speed Networking and Multimedia Computing Conference, 1994.

43. Tseng, W.H. and Huang, J.H., A high performance video server for karaoke systems, *IEEE Trans. Consumer Electr.,* 40 (3), 609–618, August, 1994.

44. Cyber-Karaoke-On-Demand, http//www.innogate.com.my/ckod.asp.

45. KARAOKE JUKEBOX; http://www.peddocko64.freeserve.co.uk.

46. StreamKaraoke; http://streamkaraoke.com

47. Streaming21, http://www.streaming21.com.

48. Lawton, G., Browsing the mobile Internet, *IEEE Comput.,* 34 (12), 18–21, 2001.

49. Kanter, T., An open service architecture for adaptive personal mobile communication, *IEEE Personal Commun.,* 8 (6), 8–17, 2001.

Index

Index

A